ELECTRICAL MACHINES, DRIVES, and POWER SYSTEMS

Second Edition

THEODORE WILDI

President, Sperika Enterprises Ltd.

Professor Emeritus of Laval University
Department of Electrical Engineering

PRENTICE HALL
Englewood Cliffs, New Jersey 07632

Library of Congress Cataloging-in-Publication Data

WILDI, THEODORE.
    Electrical machines, drives, and power systems / Theodore Wildi.—2nd Ed.

      p.   cm.
    Rev. ed. of: Electrical power technology. c1981.
    Includes bibliographical references.
    ISBN 0-13-251547-4
    1. Electrical engineering. 2. Electrical power. I. Wildi, Théodore.
Electrical power technology. II. Title.
TK145.W488   1991
621.21′042—dc20

                                        89–77774
                                        CIP

Editorial/production supervision: Mary Carnis and Barbara Cassel
Interior design: Theodore Wildi
Manufacturing buyer: Gina Brennan
Cover photographs courtesy of Siemens Electric Limited

 © 1991,1981 by Sperika Enterprises Ltd.

Printed in the United States of America
10 9 8 7 6 5 4 3 2 1

ISBN 0-13-251547-4

Prentice-Hall International (UK) Limited, *London*
Prentice-Hall of Australia Pty. Limited, *Sydney*
Prentice-Hall Canada Inc., *Toronto*
Prentice-Hall Hispanoamericana, S.A., *Mexico*
Prentice-Hall of India Private Limited, *New Delhi*
Prentice-Hall of Japan, Inc.,*Tokyo*
Simon & Shuster Asia Pte. Ltd., *Singapore*
Editora Prentice-Hall do Brasil, Ltda., *Rio de Janiero*

CONTENTS

iv

PREFACE

This Second Edition of the book (formerly entitled Electrical Power Technology), gives a broad overview of modern machines, drives and power systems. It covers the basic principles of transformers and rotating machines, transmission and distribution systems, and power electronics associated with these fields. Toward this end, the subject matter has been divided into four distinct parts: (1) Fundamentals; (2) Electrical Machines and Transformers; (3) Electrical and Electronic Drives and (4) Electric Utility Power Systems.

These major divisions, comprising 28 chapters, can be covered either separately or in sequence, in whole or in part, depending on the particular program of study. Cross references to preceding chapters and equations make it easy to return to the original source of information, when required.

Many improvements have been made in this Second Edition. Thus, a chapter has been added on Stepper Motors and another chapter discusses the development and use of the equivalent circuit of an induction motor. The chapter on high-voltage dc transmission has been updated and various topics have been streamlined and rearranged.

For example, electronic drives are assuming an evergrowing place in industry. In recognition of this fact, four important chapters are devoted to this subject. After introducing the principles of power electronics, the text goes on to cover the major aspects of ac and dc drives. They comprise the modern electronic methods of controlling medium and large-power machines.

Electric utilities also have an ongoing need for technicians and engineers that have a knowledge of electric power. To meet this objective, three chapters cover the generation, transmission and distribution of electrical energy. In addition, a fourth chapter describes the HVDC transmission of electric power as well as back-to-back converters. Major installations of this kind are presently in operation throughout the world. It is expected that dc transmission lines and dc links will become increasingly important as they complement existing ac networks.

In covering such a broad range of subjects, a conscious effort has been made to establish cohesion so that the reader will see how they all fit together. Thus, we find that the terminology and equations for synchronous machines are similar to those relating to transmission lines. Transmission lines, in turn, bring up the question of reactive power. And reactive power is important in electronic converters. Thus, knowledge gained in one area is strengthened and broadened by using it in another. As a result, the learning of electrical machines and power systems becomes a challenging, thought-provoking experience.

The subject matter requires a background in basic circuit theory, algebra and some trigonometry. Owing to its straightforward treatment of even complex topics, this book will meet the needs of a broad range of readers. First, it is appropriate for students following a two-year electrical program in community colleges, technical institutes and universities. It can also be used in those courses in an electronics program that deal with transformers or machinery and controls. Again, the text can be incorporated in a 4-year technology program, owing to its very broad coverage. Many universities have adopted the book for their electric power programs for non-electrical students.

Students following industrial training programs will also find the book contains a great deal of practical information that can be directly applied in that greatest laboratory of all – the electrical industry itself. Finally, at a time when much effort is being devoted to continuing education, this book, with its many worked-out problems, is particularly well adapted for self-study.

The exercises at the end of each chapter are organized so that readers may solve problems appropriate to their particular level of interest, background and ability. They are divided into three levels – practical, intermediate and advanced. Furthermore, to encourage the reader in solving the exercises, the answers are given at the end of the book. A 150-page solutions manual is also available.

In order to convey the real-world aspects of machinery and power systems, particular attention has been paid to such important elements as the inertia of revolving masses, the physical limitations of materials, and the problems created by heat. It is felt that this approach also falls in line with the multidisciplinary programs of many colleges and technical institutes.

A quick glance through the text shows the importance given to photographs. All equipment and systems are illustrated by diagrams and pictures, showing them in various stages of construction, or in actual use. Many students may not have had the opportunity to visit an industrial plant or to see at close hand the equipment used in the transmission and distribution of electrical energy. The photographs help convey the magnificent size of these devices and machines.

The International System of Units (SI), now widely adopted throughout the world, is used in this text. Particular care has been taken to use units and symbols according to accepted standards. The reader not yet familiar with SI will be surprised by its basic simplicity. It may take a little time to get used to such units as weber, newton, pascal and so forth, but the simple method explained in Chapter 1 facilitates the conversion from one system to any other. Also, some of the chapter-end exercises use English units, and some use SI. Thus, students can become proficient at solving problems with both systems.

The explanation of the subject matter is based on technical fact. Consequently, the reader who wishes to make an in-depth study of a particular topic will find that information given here agrees with knowledge revealed by more detailed or more advanced texts.

In summary, the book employs a theoretical, practical, multidisciplinary approach to give a broad understanding of electrical power. We hope it will stimulate renewed interest in this dynamic, exciting field.

It has been my good fortune to benefit from the technical comments and suggestions made by many instructors and professional engineers. In the preparation of this edition, I would particularly like to acknowledge the contribution of professors John Reeve, Laverne H. Hardy and Stephen Tubbs. Their recommendations, together with those of Editor Paul Berk, have been of great help in making this book what it is. Professor Tubbs, formerly of the Pennsylvania State University, has also written a laboratory manual to accompany this book.

I also want to express my appreciation to Lucie Veilleux and Anne-Marie Auclair for typing the text, to Alain Sirois for carefully checking the solutions manual, and to my son, Karl, for doing the layout and all the line drawings. And a special word of thanks to my wife, Rachel, for her encouragement and support.

Theodore Wildi

PART I

FUNDAMENTALS

1

UNITS

Units play an important role in our daily lives. In effect, everything we see and feel and everything we buy and sell is measured and compared by means of units. Units enable us to measure the distance we walk, the land we own, the time of day, and the brightness of stars – and, in fact, every feature of the environment in which we live.

Some of these units have become so familiar that we often take them for granted, seldom stopping to think how they started, or why they were given the sizes they have.

Centuries ago, the foot was defined as the length of 36 barleycorns strung end to end, and the yard was the distance from the tip of King Edgar's nose to the end of his outstretched hand.

Since then, we have come a long way in defining our units of measure more precisely. Most units are now based upon the physical laws of nature which are both invariable and reproducible. Thus, the meter and yard are measured in terms of the speed of light, and

time by the duration of atomic vibrations. This improvement in our standards of measure has gone hand in hand with the advances in technology, and the one could not have been achieved without the other.

But if the basic standards of reference are recognized by all countries of the world, the units of everyday measure are far from being universal. For example, in measuring length, some people use the inch and yard, while others use the millimeter and meter. Astronomers employ the parsec, physicists use the angstrom, and some surveyors still measure with the rod and chain. But these units of length can be compared with great accuracy because the standard of length is based upon the speed of light.

Such standards of reference make it possible to compare the units of measure in one country , or in one specialty, with the units of measure in any other. Standard units of length,mass and time are the anchors that tie together the units used in the world today.

1.1 Systems of units

Over the years, systems of units have been devised to meet the needs of commerce, industry, and science. A system of units may be described as one in which the units bear a direct numerical relationship to each other, usually expressed as a whole number. Thus, in the English system of units the inch, foot, and yard are related to each other by the numbers 12, 3, and 36.

The same correlation exists in metric systems, except that the units are related to each other by multiples of ten. Thus, the centimeter, meter and kilometer are related by the numbers 100, 1000, and 100 000. It is therefore easier to convert meters into centimeters than to convert yards into feet, and this decimal approach is one of the advantages of the metric system of units.*

Today, the officially recognized metric system is the International System of Units for which the universal abbreviation is SI. The SI was formally introduced in 1960, at the Eleventh General Conference of Weights and Measures, under the official title "Système international d'unités".

1.2 Getting used to SI

The official introduction of the International System of Units, and its adoption by most countries of the world, did not, however, eliminate the systems that were previously employed. Just like well-established habits, units become a part of ourselves, which we cannot readily let go. In effect, it is not easy to switch overnight from yards to meters and from ounces to grams. And this is quite natural, because long familiarity with a unit gives us an idea of its magnitude and how it relates to the physical world.

Nevertheless, the growing importance of SI

---

* The metric unit of length is spelled either *meter* or *metre*. In Canada, the official spelling is *metre*.

(particularly in the electrical and mechanical fields) makes it necessary to know the essentials of this measurement system. Consequently, one must be able to convert from one system to another in a simple unambiguous way. In this regard, the reader will discover that the conversion charts listed in the Appendix are particularly helpful.

The SI possesses a number of remarkable features shared by no other system of units.

1. It is a decimal system.

2. It employs many units commonly used in industry and commerce; for example, volt, ampere, kilogram, watt, etc.

3. It is a coherent system that expresses with startling simplicity some of the most basic relationships in electricity, mechanics and heat.

4. It can be used by the research scientist, the technician, the practicing engineer, and by the layman, thereby blending the theoretical and the practical world.

Despite these advantages, the SI is not the answer to everything. In specialized areas of atomic physics and even in day-to-day work, other units may be more convenient. Thus, we will continue to measure plane angles in degrees, even though the SI unit is the radian. Furthermore, "day" and "hour" will still be used, despite the fact that the SI unit of time is the second.

1.3 Base and derived units of the SI

The foundation of the International System of Units rests upon seven base units listed in Table 1A.

From these base units, we can derive other units to express quantities such as area, power, force, magnetic flux, and so on. There is really no limit to the number of units we can derive, but some occur so frequently that they have been given special names. Thus, instead of saying that the unit of pressure is the newton per square meter, we use a less cumbersome name, the pascal. Some of the derived units that have special names are listed in Table 1B.

1.4 Definitions of base units

The following are official definitions of the SI base units. Although not required for an understanding of the SI, they illustrate the extraordinary precision associated with this modern system of units. The text in italics is explanatory, and does not form part of the definition.

The **meter** (m) is the length of the path travelled by light in vacuum during a time interval of 1/299 792 458 of a second.

In 1983, the speed of light was defined to be 299 792 458 m/s exactly.

The **kilogram** (kg) is the unit of mass; it is equal to the mass of the international prototype of the kilogram.

The international prototype of the kilogram is a particular cylinder of platinum-iridium alloy that is preserved in a vault at Sèvres, France, by the International Bureau of Weights and Measures. Duplicates of the prototype exist in all important standards laboratories in the world. The platinum-iridium cylinder (90 percent platinum, 10 percent iridium) is about 4 cm high and 4 cm in diameter.

The **second** (s) is the duration of 9 192 631 770 periods of the radiation corresponding to the transition between the two hyperfine levels of the ground state of the cesium-133 atom.

A quartz oscillator, tuned to the resonant frequency of cesium atoms, produces a highly accurate and stable frequency.

The **ampere** (A) is that constant current which, if maintained in two straight parallel conductors of infinite length, of negligible circular cross

TABLE 1A   BASE UNITS

Quantity	Unit	Symbol
Length	meter	m
Mass	kilogram	kg
Time	second	s
Electric current	ampere	A
Temperature	kelvin	K
Luminous Intensity	candela	cd
Amount of substance	mole	mol

TABLE 1B   DERIVED UNITS

Quantity	Unit	Symbol
Electric capacitance	farad	F
Electric charge	coulomb	C
Electric conductance	siemens	S
Electric potential	volt	V
Electric resistance	ohm	Ω
Energy	joule	J
Force	newton	N
Frequency	hertz	Hz
Illumination	lux	lx
Inductance	henry	H
Luminous flux	lumen	lm
Magnetic flux	weber	Wb
Magnetic flux density	tesla	T
Power	watt	W
Pressure	pascal	Pa

section, and placed 1 meter apart in a vacuum, would produce between these conductors a force equal to 2×10^{-7} newton per meter of length.

The **kelvin** (K), unit of thermodynamic temperature, is the fraction 1/273.16 of the thermodynamic temperature of the triple point of water.

Pure water in an evacuated cell is cooled until ice begins to form. The resulting temperature where ice, water and water vapor co-exist is called the triple point of water and is equal to 273.16 kelvins, by definition. The triple point is equal to 0.01 degree Celsius (°C). A temperature of 0°C is therefore equal to 273.15 kelvins, exactly.

The **candela** (cd) is the luminous intensity, in a given direction, of a source that emits monochromatic radiation of frequency 540×10^{12} hertz and that has a radiant intensity in that direction of 1/683 watt per steradian.

The **mole** (mol) is the amount of substance of a system which contains as many elementary entities as there are atoms in 0.012 kilogram of carbon 12.

Note: When the mole is used, the elementary entities must be specified and may be atoms, molecules, ions, electrons, other particles or specified groups of such particles.

1.5 Definitions of derived units
Some of the more important derived units are defined as follows:

The **coulomb** (C) is the quantity of electricity transported in 1 second by a current of 1 ampere. *(Hence 1 coulomb = 1 ampere second)*

The **degree Celsius** (°C) is equal to the kelvin and is used in place of the kelvin for expressing Celsius temperature (symbol t) defined by the equation $t = T - T_o$ where T is the thermodynamic temperature and $T_o = 273.15$ K, by definition.

The **farad** (F) is the capacitance of a capacitor between the plates of which there appears a difference of potential of 1 volt when it is charged by a quantity of electricity equal to 1 coulomb. *(1 farad = 1 coulomb per volt)*

The **henry** (H) is the inductance of a closed circuit in which an electromotive force of 1 volt is produced when the electric current in the circuit varies uniformly at a rate of 1 ampere per second. *(Hence 1 henry = 1 volt second per ampere)*

The **hertz** (Hz) is the frequency of a periodic phenomenon of which the period is 1 second.

The **joule** (J) is the work done when the point of application of 1 newton is displaced a distance of 1 meter in the direction of the force. *(Hence 1 joule = 1 newton meter)*

The **newton** (N) is that force which gives to a mass of 1 kilogram an acceleration of 1 meter per second per second. *(Hence 1 newton = 1 kilogram meter per second squared)*

Although the newton is defined in terms of a mass and an acceleration, it also applies to stationary objects and to every application where a force is involved.

The **ohm** (Ω) is the electric resistance between two points of a conductor when a constant difference of potential of 1 volt, applied between these two points, produces in this conductor a current of 1 ampere, this conductor not being the source of any electromotive force. *(Hence 1 ohm = 1 volt per ampere)*

The **pascal** (P) is the unit of pressure or stress equal to one newton per square meter.

The **radian** (rad) is the unit of measure of a plane angle with its vertex at the center of a circle and subtended by an arc equal in length to the radius.

The **siemens** (S) is the unit of electric conductance equal to one reciprocal ohm. *(The siemens was formerly named the mho.)*

The **steradian** (sr) is the unit of measure of a solid angle with its vertex at the center of a sphere and enclosing an area of the spherical surface equal to that of a square with sides equal in length to the radius.

The **tesla** (T) is the unit of magnetic flux density equal to one weber per square meter.

The **volt** (V) is the difference of electric potential between two points of a conducting wire carrying a constant current of 1 ampere, when the power dissipated between these points is equal to 1 watt. *(Hence 1 volt = 1 watt per ampere)*

The **watt** (W) is the power that gives rise to the production of energy at the rate of 1 joule per second. *(Hence 1 watt = 1 joule per second.)*

The **weber** (Wb) is the magnetic flux that, linking a circuit of one turn, produces in it an electromotive force of 1 volt as it is reduced to zero at a uniform rate in 1 second. *(Hence 1 weber = 1 volt second.)*

1.6 Multiples and submultiples of SI units

Multiples and submultiples of SI units are generated by adding appropriate prefixes to the units. Thus, prefixes such as kilo, mega, nano and centi multiply the value of the unit by factors listed in Table 1 C. for example:

1 *kilo*ampere = 1000 amperes,
1 *nano*second = 10^{-9} seconds,
1 *mega*watt = 10^6 watts.

1.7 Commonly used units

Tables 1D, 1E and 1F list some common units encountered in mechanics, thermodynamics and electricity. They contain notes particularly useful to the reader who is not yet familiar with the SI.

1.8 Conversion charts and their use

Unfamiliar units can be converted to units we know well by using standard conversion tables.

TABLE 1C   PREFIXES TO CREATE MULTIPLES AND SUBMULTIPLES OF SI UNITS

Multiplier	Exponent Form	Prefix	SI Symbol
1 000 000 000 000 000 000	10^{18}	exa	E
1 000 000 000 000 000	10^{15}	peta	P
1 000 000 000 000	10^{12}	tera	T
1 000 000 000	10^{9}	giga	G
1 000 000	10^{6}	mega	M
1 000	10^{3}	kilo	k
100	10^{2}	hecto	h
10	10^{1}	deca	da
0.1	10^{-1}	deci	d
0.01	10^{-2}	centi	c
0.001	10^{-3}	milli	m
0.000 001	10^{-6}	micro	μ
0.000 000 001	10^{-9}	nano	n
0.000 000 000 001	10^{-12}	pico	p
0.000 000 000 000 001	10^{-15}	femto	f
0.000 000 000 000 000 001	10^{-18}	atto	a

But this is strictly an arithmetic process that often leaves us wondering if our calculations are correct.

The conversion charts in the Appendix eliminate this problem because they show the relative size of a unit by the position it occupies on the page. The largest unit is at the top, the smallest at the bottom, and intermediate units are ranked in between.

The units are connected by arrows, each of which bears an associated conversion factor. The conversion factor is the ratio of the larger to the smaller of the units that are connected, and hence its value is always greater than unity. The arrow always points toward the smaller unit.

In Fig. 1-1, for example, five units of length – the mile, meter, yard, inch and millimeter – are listed in descending order of size and the lines joining them bear an arrow that always points toward the smaller unit. The numbers show the relative size of the connected units; the yard is 36 times larger than the inch, the inch is 25.4 times larger than the millimeter, and so on. With this arrangement, we can convert from

TABLE 1D   COMMON UNITS IN MECHANICS

Quantity	SI Unit	Symbol	Note
Angle	radian	rad	1
Area	square meter	m^2	2
Energy (or work)	joule	J	
Force	newton	N	3
Length	meter	m	
Mass	kilogram	kg	
Power	watt	W	
Pressure	pascal	Pa	4
Speed	meter per second	m/s	
Speed of rotation	radian per second	rad/s	5
Torque	newton meter	N·m	
Volume	cubic meter	m^3	
Volume	liter	L	6

1. Although the radian is the SI unit of angular measure, we use the degree almost exclusively in this book (1 rad ~ 57.3°)
2. Most countries, including Canada (as well as many organizations in the United States), use the spelling "metre" instead of "meter".
3. The newton is a very small force, roughly equal to the force needed to press a door bell.
4. The pascal is a very small pressure equal to 1 N/m².
5. In this book, we use the revolution per minute (r/min) to designate rotational speed (1 rad/s = 9.55 r/min).
6. This unit of volume is mainly used for liquids and gases. It is spelled "liter" or "litre". The official spelling in Canada is "litre".

one unit to any other by the following simple method.

Suppose we wish to convert from yards to millimeters. Starting from *yard* in Fig. 1-1, we have to move downward and in the direction of the two arrows (36 and 25.4) until we reach *millimeter*.

Conversely, if we want to convert from millimeters to yards, we start at *millimeter* and move upward against the direction of the arrows until we reach yard. In making such conversions, we apply the following rules:

1. *If, in travelling from one unit to another, we move in the direction of the arrow, we **multiply** by the associated conversion factor.*

2. *Conversely, if we move against the arrow, we **divide**.*

Because the arrows point downward, this means that when moving down the chart we multiply, and when moving up, we divide. Fur-

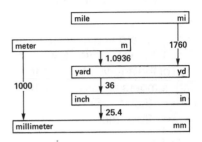

Figure 1-1
Conversion chart for units of length.

thermore, in moving from one unit to another, we can follow any path we please; the conversion result is always the same.

The rectangles bearing SI units extend slightly towards the left of the chart, to distinguish them from other units. Each rectangle bears the symbol for the unit, as well as the name of the unit written out in full.

TABLE 1E   COMMON UNITS IN THERMODYNAMICS

Quantity	SI Unit	Symbol	Note
Heat	joule	J	
Thermal power	watt	W	
Specific heat	joule per		
	(kilogram kelvin)	J/kg·K or J/kg·°C	1
Temperature	kelvin	K	2
Temperature difference	kelvin or degree Celsius	K or °C	1
Thermal conductivity	watt per		
	(meter-kelvin)	W/m·K or W/m·°C	1

1. A temperature difference of 1 K is exactly equal to a temperature difference of 1°C. The °C is now a recognized SI unit and, in practical calculations, it is often used instead of the kelvin.

2. Thermodynamic, or absolute, temperature is expressed in kelvins. On the other hand, the temperature of objects is usually expressed in °C. The absolute temperature T is related to the Celsius temperature t by the equation: $T = t + 273.15$.

Example 1-1:
Convert 2.5 yards to millimeters.

Solution:
Starting from yard and moving toward *millimeter* (Fig. 1-1), we move downwards, in the direction of the arrows. We must therefore *multiply* the numbers associated with each arrow:

2.5 yd = 2.5 ($\times$ 36) ($\times$ 25.4) millimeters
   = 2286 mm.

Example 1-2:
Convert 2000 meters into miles.

Solution:
Starting from meter and moving toward *mile*, we move once with, and once against the direction of the arrows. Consequently, we have:

2000 meters = 2000 ($\times$ 1.0936) ($\div$ 1760) miles

$$= \frac{2000 \times 1.0936}{1760}$$

$$= 1.24 \text{ mi}$$

Example 1-3:
Convert 777 calories to kilowatthours.

Solution:
Referring to the chart on ENERGY (Fig. 1-2),

TABLE 1F   COMMON UNITS IN ELECTRICITY AND MAGNETISM

Quantity	SI Unit	Symbol	Note
Capacitance	farad	F	
Conductance	siemens	S	1
Electric charge	coulomb	C	
Electric current	ampere	A	
Energy	joule	J	
Frequency	hertz	Hz	2
Inductance	henry	H	
Potential difference	volt	V	
Power	watt	W	
Resistance	ohm	Ω	
Resistivity	ohm meter	$\Omega \cdot m$	
Magnetic field strength	ampere per meter	A/m	3
Magnetic flux	weber	Wb	
Magnetic flux density	tesla	T	4
Magnetomotive force	ampere	A	5

1. Formerly called "mho".
2. 1 Hz = 1 cycle per second.
3. 1 A/m = 1 ampere turn per meter.
4. 1 T = 1 Wb/m^2.
5. What was formerly called an ampere turn is now simply called ampere: 1 A = 1 ampere turn.

and moving from calorie to kilowatthour, we first travel downwards (with the arrow 4.184) and then upwards (against the arrows 1000, 1000 and 3.6). Applying the conversion rule, we find:

$$777 \text{ calories}$$

$$= 777 \; (\times \; 4.184) \; (\div \; 1000) \; (\div \; 1000) \; (\div \; 3.6)$$

$$= 9.03 \times 10^{-4} \text{ kW·h}.$$

1.9 The per-unit system of measurement

The units we have just described enable us to specify the magnitude of any quantity. Thus, mass is expressed in kilograms, power in watts and electric potential in volts. However, we can often get a more meaningful appreciation of the size of something by comparing it with the size of something similar. For example, suppose the average weight of adults in New York is 130 lb. We can then compare the weight of any individual in terms of this average weight. Thus, a person weighing 160 lb would have a *per-unit* weight of 160 lb/130 lb = 1.23. Another person weighing 115 lb would have a *per-unit* weight of 115 lb/130 lb = 0.88.

This per-unit system of measurement has the advantage of giving the weight of a person in terms of some convenient standard, called the base of the system. Thus, knowing that a football player has a per-unit weight of 1.7 indicates that his weight is considerably above average and, furthermore, that his actual weight is 1.7 × 130 = 221 lb.

Note that when per-unit values are given they are always pure numbers. Thus, it would be incorrect to say that the football player weighs 1.7 lb. His weight is 1.7 *per-unit* where the selected "unit" is the weight of 130 lb.

To generalize, a per-unit system of measurement consists of selecting a convenient "measuring stick" and comparing similar things against it. In this book, we are particularly interested in selecting convenient measuring sticks for voltage, current, power and impedance.

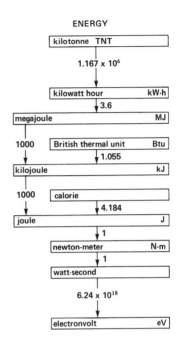

Figure 1-2
See Example 1-3.

1.10 Per-unit system with one base

If we select the size of only one quantity as our measuring stick, the per-unit system is said to have a single base. The base may be a power, a voltage, a current or a velocity. For example, suppose that three motors have power ratings of 25 hp, 40 hp and 150 hp. Let us select an arbitrary base power P_B of 50 hp. The corresponding per-unit ratings are then 25 hp/50 hp = 0.5; 40 hp/50 hp = 0.8 and 150 hp/50 hp = 3. Thus, in this per-unit world where the base is 50 hp, the three motors have power ratings of 0.5, 0.8 and 3, respectively.

We could equally well have selected a base power of 15 hp. In this case, the respective per-unit rating would be 25 hp/15 hp = 1.67, 40 hp/15 hp = 2.67 and 150 hp/15 hp = 10.

It is therefore essential to know the base of the per-unit system. If we do not know its value, the actual values of the quantities we are dealing with cannot be calculated.

The per-unit method can also be applied to impedances. Consider, for example, the circuit of Figure 1-3, composed of several resistors, capacitors and inductors having the impedances shown. If we decide to use an impedance of 1500 ohms as the base, the per-unit impedances are:

$$R_1(\text{pu}) = \frac{3500\ \Omega}{1500\ \Omega} = 2.33$$

$$R_2(\text{pu}) = \frac{450\ \Omega}{1500\ \Omega} = 0.30$$

$$X_L(\text{pu}) = \frac{4800\ \Omega}{1500\ \Omega} = 3.2$$

$$X_C(\text{pu}) = \frac{3000\ \Omega}{1500\ \Omega} = 2$$

The per-unit circuit (Fig. 1-4) contains the same elements as the real circuit but the impedances are now expressed in per-unit values. We can solve this circuit as we would any other circuit. For example, if we use complex notation, the per-unit circuit is that shown in Fig. 1-5.

1.11 Per-unit system with two bases

In electrotechnology, the per-unit system becomes particularly useful when *two* bases are used. The bases are usually a base voltage E_B and a power P_B. Thus, the selected base voltage may be 4 kV and the selected base power 500 kW.

The two base values can be selected quite independently of each other.

One interesting feature of the voltage/power per-unit system is that it automatically establishes a corresponding base current and base

Figure 1-3
Conventional circuit.

Figure 1-4
Per-unit circuit.

Figure 1-5
Per-unit circuit with j notation.

impedance. Thus, the base current I_B is

$$I_B = \frac{P_B}{E_B}$$

and the base impedance Z_B is

$$Z_B = \frac{E_B}{I_B}$$

For example, if the base voltage is 4 kV and the base power is 500 kW, the base current is

$I_\text{B} = P_\text{B}/E_\text{B} = 500\ 000/4000 = 125$ A

The base impedance is

$Z_\text{B} = E_\text{B}/I_\text{B} = 4000$ V/125 A $= 32\ \Omega$

In effect, by selecting the voltage/power per-unit system we also get a base current and a base impedance. Consequently, the so-called two-base per-unit system really gives us a 4-base per-unit system.

In order to understand the significance of this result, the reader should study the following examples. The bases are the same as before, namely:

$E_\text{B} = 4$ kV $I_\text{B} = 125$ A

$P_\text{B} = 500$ kW $Z_\text{B} = 32\ \Omega$

Example 1-4:
A 400 Ω resistor carries a current of 60 A. Calculate
a.   the per-unit resistance
b.   the per-unit current
c.   the per-unit voltage across the resistor
d.   the per-unit power dissipated in the resistor
e.   the actual E and P of the resistor

Solution:
a.   The per-unit resistance is
      $R(pu) = 400\ \Omega/32\ \Omega = 12.5$

b.   The per-unit current is
      $I(pu) = 60$ A/125 A $= 0.48$

c.   The per-unit voltage across the resistor is
      $E(pu) = I(pu) \times R(pu)$
         $= 0.48 \times 12.5$
         $= 6$

d.   The per-unit power is
      $P(pu) = E(pu) \times I(pu)$
         $= 6 \times 0.48$
         $= 2.88$

e.   The actual voltage across the resistor is
         $E = E_\text{B} \times E(pu)$
            $= 4$ kV $\times 6$
            $= 24$ kV

The actual power dissipated in the resistor is:
         $P = P_\text{B} \times P(pu)$
            $= 500$ kW $\times 2.88$
            $= 1440$ kW

Figure 1-6
See Example 1-5.

Example 1-5:
A 7.2 kV source delivers power to a 24 Ω resistor and a 400 kW motor (Fig. 1-6). Draw the equivalent per-unit circuit diagram and calculate:
a.   The per-unit $E(pu)$, $R(pu)$, $P(pu)$
b.   The per-unit current $I_2(pu)$
c.   The per-unit line current $I_\text{L}(pu)$
d.   The per-unit power absorbed by the resistor
e.   The actual power absorbed by the resistor
f.   The actual line current

Solution:
a.   The per-unit line voltage is:
      $E_1(pu) = 7.2$ kV/4 kV $= 1.8$

      The per-unit resistance is:
      $R(pu) = 24\ \Omega/32\ \Omega = 0.75$

      The per-unit power of the motor is:
      $P(pu) = 400$ kW/500 kW $= 0.8$

We can now draw the per-unit circuit (Fig. 1-7).

b.   The per-unit current I_2 is:
      $I_2(pu) = E(pu)/R(pu) = 1.8 \div 0.75$
            $= 2.4$

c. The per-unit current I_1 is:
$$I_1(pu) = P(pu)/E(pu) = 0.8/1.8$$
$$= 0.444$$
The per-unit line current I_L is:
$$I_L(pu) = I_1(pu) + I_2(pu)$$
$$= 0.444 + 2.4$$
$$= 2.844$$

d. The per-unit power in the resistor is:
$$P(pu) = E(pu) \times I_2(pu)$$
$$= 1.8 \times 2.4$$
$$= 4.32$$

Figure 1-7
Per-unit version of Figure 1-6.

e. The actual power in the resistor is:
$$P_2 = P_B \times P(pu)$$
$$= 500 \text{ kW} \times 4.32$$
$$= 2160 \text{ kW}$$

f. The actual line current is
$$I_2 = I_B \times I_L(pu)$$
$$= 125 \times 1.844 = 355.5 \text{ A.}$$

QUESTIONS AND PROBLEMS

1-1 Name the seven base units of the International System of Units.
1-2 Name five derived units of the SI.
1-3 Give the symbols of seven base units, paying particular attention to capitalization.
1-4 Why are some derived units given special names?

1-5 What are the SI units of force, pressure, energy, power and frequency?
1-6 Give the appropriate prefix for the following multipliers: 100, 1000, 10^6, 1/10, 1/100, 1/1000, 10^{-6}, 10^{-9}, 10^{15}

Express the following SI units in symbol form:

1-7	megawatt	1-21	millitesla
1-8	terajoule	1-22	millimeter
1-9	millipascal	1-23	revolution
1-10	kilohertz	1-24	megohm
1-11	gigajoule	1-25	megapascal
1-12	milliampere	1-26	millisecond
1-13	microweber	1-27	picofarad
1-14	centimeter	1-28	kilovolt
1-15	liter	1-29	megampere
1-16	milligram	1-30	kiloampere
1-17	microsecond	1-31	kilometer
1-18	millikelvin	1-32	nanometer
1-19	milliradian	1-33	milliliter

State the SI unit for the following quantities, and write the symbol:

1-34	rate of flow	1-38	density
1-35	frequency	1-39	power
1-36	plane angle	1-40	temperature
1-37	magnetic flux	1-41	mass

Give the names of the SI units that correspond to the following units:

1-42	Btu	1-53	pound force
1-43	horsepower	1-54	kilowatthour
1-44	line of flux	1-55	gallon per minute
1-45	inch	1-56	mho
1-46	angstrom	1-57	pound-force per square inch
1-47	cycle per second		
1-48	gauss	1-58	revolution
1-49	line per square inch	1-59	degree
1-50	°F	1-60	oersted
1-51	bar	1-61	ampere turn
1-52	pound mass		

Make the following conversions using the conversion charts:

1-62 10 square meters to square yards.

1-63 250 MCM to square millimeters.

1-64 1645 square millimeters to square inches.

1-65 13 000 circular mils to square millimeters.

1-66 640 acres to square kilometers.

1-67 81 000 watts to Btu per second.

1-68 33 000 foot pound-force per minute to kilowatts.

1-69 250 cubic feet to cubic meters.

1-70 10 foot pound-force to microjoules.

1-71 10 pound-force to kilogram-force.

1-72 60 000 lines per square inch to teslas.

1-73 1.2 teslas to kilogauss.

1-74 50 ounces to kilograms.

1-75 76 oersteds to amperes per meter.

1-76 5 000 meters to miles.

1-77 80 ampere hours to coulombs.

1-78 25 pound-force to newtons.

1-79 25 pounds to kilograms.

1-80 3 tonnes to pounds.

1-81 100 000 lines of force to webers.

1-82 0.3 pounds per cubic inch to kilograms per cubic meter.

1-83 2 inches or mercury to millibars.

1-84 200 pounds per square inch to pascals.

1-85 70 pounds-force per square inch to newtons per square meter.

1-86 15 revolutions per minute to radians per second.

1-87 A temperature of 120°C to kelvins.

1-88 A temperature of 200°F to kelvins.

1-89 A temperature difference of 120° Celsius to kelvins.

1-90 A resistance of 60 Ω is selected as the base resistance in a circuit. If it contains three resistors having actual values of 100 Ω, 3000 Ω and 20 Ω, calculate the per-unit value of each resistor.

1-91 A power of 25 kW and a voltage of 2400 V are selected as the base power and base voltage of a power system. Calculate the value of the base impedance and of the base current.

1-92 A resistor has a per-unit value of 5.3. If the base power is 250 kW and the base voltage is 12 470 V, calculate the ohmic value of the resistor.

1-93 A length of 4 m is selected as a base unit. Calculate
   a. the per-unit length of 1 mile
   b. the per-unit length of 1 foot
   c. the magnitude of the base area (in m^2)
   d. the magnitude of the base volume (in m^3)
   e. the per-unit value of a volume of 6000 m^3
   f. the per-unit value of an area of 2 square miles.

2

FUNDAMENTALS
OF ELECTRICITY
AND MAGNETISM

This chapter briefly reviews some of the fundamentals of electricity and magnetism. We assume the reader is already familiar with the basics covered here, including the solution of electric circuits. However, a review is useful because it focuses on those items that are particularly important in power technology. Furthermore, it establishes the notation used throughout this book to designate voltages and currents. Some of the other topics treated here will also provide the reader with a reference for subjects covered in later chapters.

2.1 Conventional and electron current flow

Consider the dry cell shown in Fig. 2-1, having one positive (+) and one negative (−) terminal. The difference of potential between them (measured in volts) is due to an excess of electrons at the negative terminal compared to the positive terminal.

If we connect a wire across the terminals, the potential difference causes an electric current to flow in the circuit. This current is composed of

a steady stream of electrons that come out of the negative terminal, move along the wire, and reenter the cell by the positive terminal (Fig. 2-2).

Before the electron theory of current flow

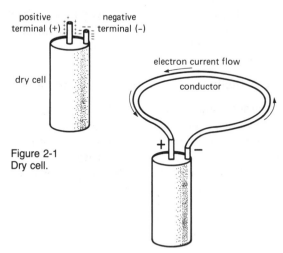

Figure 2-1
Dry cell.

Figure 2-2
Electron flow.

was fully understood, scientists of the 17th century arbitrarily decided that current in a conductor flows from the positive terminal to the negative terminal (Fig. 2-3). This so-called

conventional current flow

Figure 2-3
Conventional current flow.

conventional current flow is still used today, and is the accepted direction of current flow in electrical power technology.

In this book, we use the conventional current flow, but it is worth recalling that the actual electron flow is opposite to the conventional current flow.

2.2 Distinction between sources and loads

It is sometimes important to identify the sources and loads in an electric circuit. By definition, a *source delivers* electrical power whereas a *load absorbs* it. Every electrical device (motor, resistor, thermocouple, battery, capacitor, generator, etc.) that carries a current can be classified as either a source or a load. How can we tell the one from the other?

In order to establish a general rule, consider two black boxes A and B that are connected together by a pair of wires carrying a variable current I that is continually changing in direction (Fig. 2-4). Each box contains unknown devices and components that are connected in some way to the external terminals A_1, A_2 and B_1, B_2. A variable voltage exists between the terminals and its magnitude and polarity are also continually changing. Under such highly variable conditions, how can we tell whether A or B is a source or a load?

To answer the question, suppose we have appropriate instruments that enable us to determine the *instantaneous* polarity $(+)(-)$ of the voltage and the *instantaneous* direction of conventional current flow. The following rule then applies:

- **a device is a source whenever current flows out of the positive terminal;**

- **a device is a load whenever current flows into a positive terminal.**

If the instantaneous polarities and instantaneous current flow are as shown in Fig. 2-4, it follows from the above rule that box A is a source and box B is a load. However, if the current should reverse while the polarity remains the same, box B would become the source and box A the load.

The above rule for establishing whether a device is a source or load is very simple but it

Figure 2-4
Distinction between a source and a load.

has important applications, particularly in alternating current circuits.

Some devices, such as resistors, can behave only as loads. Other devices, such as photocells, can act only as sources. However, most devices can behave either as sources or as loads. Thus, when a battery delivers electric power it acts as a source (current flows out of the (+) terminal) but when it is being recharged it acts as a load (current flows into the (+) terminal). Electric motors act as loads on a system, but they can also behave like generators, if the electromechanical conditions are appropriate. The same thing is true of capacitors. When a capacitor is discharging it acts as a source and current flows out of the (+) terminal. On the other hand, when the capacitor is charging up it acts as a load, and current flows into the (+) terminal.

2.3 Sign notation

In arithmetic, we use the symbols (+) (−) to describe addition and subtraction. In electricity and mechanics, we broaden the meaning to indicate the *direction* of an electric current, a mechanical force, a rotational speed, etc., compared to an arbitrary chosen direction. For example, if the speed of a motor changes from + 100 r/min to − 400 r/min, it means that the direction of rotation has reversed. This interpretation of (+) and (−) is frequently met in the chapters that follow.

2.4 Sign notation for voltages

We now describe a system of notation that enables us to indicate the polarity of voltages. Figure 2-5 shows a source G having a positive terminal A and a negative terminal B. Terminal A is positive *with respect* to terminal B. Similarly, terminal B is negative *with respect* to terminal A. Note that terminal A is not positive by itself: it is only positive with respect to B.

The potential difference and the relative polarities of terminals A and B can be desig-

Figure 2-5
Double subscript notation to designate a voltage.

nated by the *double-subscript* notation, as follows:

- E_{AB} = + 100 V, which reads: **the voltage between A and B is 100 V and A is positive with respect to B.**

- E_{BA} = − 100 V, which reads: **the voltage between A and B is 100 V and B is negative with respect to A.**

Figure 2-6
If E_{21} = − 100 V terminal 2 is negative with respect to terminal 1.

As another example, if we know that the generator voltage in Fig. 2-6 has a value E_{21} = − 100 V, we know that the voltage between the terminals is 100 V and that terminal 2 is negative with respect to terminal 1. This double-subscript method of designating the magnitude and polarity of a voltage is particularly useful in analyzing three-phase circuits.

2.5 Another notation for voltages

Although we can represent the value and the polarity of voltages by the double-subscript notation (E_{12}, E_{ab}, etc.), we often prefer to use the *sign notation*. It consists of designating the voltage by a symbol (E_1, E_2, V, etc.) and identifying one of the terminals by a positive (+) sign. For example, Fig. 2-7 shows a source E_1 in which one of the terminals is *arbitrarily* marked with a positive (+) sign.

Figure 2-7
Sign notation to designate a voltage.

Figure 2-8
Circuit of Example 2-1.

With this notation, the following rules apply:

- **If we state that $E_1 = +10$ V, this means that the real polarity of the terminals corresponds to that indicated in the diagram. Furthermore, the magnitude of the voltage across the terminals is 10 V.**

- **Conversely, if $E_1 = -10$ V, the real polarity of the terminals is the reverse of that shown on the diagram. The magnitude of the voltage across the terminals is 10 V.**

Example 2-1:
The circuit of Fig. 2-8 consists of three sources V_1, V_2 and V_3, each having a terminal marked with a positive (+) sign. The sources are connected in series to a resistor R, using jumper wires A, B, C, D.

Determine the actual value and polarity of the voltage across each source, knowing that $V_1 = -4$ V, $V_2 = +10$ V and $V_3 = -40$ V.

Solution:
Using the rules just stated, we find that the true values and polarities are as shown in Fig. 2-9. However, in directing our attention to jumper A, it seems impossible that it can be both positive (+) and negative (−). However, we must remember that A is neither inherently positive or negative. It only has a polarity with respect to jumpers B and C, respectively. In effect, point A

Figure 2-9
Solution of Example 2-1.

is *negative* with respect to point B and *positive* with respect to point C. That is why A carries both a positive and a negative sign.

2.6 Graph of an alternating voltage

In the chapters that follow, we encounter sources whose voltages change polarity periodically. Such *alternating* voltages may be represented by means of a graph (Fig. 2-10). The vertical axis indicates the voltage at each instant, while the horizontal axis indicates the corresponding time. Voltages are positive when they are above the horizontal axis and negative when they are below. Figure 2-10 shows the voltage produced by the generator of Fig. 2-6.

Starting from zero, E_{21} gradually increases, attaining +100 V after 0.5 second. It then grad-

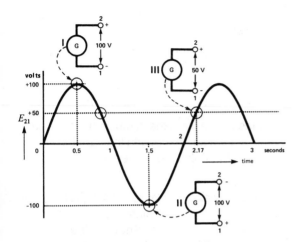

Figure 2-10
Graph of an alternating voltage having a peak of 100 V.

ually falls to zero at the end of one second. During this one-second interval, terminal 2 is positive with respect to terminal 1, because E_{21} is positive.

During the interval from 1 to 2 seconds, E_{21} is negative, therefore, terminal 2 is *negative* with respect to terminal 1. The actual voltages and polarities of the generator at 0.5, 1.5, and 2.17 seconds are shown by insets I, I I, I I I of Fig. 2-10.

2.7 Positive and negative currents

We also make use of positive and negative signs to indicate the direction of current flow. The signs are allocated with respect to a reference direction given on the circuit diagram. For

Figure 2-11
Current may flow from X to Y or Y to X.

example, the current in a circuit element such as a resistor (Fig. 2-11) may flow from X to Y or from Y to X. If one of these two directions is con-

sidered to be positive (+), the other is negative (−).

The *positive* direction is shown *arbitrarily* by means of an arrow (Fig. 2-12). Thus, if a current of 2 A flows from X to Y, it flows in the positive direction and is designated by the symbol + 2 A. Conversely, if current flows from Y to X (direction opposite to that of the arrow), it is designated by the symbol − 2 A.

Figure 2-12
Circuit element showing positive direction of current flow.

Example 2-2:
The current in a resistor R varies according to the graph shown in Fig. 2-13. Interpret the meaning of this curve.

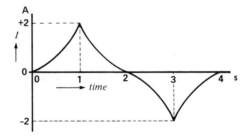

Figure 2-13
Circuit elements showing positive direction of current flow.

Solution:
According to the graph, the current increases from zero to + 2 A during the interval from 0 to

1 second. Because it is positive, the current flows from B to A in the resistor (direction of the arrow). During the interval from 1 to 2 seconds, the current decreases from $+2$ A to zero, but it still circulates from B to A in the resistor. Between 2 and 3 seconds, the current increases from zero to -2 A and, because it is negative, it really flows in a direction opposite to that of the arrow, that is, from A to B in the resistor.

2.8 Sinusoidal voltage

The ac voltage generated by commercial alternators is very nearly a perfect sine wave. It may therefore be expressed by the equation:

$$e = E_m \cos (2\pi ft + \theta) \qquad (2\text{-}1)$$

where

$e =$ instantaneous voltage [V]
$E_m =$ peak value of the sinusoidal voltage [V]
$f =$ frequency [Hz]
$t =$ time [s]
$\theta =$ a fixed angle [rad]

The expression $2\pi ft$ and θ are angles, expressed in radians. However, it is often more

Figure 2-14
Sinusoidal voltage having a peak value of 100 V and expressed by $e_{ab} = E_m \cos(360\, ft + 30°)$.

convenient to express the angle in degrees, as follows:

$$e = E_m \cos (360\, ft + \theta) \qquad (2\text{-}2)$$

or

$$e = E_m \cos (\phi + \theta) \qquad (2\text{-}3)$$

In these equations, the symbols have the same significance as before, and the variable angle ϕ ($= 360\, ft$) is also expressed in degrees.

Example 2-3:
The sine wave in Figure 2-14 represents the voltage across the terminals **a, b** of an ac motor that operates at 50 Hz. Knowing that $\theta = 30°$, and $E_m = 100$ V, calculate the voltage at t = 0 and at t = 27.144 s.

Solution:
The voltage at $t = 0$ is:
$$\begin{aligned} e_{ab} &= E_m \cos (360\, ft + \theta) \\ &= 100 \cos (360 \times 50 \times 0 + 30°) \\ &= 100 \cos 30° \\ &= 86.6 \text{ V} \end{aligned}$$

At this moment the voltage is 86.6 V and terminal **a** is positive with respect to terminal **b**.

The voltage at $t = 27.144$ s is:
$$\begin{aligned} e_{ab} &= 100 \cos (360 \times 50 \times 27.144 + 30°) \\ &= 100 \cos 488\ 622° \\ &= -20.8 \text{ V} \end{aligned}$$

Thus, at this moment the voltage is 35.3 V and terminal **a** is negative with respect to terminal **b**.

2.9 Converting cosine functions into sine functions

We can convert a cosine function of voltage or current into a sine function by adding 90° to the angle θ. Thus,

$$E_m \cos (360\, ft + \theta) = E_m \sin (360\, ft + \theta + 90°) \qquad (2\text{-}4)$$

Similarly, we can convert a sine function into a cosine function by subtracting 90° from the angle θ. Thus,

$$I_m \sin (360 \, ft + \theta) =$$
$$I_m \cos (360 \, ft + \theta - 90) \qquad (2\text{-}5)$$

2.10 Effective value of an ac voltage

Although the properties of an ac voltage are known when its frequency and peak value E_m are specified, it is much more common to use the effective value E_{eff}. For a voltage that varies sinusoidally, the relationship between E_{eff} and E_m is given by the expression:

$$E_{eff} = E_m / \sqrt{2} \qquad (2\text{-}6)$$

The effective value of an ac voltage is sometimes called the RMS (root mean square) value of the voltage. It is a measure of the heating effect of the ac voltage as compared to the same heating effect of a dc voltage. For example, an ac voltage having an effective value of 135 volts produces the same heating effect in a resistor as a dc voltage of 135 V.

The same remarks apply to the effective value of an ac current. Thus, a current that varies sinusoidally and whose peak value is I_m

possesses an effective value I_{eff} given by

$$I_{eff} = I_m / \sqrt{2} \qquad (2\text{-}7)$$

Most alternating current instruments are calibrated to show the *effective value* of voltage or current and *not the peak value* (Fig. 2-15). When the value of an alternating voltage or current is given it is understood that it is the effective value. Furthermore, the subscript in E_{eff} and I_{eff} is dropped and the effective values of voltage and current are simply represented by the symbols E and I.

Example 2-4:
A 60 Hz source having an effective voltage of 240 V delivers an effective current of 10 A to a circuit. The current lags the voltage by 30°. Draw the waveshape for E and I.

Solution:
a. The peak voltage is:
$$E_m = E \sqrt{2} = 240 \sqrt{2} = 339 \text{ V}$$

b. The peak current is:
$$I_m = I \sqrt{2} = 10 \sqrt{2} = 14.1 \text{ A}$$

Figure 2-15
Commercial voltmeters and ammeters are graduated in effective values. This range of instruments has scales ranging up to 2500 A and 9000 V. (*General Electric*)

c. Let us assume the voltage is given by:
$$e = E_m \cos (360 \times 60\ t - 90°)$$
$$= 339 \cos (21\ 600\ t - 90°)$$
$$= 339 \cos (\phi - 90°)$$

d. Owing to the phase lag of 30°, the current is given by:
$$i = I_m \cos (360\ ft - 90 - 30°)$$
$$= 14.1 \cos (21\ 600\ t - 120°)$$
$$= 14.1 \cos (\phi - 120°)$$

e. The waveshapes giving the instantaneous values of e and i are shown in Fig. 2-16.

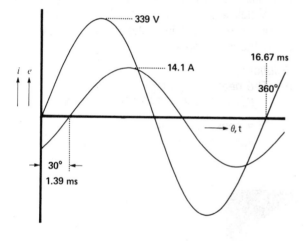

Figure 2-16
Graph showing the instantaneous values of voltage and current. The current lags 30° behind the voltage. The effective voltage is 240 V and the effective current is 10 A.

2.11 Phasor representation

In most power studies, the frequency is fixed, and so we simply take it for granted. Furthermore, we are not particularly concerned with the instantaneous voltages and currents but more with their magnitudes and the phase angle between them. And since the magnitudes are measured in terms of the effective values E rather than the peak values E_m, we are really only interested in E and θ.

This line of reasoning has given rise to the *phasor* method of representing voltages and currents.

The basic purpose of phasor diagrams is to show the magnitudes and phase angles between voltages and currents. A phasor is similar to a vector in the sense that it bears an arrow and its length is proportional to the effective value of the voltage or current it represents. The angle between two phasors is equal to the electrical phase angle between the quantities.

The following rules apply to phasors:
1. Two phasors are said to be in phase when they point in the same direction (Fig. 2-17). The phase angle between them is then zero.

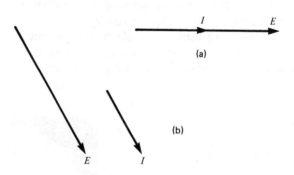

Figure 2-17
The current phasor I and voltage phasor E are in phase.

2. Two phasors are said to be out of phase when they point in different directions. The phase angle between them is the angle through which one of the phasors has to be rotated to make it point in the same direction as the other. Thus referring to Fig. 2-18, phasor I has to be rotated counterclockwise by an angle θ to make it point in the same direction as phasor E. Conversely, phasor E has to be

rotated clockwise by an angle θ to make it point in the same direction as phasor I. Consequently, whether we rotate one phasor or the other, we have to sweep through the same angle to make them line up.

Figure 2-18
Phase I lags behind phasor E by an angle of θ degrees.

3. If a phasor E has to be rotated *clockwise* to make it point in the same direction as phasor I, then phasor E is said to *lead* phasor I. Conversely, a phasor I is said to *lag* behind phasor E if phasor I has to be rotated counterclockwise to make it point in the same direction. Thus, referring to Fig. 2-18, it is clear that phasor E leads phasor I by θ degrees. But we could equally well say that I lags behind E by θ degrees.

4. Referring now to Fig. 2-19 we *could* rotate phasor I clockwise by an angle β to make it point in the same direction as phasor E. We could then say that phasor I leads phasor E by β degrees. But this is the same as saying that phasor I lags phasor E by θ degrees. In practice, we usually select the smaller phase angle between two phasors to designate the lag or lead situation.

Figure 2-19
Phasor I leads E by β degrees. But phasor I also lags E by θ degrees.

5. Phasors do not have to have a common origin, but may be entirely separate from each other, as shown in Fig. 2-20. By applying rule 3 above, we can see that E_1 is in phase with I_1 because they point in the same direction. Furthermore, phasor I_2 leads phasor E_1 by $90°$ and E_2 lags behind I_2 by $135°$.

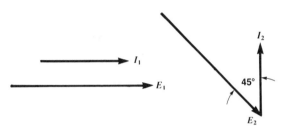

Figure 2-20
Phasors do not have to start from a common origin to show their magnitudes and phase relationships.

In the same way, the three phasors E_{ab}, E_{bc}, E_{ca} in Fig. 2-21a can be rearranged as shown in Fig. 2-21b. Note that E_{ab} in Fig. 2-21b still points in the same direction as E_{ab} in Fig. 2-21a, and the

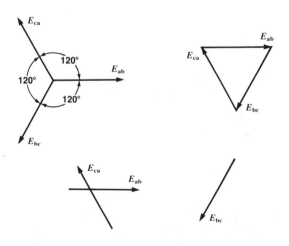

Figure 2-21
Different ways of showing the phase relationships between three voltages that are mutually displaced at $120°$.

same is true for the other phasors.

Figure 2-21c shows still another arrangement of the three phasors which does not in any way alter their magnitude or phase relationship.

The angle θ between two phasors is a measure of the time that separates their peak positive values. Knowing the frequency, we can calculate the time.

Example 2-5:
Draw the phasor diagram of the voltage and current in Fig. 2-16. Calculate the time interval between the positive peaks of E and I.

Solution:
To draw the phasor diagram, we select *any* arbitrary direction for phasor E, making its length equivalent to 240 V. Phasor I is then drawn so that it lags 30° behind E with a length equivalent to 10 A (Fig. 2-22). Knowing that the

Figure 2-22
Phasor diagram of the voltage and current given in Fig. 2-16.

frequency is 60 Hz, the time interval between the positive peaks is given by:

$$\theta = 360\ ft$$
$$30 = 360 \times 60\ t$$
$$t = 1.39\ ms$$

2.12 Harmonics

The voltages and currents in a power circuit are frequently not pure sine waves. The line voltages generally have a satisfactory waveshape but the currents are sometimes badly distorted, such as shown in Figure 2-23. This distortion can be produced by magnetic

Figure 2-23
This severely distorted 60 Hz current obtained on an electronic drive contains the following harmonics: fundamental (60 Hz) = 59 A; fifth harmonic (300 Hz) = 15.6 A; seventh harmonic (420 Hz) = 10.3 A. Higher harmonics are also present, but their amplitudes are small. (*Electro-Mécanik*).

saturation in the cores of transformers, or by the switching action of thyristors in electronic drives.

The distortion of a voltage or current can be traced to the harmonics it contains. A harmonic is any voltage whose frequency is an integral multiple of (2, 3, 4, etc., times) the line frequency.

Consider a set of sine waves in which the lowest frequency is f, and all other frequencies are integral multiples of f. By definition, the sine wave having the lowest frequency is called the *fundamental* and the other waves are called *harmonics*. For example, a set of sine waves whose frequencies are 20, 40, 100, and 380 Hz is said to possess the following components:

fundamental frequency: 20 Hz (the lowest frequency)
second harmonic: 40 Hz (2 × 20 Hz)
fifth harmonic: 100 Hz (5 × 20 Hz)
nineteenth harmonic: 380 Hz (19 × 20 Hz)

In order to understand the distorting effect of a harmonic, let us consider two sources e_1 and e_2 connected in series (Fig. 2-24a). Their frequencies are respectively 60 Hz and 180 Hz. The corresponding peak amplitudes are 100 V and 20 V. The fundamental (60 Hz) and the third harmonic (180 Hz) voltages are assumed to pass through zero at the same time, and both are perfect sine waves.

Because the sources are in series, the ter-

(a)

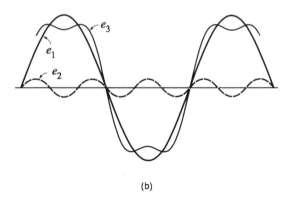

(b)

Figure 2-24
a. Two sinusoidal sources having different frequencies connected in series.
b. A fundamental and third harmonic voltage can together produce a flat-topped wave.

minal voltage e_3 is equal to the sum of the instantaneous voltages produced by each source. The resulting terminal voltage is a flat-topped wave (Fig. 2-24b). Thus, the sum of a fundamental voltage and a harmonic voltage yields a nonsinusoidal waveform whose degree of distortion depends upon the magnitude of the harmonic it contains.

We can produce a periodic voltage or current of any conceivable shape. All we have to do is to add together a fundamental component and an arbitrary set of harmonic components. For example, we can generate a square wave having an amplitude of 100 V and a frequency of 50 Hz by connecting the following sine wave sources in series:

TABLE 2 A	100 V SQUARE WAVE		
Harmonic	amplitude [V]	freq. [Hz]	relative amplitude
fundamental	127.3	50	1
third	42.44	150	1/3
fifth	25.46	250	1/5
seventh	18.46	350	1/7
ninth	14.15	450	1/9
.	.	.	.
.	.	.	.
.	.	.	.
127th	1.00	6350	1/127
.	.	.	.
.	.	.	.
.	.	.	.
n^{th}	127.3/n	50 n	1/n

A square wave is thus composed of a fundamental wave and an infinite number of harmonics. The higher harmonics have smaller and smaller amplitudes, and they are consequently less important. However, these high-frequency harmonics produce the steep sides and pointy corners of the square wave. In practice, square waves are not usually produced by adding sine waves, but the example does show that *any* wave can be built up from a fundamental wave and an appropriate number of harmonics.

Conversely, we can *decompose* a given periodic wave into its fundamental and harmonic components. However, the harmonic analysis of such waveforms is beyond the scope of this book.

Harmonic voltages and currents are usually undesirable, but in some ac circuits, they are also unavoidable. Harmonics are created by nonlinear loads, such as electric arcs and saturated magnetic circuits. They are also produced whenever voltages and currents are periodically switched, such as in power electronic circuits. All these circuits produce distorted waveshapes that are rich in harmonics.

In ac circuits, the fundamental current and fundamental voltage together produce fundamental power. This fundamental power is the useful power that causes a motor to rotate, for example. On the other hand, the product of a harmonic voltage times the corresponding harmonic current produces a harmonic power. The latter is usually dissipated as heat in the ac circuit and consequently, does no useful work. Harmonic currents and voltages should therefore be kept as small as possible.

It should be noted that the product of a fundamental voltage and a harmonic current yields zero power.

2.13 Energy in an inductor

A coil stores energy in its magnetic field when it carries a current I. The energy is given by:

$$W = \frac{1}{2} LI^2 \qquad (2\text{-}8)$$

where

W = energy stored in the coil [J]
L = inductance of the coil [H]
I = current [A]

If the current varies, the stored energy rises and falls in step with the current. Thus, whenever the current increases, the coil absorbs energy and whenever the current falls, energy is released.

The properties of an inductor are more fully discussed in Sec. 2.31.

2.14 Energy in a capacitor

A capacitor stores energy in its electric field whenever a voltage E appears across its terminals. The energy is given by:

$$W = \frac{1}{2} CE^2 \qquad (2\text{-}9)$$

where
W = energy stored in the capacitor [J]
C = capacitance of the capacitor [F]
E = voltage [V]

Example 2-6:
A coil having an inductance of 10 mH is connected in series with a 100 μF capacitor. The instantaneous current in the circuit is 40 A and the instantaneous voltage across the capacitor is 800 V. Calculate the energy stored in the electric and magnetic fields at this moment.

Solution:
The energy stored in the coil is:
$$W = 1/2\ LI^2 = 1/2 \times 10 \times 10^{-3} \times 40^2$$
$$= 8\text{J}$$

The energy stored in the capacitor is:
$$W = 1/2\ CE^2 = 1/2 \times 100 \times 10^{-6} \times 800^2$$
$$= 32\ \text{J}$$

2.15 Some useful equations

We terminate this section with a list of useful equations (Table 2B) that are frequently required when solving ac circuits. The equations are given without proof on the assumption that the student already possesses a knowledge of ac circuits in general.

2.16 Magnetic field intensity H, flux density B

Whenever a magnetic flux Φ exists in a body or component, it is due to the presence of a magnetic field intensity H, given by:

$$H = U/l \qquad (2\text{-}18)$$

where

H = magnetic field intensity [A/m]
U = magnetomotive force acting on the component [A] (or ampere turn)
l = length of the component [m]

The resulting magnetic flux density, is given by:

$$B = \Phi/A \qquad (2\text{-}19)$$

where

B = flux density [T]
Φ = flux in the component [Wb]
A = cross section of the component [m²]

There is a definite relationship between the flux density (B) and the magnetic field intensity (H) of any material. This relationship is usually expressed graphically by the B-H curve of the material.

TABLE 2B   IMPEDANCE OF SOME COMMON AC CIRCUITS

Circuit diagram	Impedance	Equation
	$X_L = 2\pi f L$	(2-10)
	$X_C = \dfrac{1}{2\pi f C}$	(2-11)
	$Z = \sqrt{R^2 + X_L{}^2}$	(2-12)
	$Z = \sqrt{R^2 + X_C{}^2}$	(2-13)
	$Z = \sqrt{R^2 + (X_L - X_C)^2}$	(2-14)
	$Z = \dfrac{R\,X_L}{\sqrt{R^2 + X_L{}^2}}$	(2-15)
	$Z = \dfrac{R\,X_C}{\sqrt{R^2 + X_C{}^2}}$	(2-16)
	$Z = \dfrac{X_C\sqrt{R^2 + X_L{}^2}}{\sqrt{R^2 + (X_L - X_C)^2}}$	(2-17)

2.17 *B-H* curve of a vacuum

In vacuum, the magnetic flux density B is proportional to the magnetic field intensity H, and is expressed by the equation:

$$B = \mu_o H \qquad (2\text{-}20)$$

where

B = flux density [T]
H = magnetic field intensity [A/m]
μ_o = magnetic constant $[= 4\pi \times 10^{-7}]*$

In the SI, the magnetic constant is fixed, by definition. It has a numerical value of $4\pi \times 10^{-7}$ or 1/800 000 approximately. This enables us to write Eq. 2-20 in the approximate form:

$$H = 800\ 000\ B \qquad (2\text{-}21)$$

The *B-H* curve of a vacuum is a straight line. A vacuum never saturates, no matter how great the flux density may be (Fig. 2-25). The curve shows that a magnetic field intensity of 800 A/m produces a flux density of 1 millitesla.

Nonmagnetic materials such as copper, paper, rubber and air have *B-H* curves almost identical to that of a vacuum.

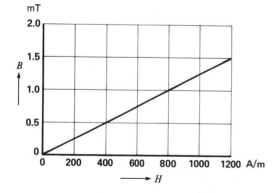

Figure 2-25
B-H curve of a vacuum and of nonmagnetic materials.

* The complete expression for μ_o is $4\pi \times 10^{-7}$ henry/meter.

2.18 *B-H* curve of a magnetic material

The flux density in a magnetic material also depends upon the magnetic field intensity. Its value is given by:

$$B = \mu_o \mu_r H \qquad (2\text{-}22)$$

where B, μ_o and H have the same significance as before, and μ_r is the *relative permeability* of the material.

The value of μ_r is not constant, but varies with the flux density in the material. Consequently, the relationship between B and H is not linear, and this makes Eq. 2-22 rather impractical to use. We usually show the relationship by means of a *B-H* saturation curve. Thus, Fig. 2-26 shows typical saturation curves of three materials commonly used in electrical machines: silicon iron, cast iron, and cast steel. The curves show that a magnetic field of 2000 A/m produces a flux density of 1.4 T in cast steel and 0.5 T in cast iron.

2.19 Determining the relative permeability

The relative permeability μ_r of a material is the ratio of the flux density in the material to the flux density that would be produced in a vacuum, using the same magnetic field intensity H.

Given the saturation curve of a magnetic material, it is easy to calculate the relative permeability using the approximate equation:

$$\mu_r \sim 800\ 000\ B/H \qquad (2\text{-}23)$$

where

B = flux density in the magnetic material [T]
H = corresponding magnetic field intensity [A/m]

Example 2-7:
Determine the permeability of silicon iron (1%) at a flux density of 1.4 T.

Solution:
Referring to the saturation curve (Fig. 2-26), we see that a flux density of 1.4 T requires a mag-

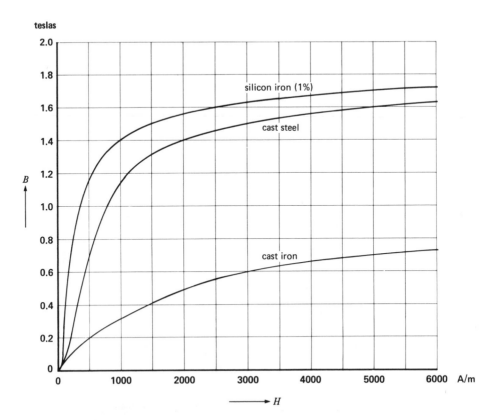

Figure 2-26
B-H saturation curves of three magnetic materials.

netic field intensity of 1000 A/m. Consequently:

$$\mu_r = 800\ 000\ B/H \qquad \text{Eq. 2-23}$$
$$= 800\,000 \times 1.4/1000 = 1120$$

At this flux density, the steel is 1120 times more permeable than a vacuum (or air). Figure 2-27 shows the saturation curves of a broad range of materials from vacuum to Permalloy®, one of the most permeable magnetic materials known. Note that as the magnetic field intensity increases the magnetic materials saturate and all the *B-H* curves eventually follow the *B-H* curve of a vacuum.

2.20 Faraday's law of electromagnetic induction

In 1831, while pursuing his experiments, Joseph Faraday made one of the most important discoveries in electromagnetism. Now known as Faraday's law of electromagnetic induction, it revealed a fundamental relationship between the voltage and flux in a circuit. Faraday's law states that:

1. **If the flux linking a loop (or turn) varies as a function of time, a voltage is induced between its terminals.**

2. **The value of the induced voltage is proportional to the rate of change of flux.**

By definition, and according to the SI system of units, when the flux inside a loop varies at the rate of 1 weber per second, a voltage of 1 V is

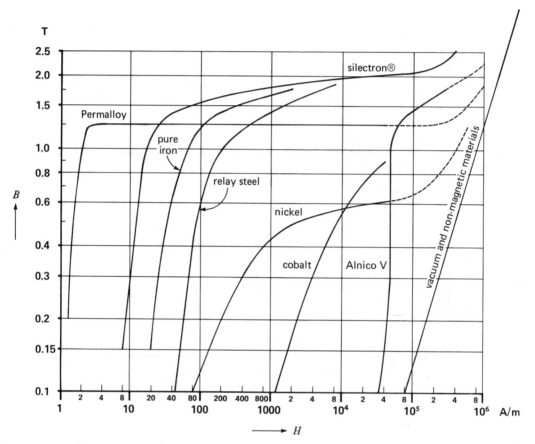

Figure 2-27
Saturation curves of magnetic and non magnetic materials.
Note that all curves become asymptotic to the *B-H* curve of
a vacuum when *H* is high.

induced between its terminals. Consequently, if the flux varies inside a coil of N turns, the voltage induced is given by:

$$E = N \frac{\Delta\Phi}{\Delta t} \qquad (2\text{-}24)$$

where

E = induced voltage [V]
N = number of turns in the coil
$\Delta\Phi$ = change of flux inside the coil [Wb]
Δt = time interval during which the flux changes [s]

Faraday's law of electromagnetic induction opened the door to a host of practical applications and established the basis of operation of transformers, generators and alternating current motors.

Example 2-8:

A coil of 2000 turns surrounds a flux of 5 mWb produced by a permanent magnet (Fig. 2-28). The magnet is suddenly withdrawn causing the flux inside the coil to drop to 2 mWb in 1/10 of a second. What is the average voltage induced?

Figure 2-28
Voltage induced by a moving magnet. Example 2-8.

Solution:
The flux change is:

$$\Delta\Phi = (5 \text{ mWb} - 2 \text{ mWb}) = 3 \text{ mWb}$$

Because this change takes place in 1/10 of a second (Δt), the average induced voltage is:

$$E = N \ \frac{\Delta\Phi}{\Delta t} = 2000 \times \frac{3}{1000 \times 1/10}$$

$$= 60 \text{ V}$$

2.21 Voltage induced in a conductor

In many motors and generators the coils move with respect to a flux that is fixed in space. The relative motion produces a change in the flux linking the coils and, consequently, a voltage is induced according to Faraday's law. However, in this special (although common) case, it is easier to calculate the induced voltage with reference to the *conductors*, rather than with reference to the coil itself. In effect, whenever a conductor "cuts" a magnetic field, a voltage is induced across its terminals. The value of the induced voltage is given by:

$$E = Blv \qquad (2\text{-}25)$$

where

$E =$ induced voltage [V]
$B =$ flux density [T]
$l =$ active length of the conductor in the magnetic field [m]
$v =$ relative speed of the conductor [m/s]

Example 2-9:
The stationary conductors of a large generator have a length of 2 m and are cut by a field of 0.6 teslas, moving at a speed of 100 m/s (Fig. 2-29). Calculate the voltage induced in each conductor.

Solution:
According to Eq. 2-25, we find:

$$E = Blv$$

$$= 0{,}6 \times 2 \times 100 = 120 \text{ V}$$

Figure 2-29
Voltage induced in a stationary conductor. Example 2-9.

2.22 Lorentz force on a conductor

When a current-carrying conductor is placed in a magnetic field, it is subjected to a force which we call *electromagnetic force*, or Lorentz force. This force is of fundamental importance because it constitutes the basis of operation of motors, of generators and of many electrical instruments. The magnitude of the force depends upon the orientation of the conductor with respect to the direction of the field. The force is greatest when the conductor is perpendicular to the field (Fig. 2-30) and zero when it is parallel to it (Fig. 2-31). Between these two extremes, the force has intermediate values.

The maximum force acting on a straight conductor is given by:

$$F = BIl \qquad (2\text{-}26)$$

where

$F =$ force acting on the conductor [N]
$B =$ flux density of the field [T]
$l =$ active length of the conductor [m]
$I =$ current in the conductor [A]

Figure 2-30
Force on a conductor.

Example 2-10:
A conductor 3 m long carrying a current of 200 A is placed in a magnetic field whose density is 0.5 T. Calculate the force on the conductor if it is perpendicular to the lines of force (Fig. 2-30).

Solution:
$$\begin{aligned} F &= BIl \\ &= 0.5 \times 3 \times 200 = 300 \text{ N} \end{aligned}$$

Figure 2-31
Force = 0

2.23 Direction of the force acting on a straight conductor

We know that when a conductor carries a current, it is surrounded by a magnetic field. For a current flowing into the page of this book, the circular lines of force have the direction shown in Figure 2-32a. The same figure shows the magnetic field created between the poles of a powerful permanent magnet.

The magnetic field does not, of course, have the shape shown in the figure because lines of force never cross each other. What, then, is the shape of the resulting field?

To answer the question, we observe that the lines of force created respectively by the conductor and by the permanent magnet act in the same direction above the conductor and in opposite directions below it. Consequently, the number of lines above the conductor must be greater than the number below. The resulting magnetic field therefore has the shape given in Figure 2-32b.

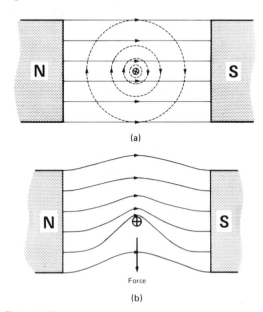

(a)

(b)

Figure 2-32
a. Magnetic field due to magnet and conductor.
b. Resulting magnetic field pushes the conductor downwards.

Recalling that lines of flux act like stretched elastic bands, it is easy to visualize that a force acts upon the conductor, tending to push it downward.

2.24 Residual flux density and coercive force

Consider the coil of Figure 2-33a which surrounds a magnetic material bent in the shape of a ring. The coil carries a current whose value and direction can be changed at will. Starting from zero, we gradually increase I, so that H and B increase. This increase traces out curve **oa** of Figure 2-33b. The flux density reaches a value B_m for a magnetic field H_m.

If the current is now gradually reduced to zero, the flux density B does not follow the original curve, but moves along a curve **ab** situated above **oa**. In effect, as we reduce the field intensity, the magnetic domains that were lined up under the influence of field H_m tend to retain their original orientation. This phenomenon is called *hysteresis*. Consequently, when H is zero, a substantial flux density remains. It is called residual flux density or *residual induction (B_r)*.

If we wish to eliminate this residual flux, we have to reverse the current in the coil and gradually increase H in the opposite direction. As we do so, we move along curve **bc**. The magnetic domains gradually change their previous orientation, until the flux density becomes zero at point **c**. The magnetic field intensity required to reduce the flux to zero is called *coercive force (H_c)*.

In reducing the flux density from B_r to zero, we also have to furnish energy. This energy is used to overcome the "frictional" resistance of the magnetic domains as they oppose the change in orientation. The energy supplied is dissipated as heat in the material. In effect, a very sensitive thermometer would indicate a slight temperature rise as the ring is being demagnetized.

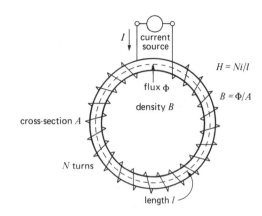

Figure 2-33a
Method of determining the *B-H* properties of a magnetic material.

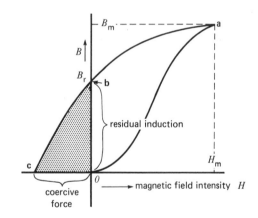

Figure 2-33b
Residual induction and coercive force.

2.25 Hysteresis loop

Transformers and most electric motors operate on alternating current. In such devices, the flux in the iron changes continuously both in value and direction. The magnetic domains are therefore oriented first in one direction, then the other, at a rate that depends upon the frequency. Thus, if the flux has a frequency of 60 Hz, the domains describe a complete cycle every 1/60 of a second, passing successively through peak densities $+B_m$ and $-B_m$ as the peak magnetic field alternates between $+H_m$ and $-H_m$. If we draw the flux density B as a function of H, we obtain a closed curve called a hysteresis loop (Fig. 2-34). The residual induction B_r and coercive force H_c have the same significance as before.

2.26 Hysteresis loss

In describing a hysteresis loop, the flux moves successively from $+B_m$, $+B_r$, 0, $-B_m$, $-B_r$, 0 and $+B_m$, corresponding respectively to points **a, b, c, d, e, f**, and **a**, of Figure 2-34. The magnetic material absorbs energy during each cycle and this energy is dissipated as heat. We can prove that the amount of heat released (expressed in J/m^3) is equal to the area of the hysteresis loop.

To reduce hysteresis losses, we select magnetic materials that have a narrow hysteresis loop, such as the grain-oriented silicon steel used in the cores of alternating-current transformers.

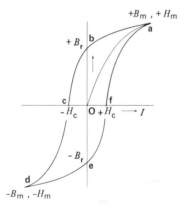

Figure 2-34
Hysteresis loop. If B is expressed in teslas and H in amperes per meter, the area of the loop is the energy dissipated per cycle, in joules per kilogram.

2.27 Hysteresis losses caused by rotation

Hysteresis losses are also produced when a piece of iron rotates in a constant magnetic field. Consider, for example, an iron armature AB that revolves in a field produced by permanent magnets N, S (Fig. 2-35). The magnetic domains in the armature tend to line up with the magnetic field, irrespective of the position of the armature. Consequently, as the armature rotates, the N poles of the domains point first towards A and then towards B. A complete reversal occurs therefore every half revolution, as can be seen in Figures 2-35a and 2-35b. Consequently, the magnetic domains in the armature reverse periodically, even though the magnetic field is constant. Hysteresis losses are produced just as they are in an ac magnetic field.

Figure 2-35
Hysteresis losses due to rotation

2.28 Eddy currents

Consider an ac flux Φ that links a rectangular-shaped conductor (Fig. 2-36). According to Faraday's law, an ac voltage E_1 is induced across its terminals.

If the conductor is short-circuited, a substantial alternating current I_1 will flow, causing the conductor to heat up. If a second conductor is placed inside the first, a smaller voltage is induced because it links a smaller flux. Consequently, the short-circuit current I_2 is less than

I_1 and so, too, is the dissipated power. Figure 2-37 shows four such concentric conductors carrying currents I_1, I_2, I_3, I_4. The currents are progressively smaller as the conductor area surrounding the flux decreases.

In Figure 2-38, the ac flux passes through a solid metal plate. It is basically equivalent to a densely-packed set of rectangular conductors touching each other. Currents swirl back and forth inside the plate, following the paths shown in the figure. These so-called *eddy currents* (or Foucault currents) can be very large owing to the low resistance of the plate. Consequently, a metal plate that is pierced by an ac flux becomes warm. In this regard, special care has to be taken in transformers so that stray leakage fluxes do not cause sections of the enclosing tank to overheat.

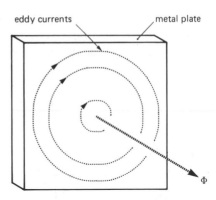

Figure 2-38
Large ac eddy currents are induced in a solid metal plate.

2.29 Eddy-currents in a stationary iron core

The eddy current problem becomes particularly important when iron has to carry an ac flux. This is the case in all ac motors and transformers. Figure 2-39a shows a coil carrying an ac current which produces an ac flux in a solid iron core. Eddy currents are set up as shown, and they flow throughout the entire length of the core. A large core could eventually become red hot (even at a frequency of 60 Hz) owing to these eddy-current losses.

We can reduce the losses by splitting the core in two along its length, taking care to insulate the two sections from each other (Fig. 2-39b). The voltage induced in each section is one-half of what it was before, with the result that the eddy currents, and the corresponding losses, are considerably reduced.

If we continue to subdivide the core, we find that the losses decrease progressively. In practice, the core is composed of stacked laminations, usually a fraction of a millimeter thick. Furthermore, a small amount of silicon is alloyed with the steel to increase its resistivity, thereby reducing the losses still more (Fig. 2-39c).

The cores of ac motors and generators are therefore always laminated. A thin coating of insulation covers each lamination to prevent

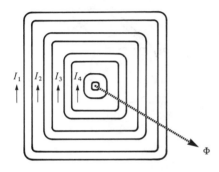

Figure 2-36
An ac flux Φ induces voltage E_1.

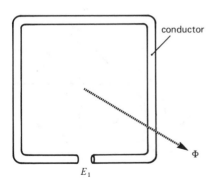

Figure 2-37
Concentric conductors carry ac currents due to ac flux Φ.

electrical contact between them. The laminations are stacked on top of each other and are tightly held in place by bolts and appropriate end-pieces.

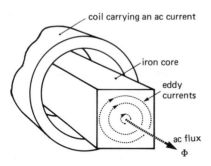

Figure 2-39a
Solid iron core carrying an ac flux.

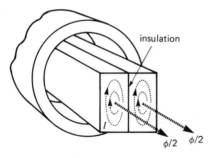

Figure 2-39b
Eddy currents are reduced by splitting the core in half.

Figure 2-39c
Core built up of thin, insulated laminations.

2.30 Eddy-current losses in a revolving core

The stationary field in direct current motors and generators produces a constant dc flux. This constant flux induces eddy currents in the revolving armature. To understand how they are induced, consider a solid cylindrical iron core that revolves between the poles of a magnet (Fig. 2-40a). As it turns, the core cuts flux lines and, according to Faraday's law, a voltage is induced along its length having the polarities shown. Owing to this voltage, large eddy currents flow in the core because its resistance is very low (Fig. 2-40b). These eddy currents produced large I^2R losses which are immediately converted into heat. The power loss is proportional to the square of the speed and the square of the flux density.

(a)

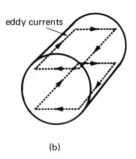

(b)

Figure 2-40
a. Voltage induced in a revolving armature and
b. large eddy currents are induced.

(a)

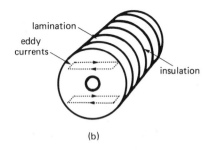

(b)

Figure 2-41
a. Armature built up of thin laminations and
b. much smaller eddy currents are induced.

To reduce the eddy-current losses, we laminate the armature using thin circular laminations that are insulated from each other. The laminations are tightly stacked with the flat side running parallel to the flux lines (Fig. 2-41).

2.31 Current in an inductor

It is well known that in an inductive circuit the voltage and current are related by the equation:

$$e = L \frac{\Delta i}{\Delta t} \qquad (2\text{-}27)$$

where

e = instantaneous voltage induced in the circuit [V]
L = inductance of the circuit [H]
$\Delta i / \Delta t$ = rate of change of current [A/s]

This equation enables us to calculate the instantaneous voltage e, when we know the

rate of change of current. However, it often happens that e is known and we want to calculate the resulting current I. We can use the same equation, but the solution requires a knowledge of advanced mathematics. To get around this problem, we can use a graphical solution, called the *volt-second method*. It yields the same results, and has the advantage of enabling us to visualize how the current in an inductor increases and decreases with time, in response to a known applied voltage.

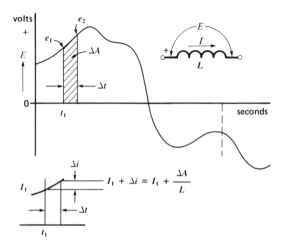

Figure 2-42
Variable voltage applied across an inductor and resulting change in current. Initial current is I_1.

Consider, for example, Fig. 2-42, in which a variable voltage E appears across an inductance L. Suppose the inductance carries a current I_1 at a time $t = t_1$. We want to determine the current a very short time later, say after an interval Δt. From Eq. 2-27 we can write

$$\Delta i = \frac{1}{L} e \Delta t$$

which means that

change in current Δi during a short interval Δt

$$= \frac{1}{L} \left(\begin{array}{l} \text{average voltage } e \text{ during the interval} \\ \times \text{ duration } \Delta t \text{ of the interval} \end{array} \right)$$

$$= \frac{1}{L} \frac{(e_1 + e_2)}{2} \times \Delta t$$

$$= \frac{1}{L} \begin{pmatrix} \text{area } \Delta A \text{ under the voltage} \\ \text{curve during the interval } \Delta t \end{pmatrix}$$

$$= \frac{1}{L} \begin{pmatrix} \text{volts-seconds across the induc-} \\ \text{tance during the interval } \Delta t \end{pmatrix}$$

Therefore, the current in the inductance after the interval Δt is:

I at time $(t_1 + \Delta t)$ = initial current + Δi

$$= I_1 + \frac{\Delta A}{L}$$

We are usually more interested in calculating the current at a time t_2, when t_2 is many Δt intervals after t_1 (Fig. 2-43). We then have to add the incremental changes in current Δi during the "long" period $(t_2 - t_1)$. As a result we find:

Figure 2-43
Volt-seconds are gained (and lost) when a variable voltage is applied across an inductor.

current I_2 at time t_2

= initial current $I_1 + (\Delta i_1 + \Delta i_2 + \Delta i_3 + \ldots)$

$$= I_1 + \frac{1}{L} (e_1 \Delta t_1 + e_2 \Delta t_2 + e_3 \Delta t_3 + \ldots)$$

$$= I_1 + \frac{1}{L} (\Delta A_1 + \Delta A_2 + \Delta A_3 + \ldots)$$

$$= I_1 + \frac{1}{L} \begin{pmatrix} \text{algebraic sum of all the little} \\ \text{areas under the voltage curve} \\ \text{between } t_1 \text{ and } t_2 \end{pmatrix}$$

$$I_2 = I_1 + \frac{1}{L} \begin{pmatrix} \text{net area } A \text{ under the voltage} \\ \text{curve between } t_1 \text{ and } t_2 \end{pmatrix}$$

The values of e_1, e_2, e_3, etc. may be (+) or (−) and therefore the little areas ΔA_1, ΔA_2, ΔA_3, etc. may be (+) or (−). The sum of these (+) and (−) values of the ΔA's gives the net area under the voltage curve between t_1 and t_2.

Thus, in Fig. 2-44 the net area A after time interval T is equal to $(A_1 - A_2)$ volt-seconds. In general, the corresponding current is then given by:

$$I = I_1 + A/L \qquad (2\text{-}28)$$

where

I_1 = current at start of interval T
I = current after time interval T [A]
A = net area under the volt-time curve during time T [V·s]
L = inductance [H]

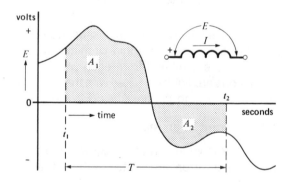

Figure 2-44
The net volt-seconds during interval T is equal to the algebraic sum of the areas A_1 and A_2.

Consider, for example an inductor L having negligible resistance, connected to a source whose voltage varies according to the curve of Fig. 2-45a. If the initial current is zero, the value at instant t_1 is:

$$I = A_1/L$$

As time goes by, the area under the curve increases progressively, and so does the current. However, the current reaches its maximum value at time t_2 because at this moment the area under the voltage curve ceases to increase any more. Beyond t_2 the voltage becomes negative and, consequently, the net area begins to diminish. At instant t_3, for example, the net area is equal to $(A_1 + A_2 - A_3)$ and the corresponding current is:

$$I = (A_1 + A_2 - A_3)/L$$

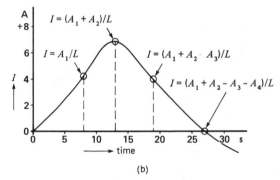

(b)

Figure 2-45
a. An inductor stores volt-seconds.
b. Current in the inductor.

At instant t_4, the negative area $(A_3 + A_4)$ is exactly equal to the positive area $(A_1 + A_2)$. The net area is zero and so the current is also zero. After instant t_4, the current becomes negative; in other words, it changes direction.

Another way of looking at the circuit is to consider that the inductor accumulates volt-seconds during the interval from 0 to t_2. As it becomes "charged up" with volt-seconds, the current increases in direct proportion to the volt-seconds received. Then during the "discharge" period from t_2 to t_4 the inductor loses volt-seconds and the current decreases accordingly. An inductor, therefore, behaves very much like a capacitor. However, instead of storing ampere-seconds (coulombs), an inductor stores volt-seconds. It is interesting to note that 1 volt-second is equal to 1 weber.

Figure 2-45b shows the instantaneous current obtained when the voltage of Fig. 2-45a is applied to an inductance of 100 H. The initial current is zero, and the current rises to a maximum of 6.9 A before again dropping to zero after a time interval of 27 s.

Important Note: If the current at the beginning of an interval T is not zero, we simply add the initial value to all the current values calculated by the volt-second method.

Example 2-11:
The voltage across the terminals of an inductor of 2 H varies according to the curve given in Fig. 2-46. a) Calculate the instantaneous current I in the circuit, knowing that the initial current is zero. b) Repeat the calculations for an initial current of 7 A.

Solution:
a. **Interval from zero to 3 s:** During this interval, the area in volt-seconds increases uniformly and progressively. Thus, after one second, the area A is 4 V·s; after two seconds, it is 8 V·s; and so forth. Using the expression $I = A/L$, the current builds up to the following respective values: 2 A, 4 A, etc., attaining a final value of 6 A after three seconds.

Interval from 3 s to 5 s: The area continues to increase, but at a slower rate, because voltage E is smaller than before.

Figure 2-46
See Example 2 - 11

When $t = 5$ s, the total surface starting from the beginning is 16 V·s; therefore the current is 16 V·s/2 H = 8 A.

Interval from 5 s to 7 s: The surface increases by 4 squares which is equivalent to 8 V·s. Consequently, the current increases by 4 A. Note that the current no longer follows a straight line because the voltage is not constant during this interval.

Interval from 7 s to 8 s: The voltage suddenly changes polarity with the result that the 8 V·s during this interval subtract from the volt-seconds that were accumulated previously. The net area from the beginning is therefore 24 V·s − 8 V·s = 16 V·s. Consequently, the current at the end of this interval is $I = 16/2 = 8$ A.

Interval from 8 s to 10 s: Because the terminal voltage is zero during this interval, the net volt-second area does not change and neither does the current (remember that we have assumed zero coil resistance).

Interval from 10 s to 14 s: The negative volt-seconds continue to accumulate and at $t = 14$ s, the negative area is equal to the positive area, and so the net current is zero. Beyond this point, the current changes direction.

b. With an initial current of $+7$ A, we have to add 7 A to each of the currents calculated previously. The new current wave is simply 7 A above the curve shown in Fig. 2-46. Thus, the current at $t = 11$ s is $6 + 7 = 13$ A.

QUESTIONS AND PROBLEMS

2-1 Three dc sources G_1, G_2, G_3 (Fig. 2-47) generate voltages as follows:
$E_{12} = -100$ V;
$E_{34} = -40$ V;
$E_{56} = +60$ V.

Show the actual polarity $(+)$ $(-)$ of the terminals in each case.

Figure 2-47
See Problems 2-1 and 2-2.

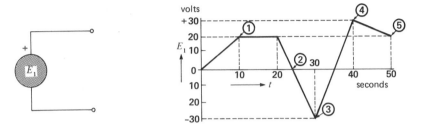

Figure 2-48.
See problem 2-3.

2-2 In Problem 2-1, if the three sources are connected in series, determine the voltage and polarity across the open terminals if the following terminals are connected together:
a. terminals 2-3 and 4-5;
b. terminals 1-4 and 3-6;
c. terminals 1-3 and 4-6.

2-3 Referring to Figure 2-48, show the voltage and the actual polarity of the generator terminals at instants 1, 2, 3, and 4.

2-4 A conductor 2 m long moves at a speed of 60 km/h through a magnetic field having a flux density of 0.6 T. Calculate the induced voltage.

2-5 A coil having 200 turns links a flux of 3 mWb, produced by a permanent magnet. The magnet is moved, and the flux linking the coil falls to 1.2 mWb in 0.2 s. Calculate the average voltage induced.

2-6 What is the SI unit of:
a. magnetic flux;
b. magnetic flux density;
c. magnetic field intensity;
d. magnetomotive force.

2-7 Referring to Figure 2-26, calculate the relative permeability of cast iron at 0.2 T, 0.6 T and 0.7 T.

2-8 We want to produce a flux density of 0.6 T in an air gap having a length of 8 mm. Calculate the mmf required.

2-9 Conductor AB in Figure 2-29 carries a current of 800 A flowing from B to A.
a. Calculate the force on the conductor.
b. Calculate the force on the moving N pole.
c. Does the force on the N pole act in the same direction as the direction of rotation?

2-10 a. Draw the waveshape of a sinusoidal voltage having a peak value of 200 V and a frequency of 5 Hz.
b. If the voltage is zero at $t = 0$, what is the voltage at $t = 5$ ms? 75 ms? 150 ms?

2-11 A sinusoidal current has an effective value of 50 A. Calculate the peak value of current.

2-12   A sinusoidal voltage of 120 V is applied to a resistor of 10 Ω. Calculate:

a.   the effective current in the resistor;
b.   the peak voltage across the resistor;
c.   the power dissipated by the resistor;
d.   the peak power dissipated by the resistor.

2-13   A distorted voltage contains an 11th harmonic of 20 V, 253 Hz. Calculate the frequency of the fundamental.

2-14   The current in a 60 Hz single-phase motor lags 36 degrees behind the voltage. Calculate the time interval between the positive peaks of voltage and current.

2-15   Referring to Fig. 2-49, determine the phase angle between the following phasors and, in each case, indicate which phasor is lagging.

a. I_1 and I_3;
b. I_2 and I_3;
c. E and I_1;

2-16   The voltage applied to an ac magnet is given by the expression $E = 160 \sin \phi$ and the current is $I = 20 \sin(\phi - 60°)$, all angles being expressed in degrees.

a.   Draw the phasor diagram for E and I, using effective values.
b.   Draw the waveshape of E and I as a function of ϕ.
c.   Calculate the peak positive power and the peak negative power in the circuit.

2-17   a.   Referring to Fig. 2-24, draw the waveshape of the distorted sine wave, if the leads of the third harmonic source are reversed.

b.   Calculate the peak voltage of the resulting waveshape.

2-18   In Figure 2-4 if terminal A_1 is (−) and current flows from A_2 to B_2 which box is the source?

2-19   The resistance of the conductors joining the two boxes in Fig. 2-4 is zero. If A_1 is (+) with respect to A_2, can B_1 be (−) with respect to B_2?

2-20   The alternating voltage e_2 in Figure 2-24a is given by the expression
$$e_2 = 20 \cos (360 ft - \theta)$$

If $\theta = 150°$ and $f = 180$ Hz, calculate the value of e_2 at $t = 0$, and at $t = 4$ min, 22 s.

Figure 2-49
See problem 2-15.

3

FUNDAMENTALS OF MECHANICS AND HEAT

In order to get a thorough grasp of electrical power technology, it is essential to know something about mechanics and heat. For example, the starting of large motors is determined not only by the magnitude of the torque, but also by the inertia of the revolving parts. And the overload capacity of an alternator is determined not only by the size of its conductors, but also by the temperature that its windings can safely withstand. And the stringing of a transmission line is determined as much by the ice-loading and mechanical strength of the conductors as by the currents they carry. And we could mention many more cases where the comprehensive approach – the electrical / mechanical / thermal approach – is essential to a thorough understanding of power technology.

For this reason, this introductory chapter covers certain fundamentals of mechanics and heat. The topics are not immediately essential to an understanding of the chapters which follow, but they constitute a valuable reference source, which the reader may wish to consult from time to time. Consequently, we recommend a quick first reading, followed by closer study of each section, as the need arises.

3.1 Force

The most familiar force we know is the force of gravity. For example, whenever we lift a stone, we exert a muscular effort to overcome the gravitational force that continually pulls it downwards. There are other kinds of forces such as the force exerted by a stretched spring or the forces created by exploding dynamite. All these forces are expressed in terms of the newton (N), which is the SI unit of force.

The magnitude of the force of gravity depends upon the mass of a body, and is given by the approximate equation:

$$F = 9.8\ m \qquad (3\text{-}1)$$

where

F = force of gravity acting on the body [N]
m = mass of the body [kg]
9.8 = an approximate constant that applies

when objects are relatively close to the surface of the earth (within 30 km)

Example 3-1:
Calculate the approximate value of the force of gravity that acts on a mass of 12 kg.

Solution:
The force of gravity is
$$F = 9.8 \, m = 9.8 \times 12 = 117.6 \text{ newtons}$$
$$= 117.6 \text{ N}$$

When using the English system of units, we have to make a distinction between the pound (symbol lb) and the pound-force (lbf). A pound is a unit of mass equal to 0.453 592 37 kg, exactly. On the other hand, a pound-force is equal to (9.806 65 × 0.453 592 37) newtons exactly, or about 4.448 N.

Example 3-2:
Calculate the approximate value of the force of gravity that acts on a mass of 140 lb. Express the result in newtons and in pounds-force.

Solution:
Using the conversion charts in the Appendix, a mass of 140 lb = 140 (÷2.205) = 63.5 kg. Using Eq. (3-1) the force of gravity is

$$F = 9.8 \, m = 9.8 \times 63.5 \text{ N} = 622.3 \text{ N}$$

Again using the conversion charts, a force of 622.3 N = 622.3 (÷4.448) = 139.9 pound-force = 139.9 lbf

Note that the force of gravity of 139.9 lbf is almost exactly equal to the mass of 140 lb. However, although the numbers are nearly the same, a force of 140 lbf is entirely different from a mass of 140 lb.

3.2 Torque

Torque is produced when a force exerts a twisting action on a body, tending to make it rotate. Torque is equal to the product of the force times the perpendicular distance between the axis of rotation and the point of application of the force. For example, suppose a string is wrapped around a pulley having a radius r (Fig. 3-1). If we pull on the string with a force F, the pulley will tend to rotate. The torque exerted on the pulley is given by:

$$T = Fr \tag{3-2}$$

where
T = torque [N·m]
F = force [N]
r = radius [m]

If the pulley is free to move it will begin to rotate around its axis.

Example 3-3:
A motor develops a starting torque of 150 N·m. If a pulley on the shaft has a diameter of 1 m, calculate the braking force needed to prevent the motor from turning.

Solution:
The radius is 0.5 m; consequently, a braking force $F = T/r = 150/0.5 = 300$ N is required. If the radius were 2 m, a braking force of 75 N would be sufficient to prevent rotation.

3.3 Mechanical work

Mechanical work is done whenever a force F moves a distance d, in the direction of the force. The work is given by:

$$W = Fd \tag{3-3}$$

where
W = work [J]
F = force [N]
d = distance the force moves [m]

Example 3-4:
A mass of 50 kg is lifted to a height of 10 m (Fig. 3-2). Calculate the work done.

Solution:
The force of gravity acting on the 50 kg mass is:

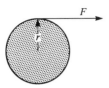

Figure 3-1
Torque $T = Fr$.

$$F = 9.8\,m = 9.8 \times 50 = 490\text{ N}$$

The work done is:
$$W = Fd = 490 \times 10 = 4900\text{ J}$$

3.4 Power

Power is the rate of doing work. It is given by the equation:

$$P = W/t \qquad (3\text{-}4)$$

where

$P = $ power [W]
$W = $ work done [J]
$t = $ time taken to do the work [s]

The unit of power is the watt (W). We often use the kilowatt (kW) which is equal to 1000 W. The power output of motors is sometimes expressed in horsepower (hp) which is an English unit equal to 746 W. It corresponds to the average power output of a dray horse.

Example 3-5:
An electric motor lifts a mass of 500 kg through a height of 30 m in 12 s (Fig. 3-3). Calculate the power developed by the motor, in kilowatts and in horsepower.

Solution:
The tension in the cable is equal to the force of gravity acting on the mass that is being lifted:

$$F = 9.8\,m = 9.8 \times 500 = 4900\text{ N}$$

The work done is:
$$W = Fd = 4900 \times 30 = 147\,000\text{ J}$$

Figure 3-2
Work $W = Fd$.

The power is:
$$P = W/t$$
$$= 147\,000/12 = 12\,250\text{ W} = 12.25\text{ kW}$$

Expressed in horsepower:
$$P = 12\,250/746 = 16.4\text{ hp}$$

3.5 Power of a motor

The mechanical power output of a motor depends upon its rotational speed and the torque it develops. The power is given by:

$$P = \frac{nT}{9.55} \qquad (3\text{-}5)$$

where

$P = $ mechanical power [W]
$T = $ torque [N·m]
$n = $ speed of rotation [r/min]
$9.55 = $ a constant to take care of units (exact value $= 30/\pi$)

Figure 3-3
Power $P = W/t$.

We can measure the power output of a motor by means of a prony brake. It consists of a stationary flat belt that presses against a pulley mounted on the motor shaft. The ends of the belt are connected to two spring scales, and the belt pressure is adjusted by tightening screw V (Fig. 3-4). As the motor turns, we can increase or decrease the power output by adjusting the tension of the belt. The mechanical power

Figure 3-4
Prony brake.

developed by the motor is entirely converted into heat by the belt rubbing against the pulley.

When the motor is disconnected, the spring scales register equal pulls and so the resulting torque is zero. However, when the motor turns clockwise (as it does in Fig. 3-4), pull P_1 exceeds pull P_2. The resultant force acting on the circumference of the pulley is therefore $(P_1 - P_2)$ newtons. If the pulley has a radius r, the net torque $T = (P_1 - P_2)r$ newton meters. Knowing the speed of rotation, we can calculate the power, using Eq. (3-5).

Example 3-6:
During a prony brake test on an electric motor, the spring scales indicate 25 N and 5 N, respectively (Fig. 3-4). Calculate the power output if the motor turns at 1700 r/min and the radius of the pulley is 0.1 m.

Solution:
The torque is:
$$T = Fr$$ Eq. 3-2
$$= (25 - 5) \times 0.1 = 2 \ \text{N·m}$$

The power is:
$$P = nT/9.55 = 356 \ \text{W}$$

The motor develops 356 W, or about 0.5 hp.

3.6 Transformation of energy

Energy can exist in one of the following forms:

1. Mechanical energy (energy stored in a coiled spring or in a moving car);

2. Thermal energy (heat released by a stove, by friction, by the sun);

3. Chemical energy (energy contained in dynamite, in coal, or in an electric storage battery);

4. Electrical energy (energy produced by a generator, or by lightning);

5. Atomic energy (energy released when the nucleus of an atom is modified).

Although energy can be neither created nor destroyed, it can be converted from one form to another by means of appropriate devices or machines. For example, the chemical energy contained in coal can be transformed into thermal energy by burning the coal in a furnace. The thermal energy contained in steam can then be transformed into mechanical energy by using a turbine. Finally, the mechanical energy can be transformed into electrical energy by means of a generator.

In the above example, the furnace, the turbine and the generator are the *machines* that do the energy transformation.

Unfortunately, whenever energy is transformed, the output is always less than the input because all machines have losses. These losses appear in the form of heat, causing the temperature of the machine to rise. Thus, the electrical energy supplied to a motor is partly dissipated as heat in the windings. Furthermore, some of its mechanical energy is also lost, due to bearing friction and air turbulence created by the fan. The mechanical losses are also transformed into heat. Consequently, the useful mechanical power output is less than the electrical input.

3.7 Efficiency of a machine

The efficiency of a machine is given by the equation:

$$\eta = \frac{P_o}{P_i} \times 100 \qquad (3\text{-}6)$$

where

η = efficiency [percent]
P_o = output power of the machine [W]
P_i = input power to the machine [W]

The efficiency is particularly low when thermal energy is converted into mechanical energy. Thus, the efficiency of steam turbines ranges from 25 to 40 percent, while that of internal combustion engines (automobile engines, diesel motors) lies between 15 and 30 percent. To realize how low these efficiencies are, we must remember that a machine having an efficiency of 20 percent loses 80 percent of the energy it receives.

Electric motors transform electrical energy into mechanical energy much more efficiently. Their efficiency ranges between 75 and 98 percent, depending on the size of the motor.

Example 3-7:
A 150 kW electric motor has an efficiency of 92 percent when it operates at full load. Calculate the losses in the machine.

Solution:
The 150 kW rating always refers to the mechanical power *output* of the motor.

The input power is:
$$P_i = P_o/\eta = 150/0.92 = 163 \text{ kW};$$

The mechanical output power is:
$$P_o = 150 \text{ kW}$$

The losses are:
$$P_i - P_o = 163 - 150 = 13 \text{ kW}.$$

Considering the high efficiency of the motor, the losses are quite moderate, but they would still be enough to continuously heat a large home in the middle of winter.

3.8 Kinetic energy of linear motion

A falling stone or a swiftly moving automobile possess kinetic energy, or energy due to motion. Kinetic energy is a form of mechanical energy given by the equation:

$$E_k = \frac{1}{2} mv^2 \qquad (3\text{-}7)$$

where
E_k = kinetic energy [J]
m = mass of the body [kg]
v = speed of the body [m/s]

Example 3-8:
A bus having a mass of 6000 kg moves at a speed of 100 km/h. If it carries 40 passengers having a total mass of 2400 kg, calculate the total kinetic energy of the vehicle. What happens to this energy when the bus is braked to a stop?

Solution:
Total mass of the bus is:
$$m = 6000 + 2400 = 8400 \text{ kg}.$$

The speed is:

$$v = 100 \text{ km/h} = \frac{100 \times 1000 \text{ m}}{3600 \text{ s}}$$
$$= 27.8 \text{ m/s}$$

The kinetic energy is:

$E_k = 1/2mv^2 = 1/2 \times 8400 \times 27.8^2$
$\quad = 3\ 245\ 928\ J = 3.25\ MJ.$

To stop the bus, the brakes are applied and the resulting frictional heat is entirely supplied at the expense of the kinetic energy. The bus will finally come to rest when all the kinetic energy (3.25 MJ) has been dissipated as heat.

3.9 Kinetic energy of rotation, moment of inertia

A revolving body also possesses kinetic energy. Its magnitude depends upon the speed of rotation and upon the mass and shape of the body. The kinetic energy of rotation is given by the equation:

$E_k = 5.48 \times 10^{-3}Jn^2$ $\hspace{2em}$ (3-8)

where

E_k = kinetic energy [J]
J = moment of inertia [kg·m²]
n = rotational speed [r/min]
5.48×10^{-3} = constant to take care of units [exact value = $\pi^2/1800$]

The moment of inertia J (sometimes simply called "inertia") depends upon the mass and shape of the body. Its value may be calculated for a number of simple shapes by using Eqs. 3-9 to 3-13 given in Table 3A. If the body has a more complex shape, it can always be broken up into two or more of the simpler shapes given in the table. The individual J's of these simple shapes are then added to give the total J of the body.

Inertia plays a very important part in rotating machines; consequently, it is worth our while to solve a few problems.

Example 3-9:
A solid 1400 kg steel flywheel has a diameter of 1 m and a thickness of 225 mm (Fig. 3-10). Calculate:
a.   its moment of inertia.

b.   the kinetic energy when the flywheel revolves at 1800 r/min.

Solution:
a.   referring to Table 3A, we find the moment of inertia is:

$J = \dfrac{mr^2}{2}$ $\hspace{3em}$ Eq. 3-10

$= \dfrac{1400 \times 0.5^2}{2} = 175\ kg·m^2$

Figure 3-10
Flywheel in Example 3-9.

b.   The kinetic energy is:
$E_k = 5.48 \times 10^{-3}Jn^2$ $\hspace{2em}$ Eq. 3-8
$\quad = 5.48 \times 10^{-3} \times 175 \times 1800^2$
$\quad = 3.1\ MJ$

Note that this relatively small flywheel possesses as much kinetic energy as the loaded bus mentioned in Example 3-8.

Example 3-10:
A flywheel having the shape given in Fig. 3-11 is composed of a ring supported by a rectangular hub. The ring and hub respectively have a mass of 80 kg and 20 kg. Calculate the moment of inertia of the flywheel.

Solution:
For the ring:

$J_1 = m\ (R_1^2 + R_2^2)/2$ $\hspace{2em}$ Eq. 3-11
$\quad = 80\ (0.4^2 + 0.3^2)/2 = 10\ kg·m^2$

TABLE 3A   MOMENT OF INERTIA J AROUND AN AXIS OF ROTATION 0

Figure 3-5
Mass m revolving at a distance r around axis 0.

$$J = mr^2 \qquad (3\text{-}9)$$

Figure 3-6
Solid disc of mass m and radius r.

$$J = \frac{mr^2}{2} \qquad (3\text{-}10)$$

Figure 3-7
Annular ring of mass m having a rectangular cross section.

$$J = \frac{m}{2}(R_1^2 + R_2^2) \qquad (3\text{-}11)$$

Figure 3-8
Straight bar of mass m pivoted on its center.

$$J = \frac{mL^2}{12} \qquad (3\text{-}12)$$

Figure 3-9
Rectangular bar of mass m revolving around axis 0.

$$J = \frac{m}{3}(R_1^2 + R_2^2 + R_1 R_2) \qquad (3\text{-}13)$$

Figure 3-11
Flywheel in Example 3-10.

For the hub:
$$J_2 = mL^2/12 \qquad \text{Eq. 3-12}$$
$$= 20 \times (0.6)^2/12 = 0.6 \text{ kg·m}^2$$

The total moment of inertia of the flywheel is:
$$J = J_1 + J_2 = 10.6 \text{ kg·m}^2$$

3.10 Torque, inertia and change in speed

There is only one way to change the speed of a revolving body and that is to subject it to a *torque* for a given period of time. The rate of change of speed depends upon the inertia, as well as on the torque. A simple equation relates these factors:

$$\Delta n = 9.55 T\Delta t/J \qquad (3\text{-}14)$$

where

Δn = change in speed [r/min]
T = torque [N·m]
Δt = interval of time during which the torque is applied [s]
J = moment of inertia [kg·m²]
9.55 = factor taking account of units [exact value = $30/\pi$]

If the torque acts in the direction of rotation, the speed rises. Conversely, if it acts against the direction of rotation, the speed falls. The term Δn may therefore represent either an increase or a decrease in speed.

Example 3-11:
The flywheel of Fig. 3-11 turns at 60 r/min. We wish to increase its speed to 600 r/min by applying a torque of 20 N·m. For how long must the torque be applied?

Solution:
The change in speed is:
$$\Delta n = (600 - 60) = 540 \text{ r/min}$$

The moment of inertia is:
$$J = 10.6 \text{ kg·m}^2$$

Substituting these values in Eq. 3-4:
$$\Delta n = 9.55 \, T\Delta t/J \qquad \text{Eq. 3-14}$$
$$540 = 9.55 \times 20\Delta t/10.6$$

therefore,
$$\Delta t = 30 \text{ seconds}$$

3.11 Speed of a motor/load system

In electrical power technology, it often happens that an electric motor drives a mechanical load. In such a system there are three main factors to consider: the torque developed by the motor, the torque exerted by the load, and the speed. We now explain how they interreact.

Consider a load coupled to a motor by means of a shaft (Fig. 3-12). The load exerts a constant torque T_L that acts in a counterclockwise direction. On the other hand, the torque T_M developed by the motor acts clockwise, and it can be varied by increasing or decreasing the electric current I. Suppose the system is initially at rest and that $T_M = T_L$. Because the torques are equal and opposite, the net torque acting on the shaft is zero, and so it has no tendency to rotate.

As a result of the opposing torques, the shaft twists and becomes slightly deformed, but otherwise nothing happens.

Suppose we want the load to turn clockwise at a speed n_1. To do this, we increase the motor current so that T_M exceeds T_L. The net torque on the shaft acts clockwise, and so it begins to

Figure 3-12
Shaft is stationary $T_M = T_L$.

Figure 3-13
Shaft turns cw $T_M = T_L$.

Figure 3-14
Shaft turns ccw $T_M = T_L$.

rotate clockwise. The speed increases progressively with time but as soon as the desired speed n_1 is reached, we reduce the motor current so that T_M is again exactly equal to T_L. The net torque acting on the system is now zero and the speed n_1 will neither increase or decrease any more (Fig. 3-13).

This brings us to a very important conclusion. The speed of a mechanical load remains constant when the torque T_M developed by the motor is *equal* and opposite to the torque T_L exerted by the load. At first, this conclusion is rather difficult to accept, because we are inclined to believe that when $T_M = T_L$, the system should simply stop. But this is not so, as our reasoning (and reality) shows. We repeat: the speed of a motor remains constant when the motor torque is *exactly* equal and opposite to the load torque. In effect, the motor/load system is then in a state of dynamic equilibrium.

With the load now running clockwise at a speed n_1, suppose we reduce T_M so that it is less than T_L. The net torque on the shaft now acts counterclockwise. Consequently, the speed decreases and will continue to decrease as long as T_L exceeds T_M. If the imbalance between T_L and T_M lasts long enough, the speed will eventually become zero and then reverse. If we control the motor torque so that

$T_M = T_L$ when the reverse speed reaches a value n_2, the system will continue to run indefinitely at this new speed (Fig. 3-14).

In conclusion, torques T_M and T_L are *identical* in Figs. 3-12, 3-13, and 3-14, and yet the shaft may be turning clockwise, counterclockwise, or not at all. The actual steady-state speed depends upon whether T_M was greater or less than T_L for a certain period of time *before* the actual steady-state condition was reached. The reader should ponder a few moments over this statement.

Whenever the motor torque T_M and load torque T_L are not equal and opposite, the speed changes gradually. The rate of change depends upon the inertia of the rotating parts, and this aspect is covered in more detail in Sec. 3-13.

3.12 Power flow in a mechanically coupled system

Returning again to Fig. 3-13, we see that motor torque T_M acts in the same direction (clockwise) as speed n_1. This means that the motor *delivers* mechanical power to the shaft. Conversely, load torque T_L acts opposite to speed n_1. Consequently, the load *receives* mechanical power from the shaft. We can therefore state the following general rule:

When the torque developed by a motor acts in the same direction as the speed, the motor delivers power to the load. For all other conditions, the motor receives power from the load.

In Fig. 3-14, for example, the motor receives power from the load because T_M acts opposite to n_2. Although this is an unusual condition, it occurs for brief periods in electric trains and electric hoists. The behavior of the motor under these conditions will be examined in later chapters.

3.13 Motor driving a load having inertia

When a motor drives a mechanical load, the speed is usually constant. In this state of dynamic equilibrium, the torque T_M developed by the motor is exactly equal to the torque T_L imposed by the load. The inertia of the revolving parts does not come into play under these conditions. However, if the motor torque is raised so that it exceeds the load torque, the speed will increase, as we already know. Conversely, when the motor torque is less than that of the load, the speed drops. The increase or decrease in speed (Δn) is still given by the Eq. 3-14, except that torque T is now replaced by the *net* torque ($T_M - T_L$):

$$\Delta n = 9.55 \, (T_M - T_L)\Delta t/J \qquad (3\text{-}15)$$

where

Δn = change in speed [r/min]
T_M = motor torque [N·m]
T_L = load torque [N·m]
Δt = time interval during which T_M and T_L are acting [s]
J = moment of inertia of all revolving parts [kg·m²]

Example 3-12:

A large reel of paper installed at the end of a paper machine has a diameter of 1.8 m, a length of 5.6 m and a moment of inertia of 4500 kg·m². It is driven by a directly coupled variable-speed dc motor turning at 120 r/min. The paper is kept at a constant tension of 6000 N.

a. Calculate the power of the motor when the reel turns at a constant speed of 120 r/min,

b. If the speed has to be raised from 120 r/min to 160 r/min in 5 seconds, calculate the torque that the motor must develop during this interval,

c. Calculate the power of the motor after it has reached the desired speed of 160 r/min.

Solution:

a. The torque exerted on the reel is:

$$T = Fr = 6000 \times \frac{1.8}{2} = 5400 \text{ N·m}$$

The power developed by the reel motor is:

$$P = \frac{nT}{9.55} = \frac{120 \times 5400}{9.55} \qquad \text{Eq. 3-5}$$

$$= 67.85 \text{ kW (about 91 hp)}$$

b. As the speed increases from 120 r/min to 160 r/min the load torque (5400 N·m) stays constant because the tension in the paper remains unchanged . Let the required motor torque be T_M. It must be greater than the load torque in order for the speed to increase.

We have:
$\Delta n = 160 - 120 = 40 \text{ r/min}$
$J = 4500 \text{ kg·m}^2$
$\Delta t = 5 \text{ s}$

$$\Delta n = \frac{9.55 \, (T_M - T_L)\Delta t}{J}$$

$$40 = \frac{9.55 \, (T_M - 5400)5}{4500}$$

Thus,
$$T_M - 5400 = 3770$$
$$T_M = 9170$$

The motor must therefore develop a constant torque of 9170 N·m during the acceleration period.

The mechanical power of the reel motor at 160 r/min is:

$$P = \frac{nT}{9.55} = \frac{160 \times 9170}{9.55}$$

$$= 153.6 \text{ kW (equivalent to}$$

$$206 \text{ hp)}$$

c. As soon as the desired speed (160 r/min) is reached, the motor only has to develop a torque equal to the load torque (5400 N·m). The power of the motor is therefore reduced to:

$$P = \frac{nT}{9.55} = \frac{160 \times 5400}{9.55}$$

$$= 90.5 \text{ kW (equivalent to}$$

$$121 \text{ hp)}$$

3.14 Electric motors driving linear motion loads

Rotating loads such as fans, pumps and machine tools are well suited for direct mechanical coupling to electric motors. On the other hand, loads that move in a straight line, such as hoists, trains, wire-drawing machines, etc., must be equipped with a "motion converter" before they can be connected to a rotating machine. The motion converter may be a rope-pulley arrangement, a rack and pinion mechanism, or simply a wheel moving over a track. These converters are so utterly simple that we seldom think of the important part they play.

Straight-line motion involves a linear speed v and a force F, while rotary motion involves a rotational speed n and a torque T. How are these quantities related when a motion converter is used?

Consider a jack driven by a motor that rotates at a speed n while exerting a torque T (Fig. 3-15). This causes a vertical ram to exert a powerful force F while moving at a linear speed v. The power supplied in raising the load is given by:

$$P_o = Fv$$

On the other hand, the power input to the jack is given by:

$$P_i = \frac{nT}{9.55} \qquad \text{Eq. 3-5}$$

Figure 3-15
Converting rotary motion into linear motion.

Assuming there are no losses in the motion converter, we have:

$$P_i = P_o$$

Consequently:

$$nT = 9.55Fv \qquad (3\text{-}16)$$

where

$n =$ rotational speed [r/min]
$T =$ torque [N·m]
$F =$ force [N]
$v =$ linear speed [m/s]
$9.55 =$ a constant [exact value $= 30/\pi$]

Example 3-13:

A force of 25 kN is needed to pull an electric train at a speed of 90 km/h. The motor on board the locomotive turns at 1200 r/min. Calculate the torque developed by the motor.

Solution:

$$nT = 9.55Fv$$
$$1200T = 9.55 \times 25\ 000 \times (90\ 000/3600)$$
$$T = 4974\ \text{N·m} = 5\ \text{kN·m}$$

example, if we add 100 kJ of heat to 1 kg of water, the temperature rises by 24°C. The same amount of heat supplied to 1 kg of copper raises *its* temperature by 263°C. It is therefore obvious that heat and temperature are two quite different things.

If we remove heat from a body, its temperature drops. However, the temperature cannot fall below a lower limit. This limit is called abso-

Figure 3-16
Temperature scales.

Figure 3-17
Electric water heater.

3.15 Heat and temperature

Whenever we apply heat to a body, we supply it with thermal energy. Heat is therefore a form of energy and the SI unit is the joule.

What happens when a body receives this type of energy? First, the atoms of the body vibrate more intensely. Second, its temperature increases, a fact we can verify by touching it or by observing the reading of a thermometer.

For a given amount of heat, the increase in temperature depends upon the mass of the body and the material of which it is made. For

lute zero. It corresponds to a temperature of 0 kelvin or −273.15°C. At absolute zero, all atomic vibrations cease and the only motion which subsists is that of the moving electrons.

3.16 Temperature scales

The kelvin and the degree Celsius are the SI units of temperature. Figure 3-16 shows some interesting relationships between the Kelvin, Celsius and Fahrenheit temperature scales. Thus, iron melts at 1808 K, or 1533°C or 2791°F.

3.17 Heat required to raise the temperature of a body

The temperature rise of a body depends upon the heat it receives, its mass, and the nature of the material. The relationship between these quantities is given by the equation:

$$Q = mc\Delta t \qquad (3-17)$$

where

Q = quantity of heat added to (or removed from) a body [J]

m = mass of the body [kg]

c = specific heat capacity of the material making up the body [J/(kg·°C)]

Δt = change in temperature [°C]

The specific heat capacity of several materials is given in the Appendix.

Example 3-14:
Calculate the heat required to raise the temperature of 200 L of water from 10°C to 70°C, assuming the tank is perfectly insulated (Fig. 3-17). The specific heat capacity of water is 4180 J/kg·°C, and one liter weighs 1 kg.

Solution:
The mass of water is 200 kg, and so the heat required is:

$$\begin{aligned} Q &= mc\Delta t \\ &= 200 \times 4180 \times (70 - 10) \\ &= 50.2 \text{ MJ} \end{aligned}$$

Referring to the conversion table for Energy (see Appendix), we find that 50.2 MJ is equal to 13.9 kW·h.

3.18 Transmission of heat

Many problems in electric power technology are related to the adequate cooling of devices and machines. This, in turn, requires a knowledge of the mechanism by which heat is transferred from one body to another. In the sections that follow, we briefly review the ele-

mentary physics of heat transmission. We also include some simple but useful equations, enabling us to determine, with reasonable accuracy, the heat loss, temperature rise, etc., of electrical equipment.

3.19 Heat transfer by conduction

If we bring a hot flame near one end of an iron bar, its temperature rises owing to the increased vibration of its atoms (Fig. 3-18). This atomic vibration is transmitted from one atom to the next, to the other end of the bar. Consequently, the end opposite the flame also warms up, an observation we have all made at

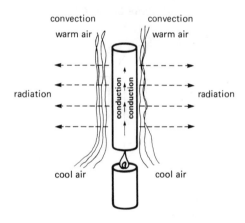

Figure 3-18
Heat transmission by convection, conduction and radiation.

one time or another. In effect, heat is transferred along the bar by a process called *conduction*.

The rate of heat transfer depends upon the *thermal conductivity* of the material. Thus, copper is a better thermal conductor than steel is, and insulators are especially poor conductors of heat.

The SI unit of thermal conductivity is the watt per meter degree Celsius (W/(m·°C)). The thermal conductivity of several common materials is given in Tables AX1 and AX2 in the Appendix.

Referring to Fig. 3-19, we can calculate the rate of heat flow through a body by using the equation:

$$P = \frac{\lambda A(t_1 - t_2)}{d} \qquad (3\text{-}18)$$

where

P = power (heat) transmitted [W]
λ = thermal conductivity of the body [W/(m·°C)]
A = surface area of the body [m²]
$(t_1 - t_2)$ = difference of temperature between opposite faces [°C]
d = thickness of the body [m]

Figure 3-19
Heat transmission by conduction.

Example 3-15:
The temperature difference between two sides of a sheet of mica is 50°C (Fig. 3-20). If its area is 200 cm² and thickness is 3 mm, calculate the heat flowing through the sheet, in watts.

Solution:
According to Table AX1, the thermal conductivity of mica is 0.36 W/m·°C. The heat transmitted is therefore:

$$P = \frac{\lambda A(t_{1,} - t_2)}{d} \qquad \text{Eq. 3-18}$$

$$= \frac{0.36 \times 0.02 \, (120 - 70)}{0.003} = 120 \text{ W}$$

Figure 3-20
Mica sheet, Example 3-15.

3.20 Heat transfer by convection

In Fig. 3-18, the air in contact with the hot steel bar warms up and, becoming lighter, rises like smoke in a chimney. As the hot air moves upwards, it is replaced by cooler air which, in turn, also warms up. A continual current of air is therefore set up around the bar, removing its heat by a process called *natural convection*.

The convection process can be accelerated by employing a fan to create a rapid circulation of fresh air. Heat transfer by *forced convection* is used in most electric motors to obtain efficient cooling.

Natural convection also takes place when a hot body is immersed in a liquid, such as oil. The oil in contact with the body heats up,

Figure 3-21
Convection currents in oil.

creating convection currents which follow the path shown in Fig. 3-21. When the warm oil comes in contact with the cooler tank, it chills, becomes heavier, sinks to the bottom, and moves upward again to replace the warmer oil now moving away. The heat dissipated by the body is therefore carried away by convection to the external tank. The tank, in turn, loses *its* heat by natural convection to the surrounding air.

3.21 Calculating the losses by convection

The heat loss by natural convection in air is given by the approximate equation:

$$P = 3A(t_1 - t_2)^{1.25} \qquad (3\text{-}19)$$

where

P = heat loss by natural convection [W]
A = surface of the body [m]
t_1 = surface temperature of the body [°C]
t_2 = ambient temperature of the surrounding air [°C]

In the case of *forced* convection such as that produced by a blower, the heat carried away is given by:

$$P = 1280 \, V_a(t_2 - t_1) \qquad (3\text{-}20)$$

where

P = heat loss by forced convection [W]
V_a = volume of cooling air [m³/s]
t_1 = temperature of the incoming (cool) air [°C]
t_2 = temperature of the outgoing (warm) air [°C]

Equation 3-20 also applies if hydrogen is used as the cooling medium.

Example 3-16:

A totally enclosed motor has an external surface area of 1.2 m². When it operates at full load, the surface temperature rises to 60°C in an ambient of 20°C (Fig. 3-22). Calculate the heat loss by natural convection.

Solution:

$$\begin{aligned} P &= 3A \, (t_1 - t_2)^{1.25} \\ &= 3 \times 1.2 \, (60 - 20)^{1.25} = 362 \text{ W} \end{aligned}$$

Example 3-17:

A fan rated at 3.75 kW blows 240 m³/min of air through a 750 kW motor to carry away the heat. If the inlet temperature is 22°C and the outlet temperature is 31°C, estimate the losses in the motor.

Solution:

The losses are carried away by the circulating air. Consequently, the losses are:

$$\begin{aligned} P &= 1280 \, V_a \, (t_2 - t_1) \\ &= 1280 \times 240/60 \, (31 - 22) = 46 \text{ kW} \\ & \quad \text{(approximate)} \end{aligned}$$

Figure 3-22
Convection and radiation losses in a totally-enclosed motor.

3.22 Heat transfer by radiation

We have all basked in the warmth produced by the sun's rays. This radiant heat energy possesses the same properties as light, and it readily passes through the empty space between the sun and the earth. Solar energy is only converted to heat when the sun's rays meet a solid body, such as the physical objects and living things on the surface of the earth. Scientists have discovered that all bodies radiate heat, even those that are very cold. The amount of energy given off depends upon the temperature of the body.

On the other hand, all bodies *absorb* radiant energy from the objects which surround them. The energy absorbed depends upon the temperature of the surrounding objects. There is, therefore, a continual exchange of radiant energy between material bodies, as if each were a miniature sun. Equilibrium sets in when the temperature of a body is the same as that of its surroundings. The body then radiates as much energy as it receives and the net radiation is zero. On the other hand, if the body is hotter than its environment, it will continually lose heat by radiation, even if it is located in a vacuum.

3.23 Calculating radiation losses

The heat which a body loses by radiation is given by the equation:

$$P = kA \ (T_1{}^4 - T_2{}^4) \qquad (3\text{-}21)$$

where

$P =$ heat radiated [W]
$A =$ surface area of the body [m²]
$T_1 =$ absolute temperature of the body [K]
$T_2 =$ absolute temperature of the surrounding objects [K]
$k =$ a constant, which depends upon the nature of the body surface

Table 3B gives the values of k for surfaces commonly encountered in electrical equipment.

Example 3-18:
The motor in Example 3-16 is coated with a non metallic enamel. Calculate the heat lost by radiation, knowing that all surrounding objects are at an ambient temperature of 20°C.

Solution:

$T_1 =$ surface temperature = 60°C
   or (273.15 + 60) = 333 K

$T_2 =$ surrounding temperature = 20°C
   or (273.15 + 20) = 293 K

From Table 3B, $k = 5 \times 10^{-8}\,\text{W/(m}^2\cdot\text{K}^4)$

The power lost by radiation is therefore:

$$P = kA \ (T_1{}^4 - T_2{}^4) \qquad \text{Eq. 3-21}$$
$$= 5 \times 10^{-8} \times 1.2 \ (333^4 - 293^4)$$
$$= 296 \text{ W (approximate)}$$

It is interesting to note that the motor dissipates almost as much heat by radiation (296 W) as it does by convection (362 W).

TABLE 3B   RADIATION CONSTANTS

Type of surface	Constant k W/(m²·K⁴)
polished silver	0.2 x 10⁻⁸
bright copper	1 x 10⁻⁸
oxidized copper	3 x 10⁻⁸
aluminum point	3 x 10⁻⁸
oxidized nichrome	2 x 10⁻⁸
tungsten	2 x 10⁻⁸
oxidized iron	4 x 10⁻⁸
insulating materials	5 x 10⁻⁸
paint or non metallic enamel	5 x 10⁻⁸
perfect emitter (blackbody)	5.669 x 10⁻⁸

QUESTIONS AND PROBLEMS

Practical level

3-1 A cement block has a mass of 40 kg. What is the force of gravity acting on it? What force is needed to lift it?

3-2 How much energy is needed to lift a sack of flour weighing 75 kg to a height of 4 m?

3-3 Give the SI unit and the corresponding SI symbol for the following quantities:
   force   work
   pressure   area
   mass   temperature
   thermal energy   thermal power
   mechanical energy mechanical power
   electrical energy   electrical power

3-4 In tightening a bolt, a mechanic exerts a force of 200 N at the end of a wrench having a length of 0.3 m. Calculate the torque he exerts.

3-5 An automobile engine develops a torque of 600 N·m at a speed of 4000 r/min. Calculate the power output in watts and in horsepower.

3-6 A crane lifts a mass of 600 lb to a height of 200 ft in 15 s. Calculate the power in watts and in horsepower.

3-7 An electric motor draws 120 kW from the line and has losses equal to 20 kW. Calculate:
a. the power output of the motor [kW] and [hp];
b. the efficiency of the motor;
c. the amount of heat released [Btu/h].

3-8 A large flywheel has a moment of inertia of 500 lb·ft^2. Calculate its kinetic energy when it rotates at 60 r/min.

3-9 The rotor of an induction motor has a moment of inertia of 5 kg·m^2. Calculate the energy needed to bring the speed:
a. from zero to 200 r/min;
b. from 200 r/min to 400 r/min;
c. from 3000 r/min to 400 r/min.

3-10 Name the three ways whereby heat is carried from one body to another.

3-11 A motor develops a cw torque of 60 N·m and the load develops a ccw torque of 50 N·m.
a. If this situation persists for some time, will the direction of rotation eventually be cw or ccw?
b. What value of motor torque is needed to keep the speed constant?

3-12 A motor drives a load at cw speed of 1000 r/min. The motor develops a cw torque of 12 N·m, and the load exerts a ccw torque of 15 N·m.
a. Will the speed increase or decrease?
b. If this situation persists for some time, in what direction will the shaft eventually rotate?

3-13 Referring to Fig. 3-12, if T_M = 40 N·m, what is the power delivered by the motor?

3-14 Referring to Fig. 3-13, if T_M = 40 N·m and n_1 = 50 r/min, calculate the power delivered by the motor.

3-15 Referring to Fig. 3-14, if T_M = 40 N·m and n_2 = 50 r/min, calculate the power received by the motor.

Intermediate level

3-16 During a prony brake test on a motor (see Fig. 3-4), the following scale and speed readings were noted:

P_2 = 5 lbf, P_1 = 28 lbf,
n = 1160 r/min.

If the diameter of the pulley is 12 inches, calculate the power output of the motor in kilowatts and in horsepower.

3-17 A motor drives a flywheel having a moment of inertia of 5 kg·m^2. The speed increases from 1600 r/min to 1800 r/min in 8 s. Calculate:

a. the torque developed by the motor [N·m];

b. the energy in the flywheel at 1800 r/min [kJ];

c. the motor power [W] at 1600 r/min;

d. the power input [W] to the flywheel at 1750 r/min.

3-18 A dc motor coupled to a large grinder develops 120 hp at a constant speed of 700 r/min. The moment of inertia of the revolving parts is 2500 lb·ft^2.

a. Calculate the torque developed by the motor;

b. Calculate the motor torque needed so that the speed will increase to 750 r/min in 5 s. Note: The torque exerted by the grinder remains the same.

3-19 The electric motor in a trolley bus develops a power output of 80 hp at 1200 r/min as the bus moves up a hill at a speed of 30 miles per hour. Assuming that the gear losses are negligible, calculate:

a. the torque developed by the motor [N·m];

b. the force opposing the motion of the bus [N].

3-20 Calculate the heat [MJ] required to raise the temperature of 100 kg of copper from 20°C to 100°C.

3-21 Repeat Problem 3-20 for 100 kg of aluminum.

Figure 3-23
Electric hoist, Problem 3-22.

3-22 The motor in Fig. 3-23 drives a hoist, raising a mass m of 800 kg at a uniform rate of 5 m/s. The winch has a radius of 20 cm. Calculate the torque [N·m] and speed [r/min] of the motor.

3-23 If the hoisting rate in Problem 3-22 is reduced to 1 m/s, calculate the new speed [r/min] and torque [ft·lbf] of the motor.

PART II

ELECTRICAL MACHINES AND TRANSFORMERS

4

DIRECT CURRENT GENERATORS

We begin our study of rotating machinery with the direct current generator. Direct current generators are not as common as they used to be, because direct current, when required, is mainly produced by electronic rectifiers. These rectifiers can convert the current of an ac system into direct current without using any moving parts. Nevertheless, an understanding of dc generators is important because it represents a logical introduction to the behavior of dc motors. Indeed, many dc motors in industry actually operate as generators for brief periods.

Commercial dc generators and motors are built the same way; consequently, any dc generator can operate as a motor and vice versa. Owing to their similar construction, the fundamental properties of generators and motors are identical. Consequently, anything we learn about a generator can be directly applied to the dc motor.

In this chapter, we begin with the basic principles of a 2-pole generator when it operates at no-load. We show the importance of brush position and define what is meant by the neutral point. We show how the induced voltage is generated and what determines its magnitude.

This is followed by studying the behavior of the generator under load. Mechanical torque, direction of current flow and the importance of armature reaction are discussed. The need for commutating poles and the problem of pole-tip saturation are then covered.

We then go on to discuss the major types of dc generators and their voltage-regulation characteristics.

The chapter ends with a description of the actual physical construction of direct-current machines, including multipole designs.

4.1 Generating an ac voltage

Irrelevant as it may seem, the study of a direct current (dc) generator has to begin with a knowledge of the alternating current (ac) generator. The reason is that the voltage generated in any dc generator is inherently alternating, and it only becomes dc after it has been rectified.

Figure 4-1 shows an elementary ac generator composed of a coil that revolves at 60 r/min between the N, S poles of a permanent magnet. The rotation is due to an external driving force, such as a motor (not shown). The coil is connected to two slip rings which, in turn, are connected to an external load by means of two stationary brushes X, Y.

Figure 4-1
Elementary ac generator turning at 1 revolution per second.

As the coil rotates, a voltage is induced between its terminals A, D, and this voltage appears between the brushes, and therefore across the load. The voltage is generated because the conductors of the coil cut across the flux produced by the N, S poles. The induced voltage is therefore maximum (20 V) when the coil is momentarily in the horizontal position, as shown. No flux is cut when the coil is momentarily in the vertical position; consequently the voltage at these instants is zero. Another feature of the voltage is that its polarity changes every time the coil makes half a turn. The voltage can therefore be represented as a function of the angle of rotation (Fig. 4-2).

The coil in our example revolves at uniform speed, therefore each angle of rotation corre-

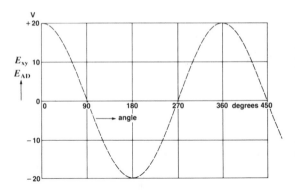

Figure 4-2
Voltage induced in the ac generator as a function of the angle of rotation.

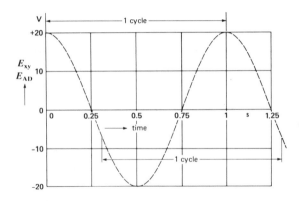

Figure 4-3
Voltage induced as a function of time.

sponds to a specific interval of time. Because the magnet makes one turn per second, the angle of 360° in Fig. 4-2 corresponds to an interval of one second. Consequently, we can also represent the induced voltage as a function of time (Fig. 4-3).

4.2 Direct current generator

If the brushes in Fig. 4-1 could be switched from one slip ring to the other every time the polarity was about to change, we would obtain a voltage of constant polarity across the load. Brush X would always be positive and brush Y, negative. We can obtain this result by using a *commutator* (Fig. 4-4). A commutator in its most simple form is composed of a slip ring that is cut in half, with each segment insulated from the other, as well as from the shaft. One segment is connected to coil-end A and the other to coil-end D. The commutator revolves with the coil and the induced voltage between the segments is picked up by two stationary brushes X and Y.

Figure 4-4
Elementary dc generator is simply an ac generator equipped with a mechanical rectifier called a commutator.

The voltage between brushes X and Y pulsates but never changes polarity (Fig. 4-5). In effect, the alternating voltage in the coil is *rectified* by the commutator, which acts as a mechanical reversing switch.

Owing to the constant polarity between the brushes, the current in the external load always flows in the same direction. The machine represented in Fig. 4-4 is called a direct current generator, or dynamo.

4.3 Difference between ac and dc generators

The elementary ac and dc generators studied so far are essentially built the same way. In each case, a coil rotates between the poles of a magnet and an ac voltage is induced in the coil.

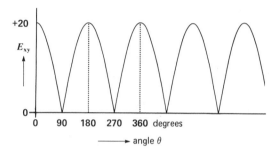

Figure 4-5
The elementary dc generator produces a pulsating dc voltage.

The machines only differ in the way the coils are connected to the external circuit: ac generators carry slip rings while dc generators require a commutator (Fig. 4-6). We sometimes build small machines which carry both slip rings and a commutator. Such machines can function simultaneously as ac and dc generators (Fig. 4-6).

(a)   (b)   (c)

Figure 4-6
The three armatures (a), (b), (c) have identical windings. Depending upon how they are connected (to slip rings or a commutator) an ac or dc voltage is obtained.

4.4 Improving the waveshape

Returning to the dc generator, we can improve the pulsating dc voltage by using four coils and four segments, as shown in Fig. 4-7. The resulting waveshape is given in Fig. 4-8. The voltage still pulsates but it never falls to zero; it is much closer to a steady dc voltage.

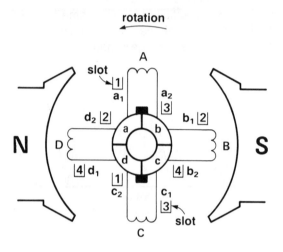

Figure 4-7
Schematic diagram of a dc generator having 4 coils and 4 commutator bars.

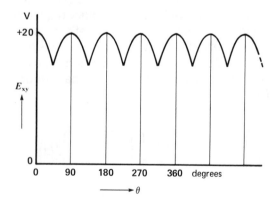

Figure 4-8
The voltage between the brushes is more uniform than in Fig. 4-5.

By increasing the number of coils and segments, we can obtain a dc voltage that is very smooth. Modern dc generators produce voltages having a ripple of less than 5 percent.

It is important to understand the physical meaning of Fig. 4-7, because we will be using similar drawings to explain the behavior of dc machines. The four coils in the figure are identical to the coil shown in Fig. 4-1. At the instant shown, coil A is not cutting any flux and neither is coil C. The reason is that the coil sides of

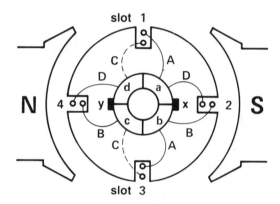

Figure 4-9
The actual physical construction of the generator shown in Fig. 4-7. The armature has 4 slots, 4 coils and 4 commutator bars.

these two coils are midway between the poles. On the other hand, coils B and D are cutting flux coming from the center of the N, S poles. Consequently, the voltage induced in these coils is at its maximum possible value (20 V).

A schematic diagram such as Fig. 4-7, tells us where the coil sides of the individual coils are located: between the poles, under the poles, near the pole-tips, and so on. But we must remember that the coil sides (a_1, a_2; b_1, b_2; etc.) of each coil are actually located at 180° to each other, and not side by side as Fig. 4-7 tends to indicate.

The actual construction of this so-called *armature* is shown in Fig. 4-9. The four coils are placed in four slots. Each coil has two coil sides and so there are two coil-sides per slot. Thus, each slot contains the conductors of two coils.

For reasons of symmetry, the coils are wound so that one coil-side is at the bottom of a slot and the other is at the top. For example, in Fig. 4-7 coil side a_1 is in the top of slot 1 while coil side a_2 is in the bottom of slot 3. The coil connections to the commutator segments are easy to follow in this simple armature. The reader should compare these connections with those in Fig. 4-9 to verify that they are the same. Note also the actual and schematic position of the brushes with respect to the poles.

Figure 4-10 shows the position of the coils when the armature has moved through 45°. The sides a_1, a_2 of coil A are now sweeping

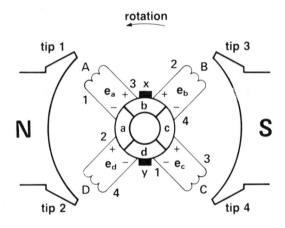

Figure 4-10
Position of the coils when the armature of Fig. 4-9 has rotated through 45°.

past pole tip 1 and pole tip 4. The sides of coil C are experiencing the same flux because they are in the same slots as coil A. Consequently, the voltage e_a induced in coil A is exactly the same as the voltage e_c induced in coil C. Note, however, that coil A is moving downwards, while coil C is moving upwards. The polarities of e_a and e_c are therefore opposite, as shown.

The same reasoning leads us to conclude that e_b and e_d are equal and opposite in polarity.

This means that $e_a + e_b + e_c + e_d = 0$ at all times. Consequently, no current will flow in the closed loop formed by the four coils. This is most fortunate, because any such circulating current would produce I^2R losses.

The voltage between the brushes is not zero, but is equal to $e_b + e_c$ (or $e_a + e_d$) at the instant shown.

The armature winding we have just discussed is called a *lap winding*. It is the most common type of winding used in direct current generators and motors.

4.5 Induced voltage

Figures 4-11a and 4-11b show a more realistic armature having 12 coils and 12 slots instead of only 4. When the armature rotates, the voltage E induced in each conductor depends upon the flux density which it cuts. This fact is based upon the equation:

$$E = Blv \qquad \text{Eq. 2-25}$$

Because the density in the air gap varies from point to point, the induced voltage per coil depends upon its instantaneous position. Consider, for example, the voltages induced in the armature when it occupies the position shown in Fig. 4-11. The conductors in slots 1 and 7 are exactly between the poles where the flux density is zero. The voltage induced in the two coils lodged in slots 1 and 7 is therefore zero. On the other hand, the conductors in slots 4 and 10 are directly under the center of the poles, where the flux density is greatest. The voltage induced in the two coils lodged in *these* slots is therefore maximum. Finally, owing to magnetic symmetry, the voltage induced in the coils lodged in slots 3 and 9 is the same as that induced in the coils lodged in slots 5 and 11.

Figure 4-11b shows the instantaneous voltage induced in each of the 12 coils of the armature. They are 0-7-18 and 20 V, respectively. Note that the brushes short-circuit the coils in which the voltage is momentarily zero.

Taking polarities into account, we can see

Figure 4-11a
Physical construction of a dc generator having 12 coils, 12 slots and 12 commutator bars.

Figure 4-11b
Schematic diagram of the armature and the voltages induced in the 12 coils.

that the voltage between the brushes is (7 + 18 + 20 + 18 + 7) = 70 V, and brush **x** is positive with respect to brush **y**. This voltage remains essentially constant as the armature rotates, because the number of coils between the brushes is always the same, irrespective of armature position.

If we were to shift the brush yoke by 30° (Fig. 4-12), the voltage between the brushes would become (0 + 7 + 18 + 20 + 18) = 63 V.

Thus, by shifting the brushes, the output voltage decreases. Furthermore, in this position, the brushes continually short-circuit coils that generate 7 V. Large currents will flow in the short-circuited coils and brushes, and sparking will result. Thus, shifting the brushes off the neutral position reduces the voltage between the brushes and causes sparking.

4.6 Neutral zones

Neutral zones are those places on the surface of the armature where the flux density is zero. When the generator operates at no-load, the

neutral zones are located exactly between the poles. No voltage is induced in a coil that cuts through the neutral zone. We always try to set the brushes so that they are in contact with coils that are momentarily in a neutral zone.

Figure 4-12
Moving the brushes off the neutral point reduces the output voltage and produces sparking.

4.7 Value of the induced voltage

The voltage induced in a dc generator having a lap winding is given by the equation:

$$E_\mathrm{o} = Zn\Phi/60 \qquad (4\text{-}1)$$

where

E_o = voltage between the brushes [V]
Z = total number of conductors on the armature
n = speed of rotation [r/min]
Φ = flux per pole [Wb]

This important equation shows that for a given generator, the voltage is directly proportional to the flux per pole and to the speed of rotation. The equation only holds true if the brushes are on the neutral position. If the brushes are shifted off neutral, the effect is equivalent to reducing the number of conductors Z.

Example 4-1:
The armature of a six-pole, 600 r/min generator, has 90 slots. If each coil has 4 turns and the flux per pole is 0.04 Wb, calculate the value of the induced voltage.

Solution:
Each turn has two active conductors, and 90 coils are required to fill the 90 slots. The total number of armature conductors is:

Z = 90 coils × 4 turns/coil × 2 conductors/turn
= 720

The speed is:
n = 600 r/min

consequently,

E_o = $Zn\Phi/60$ = 720 × 600 × 0.04/60
= 288 V

The voltage between the brushes at no-load is therefore 288 V, provided the brushes are on neutral.

4.8 Generator under load: the energy conversion process

When a direct-current generator is under load, some fundamental flux and current relationships take place that are directly related to the mechanical-electrical energy conversion process. Consider for example, a two-pole generator that is driven counterclockwise while delivering current to a load (Fig. 4-13)

The current delivered by the generator also flows through all the armature conductors. If we could look inside the machine, we would discover that current always flows in the same direction in those conductors that are momentarily under a N pole. The same is true for conductors that are momentarily under a S pole. However, the currents under the N pole flow in

Figure 4-13
The energy conversion process. The electromagnetic torque due to *F* must be balanced by the applied mechanical torque.

the opposite direction to those under a S pole. Referring to Figure 4-13, the armature conductors under the S pole carry currents that flow *into* the page, away from the reader. Conversely, the armature currents under the N pole flow *out* of the page, towards the reader.

Because the conductors lie in a magnetic field, they are subjected to a force, according to Lorentz's law. If we examine the direction of current flow and the direction of flux, we find that the individual forces F on the conductors all act clockwise. In effect, they produce a torque which acts *opposite* to the direction in which the generator is being driven. To keep the generator going, we must exert a torque on the shaft to overcome this opposing electromagnetic torque. The resulting mechanical power is converted into electrical power, which is delivered to the generator load. That is how the energy conversion process takes place.

4.9 Armature reaction

Up till now, we have assumed that the only mmf acting in a dc generator is that due to the field. However, the current flowing in the armature coils also creates a powerful magnetomotive force that distorts and weakens the flux coming from the poles. This distortion and field weakening takes place in both motors and generators. The effect produced by the armature mmf is called *armature reaction*.

To understand the effect of the armature mmf, we return to the generator under load (Fig. 4-13). If we consider the armature alone, it will produce a magnetic field as shown in Fig. 4-14. This field acts at right angles to the field produced by the N, S poles. The intensity of the armature flux depends upon its mmf, which in turn depends upon the current carried by the armature. Thus, contrary to the field flux, the armature flux is not constant but varies with the load.

We can immediately foresee a problem which the armature flux will produce. Figure 4-14 shows that the flux in the neutral

Figure 4-14
Magnetic field produced by the current flowing in the armature conductors.

zone is no longer zero, and consequently a voltage will be induced in the coils that are short-circuited by the brushes. As a result, severe sparking may occur. The intensity of the sparking will depend upon the armature flux and hence upon the load current delivered by the generator.

The second problem created by the armature mmf is that it distorts the flux produced by the poles. In effect, the combination of the armature mmf and field mmf produces a magnetic field whose shape is illustrated in Fig. 4-15. The neutral zones have shifted in the direction of rotation of the armature. This occurs in all dc generators.

The flux distorsion produces still another effect: the higher flux density in pole tips 2, 3 causes saturation to set in. Consequently, the increase in flux under pole tips 2, 3 is less than the decrease in flux under pole tips 1, 4. As a result, the total flux produced by the N, S poles

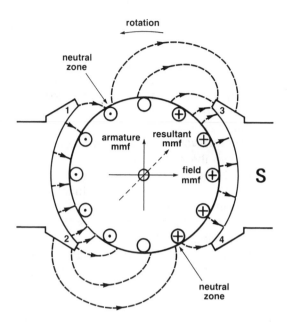

Figure 4-15
Armature reaction distorts the field produced by the N, S poles.

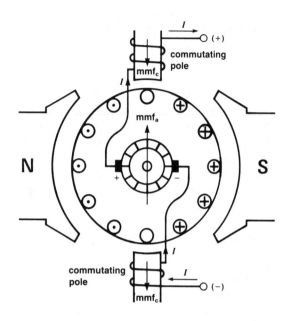

Figure 4-16
Commutating poles produce a mmf$_c$ that opposes the mmf$_a$ of the armature.

is less than it was when the generator was running at no-load. This causes a corresponding reduction in the induced voltage given by Eq. 4-1. For large machines, the decrease in flux may be as much as 10 percent.

It is important to note that the orientation of the armature flux remains fixed in space; it does not rotate with the armature.

4.10 Shifting the brushes to improve commutation

Owing to the shift in the neutral zone when the generator is under load, we could move the brushes to reduce the sparking.

For generators, the brushes are shifted to the new neutral zone by moving them in the direction of rotation. For motors, the brushes are shifted against the direction of rotation.

As soon as the brushes are moved, the commutation improves. However, if the load fluctuates, the armature mmf rises and falls and so

the neutral zone shifts back and forth between the no-load and full-load positions. We would therefore have to move the brushes back and forth to obtain sparkless commutation. This procedure is not practical and other means are used to resolve the problem. For small dc machines, however, the brushes are set in an intermediate position to produce acceptable commutation at all loads.

4.11 Commutating poles

To counter the effect of armature reaction in medium- and large-power machines, we always place a set of *commutating* poles between the main poles of a dc machine (Fig. 4-16).* These narrow poles develop a magnetomotive force mmf$_c$ equal and opposite to the magnetomotive force mmf$_a$ of the arma-

---

* Commutating poles are sometimes called interpoles.

ture. In effect, the two magnetomotive forces rise and fall together as the load current varies. By nullifying the armature mmf in this way, we no longer have to shift the brushes.

Figure 4-16 shows how the commutating poles of a two-pole motor are connected. Clearly, the mmf of the commutating poles acts opposite to the mmf of the armature and therefore neutralizes its effect. However, the neutralization is restricted to the narrow brush zone where commutation takes place. The distorted flux distribution under the pole face unfortunately remains the same.

4.12 Separately excited generator

Now that we have learned some basic facts about dc generators, we can study the various types and their properties. Thus, instead of using permanent magnets to create the magnetic field, we can use a pair of electromagnets as shown in Fig. 4-17. When the dc field current in such a generator is supplied by an independent source (such as a storage battery or another generator, called *exciter*), the generator is said to be *separately excited*. Thus, in Fig. 4-17, the dc source connected to terminals **a** and **b** causes an exciting current I_x to flow. If the armature is driven by a motor or a diesel engine, a voltage E_o appears between brush terminals **x** and **y**.

4.13 No-load operation and saturation curve

When a separately excited dc generator runs at no-load (armature circuit open), a change in the exciting current causes a corresponding change in the induced voltage. We now examine the relationship between the two.

Field flux vs exciting current. Let us gradually raise the exciting current I_x, so that the mmf of the field increases, which increases the flux Φ per pole. If we plot Φ as a function of I_x, we obtain the *saturation curve* of Fig. 4-18a. This curve is obtained whether or not the generator is turning.

When the exciting current is relatively small, the flux is small, and the iron in the machine is

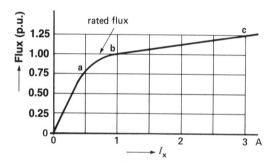

Figure 4-18a
Flux per pole versus exciting current.

unsaturated. Very little mmf is needed to force the flux through the iron, with the result that the mmf developed by the field coils is almost entirely available to force the flux through the air gap. Because the permeability of air is constant, the flux increases in direct proportion to the exciting current, as shown by the portion **oa** of the saturation curve.

However, as we continue to raise the exciting current, the iron in the field and armature begins to saturate. A large mmf is now required to produce a small increase in flux, as shown by portion **bc** of the curve. The machine is now said to be saturated. Saturation of the iron

Figure 4-17
Separately excited 2-pole generator.

begins to be important when we reach the so-called "knee" **ab** of the saturation curve.

How does the saturation curve relate to the induced voltage E_o? If we drive the generator at constant speed, E_o is directly proportional to the flux Φ. Consequently, by plotting E_o as a function of I_x, we obtain a curve whose shape is identical to the saturation curve of Fig. 4-18a. The result is shown in Fig. 4-18b; it is called the *no-load saturation curve* of the generator.

The nominal voltage of a dc generator is usually a little above the knee of the curve. In Fig. 4-18b, for example, the nominal voltage is 120 V. By varying the exciting current, we can vary the induced voltage as we please. Furthermore, by reversing the current, the flux will

4.14 Shunt generator

A shunt-excited generator is a machine whose so-called *shunt-field* coils are connected in parallel with the armature terminals, so that the generator can be self-excited (Fig. 4-19). The principal advantage of this connection is that it eliminates the need for an external source of excitation.

How is self-excitation achieved? When a shunt generator is started up, a small voltage is induced in the armature owing to the remanent flux in the poles. This voltage produces a small exciting current I_x in the shunt field. The

Figure 4-18b
Saturation curve of a dc generator.

reverse and so, too, will the polarity of the induced voltage.

Induced voltage vs speed. For a given exciting current, the induced voltage increases in direct proportion to the speed, a result that follows from Eq. 4-1.

If we reverse the direction of rotation, the polarity of the induced voltage also reverses. However, if we reverse both the exciting current *and* the direction of rotation, the polarity of the induced voltage remains the same.

(a)

(b)

Figure 4-19
a. Self-excited shunt generator.
b. Schematic diagram of a shunt generator.

resulting small mmf acts in the same direction as the remanent flux, causing the flux per pole to increase. The increased flux increases E_o, which increases I_x, which increases the flux still more, which increases E_o even more . . . and this progressive buildup continues until E_o reaches a maximum value determined by the field resistance and the degree of saturation.

4.15 Controlling the voltage of a shunt generator

It is easy to control the induced voltage of a shunt-excited generator. We simply vary the exciting current by means of a rheostat connected in series with the shunt field (Fig. 4-20).

To understand how the output voltage varies, suppose that E_o is 120 V when the movable contact **p** is in the center of the rheostat. If we move the contact towards extremity m, the

Figure 4-20
Controlling the generator voltage with a field rheostat.

resistance between points a and b diminishes, which causes the exciting current to increase. This increases the flux and, consequently, the induced voltage E_o. On the other hand, if we move the brush towards extremity n, the resistance increases, the exciting current diminishes, the flux diminishes, and E_o will fall.

We can determine the no-load value of E_o if we know the saturation curve of the generator and the total resistance R_t of the shunt field circuit. We simply draw a straight line

corresponding to the resistance R_t, superimposing it on the saturation curve (Fig. 4-21). The line passes through the origin, and its slope is equal to R_t. The point where the line intersects the curve, corresponds to the induced voltage.

For example, if the shunt field has a resistance of 50 Ω, and the rheostat is set to zero, then $R_t = 50\ \Omega$. The slope of the line corresponding to R_t must pass through the point $E = 50\ V, I = 1\ A$. This line intersects the saturation curve at a point corresponding to a voltage of 150 V (Fig. 4-21). This is the maximum voltage which the shunt generator can produce. By changing the setting of the rheostat, the total resistance of the field circuit increases, causing E_o to decrease progressively. For example, if R_t is increased to 120 Ω, the line cuts the curve at a voltage E_o of 120 V.

If we continue to increase R_t we will reach a

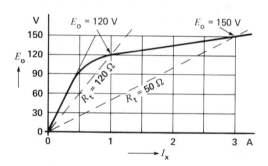

Figure 4-21
The no-load voltage depends upon the resistance of the shunt-field circuit.

critical value where the slope of the resistance line is equal to that of the saturation curve. When this point is reached, the induced voltage drops to zero. In Fig. 4-21, this critical resistance is equal to 200 Ω.

4.16 Equivalent circuit

We have seen that the armature winding contains a set of identical coils, all of which possess a certain resistance. The total armature resistance R_o is that which exists between the armature terminals when the machine is stationary. It is measured on the commutator surface between those segments that lie under the (+) and (−) brushes. The resistance is usually very small, often less than one hundredth of an ohm. Its value depends mainly upon the power and voltage of the generator. To simplify the generator circuit, we can represent R_o as if it were in series with one of the brushes. If the machine has interpoles, the resistance of these windings is included in R_o.

The equivalent circuit of a generator is thus composed of a resistance R_o in series with a voltage E_o (Fig. 4-22). The latter is the voltage induced in the revolving conductors. Terminals

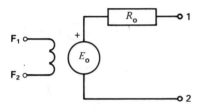

Figure 4-22
Equivalent circuit of a dc generator.

Figure 4-23
Separately excited generator under load.

Figure 4-24
Load characteristic of a separately excited generator.

1, 2 are the external armature terminals of the machine, and F_1, F_2 are the field winding terminals. Using this circuit, we shall now study the more common types of direct current generators and their behavior under load.

4.17 Separately excited generator under load

Let us consider a separately excited generator that is driven at constant speed and whose field is excited by a battery (Fig. 4-23). Induced voltage E_o is therefore fixed. When the machine operates at no-load, terminal voltage E_{12} is equal to the induced voltage E_o because the voltage drop in the armature resistance is zero. However, if we connect a load across the armature (Fig. 4-23), the resulting load current I produces a voltage drop across resistance R_o. Terminal voltage E_{12} is now less than the induced voltage E_o. As we increase the load, the terminal voltage diminishes progressively, as shown in Fig. 4-24. The graph of terminal voltage as a function of load current is called the *load curve* of the generator.

In practice, the induced voltage E_o also decreases slightly with increasing load because pole-tip saturation tends to decrease the field flux. Consequently, the terminal voltage E_{12} falls off more rapidly than can be attributed to armature resistance alone.

4.18 Shunt generator under load

The terminal voltage of a self-excited shunt generator falls off more sharply with increasing load than that of a separately excited generator. The reason is that the field current in a separately excited machine remains constant. This is not so for a self-excited generator, because its exciting current falls as the terminal voltage drops. For a self-excited generator, the drop in voltage from no-load to full-load is about 15 percent of the no-load voltage, whereas for a separately excited generator it is usually less than 10 percent.

4.19 Compound generator

The compound generator was developed to prevent the terminal voltage of a dc generator from decreasing with increasing load. Thus, although we can usually tolerate a reasonable drop in terminal voltage as the load increases, this has a serious effect on lighting circuits. For example, the distribution system of a ship supplies power to both dc machinery and incandescent lamps. The current delivered by the generator fluctuates continually, in response to the varying loads. These current variations produce corresponding changes in the generator terminal voltage, causing the lights to flicker. Compound generators eliminate this problem.

A compound generator (Fig. 4-25) is similar to a shunt generator, except that it has additional field coils connected in series with the armature. These *series field* coils are composed of a few turns of heavy wire, big enough to carry the armature current. The total resistance of the series coils is therefore small. Figure 4-25b is a schematic diagram showing the shunt and series field connections.

When the generator runs at no-load, the current in the series coils is zero. The shunt coils, however, carry exciting current I_x which produces the field flux, just as in a standard self-excited shunt generator. As the generator is loaded, the terminal voltage tends to drop, but load current I_c now flows through the series field coils. The mmf developed by these coils acts in the same direction as the mmf of the shunt field. Consequently, the field flux under

load rises above its original no-load value, which raises the value of E_o. If the series coils are properly designed, the terminal voltage remains practically constant from no-load to full-load. In effect, the rise in the induced voltage compensates for the armature IR drop.

In some cases, we have to compensate not only for the armature voltage drop, but also for the IR drop in the feeder line between the gen-

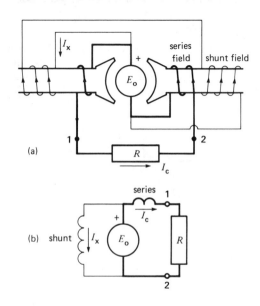

Figure 4-25
a. Compound generator under load.
b. Schematic diagram.

erator and load. The generator manufacturer then adds one or two extra turns on the series winding so that the terminal voltage *increases* as the load current rises. Such machines are called *over-compound generators*. If the compounding is too strong, a low resistance can be placed in parallel with the series field. This reduces the current in the series field and has the same effect as reducing the number of turns. For example, if the value of the so-called *diverter* resistance is equal to that of the series field, the current in the latter is reduced by half.

4.20 Differential compound generator

In a *differential compound generator* the mmf of the series field acts opposite to the shunt field. As a result, the terminal voltage falls drastically with increasing load. We can make such a generator by simply reversing the series field of a standard compound generator. Differential compound generators were formerly used in dc arc welders, because they tended to limit the short-circuit current and to stabilize the arc during the welding process.

4.21 Load characteristics

The load characteristics of some shunt and compound generators are given in Fig. 4-26. The voltage of an over-compound generator increases by 10 percent when full load is applied, whereas that of a flat-compound generator remains constant. On the other hand, the full-load voltage of a shunt generator is 15 percent below its no-load value, while that of a differential-compound generator is 30 percent lower.

Figure 4-26
Typical load characteristics of dc generators.

4.22 Generator specifications

The nameplate of a generator indicates the power, voltage, speed and other details about the machine. These specifications, or *nominal characteristics*, are the values guaranteed by the manufacturer. For example, the following information is punched on the nameplate of a 100 kW generator:

Power	100 kW	Speed	1200 r/min
Voltage	250 V	Type	Compound
Exciting current	20 A	Class	B
Temperature rise	50°C		

These specifications tell us that the machine can deliver, continuously, a power of 100 kW at a voltage of 250 V, without exceeding a temperature rise of 50°C. It can therefore supply a load current of 100 000/250 = 400 A. It possesses a series winding, and the current in the shunt field, is 20 A. In practice, the terminal voltage is adjusted to a value close to its rating of 250 V. We may draw any amount of power from the generator, as long as it does not exceed 100 kW. The class B designation refers to the class of insulation used in the machine.

CONSTRUCTION OF DIRECT CURRENT GENERATORS

We have described the basic features and properties of direct current generators. We now look at the mechanical construction of these machines, directing our attention to the field, the armature, the commutator and the brushes.

4.23 Field

The field produces the magnetic flux in the machine. It is basically a stationary electromagnet composed of a set of salient poles bolted to the inside of a circular frame (Figs. 4-27, 4-28). Field coils, mounted on the poles, carry the dc exciting current. The frame is usually made of solid cast steel, whereas the pole pieces are composed of stacked iron laminations. In some generators the flux is created by permanent magnets.

Figure 4-27
Cross section of a 2-pole generator.

Figure 4-28
Cutaway view of a 4-pole shunt generator. It has 3 brushes per brush set.

In our discussions so far we have considered only 2-pole generators. However, in practice, a dc generator or motor may have 2, 4, 6, to as many as 24 poles. The number of poles depends upon the physical size of the machine; the bigger it is, the more poles it will have. By using a multi-pole design, we can reduce the dimensions and cost of large machines, and also improve their performance.

The field coils of a multipole machine are connected together so that adjacent poles have opposite magnetic polarities (Fig. 4-29). The shunt field coils are composed of several hun-

dred turns of wire carrying a relatively small current. The coils are insulated from the pole pieces to prevent short-circuits.

The mmf developed by the coils produces a magnetic field that passes through the pole pieces, the frame, the armature and the air gap. The air gap is the short space between the armature and the pole pieces. It increases

Figure 4-29
Adjacent poles of multipole generators have opposite magnetic polarities.

approximately from 1.5 to 5 mm as the generator rating increases from 1 kW to 100 kW.

Because the armature and field are composed of magnetic materials having excellent permeability, most of the mmf is used to force the flux across the air gap. Consequently, by reducing its length, we can diminish the size of the shunt field coils. However, the air gap cannot be made too short otherwise the armature reaction effect becomes too great.

If the generator has a series field, the coils are wound on top of the shunt-field coils. The conductor size must be large enough so that the winding does not overheat when it carries the full-load current of the generator.

4.24 Armature

The armature is the rotating part of a dc generator. It consists of a commutator, an iron core and a set of coils (Fig. 4-30). The armature is

Figure 4-30
Armature of a dc generator showing the commutator, stacked laminations, slots and shaft. (*General Electric Company, USA*)

Figure 4-31
Armature lamination with tapered slots.

keyed to a shaft and revolves between the field poles. The iron core is composed of slotted, iron laminations that are stacked to form a solid cylindrical core. The laminations are individually coated with an insulating film so that they do not come in electrical contact with each other. As a result, eddy-current losses are reduced. The slots are lined up to provide space to insert the armature conductors.

The armature conductors carry the load current delivered by the generator. They are insulated from the iron core by several layers of paper or mica and are firmly held in place by fiber slot sticks. If the armature current is below 10 A, round wire is used, but for currents exceeding 20 A, rectangular conductors are preferred because they make better use of the available slot space. The lamination of a small armature is shown in Fig. 4-31. A cross section view of the slot of a large armature is shown in Fig. 4-32.

Figure 4-32
Cross section of a slot containing 4 conductors.

4.25 Commutator and brushes

The commutator is composed of an assembly of tapered copper segments insulated from each other by mica sheets, and mounted on the shaft of the machine (Fig. 4-33). The armature

mica

segment

Figure 4-33
Commutator of a dc machine.

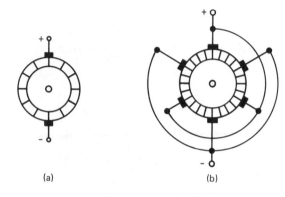

(a)

(b)

Figure 4-34
a. Brushes of a 2-pole generator
b. Brushes and connections of a 6-pole generator.

conductors are connected to the commutator in a manner we explain below (Sec. 4.26).

Great care is taken in building the commutator because any eccentricity will cause the brushes to bounce, producing unacceptable sparking. The sparks burn the brushes and overheat and carbonize the commutator.

A two-pole generator has two fixed brushes diametrically opposite each other (Fig. 4-34a). They slide on the commutator and ensure good electrical contact between the revolving armature and the stationary external load.

Multipole machines possess as many brush sets as they have poles. The brush sets, in turn, are composed of one or more brushes, depending upon the current that has to be carried. In Fig. 4-35c, for example, two brushes make up the brush set. The brush sets are spaced at equal intervals around the commutator. They are supported by a movable brush yoke that permits the entire brush assembly to be rotated through an angle and then locked in the neutral position. In going around the commutator, the successive brush sets have positive and negative polarities. Brushes having the same polarity are connected together and the leads are brought out to one positive and one

(a)

(b)

spring

(c)

Figure 4-35
a. Carbon brush and ultra-flexible copper lead.
b. Brush holder and spring to exert pressure.
c. Brush set composed of two brushes, mounted on rocker arm. (*General Electric Company, USA*).

negative terminal (Fig. 4-34b).

The brushes are made of carbon because it has good electrical conductivity and its softness does not score the commutator. To improve the conductivity, a small amount of copper is sometimes mixed with the carbon. The brush pressure is set by means of adjustable springs. If the pressure is too great, the friction produces excessive heating of the commutator and brushes; on the other hand, if it is too weak, the imperfect contact may produce sparking. The pressure is usually about 15 kPa ($\sim$ 2 lb/in^2) and the permissible current density is approximately 10 A/cm^2 ($\sim$ 65 A/in^2). Thus, a typical brush having a cross section of 3 cm $\times$ 1 cm ($\sim$ 1.2 in $\times$ 0.4 in) exerts a pressure of 4.5 N ($\sim$ 1 lb) and can carry a current of about 30 A.

Figure 4-36 shows the construction of a modern 4-pole dc generator. In order to appreciate the progress that has been made, Fig. 4-37 shows a generator that was built in 1889.

Figure 4-36
Sectional view of a 100 kW, 250 V, 1750 r/min 4-pole dc generator. (*General Electric Company, USA*)

Figure 4-37
This direct-current Thompson generator was first installed in 1889, to light the streets of Montreal. It delivered a current of 250 A at a voltage of 110 V. Other properties of this pioneering machine:

Speed	1300 r/min
Total weight	2390 kg
Armature diameter	292 mm
Stator internal diameter	330 mm
Number of commutator bars	76
Armature conductor size	# 4
Shunt field conductor size	# 14

A modern generator having the same power and speed weighs 7 times less and occupies only 1/3 the floor space.

4.26 Details of a multipole generator

In order to get a better understanding of multipole generators, we shall examine the construction of a 12-pole machine. Figure 4-38a is the schematic diagram of such a machine having 72 slots on the armature, 72 segments on the commutator, and 72 coils. The armature has a lap winding, and the reader should note how similar it is to the schematic

diagram of a 2-pole machine (Fig. 4-11b). Coils A and C are momentarily in the neutral zone, while coil B is cutting the flux coming from the center of the poles.

The coil width (known as coil pitch) is such that the coil sides cut the flux coming from adjacent N, S poles. Thus, the coil sides of coil B lie under the center of pole 2 and the center of pole 3. Similarly, the coil sides of coil A are in the neutral zones between poles 1, 2 and poles 2, 3.

The voltage generated between brushes X and Y is equal to the sum of the voltages generated by the 5 coils connected to commutator segments 1-2, 2-3, 3-4, 4-5 and 5-6. The voltages between the other brush sets are similarly generated by five coils.

The (+) brush sets are connected together to form the (+) terminal. The (−) brush sets are similarly connected to form the (−) terminal. These connections are not shown on the diagram. For similar reasons of clarity, we do not show the interpoles which are placed between the N, S poles.

Figure 4-38b gives a detailed view of the armature coils lying between brushes X and Y. Only the three coils A, B, C are shown so as not to confuse the diagram. Coil A has its coil-sides in slots 1 and 7, while those of coil B are in slots 4 and 10. Furthermore, coil A is connected to commutator segments 72 and 1, while coil B is connected to segments 3 and 4.

In the position shown, the coil-sides of coil A are in the neutral zone between the poles. Consequently, no voltage is induced in coil A. On the other hand, the coil-sides of B are directly under the N and S poles. The voltage in coil B is maximum at this moment. Consequently, the voltage between adjacent commutator segments 3 and 4 is maximum.

The voltage in coil C is also zero because its coil-sides are sweeping across the neutral zone. Note that the positive and negative brushes short-circuit coils having zero voltage.

Figure 4-38a
Schematic diagram of a 12-pole, 72-coil dc generator.

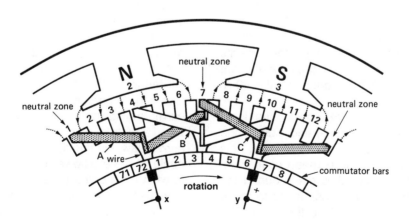

Figure 4-38b
Closeup view of the armature coils between adjacent
brushes.

Example 4-2:

The generator in Fig. 4-38 generates 240 V between adjacent brushes and delivers a current of 2400 A to the load. Calculate:

a. the current delivered per brush set;
b. the current flowing in each coil;
c. the average voltage induced per coil.

Solution:

a. A current of 2400 A flows out of the (+) terminal and back into the (−) terminal of the generator. There are 12 brush sets, six positive and six negative.

The current per brush set is:
$$I = 2400/6 = 400 \text{ A}$$

b. Each positive brush set gathers current from the coils to the right and to the left of the brush. Consequently, the current in each coil is:
$$I = 400/2 = 200 \text{ A}.$$

c. There are six coils between adjacent brush sets. The average voltage per coil is:
$$E_{\text{avge}} = 240/6 = 40 \text{ V}$$

4.27 The ideal commutation process

When a generator is under load, the individual coils on the armature carry one-half the load current that enters into (or flows out of) one brush. In the case of the 12-pole generator discussed in the previous section, the currents flowing in the windings next to a positive brush are shown in Fig. 4-39a. Note that the currents in the coils flow towards the brush, both from the right and the left. If the load current is 80 A, the coils all carry 40 A.

However, if the armature turns counterclockwise, the coils on the right-hand side of the brush will soon be on the left-hand side. This means that the current in these coils must reverse. The reversal takes place during the millisecond interval that a coil takes to move from one end of the brush to the other. The process whereby the current changes direction in this brief interval is called *commutation*.

To understand how commutation takes place, we refer to Figs. 4-39a to 4-39e.

In Fig. 4-39a, the brush is in the middle of segment 1, and the 40 A from the coils on the right and the left of the brush unite to give the 80 A output. The contact resistance between the segment and brush produces a voltage drop of about 1 V.

In Fig. 4-39b, the commutator has moved a short distance and 25 percent of the brush surface is now in contact with segment 2 while 75 percent is in contact with segment 1. Owing to the contact resistance, the *conductivity* between the brush and commutator is proportional to the contact area. The area in contact with segment 2 is only one-fourth of the total contact area, and so the current from segment 2 is only one-fourth of the total current namely 0.25 × 80 = 20 A. By the same token, the current from segment 1 to the brush is 0.75 × 80 = 60 A.

If we now apply Kirchhoff's current law, we discover that the current flowing in coil 1 must be 20 A. Thus, by coming in contact with the brush, the current in this coil has dropped from 40 A to 20 A.

In Fig. 4-39c, the commutator has moved a little further and the brush area in contact with segments 1 and 2 is now the same. Consequently, the conductivities are the same and so the currents are equal. This means that the current in coil 1 is zero at this instant.

In Fig. 4-39d, the commutator has moved still farther to the left. Segment 2 is now in contact with 75 percent of the brush, and so the currents divide accordingly: 60 A from segment 2 and 20 A from segment 1. Applying Kirchoffs' current law, we find that the current in coil 1 is again 20 A, but it flows in the opposite direction to what it did before! We can now understand how the brush contact resistance forces a progressive reversal of the current as the segments slide over the brush.

In Fig. 4-39e the current reversal in coil 1 is

(a)

(b)

(c)

(d)

(e)

Figure 4-39
Commutation of the current in coil 1. Inductive effects are neglected and current reversal is caused by the brush contact resistance.

complete and the current in coil 2 is about to be reversed.

In this ideal commutation process, it is important to note that the current density (amperes per square inch) remains the same at every point across the brush face. Thus, the heat produced by the contact resistance is spread uniformly across the brush surface. Unfortunately, such ideal commutation is not possible in practical machines, and we now investigate the reason why.

4.28 The practical commutation process

The problem with commutation is that it takes place in a very short time; consequently, the current cannot reverse as quickly as it should. The reason is that the armature coils have inductance and it strongly opposes a rapid change in current.

Suppose, for example, that the commutator in Fig. 4-39 turns at 600 r/min. One revolution is therefore completed in 1/10 of a second and during this short period 72 commutator bars sweep past the brush. Thus, the time available to reverse the current in coil 1 is only $1/10 \times 1/72 = 1/720$ s or 1.39 ms!

The voltage induced by self-induction is given by

$$e = L\Delta I/\Delta t \qquad (4\text{-}2)$$

in which

$e =$ induced voltage [V]

$L =$ inductance of the coil [H]

$\Delta I/\Delta t =$ rate of change of current [A/s]

If coil 1 has an inductance of, say, 100 μH, the induced voltage is

$$e = L\,\Delta I/\Delta t$$

$$= \frac{100 \times 10^{-6} \times (+40 - (-40))}{1.39 \times 10^{-3}}$$

$$= 5.75\ \text{V}$$

It is the presence of this induced voltage that opposes the change in current.

Figs. 4-40a to 4-40e illustrate the new currents flowing in coil 1 when the self-inductance of the coil is considered. We have assumed some plausible values so as to determine the current flows in the brush. The currents should be compared with those in Fig. 4-39.

In Fig. 4-40a, the brush is in the middle of segment 1, and the currents in all the coils are stable. As a result, the coil inductance does not come into play.

In Fig. 4-40b, the current is changing due to the contact resistance effect. However, the induced voltage *e* prevents the current in coil 1 from dropping to its ideal value of 20 A. Indeed, the coil current is 35 A and the currents flowing from segments 1 and 2 into the brush are now 75 A and 5 A, instead of 60 A and 20 A. Note that the current density is no longer uniform over the brush face. The density is low where the brush touches segment 2, and high where it touches segment 1.

In Fig. 4-40d the brush is beyond the midpoint of segment 1 and the current has *still* not reversed.

Consequently, the current flowing from segment 1 to the brush is very high (60 A) despite the fact that the contact area is small. The high current density causes the brush to overheat at the tip. Because 720 coils are being commutated every second, this overheating raises the brush tip to the incandescent point and serious sparking is the result.

In designing dc motors and generators, every effort is made to reduce the inductance of the coils. One of the most effective ways is to reduce the number of turns per coil. But for a given output voltage, this means that the number of coils must be increased. And more coils implies more commutators bars. Thus, we end up with direct-current generators that have a large number of commutator bars − not to reduce the ripple in the output voltage, but to reduce the problem of commutation.

(a)

(b)

(c)

(d)

(e)

Figure 4-40
Commutation of the current in coil 1. The coil inductance opposes the reversal of current.

QUESTIONS AND PROBLEMS

Practical level

4-1 Sketch the main components of a dc generator.

4-2 Why are the brushes of a dc machine always placed at the neutral points?

4-3 Describe the construction of a commutator.

4-4 How is the induced voltage of a separately excited dc generator affected if:
a. the speed increases;
b. the exciting current is reduced?

4-5 How do we adjust the voltage of a shunt generator?

4-6 The terminal voltage of a shunt generator decreases with increasing load. Explain.

4-7 Explain why the output voltage of an over-compound generator increases as the load increases.

4-8 Explain the difference between shunt, compound and differential compound generators:
a. as to construction;
b. as to electrical properties.

Intermediate level

4-9 A separately excited dc generator turning at 1400 r/min produces an induced voltage of 127 V. The armature resistance is 2 Ω and the machine delivers a current of 12 A. Calculate:
a. the terminal voltage [V];
b. the heat dissipated in the armature [W];
c. the braking torque exerted by the armature [N·m].

4-10 A separately excited dc generator produces a no-load voltage of 115 V. What happens if:

a. the speed is increased by 20 percent?
b. the direction of rotation is reversed?
c. the exciting current is increased by 10 percent?
d. the polarity of the field is reversed?

4-11 Each pole of a 100 kW, 250 V compound generator has a shunt field of 2000 turns and a series field of 7 turns. If the total shunt-field resistance is 100 Ω, calculate the mmf when the machine operates:
a. at no-load;
b. at full-load.

4-12 Figure 4-18b shows the no-load saturation curve of a separately excited dc generator when it revolves at 1500 r/min. Calculate the exciting current needed to generate 120 V at 1330 r/min.

4-13 Referring to Fig. 4-10, the induced voltage in coil D is momentarily 18 V, in the position shown. Calculate the voltages induced in coils A, B, and C at the same instant.

4-14 Referring to Fig. 4-11b, calculate the voltage induced in coil A when the armature has rotated by 90°; by 120°.

4-15 Brush **x** is positive with respect to brush **y** in Fig. 4-11b. Show the polarity of each of the 12 coils. Does the polarity reverse when a coil turns through 180°?

4-16 The generator of Fig. 4-38 revolves at 960 r/min and the flux per pole is 20 mWb. Calculate the no-load armature voltage if each armature coil has 6 turns.

4-17 a. How many brush sets are needed for the generator in Fig. 4-38?
b. If the machine delivers a load current of 1800 A, calculate the current flowing in each armature coil.

Advanced level

4-18   The voltage between brushes X and Y is 240 V in the generator shown in Fig. 4-38. Why can we say that the voltage between segments 3 and 4 *must* be greater than 40 V?

4-19   Referring to Fig. 4-10, determine the polarity of E_{xy} when the armature turns counterclockwise.

4-20   a.   In Fig. 4-38, determine the polarity of E_{34} between commutator segments 3 and 4, knowing that the armature is turning clockwise.
       b.   At the same instant, what is the polarity of segment 35 with respect to segment 34?

4-21   The armature shown in Fig. 5-4 has 81 slots and the commutator has 243 segments. It will be wound to give a 6-pole lap winding having 1 turn per coil. If the flux per field pole is 30 mWb, calculate:
       a.   the induced voltage at a speed of 1200 r/min;
       b.   the average flux density per pole;
       c.   the time needed to reverse the current in each armature coil, knowing that the brushes are 15 mm wide and that the diameter of the commutator is 450 mm.

4-22   A 200W, 120V, 1800 r/min dc generator has 75 commutator bars. The brush width is such as to cover 3 commutator segments. Show that the duration of the commutator process is equal to 1.33 ms.

4-23   A 4-pole 250 kW, 750V dc generator has a lap winding on the armature. Calculate:
       a.   the full load current of the generator.
       b.   the current carried by the armature coils.

5

DIRECT CURRENT MOTORS

Now that we have a good understanding of dc generators, we can begin our study of dc motors. Direct current motors transform electrical energy into mechanical energy. They drive devices such as hoists, fans, pumps, calenders, punch-presses, and cars. These devices may have a definite torque-speed characteristic (such as a pump or fan) or a highly variable one (such as hoist or car). The torque-speed characteristic of the motor must be adapted to the type of the load it has to drive, and this requirement has given rise to three basic types of motors:

1. shunt motors,
2. series motors,
3. compound motors.

Direct current motors are seldom used in ordinary industrial applications because all electric utility systems furnish alternating current. However, for special applications such as in steel mills, mines and electric trains, it is advantageous to transform the alternating current into direct current in order to use dc motors. The reason is that the torque-speed characteristics of dc motors can be varied over a wide range while retaining high efficiency.

5.1 Counter-electromotive force (cemf)

Direct current motors are built the same way as generators are; consequently, a dc machine can operate either as a motor or as a generator. To illustrate, consider a dc generator in which the armature, initially at rest, is connected to a dc source E_s by means of a switch (Fig. 5-1). The armature has a resistance R and the field is created by a set of permanent magnets.

As soon as the switch is closed, a large current flows in the armature because its resistance is very low. The individual armature conductors are immediately subjected to a force because they are immersed in the magnetic field created by the permanent magnets. These forces add up to produce a powerful torque, causing the armature to rotate.

On the other hand, as soon as the armature begins to turn, a second phenomenon takes

Figure 5-1
Starting a dc motor across the line.

place: the generator effect. We know that a voltage E_o is induced in the armature conductors as soon as they cut a magnetic field (Fig. 5-2). This is always true, *no matter what causes the rotation*. The value and polarity of the induced voltage are the same as those obtained when the machine operates as a generator. The induced voltage E_o is therefore proportional to

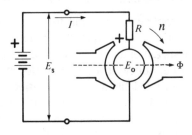

Figure 5-2
Counter-electromotive force (cemf) in a dc motor.

the speed of rotation n of the motor and to the flux Φ per pole, as previously given by Eq. 4-1:

$$E_o = Zn\Phi/60 \qquad \text{Eq. 4-1}$$

As in the case of a generator, Z is a constant which depends upon the number of turns on the armature and the type of winding. For lap windings, Z is equal to the number of armature conductors.

In the case of a motor, the induced voltage

E_o is called *counter-electromotive force* (cemf) because its polarity always acts "against" the source voltage E_s. It acts against the voltage in the sense that the net voltage acting in the series circuit of Fig. 5-2 is equal to $(E_s - E_o)$ volts and not $(E_s + E_o)$ volts.

5.2 Acceleration of the motor

The net voltage acting in the armature circuit in Fig. 5-2 is $(E_s - E_o)$ volts. The resulting armature current I is limited only be the armature resistance R, and so

$$I = (E_s - E_o)/R \qquad (5\text{-}1)$$

When the motor is at rest, the induced voltage $E_o = 0$, and so the starting current is:

$$I = E_s/R$$

The starting current may be 20 to 30 times greater than the nominal full-load current of the motor. The large forces acting on the armature conductors produce a powerful starting torque and a consequent rapid acceleration of the armature.

As the speed increases, the counter-emf E_o increases, with the result that the value of $(E_s - E_o)$ diminishes. It follows from Eq. 5-1 that the armature current I drops progressively as the speed increases.

Although the armature current decreases, the motor continues to accelerate until it reaches a definite, maximum speed. At no-load this speed produces a counter-emf E_o slightly less than the source voltage E_s. In effect, if E_o were equal to E_s, the net voltage $(E_s - E_o)$ would become zero and so, too, would the current I. The driving forces would cease to act on the armature conductors, and the mechanical drag produced by the fan and the bearings would immediately cause the motor to slow down. As the speed decreases the net voltage $(E_s - E_o)$ increases, and so does the current I. The speed will cease to fall as soon as the torque developed by the armature current is equal to the dragging torque. Thus, when a motor runs at no-load, the counter-emf must be slightly less than E_s so as to enable a small current to flow, sufficient to produce the required torque.

Example 5-1:

The armature of a permanent-magnet dc generator has a resistance of 1 Ω and generates a voltage of 50 V when the speed is 500 r/min. If the armature is connected to a source of 150 V, calculate:

a. the starting current;

b. the counter-emf when the motor runs at 1000 r/min? at 1460 r/min?

c. the armature current at 1000 r/min? at 1460 r/min?

Solution:

a. At the moment of start-up, the armature is stationary, so $E_o = 0$ V (Fig. 5-3a). The

The corresponding armature current is:
$$I = (E_s - E_o)/R$$
$$= 50/1 = 50 \text{ A (Fig. 5-3b)}$$

When the motor speed reaches 1460 r/min, the cemf will be 146 V, almost equal to the source voltage. Under these conditions, the armature current is only
$$I = (E_s - E_o)/R = (150 - 146)/1$$
$$= 4 \text{ A}$$

and the corresponding motor torque is much smaller than before (Fig. 5-3c).

5.3 Mechanical power and torque

The power and torque of a dc motor are two of its most important properties. We now derive two simple equations that enable us to calculate them.

1. According to Eq. 4-1, the cemf induced in a

(a)   (b)   (c)

Figure 5-3
See Example 5-1.

starting current is limited only by the armature resistance:
$$I = E_s/R = 150 \text{ V}/1 \ \Omega = 150 \text{ A}$$

b. Because the generator voltage is 50 V at 500 r/min, the cemf of the motor will be at 100V at 1000 r/min and 146 V at 1460 r/min.

c. The net voltage in the armature circuit at 1000 r/min is:
$$E_s - E_o = 150 - 100 = 50 \text{ V}$$

lap-wound armature is given by

$$E_o = Zn\Phi/60 \qquad \text{Eq. 4-1}$$

Referring to Fig. 5-2, the electrical power P_a supplied to the armature is equal to the supply voltage E_s multiplied by the armature current I:

$$P_a = E_sI \qquad (5-2)$$

However, E_s is equal to the sum of E_o plus the IR drop in the armature:

$$E_s = E_o + IR \qquad (5\text{-}3)$$

It follows that

$$\begin{aligned}P_a &= E_s I \\ &= (E_o + IR)\, I \\ &= E_o I + I^2 R \qquad (5\text{-}4)\end{aligned}$$

The I^2R term represents heat dissipated in the armature, but the very important term $E_o I$ is the electrical power that is converted into mechanical power. The mechanical power of the motor is therefore exactly equal to the product of the cemf multiplied by the armature current:

$$P = E_o I \qquad (5\text{-}5)$$

where

P = mechanical power developed by the motor [W]

E_o = induced voltage in the armature (cemf) [V]

I = total current supplied to the armature [A]

2. Turning our attention to torque T, we know that the mechanical power P is given by the expression

$$P = nT/9.55 \qquad \text{Eq. 3-5}$$

where n is the speed of rotation.

Combining Eqs. 3-5, 4-1 and 5-5, we obtain

$$\begin{aligned}nT/9.55 &= E_o I \\ &= Zn\Phi I/60\end{aligned}$$

whence $\qquad T = Z\Phi I/6.28$

The torque developed by a lap-wound motor is therefore given by the expression

$$T = Z\Phi I/6.28 \qquad (5\text{-}6)$$

where

T = torque [N·m]

Z = number of conductors on the armature

Φ = effective flux per pole [Wb]*

I = armature current [A]

6.28 = constant, to take care of units [exact value = 2π]

Equation 5-6 shows that we can raise the torque of a motor either by raising the armature current or by raising the flux created by the poles.

Example 5-2:
The following details are given on a 225 kW ($\sim$ 300 hp), 250 V, 1200 r/min dc motor (See Figs. 5-4 and 5-5):

armature coils	243
turns per coil	1
type of winding	lap
armature slots	81
commutator segments	243
field poles	6
diameter of armature	559 mm
axial length of armature	235 mm

Calculate:
a. the armature current;
b. the number of conductors per slot;
c. the flux per pole.

Solution:
a. We can assume that the induced voltage E_o is nearly equal to the applied voltage (250 V).

The rated armature current is:
$$\begin{aligned}I &= P/E_o = 225\,000/250 \\ &= 900 \text{ A}\end{aligned}$$

b. Each coil is made up of 2 conductors, so altogether there are $243 \times 2 = 486$ conductors on the armature.
   Conductors per slot = $486/81 = 6$
   Coil-sides per slot = 6

c. The motor torque is:
$$\begin{aligned}T &= 9.55\, P/n \qquad \text{Eq. 3-5} \\ &= 9.55 \times 225\,000/1200 \\ &= 1790 \text{ N·m}\end{aligned}$$

The flux per pole is:
$$\begin{aligned}\Phi &= 6.28\, T/ZI \qquad \text{Eq. 5-6} \\ &= (6.28 \times 1790)/(486 \times 900) \\ &= 25.7 \text{ mWb}\end{aligned}$$

* The effective flux is given by $\Phi = 60\, E_o/Zn$.

Figure 5-4
Bare armature and commutator of a dc motor rated 225 kW, 250 V, 1200 r/min. The armature core has a diameter of 559 mm and an axial length of 235 mm. It is composed of 400 stacked laminations 0.56 mm thick. The armature has 81 slots and the commutator has 243 bars. (*H. Roberge*)

(a)

(b)

(c)

(d)

Figure 5-5
a. Armature of Fig. 5-4 in the process of being wound: coil-forming machine gives the coils the desired shape

b. One of the 81 coils ready to be placed in the slots;
c. Connecting the coil ends to the commutator bars;
d. Commutator connections ready for brazing. (*H. Roberge*)

5.4 Speed of rotation

When a dc motor runs between no-load and full-load, the IR drop due to armature resistance is always small compared to the supply voltage E_s. This means that the counter-emf E_o is very nearly equal to E_s.

On the other hand, we have already seen that E_o may be expressed by the equation:

$$E_o = Zn\Phi/60 \qquad \text{Eq. 4-1}$$

Replacing E_o by E_s, we obtain

$$E_s = Zn\Phi/60$$

That is, $n = \dfrac{60E_s}{Z\Phi}$ (approx)   (5-7)

where

    n = speed of rotation [r/min]
    E_s = armature voltage [V]
    Z = total number of armature conductors

This important equation shows that motor speed is directly proportional to the armature supply voltage, and inversely proportional to the flux per pole. We will now study how it is applied.

5.5 Armature speed control

According to Eq. 5-7, if the flux per pole Φ is kept constant (permanent magnet field or field with fixed excitation), the speed depends only upon the armature voltage E_s. By raising or lowering E_s, the motor speed will rise and fall in proportion.

In practice, we can vary E_s by connecting the motor armature M to a separately-excited variable-voltage dc generator G (Fig. 5-6). The field excitation of the motor is kept constant, but the generator excitation I_x can be varied from zero to maximum, and even reversed. The generator output voltage E_s can therefore be varied from zero to maximum, with either positive or negative polarity. Consequently, the motor speed can be varied from zero to maximum, in either direction. Note that the generator is driven by an ac motor connected to a 3-phase line. This method of speed control, known as the Ward-Leonard system, is employed in steel mills, high-rise elevators, mines and paper mills.

In modern installations, the generator is often replaced by a high-power electronic converter which changes the ac power of the electrical utility to dc, by electronic means.

The Ward-Leonard system is more than just a simple way of applying a variable dc voltage to the armature of a dc motor. It can actually force the motor to develop the torque and speed required by the load. For example, suppose E_s is adjusted to be slightly higher than the cemf E_o of the motor. Current will then flow in the direction shown in Fig. 5-6, and the motor develops a positive torque. The armature of the motor absorbs power because I flows into the positive terminal.

Now, suppose we reduce E_s by reducing the generator excitation Φ_G. As soon as E_s becomes less than E_o, current I reverses. As a

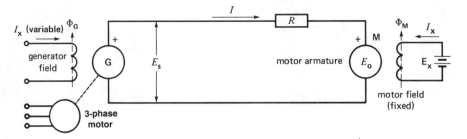

Figure 5-6
Ward-Leonard speed control system.

result, (a) the motor torque reverses and (b) the armature of the motor *delivers* power to generator G. In effect, the dc motor suddenly becomes a generator and generator G suddenly becomes a motor. The electric power which the dc motor now delivers to G is derived at the expense of the kinetic energy of the rapidly decelerating armature and its connected mechanical load. As a result of reducing E_s, the motor is suddenly forced to slow down.

What happens to the dc power received by generator G? When G receives electric power, it operates as a motor, driving its own ac motor as an asynchronous generator!* The result is that ac power is fed back into the line that normally feeds the ac motor. The fact that power can be recovered this way makes the Ward-Leonard system very efficient, and constitutes another of its advantages.

Example 5-3:
A 2000 kW, 500 V, variable-speed motor is driven by a 2500 kW generator, using a Ward-Leonard control system shown in Fig. 5-6. The total resistance of the motor and generator armature circuit is 10 mΩ. The motor turns at a nominal speed of 300 r/min, when E_o is 500 V.

Calculate:
a. the motor torque and speed when
    E_s = 400 V and E_o = 380 V;

b. the motor torque and speed when
    E_s = 350 V and E_o = 380 V

Solution:
a. The armature current is:
    $$I = (E_s - E_o)/R = (400 - 380)/0.01$$
    $$= 2000 \text{ A}$$

The power to the motor armature is:
$$P = E_o I = 380 \times 2000 = 760 \text{ kW}$$

The motor speed is:
$$n = (380 \text{ V}/500 \text{ V}) \times 300 = 228 \text{ r/min}$$

The motor torque is:
$$T = 9.55P/n \qquad \text{Eq. 3-5}$$
$$= (9.55 \times 760\ 000)/228$$
$$= 47.8 \text{ kN·m}$$

b. Because E_o = 380 V, the motor speed is still 228 r/min.

The armature current is:
$$I = (E_s - E_o)/R = (350 - 380)/0.01$$
$$= -3000 \text{ A}$$

The current is negative and so it flows in reverse; consequently, the motor torque also reverses.

Power returned by the motor to the generator and the 10 mΩ resistance:
$$P = E_o I = 380 \times 3000 = 1140 \text{ kW}$$

Breaking torque developed by the motor:
$$T = 9.55P/n$$
$$= (9.55 \times 1\ 140\ 000)/228$$
$$= 47.8 \text{ kN·m}$$

The speed of the motor and its connected mechanical load will rapidly drop under the influence of this electromechanical braking torque.

Another way to control the speed of a dc motor is to place a rheostat in series with the armature (Fig. 5-7). The current in the rheostat produces a voltage drop which subtracts from the fixed source voltage E_s, yielding a smaller supply voltage across the armature. This method enables

Figure 5-7
Armature speed control using a rheostat.

---

*   The asynchronous generator is explained in Chapter 14.

us to *reduce* the speed below its nominal speed. It is only recommended for small motors because a lot of power and heat is wasted in the rheostat, and the overall efficiency is low. Furthermore, the speed regulation is poor, even for a fixed setting of the rheostat. In effect, the *IR* drop across the rheostat increases as the armature current increases. This produces a substantial drop in speed with increasing mechanical load.

5.6 Field speed control

According to Eq. 5-7, we can also vary the speed of a dc motor by varying the field flux Φ. Let us now keep the armature voltage E_s constant so that the numerator in Eq. 5-7, is constant. Consequently, the motor speed now changes in inverse proportion to the flux Φ: if we increase the flux, the speed will drop and vice versa.

This method of speed control is frequently used because it is simple and inexpensive. To control the flux (and hence, the speed), we connect a rheostat R_f in series with the field (Fig. 5-8a).

To understand this method of speed control, suppose that the motor in Fig. 5-8a is initially running at constant speed. The counter-emf E_o is slightly less than the armature supply voltage E_s, owing to the *IR* drop in the armature. If we suddenly increase the resistance of the rheostat, both the exciting current I_x and the flux Φ will diminish. This immediately reduces the cemf E_o, causing the armature current I to jump to a much higher value. The current changes dramatically because its value depends upon the very small *difference* between E_s and E_o. Despite the weaker field, the motor develops a greater torque than before. It will accelerate until E_o is again almost equal to E_s.

Clearly, to develop the same E_o with a weaker flux, the motor must turn faster. We can therefore raise the motor speed above its nomi-

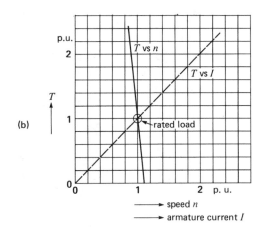

Figure 5-8
a. Schematic diagram of a shunt motor including the field rheostat.
b. Torque-speed and torque-current characteristic of a shunt motor.

nal value by introducing a resistance in series with the field. For shunt-wound motors, this method of speed control enables high-speed/low-speed ratios as high as 3 to 1. Broader speed ranges tend to produce instability and poor commutation.

Under certain abnormal conditions, the flux may drop to dangerously low values. For example, if the exciting current of a shunt motor is interrupted accidentally, the only flux remaining is that due to the remanent magnetism in the poles.* This flux is so small that the motor has to rotate at a dangerously high speed to induce the required cemf. Safety devices are introduced to prevent such runaway conditions.

5.7 Shunt motor under load

Consider a dc motor running at no-load. If a mechanical load is suddenly applied to the shaft, the small no-load current does not produce enough torque to carry the load and the motor begins to slow down. This causes the cemf to diminish, resulting in a higher current and a corresponding higher torque. When the torque developed by the motor is exactly equal to the torque imposed by the mechanical load, then, and only then, will the speed remain constant (see Sec. 3-11). To sum up, as the mechanical load increases, the armature current rises and the speed drops.

The speed of a shunt motor stays relatively constant from no-load to full-load. In small motors, it only drops by 10 to 15 percent when full load is applied. In big machines, the drop is even less, owing, in part, to the very low armature resistance. By adjusting the field rheostat, the speed can, of course, be kept absolutely constant as the load changes.

Typical torque-speed and torque-current characteristics of a shunt motor are shown in Fig. 5-8b. The speed, torque and current are given in per-unit values. The torque is directly proportional to the armature current. Furthermore, the speed changes only from 1.1 pu to 0.9 pu as the torque increases from 0 pu to 2 pu.

Example 5-4:
A shunt motor rotating at 1500 r/min is fed by a 120 V source. (Fig. 5-9a). The line current is

---

* The term *residual* magnetism is also used. However, the IEEE Standard Dictionary of Electrical and Electronic Terms states "... If there are no air gaps... in the magnetic circuit, the remanent induction will equal the residual induction; if there are air gaps ... the remanent induction will be less that the residual induction".

---

51 A and the shunt field resistance is 120 Ω. If the armature resistance is 0.1 Ω, calculate:

a. the current in the armature;
b. the counter-emf;
c. the mechanical power developed by the motor.

Solution:
a. The field current (Fig. 5-9b) is:
$$I_x = 120 \text{ V}/120 \text{ } \Omega = 1 \text{ A}$$

The armature current is:
$$I = 51 - 1 = 50 \text{ A}$$

b. The voltage across the armature is:
$$E = 120 \text{ V}$$

Figure 5-9
See Example 5-4.

Voltage drop due to armature resistance is:
$$IR = 50 \times 0.1 = 5 \text{ V}$$

The cemf generated by the armature is:
$$E_o = 120 - 5 = 115 \text{ V}$$

c. The total power supplied to the motor is:
$$P_i = EI = 120 \times 51 = 6120 \text{ W}$$

Power absorbed by the armature is:
$$P_a = EI = 120 \times 50 = 6000 \text{ W}$$

Power dissipated in the armature is:
$$P = IR^2 = 50^2 \times 0.1 = 250 \text{ W}$$

Mechanical power developed by the armature is:

$P = 6000 - 250 = 5750\,\text{W}$ (equivalent to $5750/746 = 7.7\,\text{hp}$)

The actual mechanical output is slightly less than 5750 W because some of the mechanical power is dissipated in bearing friction losses, in windage losses and in armature iron losses.

5.8 Series motor

A series motor is identical in construction to a shunt motor except for the field. In effect, the field is connected in series with the armature, and must therefore carry the full armature current (Fig. 5-10a). This so-called *series field* is composed of a few turns of wire having a cross section sufficiently large to carry the current.

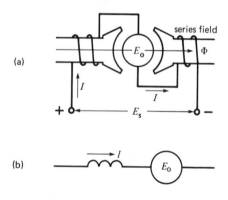

Figure 5-10
a. Series motor connection diagram.
b. Schematic diagram of a series motor.

Although the construction is similar, the properties of a series motor are completely different from those of a shunt motor. In a shunt motor, the flux Φ per pole is constant at all loads because the shunt field is connected to the line. But in a series motor the flux per pole depends upon the armature current, and hence upon the load. When the current is large, the flux is large

and vice versa. Despite these differences, the same basic principles and equations apply to both machines.

When a series motor operates at full load, the flux per pole is the same as that of a shunt motor of identical power and speed. However, when the series motor starts up, the armature current is higher than normal, with the result that the flux per pole is also greater than normal. It follows that the starting torque of a series motor is considerably greater than that of a shunt motor.

On the other hand, if the motor operates at less than full-load, the armature current and the flux per pole are smaller than normal. The weaker field causes the speed to rise, in the same way as it would for a shunt motor with a weak shunt field. For example, if the load current of a series motor drops to half its normal value, the flux diminishes by half and so the speed doubles. Obviously, if the load is small, the speed may rise to dangerously high values.

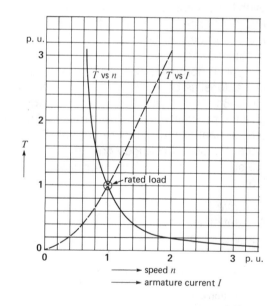

Figure 5-11
Typical torque-speed and torque-current characteristic of a series motor.

For this reason, we never permit a series motor to operate at no-load. It tends to run away, and the resulting centrifugal forces may tear the windings out of the armature and destroy the machine.

5.9 Series motor speed control

When a series motor carries a load, its speed may have to be adjusted slightly. Thus, the speed can be increased by placing a low resistance in parallel with the series field. The field current is then smaller than before, which produces a drop in flux and an increase in speed.

Conversely, the speed may be lowered by connecting an external resistor is series with the armature and the field. The total IR drop across the resistor and field reduces the armature supply voltage, and so the speed must fall.

Typical torque-speed and torque-current characteristics are shown in Fig. 5-11. They are quite different from the shunt motor characteristics given in Fig. 5-8b.

Example 5-5:
A 15 hp, 240 V, 1780 r/min dc series motor has a full-load rated current of 54 A. Its operating characteristics are given by the per-unit curves of Fig. 5-11. Calculate
a. the current and speed when the load torque is 24 N·m
b. the efficiency under these conditions.

Solution:
a. We first establish the base power, base speed and base current of the motor. They correspond to the full-load ratings as follows:
$$P_B = 15 \text{ hp} = 15 \times 746 = 11\ 190 \text{ W}$$
$$n_B = 1780 \text{ r/min}$$
$$I_B = 54 \text{ A}.$$

The base torque is therefore:
$$T_B = \frac{9.55 P_B}{n_B} = 9.55 \times 11\ 190/1\ 780$$
$$= 60 \text{ N·m}$$

A load torque of 24 N·m corresponds to a per-unit torque of
$$T(\text{pu}) = 24/60 = 0.4$$
Referring to Fig. 5-11, a torque of 0.4 pu is attained at a speed of 1.4 pu. Thus, the speed is:
$$n = n(\text{pu}) \times n_B = 1.4 \times 1780$$
$$= 2492 \text{ r/min}.$$

From the T vs I curve, a torque of 0.4 pu requires a current of 0.6 pu. Consequently, the load current is
$$I = I(\text{pu}) \times I_B = 0.6 \times 54 = 32.4 \text{ A}.$$

b. To calculate the efficiency, we have to know P_o and P_i.
$$P_i = EI = 240 \times 32.4 = 7776 \text{ W}$$
$$P_o = nT/9.55 = 2492 \times 24/9.55$$
$$= 6262 \text{ W}$$
$$\eta = P_o/P_i = 6262/7776 = 0.805 \text{ or } 80.5\%$$

5.10 Applications of the series motor

Series motors are used on equipment requiring a high starting torque. They are also used to drive devices which must run at high speed at light loads. The series motor is particularly well adapted for traction purposes, such as in electric trains. Acceleration is rapid because the torque is high at low speeds. Furthermore, the series motor automatically slows down as the train goes up a grade, yet turns at top speed on flat ground. The power of a series motor tends to be constant because high torque is accompanied by low speed and vice versa. Series motors are also used in electric cranes and hoists: light loads are lifted quickly and heavy loads more slowly.

5.11 Compound motor

A compound dc motor carries both a series field and a shunt field. In a *cumulative compound* motor, the mmf of the two fields add. The shunt field is always stronger than the series field.

Figure 5-12 shows the connection and schematic diagrams of a compound motor. When the motor runs at no-load, the armature current I in the series winding is low and the mmf of the series field is negligible. However, the shunt field is fully excited by current I_x and so the motor behaves like a shunt machine: it does not tend to run away at no load.

(a)

(b)

Figure 5-12
a. Connection diagram of a dc compound motor.
b. Schematic diagram of the motor.

As the load increases, the mmf of the series field increases, but the shunt field strength remains constant. The total mmf (and the resulting flux per pole) is therefore greater under load than at no-load. The motor speed falls with increasing load and the speed drop from no-load to full-load is generally between 10 percent and 30 percent.

If the series field is connected so that it opposes the shunt field, we obtain a *differential compound* motor. In such a motor, the total mmf decreases with increasing load. The speed rises as the load increases, and this may lead to instability. The differential compound motor has very few applications.

Figure 5-13 shows the typical torque-speed curves of shunt, compound and series motors on a per-unit basis. Figure 5-14 shows a typical application of dc motors in a steel mill.

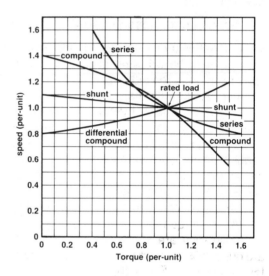

Figure 5-13
Typical speed versus torque characteristics of various dc motors.

5.12 Reversing the direction of rotation

To reverse the direction of rotation of a dc motor, we must reverse either (a) the armature connections, or (b) both the shunt and series field connections. The interpoles are considered to form part of the armature. The change in connections is shown in Fig. 5-15.

5.13 Starting a shunt motor

If we apply full voltage to a stationary shunt motor, the starting current in the armature will be very high, and we run the risk of:
   a. burning out the armature;
   b. damaging the commutator and brushes, owing to heavy sparking;
   c. overloading the feeder;
   d. snapping off the shaft due to mechanical shock;
   e. damaging the driven equipment because of the sudden mechanical hammerblow.

Figure 5-14
Hot strip finishing mill composed of 6 stands, each driven by a 2500 kW dc motor. The wide steel strip is delivered to the runout table (left foreground) driven by 161 dc motors, each rated 3 kW. (*General Electric*)

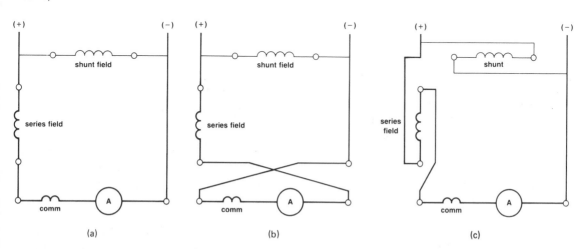

Figure 5-15
a. Original connections of a compound motor.
b. reversing the armature connections to reverse the direction of rotation.
c. Reversing the field connections to reverse the direction of rotation.

All dc motors must therefore be provided with a means to limit the starting current to reasonable values, usually between 1.5 and twice full-load current. One solution is to connect a rheostat in series with the armature. The resistance is gradually reduced as the motor accelerates and is eventually eliminated entirely, when the machine has attained full speed.

Today, electronic methods are often used to limit the starting current and to provide speed control.

5.14 Face-plate starter

Figure 5-16 shows the schematic diagram of a manual face-plate starter for a shunt motor. Bare copper contacts are connected to current-

more, the arm is pulled to the next contact, thereby removing resistor R_1 from the armature circuit. The current immediately jumps to a higher value and the motor quickly accelerates to the next higher speed. When the speed is again stable, we move to the next contact, and so forth, until the arm finally touches the last contact. The arm is magnetically held in this position by a small electromagnet **4**, which is in series with the shunt field.

If the supply voltage is suddenly interrupted, or if the field excitation should accidentally be cut, the electromagnet releases the arm, allowing it to return to its dead position, under the pull of spring **3**. This safety feature prevents the motor from restarting unexpectedly when

Figure 5-16
Manual face-plate starter for a shunt motor.

limiting resistors R_1, R_2, R_3 and R_4. Conducting arm **1** sweeps across the contacts when it is pulled to the right by means of insulated handle **2**. In the position shown, the arm touches dead copper contact **M** and the motor circuit is open. As we draw the handle to the right, the conducting arm first touches fixed contact **N**.

The supply voltage E_s immediately causes full field current I_x to flow, but the armature current I is limited by the four resistors in the starter box. The motor begins to turn and, as the cemf builds up, the armature current gradually falls. When the motor speed ceases to rise any

the supply voltage is reestablished.

5.15 Stopping a motor

One is inclined to believe that stopping a dc motor is a simple, almost trivial, operation. Unfortunately, this is not always true. When a large dc motor is coupled to a heavy inertia load, it may take an hour or more for the system to come to a halt. For many reasons, such a lengthy deceleration time is often unacceptable and, under these circumstances, we must apply a braking torque to ensure a rapid stop. One way to brake the motor is by simple mechanical friction, in the same way we stop a car. A more elegant method consists of circulating a reverse current in the armature, so as to brake the motor electrically. Two methods are employed to create such an electromechanical brake: (a) dynamic braking and (b) plugging.

5.16 Dynamic braking

Consider a shunt motor whose field is directly connected to a source E_s, and whose armature is connected to the same source by means of a double-throw switch. The switch connects the armature to either the line or to an external resistor R (Fig. 5-17).

When the motor is running normally, the direction of the armature current I_1 and the polarity of the cemf E_o are as shown in Fig 5-17a. Neglecting the armature IR drop, E_o is equal to E_s.

If we suddenly open the switch (Fig. 5-17b), the motor continues to turn, but its speed will gradually drop owing to friction and windage losses. On the other hand, because the shunt field is still excited, induced voltage E_o continues to exist, falling at the same rate as the speed. In essence, the motor is now a generator whose armature is on open-circuit.

Let us close the switch on the second set of contacts so that the armature is suddenly connected to the external resistor (Fig. 5-17c). Voltage E_o will immediately produce an armature current I_2. However, this current flows in the *opposite* direction to the original current I_1. It follows that a reverse torque is developed whose magnitude depends upon I_2. The reverse torque brings the machine to a rapid, but very smooth stop.

In practice, resistor R is chosen so that the initial braking current is about twice the rated motor current. The initial braking torque is then twice the normal torque of the motor.

As the motor slows down, the gradual decrease in E_o produces a corresponding decrease in I_2. Consequently, the braking torque becomes smaller and smaller, finally

Figure 5-17a
Armature connected to a dc source E_s.

Figure 5-17b
Armature on open circuit generating a voltage E_o.

Figure 5-17c
Dynamic braking.

becoming zero when the armature ceases to turn. The speed drops quickly at first and then more slowly, as the armature comes to a halt. The speed decreases exponentially, like the voltage across a discharging capacitor. Consequently, the speed decreases by half in equal intervals of time T_o. To illustrate the usefulness of dynamic braking, Fig. 5-18 compares the speed-time curves for a motor equipped with dynamic braking and one which simply coasts to a stop.

Figure 5-19a
Armature connected to dc source E_s

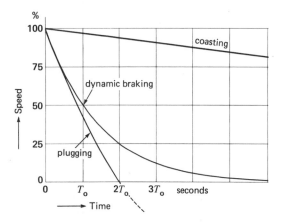

Figure 5-18
Speed versus time curves for various braking methods.

Figure 5-19b
Plugging.

5.17 Plugging

We can stop the motor even more rapidly by using a method called *plugging*. It consists of suddenly reversing the armature current by reversing the terminals of the source (Fig. 5-19).

Under normal motor conditions, armature current I_1 is given by

$$I_1 = (E_s - E_o)/R_o$$

where R_o is the armature resistance. If we suddenly reverse the terminals of the source, the net voltage acting on the armature circuit becomes $(E_o + E_s)$. The so-called "counter" emf E_o of the armature is no longer counter to anything, but actually *adds* to the supply voltage E_s. This net voltage would produce an enormous reverse current, perhaps 100 times greater than the full-load armature current. This current would initiate an arc around the commutator, destroying segments, brushes and

supports, even before the line circuit-breakers could open.

To prevent such a catastrophe, we must limit the reverse current by introducing a resistor R in series with the reversing circuit (Fig. 5-19b). As in dynamic braking, the resistor is designed to limit the initial braking current I_2 to about twice full-load current. With this plugging circuit, a reverse torque is developed even when the armature has come to a stop. In effect, at zero speed, $E_o = 0$, but $I_2 = E_s/R$, which is about one-half its initial value. As soon as the motor stops, we must immediately open the armature circuit, otherwise it will begin to run in reverse. Circuit interruption is usually controlled by an automatic null-speed device, mounted on the motor shaft.

The curves of Fig. 5-18 enable us to compare plugging and dynamic braking for the same initial braking current. Note that plugging stops the motor completely after an interval $2T_o$. On the other hand, if dynamic braking is used, the speed is still 25 percent of its original value at this time. Nevertheless, the comparative simplicity of dynamic braking renders it more popular in most applications.

5.18 Dynamic braking and mechanical time constant

We mentioned that the speed decreases exponentially with time when a dc motor is stopped by dynamic braking. We can therefore speak of a mechanical time constant T in much the same way we speak of the electrical time constant of a capacitor that discharges into a resistor.

In essence, T is the time it takes for the speed of the motor to fall to 36.8 percent of its initial value. However, it is much easier to draw the speed-time curves by defining a new time constant T_o which is the time for the speed to decrease to 50 percent of its original value. There is a direct mathematical relationship between the conventional time constant T and the "half" time constant T_o. It is given by:

$$T_o = 0.693T \qquad (5\text{-}8)$$

We can prove that this mechanical time constant is given by:

$$T_o = \frac{Jn_1{}^2}{131.5P_1} \qquad (5\text{-}9)$$

where

$T_o^/$ = time for the motor speed to fall to one-half its previous value [s]

J = moment of inertia of the rotating parts, referred to the motor shaft [kg·m²]

n_1 = initial speed of the motor when braking starts [r/min]

P_1 = initial power delivered by the motor to the braking resistor [W]

131.5 = a constant [exact value = $(30/\pi)^2/\log_e 2$]

0.693 = a constant [exact value = $\log_e 2$]

This equation is based upon the assumption that the braking effect is entirely due to the energy dissipated in the braking resistor. In general, the motor is subjected to an extra braking torque due to windage and friction and so the braking time will be less than that given by Eq. 5-9.

Example 5-6:
A 225 kW (~ 300 hp), 250 V, 1280 r/min dc motor has windage, friction, and iron losses of 8 kW. It drives a large flywheel and the total moment of inertia of the flywheel and armature is 177 kg·m². The motor is connected to a 210 V dc source, and its speed is 1280 r/min just before the armature is switched across a braking resistor of 0.2 Ω. Calculate:

a. the mechanical time constant T_o of the braking system;
b. the time for the motor speed to drop to 20 r/min;
c. the time for the motor speed to drop to 20 r/min, if the only braking force is that due to the windage, friction and iron losses.

Solution:
a. We note that the armature voltage is 210 V and the speed is 1280 r/min.

When the armature is switched to the braking resistor, the induced voltage is still very close to 210 V. The initial power delivered to the resistor is:

$$P_1 = E^2/R = 210^2/0.2 = 220\,500 \text{ W}$$

The time constant T_o is:

$$T_o = Jn_1{}^2/(131.5P_1) \qquad \text{Eq. 5-9}$$
$$= \frac{177 \times 1280^2}{131.5 \times 220\,500}$$
$$= 10 \text{ s}$$

b. The motor speed drops by 50 percent every 10 s. The speed versus time curve follows the sequence given below:

speed r/min	time s
1280	0
640	10
320	20
160	30
80	40
40	50
20	60

The speed of the motor drops to 20 r/min after an interval of 60 s.

c. The initial windage, friction and iron losses are 8 kW. These losses do not vary in exactly the same way as do the losses in a braking resistor. However, the behavior is comparable, which enables us to make a rough estimate of the braking time. We have:

$$n_1 = 1280 \qquad P_1 = 8000$$

The new time constant is:

$$T_o = Jn_1^2/(131.5P_1)$$
$$= (177 \times 1280^2)/(131.5 \times 8000)$$
$$= 276 \text{ s} = 4.6 \text{ min}$$

The stopping time increases in proportion to the time constant. Consequently, the time to reach 20 r/min is approximately:

$$t = (276/10) \times 60 = 1656 \text{ s}$$
$$= 28 \text{ min}$$

This braking time is 28 times longer than when dynamic braking is used.

Theoretically, a motor which is dynamically braked never comes to a complete stop. In practice, however, we can assume that the machine stops after an interval equal to 5 T_o seconds.

If the motor is plugged, the stopping time has a definite value given by

$$t_s = 2T_o \qquad (5\text{-}10)$$

where

t_s = stopping time using plugging [s]
T_o = time constant as given in Eq. 5-9 [s]

Example 5-7:
The motor in Example 5-6 is plugged, and the braking resistor is increased to 0.4 Ω, so that the initial braking current is the same as before. Calculate:

a. the initial braking current and braking power;
b. the stopping time.

Solution:
a. The net voltage acting across the resistor is:

$$E = E_o + E_s = 210 + 210 = 420 \text{ V}$$

The initial braking current is:
$$I_1 = E/R = 420/0.4 = 1050 \text{ A}$$

The initial braking power is:
$$P_1 = E_oI_1 = 210 \times 1050 = 220.5 \text{ kW}$$

According to Eq. 5-10, T_o has the same value as before:
$$T_o = 10 \text{ s}$$

The time to come to a complete stop is:
$$t_s = 2T_o = 20 \text{ s}$$

5.19 Armature reaction

Up till now, we have assumed that the only mmf acting in a dc motor is that due to the field. However, the current flowing in the armature conductors also creates a magnetomotive force that distorts and weakens the flux coming from the poles. This distortion and field weakening takes place in motors as well as in generators. We recall that the magnetic action of the armature mmf is called *armature reaction*.

5.20 Flux distortion due to armature reaction

When a motor runs at no-load, the small current flowing in the armature does not appreciably affect the flux Φ_1 coming from the poles (Fig. 5-20). But when the armature carries its normal current, it produces a strong magnetomotive force which, if it acted alone, would create a flux Φ_2 (Fig. 5-21). By superimposing Φ_1 and Φ_2, we obtain the resulting flux Φ_3 (Fig. 5-22). In our example, the flux density increases under the left half of the pole and it decreases under the right half. This unequal distribution produces two important effects. First, the neutral zone shifts towards the left (against the direction or rotation). Second, owing to the higher flux density in pole tip A, saturation sets in. Consequently, the increase of flux under the left-hand side of the pole is less than the decrease under the right-hand side. Flux Φ_3 at full-load is therefore slightly less than flux Φ_1 at no-load. For large machines, the decrease in flux may be as much as 10 percent and it causes the speed to increase with load. Such a condition tends to be unstable; to eliminate the problem, we sometimes add a series field of one or two turns to increase the flux under load. Such motors are said to have a *stabilized-shunt winding.*

5.21 Commutating poles

To counter the effect of armature reaction in medium- and large-power machines, we always place a set of *commutating* poles between the main poles of a dc machine (Fig. 5-23). As in the case of a dc generator, these narrow poles develop a magnetomotive force equal and opposite to the mmf of the armature so that the respective magnetomotive forces rise and fall together as the load current varies. As a result, commutation is improved. However, the neutralization is restricted to the narrow zone where commutation takes place. The distorted flux distribution under the pole face unfortunately remains the same.

5.22 Compensating winding

Some dc motors in the 100 kW to 10 MW, ($\sim$ 134 hp to 13 400 hp) range employed in

Figure 5-20
Flux distribution in a motor running at no-load.

Figure 5-21
Flux created by the full-load armature current.

Figure 5-22
Resulting flux distribution in a motor running at full load.

steel mills, perform a series of rapid, heavy-duty operations. They accelerate, decelerate, stop, reverse, all in a matter of seconds. The corresponding armature current increases, decreases, reverses in stepwise fashion, producing very sudden changes in armature reaction.

For such motors, the commutating poles and series stabilizing windings do not adequately neutralize the armature mmf. Torque and speed control is difficult under such transient

Figure 5-23
The narrow commutating poles are placed between the main poles of this 6-pole motor.

conditions and flash-overs may occur across the commutator. To eliminate this problem, additional *compensating windings* are connected in series with the armature. They are distributed in slots, cut into the pole faces of the main field poles (Fig. 5-24). Like commutating poles, these windings produce a mmf equal and opposite to the mmf of the armature. However, because the windings are distributed across the pole faces, the armature mmf is bucked from point to point, which eliminates the field distortion shown in Fig. 5-22. With compensating windings, the field distribution remains essentially undisturbed from no-load to full-load, retaining the general shape shown in Fig. 5-20.

The addition of compensating windings has

a profound effect on the design and performance of a dc motor;

1. A shorter air gap can be used because we no longer have to worry about the demagnetizing effect of the armature. A shorter gap means that the shunt field strength can be reduced and hence the coils are smaller.

2. The inductance of the armature circuit is reduced by a factor of 4 or 5; consequently, the armature current can change more quickly and the motor gives a much better response. This is particularly true in big machines.

3. A motor equipped with compensating windings can briefly develop 3 to 4 times its rated torque. The peak torque of an uncompensated motor is much lower when the armature current is large. The reason is that the effective flux in the air gap falls off rapidly with increasing current because of armature reaction.

We conclude that compensating windings are essential in large motors subjected to severe duty cycles.

QUESTIONS AND PROBLEMS

Practical level

5-1   Name three types of dc motors and make a sketch of the connections.

5-2   Explain what is meant by the generator effect in a motor.

5-3   What determines the magnitude and polarity of the counter-emf in a dc motor?

5-4   The counter-emf of a motor is always slightly less than the applied armature voltage. Explain.

5-5   Name two methods that are used to vary the speed of a dc motor.

Figure 5-24
Six-pole dc motor having a compensating winding distributed in slots in the main poles. The machine also has 6 commutating poles. (*General Electric Company*)

5-6 Explain why the armature current of a shunt motor decreases as the motor accelerates.

5-7 Why is a starting resistor needed to bring a motor up to speed?

5-8 Show one way to reverse the direction of rotation of a compound motor.

5-9 A 230 V shunt motor has a nominal armature current of 60 A. If the armature resistance is 0.15 Ω, calculate:
a. the counter-emf [V];
b. the power supplied to the armature [W];
c. the mechanical power developed by the motor, [kW] and [hp].

5-10 a. In Problem 5-9, calculate the initial starting current if the motor is directly connected across the 230 V line;

b. Calculate the value of the starting resistor needed, to limit the initial current to 115 A.

Intermediate level

5-11 The compound motor of Fig. 5-12 has 1200 turns on the shunt winding and 25 turns on the series winding, per pole. The shunt field has a total resistance of 115 Ω, and the nominal armature current is 23 A. If the motor is connected to a 230 V line, calculate:
a. the mmf per pole at full load;
b. the mmf at no-load.

5-12 A separately-excited dc motor turns at 1200 r/min when the armature is connected to a 115 V source. Calculate the armature voltage required so that the motor runs at 1500 r/min? 100 r/min?

5-13 The following details are known about a 250 hp, 230 V, 435 r/min dc shunt motor:

> nominal full-load current – 862 A
> insulation class – H
> weight – 3400 kg
> external diameter of the frame – 915 mm
> length of frame – 1260 mm

a. Calculate the total losses and efficiency at full load.
b. Calculate the approximate shunt field exciting current if the shunt field causes 20% of the total losses.
c. Calculate the value of the armature resistance, as well as the counter-emf, knowing that 50% of the total losses at full load are due to armature resistance.
d. If we wish to attain a speed of 1100 r/min, what should be the approximate exciting current?

5-14 We wish to stop a 120 hp, 240 V, 400 r/min motor by using the dynamic braking circuit shown in Fig. 5-17. If the nominal armature current is 400 A, calculate:
a. the value of the braking resistor R if we want to limit the maximum braking current to 125% of its nominal value;
b. the braking power [kW], when the motor has decelerated to 200 r/min? 50 r/min? 0 r/min?

5-15 a. The motor in Problem 5-14 is now stopped by using the plugging circuit of Fig. 5-19. Calculate the new braking resistor R so that the maximum braking current is 500 A.
b. Calculate the braking power [kW] when the motor has decelerated to 200 r/min? 50 r/min? 0 r/min?
c. Compare the braking power developed at 200 r/min to the instantaneous power dissipated in resistor R.

Advanced level

5-16 The armature of a 225 kW, 1200 r/min motor has a diameter of 559 mm and an axial length of 235 mm. Calculate:
a. the approximate moment of inertia, knowing that iron has a density of 7900 kg/m^3;
b. the kinetic energy of the armature alone when it turns at 1200 r/min;
c. the total kinetic energy of the revolving parts at a speed of 600 r/min, if the J of the windings and commutator is equal to the J calculated in (a).

5-17 If we reduce the normal exciting current of a practical shunt motor by 50%, the speed increases, but it never doubles. Explain why, bearing in mind the saturation of the iron under normal excitation.

5-18 The speed of a series motor drops with rising temperature, while that of a shunt motor increases. Explain.

6

EFFICIENCY AND HEATING OF ELECTRICAL MACHINES

Whenever a machine transforms energy from one form to another, there is always a certain loss. The loss takes place in the machine itself, causing (a) an increase in temperature and (b) a reduction in efficiency.

From the standpoint of losses, electrical machines may be divided into two groups: those which have revolving parts (motors, generators, etc.) and those which do not (transformers, reactors, etc.). Electrical and mechanical losses are produced in rotating machines, while only electrical losses are produced in stationary machines.

In this chapter, we analyze the losses in dc machines, but the same losses are also found in most machines operating on alternating current. The study of power losses is important because it gives us a clue as to how they may be reduced.

We also cover the important topics of service life and temperature rise. We show that both are related to the class of insulation used and that these insulation classes have been standardized.

6.1 Mechanical losses

Mechanical losses are due to bearing friction, brush friction and windage. The friction losses depend upon the speed of the machine, and upon the design of the bearings, brushes, commutator and slip rings. Windage losses depend on the speed and design of the cooling fan, and on the turbulence produced by the revolving parts. In the absence of prior information, we usually conduct tests on the machine itself to determine the value of these mechanical losses.

Rotating machines are usually cooled by an internal fan mounted on the motor shaft. It draws in cool air from the surroundings, blows it over the windings, and expels it again through suitable vents. In hostile environments, special cooling methods are sometimes used, as illustrated in Fig. 6-1.

Figure 6-1
Totally enclosed, water-cooled, 450 kW, 3600 r/min motor for use in a hostile environment. Warm air inside the machine is blown up-wards and through a water-cooled heat exchanger, situated immediately above the Westinghouse nameplate. After releasing its heat to a set of water cooled pipes, the cool air re-enters the machine by way of two rectangular pipes leading into the end bells. The cooling air therefore moves in a closed circuit, and the surrounding contaminated atmosphere never reaches the motor windings. The circular capped pipes located diagonally on the heat exchanger serve as cooling-water inlet and outlet respectively. (*Westinghouse*)

6.2 Electrical losses

Electrical losses are composed of:
1) conductor I^2R losses (sometimes called *copper* losses);
2) brush losses;
3) iron losses.

1. Conductor losses. In dc motors and generators, copper losses occur in the armature, the series field, the shunt field, the commutating poles and the compensating winding. These I^2R losses show up as heat, causing the conductor temperatures to rise above ambient temperature.

Instead of using the I^2R equation, we sometimes prefer to express the losses in terms of the number of watts per kilogram of conductor material. The losses are then given by the equation:

$$P_c = 1000J^2\rho/\zeta \qquad (6\text{-}1)$$

where
P_c = specific conductor power loss [W/kg]
J = current density [A/mm^2]
ρ = resistivity of the conductor [n$\Omega\cdot$m]
ζ = density of the conductor [kg/m^3]
1000 = constant, to take care of units

According to this equation, the loss per unit mass is proportional to the square of the current density. For copper conductors, we use densities between 1.5 A/mm^2 and 6 A/mm^2. The corresponding losses vary from 5 W/kg to 90 W/kg (Fig. 6-2). The higher densities require an efficient cooling system to prevent an excessive temperature rise.

2. Brush losses. The I^2R losses in the brushes are negligible because the current density is only about 0.1 A/mm^2 which is far less than that used in copper. However, the *contact* voltage drop between the brushes and commutator may produce significant losses. The drop varies from 0.8 V to 1.3 V, depending on the type of brush, the applied pressure and the brush current (Fig. 6-3).

3. Iron losses. Iron losses are produced in the armature of a dc machine. They are due to hysteresis and eddy currents, as previously explained in Secs. 2-27 and 2-30. Iron losses depend upon the flux density, the speed of rotation, the quality of the steel and the size of the armature. They typically range from 0.5 W/kg to 20 W/kg. The higher values occur in the armature teeth, where the flux density may be as high as 1.7 T. The losses in the armature core are usually much lower. The losses can be further reduced by annealing the steel (Fig. 6-4).

Some iron losses are also produced in the

Figure 6-2
Copper losses may be expressed in watts per kilogram.

Figure 6-3
Brush contact voltage drop occurs between the brush face and commutator.

Figure 6-4
This 150 kW electric oven is used to anneal punched steel laminations. This industrial process, carried out in a controlled atmosphere of 800°C, significantly reduces the iron losses. The laminations are seen as they leave the oven. (*General Electric*)

pole faces. They are due to flux pulsations created as successive armature teeth and slots sweep across the pole face.

Strange as it may seem, iron losses impose a mechanical drag on the armature, producing the same effect as mechanical friction.

Example 6-1:
A dc machine turning at 875 r/min carries an armature winding whose total weight is 40 kg. The current density is 5 A/mm^2 and the operating temperature is 80°C. The total iron losses in the armature amount to 1100 W. Calculate:
a. the copper losses;
b. the mechanical drag [N·m] due to the iron losses.

Solution:
a. The resistivity of copper at 80°C is:
$$\rho = \rho_o (1 + \alpha t)$$
$$= 15.88 (1 + 0.004\ 27 \times 80)$$
$$= 21.3\ n\Omega{\cdot}m$$

The density of copper is:
8890 kg/m^3

The specific power loss is:
$$P_c = 1000 J^2 \rho / \zeta \qquad \text{Eq. 6-1}$$
$$= 1000 \times 5^2 \times 21.3/8890$$
$$= 60\ W/kg$$

Total copper loss is:
$$P = 60 \times 40 = 2400\ W$$

b. The braking torque due to iron losses can be calculated from:
$$P = nT/9.55 \qquad \text{Eq. 3-5}$$
$$1100 = 875T/9.55$$
$$T = 12 N{\cdot}m \text{ or approximately}$$
$$8.85\ ft{\cdot}lbf.$$

6.3 Losses as a function of load

A dc motor running at no-load develops no useful power. However, it must absorb some power from the line to continue to rotate. This no-load power overcomes the friction, windage, and iron losses, and provides for the copper losses in the shunt field. The I^2R losses in the armature, series field and commutating field are negligible because the no-load current is seldom more than 10% of the nominal full-load current.

As we load the machine the current increases in the armature circuit. Consequently, the I^2R losses in the armature circuit will rise. On the other hand, the no-load losses, mentioned above, remain essentially constant as the load increases, unless the speed of the machine changes appreciably. It follows that the total losses increase with load. Because they are converted into heat, the temperature of the machine rises progressively as the load increases.

However, the temperature must not exceed the limiting temperature of the insulation used in the machine. Consequently, there is a limit to the power which the machine may deliver. This temperature-limited power enables us to establish the *nominal* or *rated power* of the machine. A machine loaded beyond its nominal rating will usually overheat. The insulation deteriorates more rapidly, which inevitably shortens the service life of the machine.

If a machine runs *intermittently,* it can carry heavy overloads without overheating, provided that the operating time is short. Thus, a motor having a nominal rating of 10 kW can easily carry a load of 15 kW, if it only operates a few minutes per hour. However, for higher loads, the capacity is limited by other factors, usually electrical. For instance, it is physically impossible for a machine rated at 10 kW to yield an output of 100 kW.

6.4 Efficiency curve

The efficiency of a machine is the ratio of the useful output power P_o to the input power P_i (Sec. 3-7). Furthermore, input power is equal to useful power plus the losses p. We can therefore write:

$$\eta = \frac{P_o}{P_i} \times 100 = \frac{P_o}{P_i + p} \times 100 \quad (6\text{-}2)$$

where

η = efficiency [%]
P_o = output power [W]
P_i = input power [W]
p = losses [W]

The following example shows how to calculate the efficiency of a dc machine.

Example 6-2:
A dc compound motor having a rating of 10 kW, 1150 r/min, 230 V, 50 A, has the following losses *at full load*:

bearing friction loss	=	40 W
brush friction loss	=	50 W
windage loss	=	200 W
(1) total mechanical losses	=	290 W
(2) iron losses	=	420 W
(3) copper loss in the shunt field	=	120 W

copper losses at full load:

a. in the armature	=	500 W
b. in the series field	=	25 W
c. in the commutating winding	=	70 W
(4) total copper loss in the armature circuit at full load	=	595 W

Calculate the losses and efficiency at no-load and at 25, 50, 75, 100 and 150 percent of the nominal rating of the machine. Draw a graph showing efficiency as a function of mechanical load (neglect the losses due to brush contact drop).

Solution:
No-load. The copper losses in the armature circuit are negligible at no-load. Consequently, the no-load losses are equal to the sum of the mechanical losses (1), the iron losses (2), and the shunt field losses (3):

no-load losses = 290 + 420 + 120 = 830 W

These losses remain essentially constant as the load varies.

The efficiency is zero at no-load because no useful power is developed by the motor.

25% load. When the motor is loaded to 25% of its nominal rating, the armature current is approximately 25% (or 1/4) of its full-load value. Because the copper losses vary as the square of the current, we have:

copper losses in the armature circuit
$= (1/4)^2 \times 595 = 37$ W

no-load losses
$= 830$ W

total losses
$= 37 + 830 = 867$ W

Useful power developed by the motor at 25% load:
$$P_o = 10\ \text{kW} \times (1/4) = 2500\ \text{W}\ (\sim 3.35\ \text{hp})$$

power supplied to the motor is:
$$P_i = 2500 + 867 = 3367\ \text{W}$$

The efficiency is:
$$\begin{aligned} v &= (P_o / P_i) \times 100 \qquad \text{Eq. 6-2} \\ &= (2500/3367) \times 100 = 74\% \end{aligned}$$

In the same way, we find the losses at 50, 75, 100 and 150 percent of the nominal load:

At 50% load, the losses are
$(1/2)^2 \times 595 + 830 = 980$ W

At 75% load, the losses are
$(3/4)^2 \times 595 + 830 = 1165$ W

At 100% load, the losses are
$595 + 830 = 1425$ W

At 150% load, the losses are
$(1.5)^2 \times 595 + 830 = 2170$ W

The efficiency calculations for the various loads are listed in Table 6A and the results are shown graphically in Fig. 6-5.

TABLE 6A		LOSSES AND EFFICIENCY OF A DC MOTOR		
Load [%]	Total Losses [W]	Output Power P_0 [%]	Input Power P_i [W]	Efficiency [%]
0	830	0	830	0
25	867	2 500	3 367	74
50	980	5 000	5 980	83.6
75	1 165	7 500	8 665	86.5
100	1 425	10 000	11 425	87.5
150	2 170	15 000	17 170	87.4

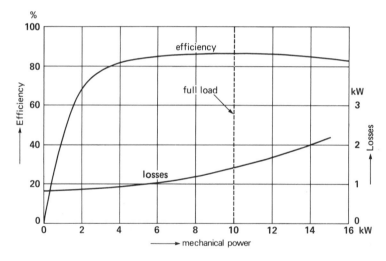

Figure 6-5
Losses and efficiency as a function of mechanical power. See Example 6-2.

It is important to remember that at light loads the efficiency of any motor is poor. Consequently, when selecting a motor to do a particular job, we always try to choose one having a power rating roughly equal to the load it has to drive.

We can prove that the efficiency of any dc machine reaches a maximum at that load where the armature circuit copper losses are equal to the no-load losses. In our example, this corresponds to a total loss of (830 + 830) = 1660 W, an output of 11 811 W (15.8 hp) and an efficiency of 87.68 percent. The reader may wish to check these results.

6.5 Temperature rise

The temperature rise of a machine is the difference between the temperature of its warmest accessible part and the ambient temperature. It may be measured by simply using two thermometers. However, owing to the practical difficulty of placing a thermometer close to the really warmest spot inside the machine, this method is seldom used. We usually rely upon more sophisticated methods, described in the following sections.

Temperature rise has a direct bearing on the power rating of a machine or device. It also has a direct bearing on the service life of a machine. Consequently, it is a very important quantity.

by half every time the temperature increases by 10°C. This means that if a motor has a normal life expectancy of eight years at a temperature of 105°C, it will have a service life of only four years at a temperature of 115°C, of two years at 125°C, and of only one year at 135°C!

The factors that contribute most to the deterioration of insulators are: 1. heat; 2. humidity; 3. vibration; 4. acidity; 5. oxidation and 6. time (Fig. 6-6). Because of these factors, the state of the insulator changes gradually; it slowly begins to crystallize and the transformation takes place more rapidly as the temperature rises.

In crystallizing, the organic insulator becomes hard and brittle. Eventually, the

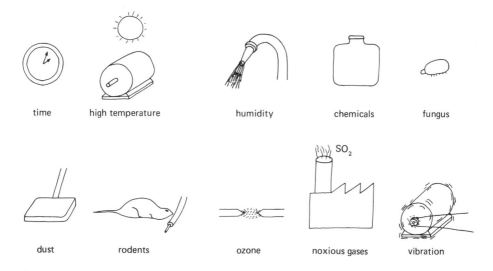

time   high temperature   humidity   chemicals   fungus

dust   rodents   ozone   noxious gases   vibration

Figure 6-6
Factors which may shorten the service life of an insulator.

6.6 Life expectancy of electric equipment

Apart from accidental electrical and mechanical failures, the life expectancy of electrical apparatus is limited by the temperature of its insulation: the higher the temperature, the shorter its life. Tests made on many insulating materials have shown that the service life of electrical apparatus diminishes approximately

slightest shock or mechanical vibration will cause it to break. Under normal conditions, with respect to the above factors, most organic insulators have a life expectancy of eight to ten years provided that their temperature does not exceed 100°C. On the other hand, some synthetic polymers can withstand temperatures as high as 200°C for the same length of time.

Low temperatures are just as harmful as

high temperatures are, because the insulation tends to freeze and crack. Special synthetic organic insulators have been developed, however, which retain their flexibility at temperatures as low as −60°C.

6.7 Thermal classification of insulators

Committees and organizations that set standards* have grouped insulators into five classes, depending upon their ability to withstand heat. These classes correspond to the maximum temperature levels of: 105°C, 130°C, 155°C, 180°C and 220°C (formerly represented by the letters A, B, F, H and R). This thermal classification (Table 6B) is a cornerstone in the design and manufacture of electrical apparatus.

6.8 Limiting ambient temperature and hot-spot temperature rise

Standards organizations have also established a maximum *ambient temperature*, which is usually 40°C. This limiting temperature was established for the following reasons:

1. It enables electrical manufacturers to foresee the worst ambient temperature conditions which their machines are likely to encounter. Consequently,
2. it enables them to standardize the size of their machines and to give performance guarantees. The temperature of a machine varies from point to point, but there are places where the temperature is warmer than anywhere else. This so-called *hottest-spot temperature* must not exceed the temperature limit for the particular class of insulation used.

Figure 6-7 shows the hot-spot temperature limits for class A, B, F, and H insulation (curve 1). They are the temperature limits previously mentioned in Sec. 6.7. The limiting ambient temperature of 40°C is also shown (curve 3). The temperature difference between these two

curves gives the *maximum permissible temperature rise* for each insulation class. This limiting temperature rise enables the manufacturer to establish the physical size of the motor, relay, and so forth, he intends to put on the market. Thus, for class B insulation, the limiting temperature rise is (130 − 40) = 90°C.

To show how the temperature rise affects the size, suppose a manufacturer has designed and built a 10 kW motor using class B insulation. To test the motor, he places it in a constant ambient temperature of 40°C and loads it up until it delivers 10 kW of mechanical power. Special temperature detectors, located at strategic points throughout the machine, record the temperature of the windings. After the temperatures have stabilized (which may take several hours), the hottest temperature is noted, and this is called the hot-spot temperature. If the hot-spot temperature so recorded is, say 147°C, the manufacturer is not permitted to sell his product. The reason is that the temperature rise (147° − 40°) = 107°C exceeds the maximum permissible rise of 90°C for class B insulation.

On the other hand, if the hottest-spot temperature is only 100°C, the temperature rise is only (100° − 40°) = 60°C. The manufacturer immediately perceives that he can make a more economical design and still remain within the permissible temperature rise limits. For instance, he can reduce the conductor size until the hot-spot temperature rise is very close to 90°C. Obviously, this reduces the weight and cost of the windings. But the manufacturer also realizes that the reduced conductor size now enables him to reduce the size of the slots. This, in turn, reduces the amount of iron. By thus redesigning the motor, the manufacturer ultimately ends up with a machine that operates within the permissible temperature rise limits and has the smallest possible physical size.

In practice, it it not convenient to carry out performance tests in a controlled ambient tem-

---

*   Such as IEEE, Underwriters Laboratories, Canadian Standards Association.

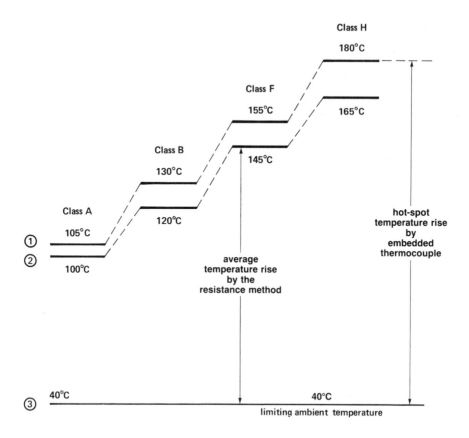

Figure 6-7
Typical limits of some dc and ac industrial machines, according to the insulation class:
(1) shows the maximum permissible temperature of the insulation to obtain a reasonable service life;
(2) shows the maximum permissible temperature using the resistance method;
(3) shows the limiting ambient temperature.

perature of 40°C. The motor is usually loaded to its rated capacity in much lower (and more comfortable) ambient temperatures. The hottest-spot temperature is recorded as before. If the temperature rise under these conditions is equal to or less than 90°C (for class B insulation), the manufacturer is allowed to sell his product. There is, however, one further condition: In carrying out the test, the ambient temperature must lie between 10°C and 40°C. These upper and lower ambient temperature limits are set by standards.

Example 6-3:
A 75 kW motor, insulated class F, operates at full-load in an ambient temperature of 32°C. If the hot-spot temperature is 125°C, does the motor meet the temperature standards?

Solution:
The hot-spot temperature rise is
$(125° − 32°) = 93°C$.
According to Fig. 6-7, the permissible hot-spot temperature rise is $(155° − 40°) = 115°C$.
The motor easily meets the temperature standards. The manufacturer could reduce the

size of the motor and thereby produce a more competitive product.

6.9 Temperature rise by the resistance method

The hot-spot temperature rise is rather difficult to measure because it has to be taken at the very inside of the winding. This can be done by embedding a small temperature detector, such as a thermocouple or thermistor. However, this direct method of measuring hot-spot temperature is costly, and is only justified for large machines.

TABLE 6B   CLASSES OF INSULATION SYSTEMS

Class	Illustrative Examples
105°C A	Materials or combinations of materials such as cotton, silk, and paper when suitably impregnated or coated or when immersed in a dielectric liquid such as oil. Other materials or combinations of materials may be included in this class if by experience or accepted tests they can be shown to have comparable thermal life at 105°C.
130°C B	Materials or combinations of materials such as mica, glass fiber, asbestos, etc., with suitable bonding substances. Other materials or combinations of materials may be included in this class if by experience or accepted tests they can be shown to have comparable thermal life at 130°C.
155°C F	Materials or combinations of materials such as mica, glass fiber, asbestos, etc., with suitable bonding substances. Other materials or combinations of materials may be included in this class if by experience or accepted tests they can be shown to have comparable life at 155°C.
180°C H	Materials or combinations of materials such as silicone elastomer, mica, glass fiber, asbestos, etc., with suitable bonding substances such as appropriate silicone resins. Other materials or combinations of materials may be included in this class if by experience or accepted tests they can be shown to have comparable life at 180°C.
200°C N	Materials or combinations of materials which by experience or accepted tests can be shown to have the required thermal life at 200°C.
220°C R	Materials or combinations of materials which by experience or accepted tests can be shown to have the required thermal life at 220°C.
240°C S	Materials or combinations of materials which by experience or accepted tests can be shown to have the required thermal life at 240°C.
above 240°C C	Materials consisting entirely of mica, porcelain, glass, quartz, and similar inorganic materials. Other materials or combinations of materials may be included in this class if by experience or accepted tests they can be shown to have the required thermal life at temperatures above 240°C.

The above insulation classes indicate a normal life expectancy of 20 000 h to 40 000 h at the stated temperature. This implies that electrical equipment insulated with a class A insulation system would probably last for 2 to 5 years if operated continously at 105°C. Note that this classification assumes that the insulation system is not in contact with corrosive, humid, or dusty atmospheres.

For a complete explanation of insulation classes, insulation systems and temperature indices, see IEEE Std 1-1969, and the companion IEEE Standards Publications Nos 96, 97, 98, 99 and 101. See also IEEE Std 117-1974 and Underwriters Laboratories publication on insulation systems UL 1446, 1978.

To simplify matters, accepted standards permit a second method of determining temperature rise. It is based upon the *average* winding temperature rather than the hot-spot temperature. The maximum average winding temperatures for the various insulation classes are shown in curve 2, Fig. 6-7. For example, in the case of class B insulation, an *average* winding temperature of 120°C is assumed to correspond to a hot-spot temperature of 130°C. Consequently, an average temperature rise of (120° − 40°) = 80°C is assumed to correspond to a hot-spot temperature rise of (130° − 40°) = 90°C.

The average temperature of a winding is found by the so-called *resistance method*. It consists of measuring the winding resistance at a known temperature, and measuring it again when the machine is hot. For example, if the winding is made of copper, we can use the following equation to determine its average temperature:

$$t_2 = \frac{R_2}{R_1}(234 + t_1) - 234 \qquad (6\text{-}3)$$

where

t_2 = average temperature of the winding when hot [°C]

234 = a constant equal to $1/\alpha = 1/0.004\,27$

R_2 = hot resistance of the winding [Ω]

R_1 = cold resistance of the winding [Ω]

t_1 = temperature of the winding when cold [°C]

Knowing the hot winding temperature by the resistance method, we can immediately calculate the corresponding temperature rise by subtracting the ambient temperature. If this temperature rise falls within the permissible limit (80°C for class B insulation), the product is acceptable, from a standards point of view. Note that when performance tests are carried out using the resistance method, the ambient temperature must lie between 10°C and 40°C.

Example 6-4:

A dc motor that has been idle for several days in an ambient temperature of 19°C, is found to have a shunt-field resistance of 22 Ω. The motor then operates at full-load and, when temperatures have stabilized, the field resistance is found to be 30 Ω. The corresponding ambient temperature is 24°C. If the motor is built with class B insulation, calculate:

a. the average temperature of the winding, when hot;

b. the temperature rise by the resistance method;

c. whether the motor meets the temperature standards.

Solution:

a. The average temperature of the hot winding is:

$$t_2 = (R_2/R_1)\,(234 + t_1) - 234$$
$$= (30/22)\,(234 + 19) - 234$$
$$= 111°C$$

b. The average temperature rise is:

$$111° - 24° = 87°C$$

c. The temperature rise permitted for class B insulation is (120 − 40) = 80°C. Consequently, the motor does not meet the standards. Either its rating will have to be reduced, or the cooling system improved, before it can be put on the market. Alternatively, it may be rewound using class F insulation. As a very last resort, its size may have to be increased.

A final word of caution: temperature rise standards depend not only on the class of insulation, but also on the type of apparatus (motor, transformer, relay, etc.), the type of construction (drip-proof, totally enclosed, etc.) and the field of application of the apparatus (commercial, industrial, naval, etc.) Consequently, the pertinent standards must always be consulted before conducting a heat-run test on a specific machine or device.

6.10 Relationship between the speed and size of a machine

Although temperature limits establish the nominal power rating of a machine, its basic physical size depends upon the power and speed of rotation.

Consider a 100 kW, 250 V, 2000 r/min generator shown in Fig. 6-8. Suppose we have to build another generator having the same power and voltage, but running at half the speed.

To generate the same voltage at half the speed, we either have to double the number of conductors on the armature, or double the flux from the poles. Consequently, we must either increase the size of the armature, or increase the size of the poles. In practice, we increase both. We conclude that for a given power output, a low-speed machine is always bigger than a high-speed machine (Fig. 6-9). This is true for both ac and dc machines.

Figure 6-8
100 kW, 2000 r/min motor, mass: 300 kg.

Figure 6-9
100 kW, 1000 r/min motor, mass: 500 kg.

Basically, the size of a machine depends uniquely upon its torque. Thus, a 100 kW, 2000 r/min motor has about the same physical size as a 10 kW motor running at 200 r/min because they develop the same torque.

Low-speed motors are therefore more costly than high-speed motors of equal power. Consequently, for low-speed drives, it is often cheaper to use a small high-speed motor with a gear box, than to use a large low-speed motor directly coupled to its load.

QUESTIONS AND PROBLEMS

Practical level

6-1   Name the losses in a dc motor.

6-2   What causes iron losses and how can they be reduced?

6-3   Explain why the temperature of a machine increases as the load increases.

6-4   What determines the power rating of a machine?

6-5   If we cover up the vents in a motor, its output power must be reduced. Explain.

6-6   If a motor operates in a cold environment, may we load it above its rated power? Why?

6-7   Name some of the factors which contribute to the deterioration of organic insulators.

6-8   A motor is built with class H insulation. What maximum hot-spot temperature can it withstand?

Intermediate level

6-9 A dc motor connected to a 240 V line produces a mechanical output of 160 hp. Knowing that the losses are 12 kW, calculate the input power and the line current.

6-10 A 115 V dc generator delivers 120 A. If it has an efficiency of 81%, calculate the mechanical power needed to drive it [hp].

6-11 Calculate the full-load current of a 250 hp, 230 V dc motor having an efficiency of 92%.

6-12 A machine having class B insulation, attains a temperature of 208°C (by resistance) in an ambient temperature of 180°C.
a. What is the temperature rise?
b. Is the machine running too hot and, if so, by how much?

6-13 The efficiency of a motor is always low when it operates at 10% of its nominal power rating. Explain.

6-14 Calculate the efficiency of the motor in Example 6-2 when it delivers an output of 40 hp.

6-15 An electric motor driving a skip hoist withdraws 1.5 metric tons of minerals from a trench 20 m deep, every 30 seconds. If the hoist has an overall efficiency of 94%, calculate the power output of the motor, in horsepower and in kilowatts.

6-16 Thermocouples are used to measure the internal winding temperature of a 1200 kW ac motor, insulated class F. If the motor runs at full load, what is the maximum temperature these detectors should indicate in an ambient temperature of 40°C? 30°C? 14°C?

6-17 A 60 hp ac motor with class F insulation has a cold winding resistance of 12 Ω at 23°C. When it runs at rated load in an ambient temperature of 31°C, the hot winding resistance is found to be 17.4 Ω.
a. Calculate the hot winding temperature;
b. Calculate the temperature rise of the motor;
c. Could the manufacturer increase the nameplate rating of the motor? Explain.

6-18 An electric motor has a normal life of eight years when the ambient temperature is 30°C. If it is installed in a location where the ambient temperature is 60°C, what is the new probable service life of the motor?

6-19 A No. 10 round copper wire 210 m long carries a current of 12 A. Knowing that the temperature of the conductor is 105°C, calculate:
a. the current density [A/mm^2];
b. the specific copper losses [W/kg].

Advanced level

6-20 An aluminum conductor operates at a current density of 2 A/mm^2
a. If the conductor temperature is 120°C, calculate the specific losses [W/kg].
b. Express the current density in circular mils per ampere.

6-21 The temperature rise of a motor is roughly proportional to its losses. On the other hand, its efficiency is reasonably constant in the range between 50% and 150% of its nominal rating (see, for example, Fig. 6-5). Based on these facts, if a 20 kW motor has a full-load temperature rise of 80°C, what power can it deliver at a temperature rise of 105°C?

6-22 An electromagnet (insulated class A) has a normal service life of two years. What is its new life span if it is rewound using class F insulation? (Assume the temperature of the magnet remains the same).

6-23 An 11 kW ac motor having class B insulation would normally have a service life of 20 000 h provided the winding temperature by resistance does not exceed 120°C. By how many hours is the service life reduced if the motor runs for 3 h at a temperature (by resistance) of 200°C?

7

ACTIVE, REACTIVE, AND APPARENT POWER

The concept of active, reactive, and apparent power plays a major role in electric power technology. In effect, the transmission of electrical energy and the behavior of ac machines are often easier to understand by working with power, rather than dealing with voltages and currents. The reader is therefore encouraged to pay particular attention to this chapter.

The terms active, reactive, and apparent power apply to steady-state alternating current circuits in which the voltages and currents are sinusoidal. They cannot be used to describe transient-state behavior, nor can we apply them to dc circuits.

Our study begins with an analysis of the instantaneous power in an ac circuit. We then go on to define the meaning of active and reactive power and how to identify sources and loads. This is followed by a definition of apparent power, power factor and the power triangle. We then show how ac circuits can be solved using these power concepts. In conclusion, complex notation is used to determine the active and reactive power in an ac circuit.

7.1 Instantaneous power

The instantaneous power supplied to a load is simply the product of the instantaneous voltage across its terminals times the instantaneous current that flows through it.

Instantaneous power is always expressed in watts, irrespective of the type of circuit used. The instantaneous power may be positive or negative. A positive value means that power flows from the source to the load. Conversely, a negative value indicates that power is flowing in reverse, from the load to the source.

Example 7-1:
A sinusoidal voltage having a peak value of 162 V and a frequency of 60 Hz is applied to the terminals of an ac motor. The resulting current has a peak value of 7.5 A and lags 50° behind the voltage.

a. Express the voltage and current in terms of the electrical angle ϕ;

b. Calculate the value of the instantaneous voltage and current at an angle of 120°;

c. Calculate the value of the instantaneous power at 120°;

d. Plot the curve of the instantaneous power delivered to the motor.

Solution:

a. Let us assume that the voltage starts at zero, and increases positively with time. We can therefore write:

$$e = E_m \cos (\phi - 90°)$$
$$= E_m \sin \phi = 162 \sin \phi$$

The current lags behind the voltage by an angle $\theta = 50°$; consequently, we can write:

$$i = I_m \sin (\phi - \theta) = 7.5 \sin (\phi - 50°)$$

b. At $\phi = 120°$, we have:

$$e = 162 \sin 120° = 162 \times 0.866$$
$$= 140.3 \text{ V}$$
$$i = 7.5 \sin (120° - 50°) = 7.5 \sin 70°$$
$$= 7.5 \times 0.94$$
$$= 7.05 \text{ A}$$

c. The instantaneous power at 120° is:

$$p = ei = 140.3 \times 7.05 = +989 \text{ W}$$

Because the power is positive, it flows at this instant from the source to the motor.

d. In order to plot the curve of instantaneous power we repeat procedures (b) and (c) above for angles ranging from $\phi = 0$ to $\phi = 360°$. Table 7A lists part of the data used.

The voltage, current and instantaneous power are plotted in Fig. 7-1. The power attains a positive peak of $+1000$ W, and a negative peak of -218 W. The negative power means that power is actually flowing from the load (motor) to the source. This occurs during the intervals 0-50°, 180°-230° and 360°-410°. Although a power flow from a load to a source may seem to be impossible, it happens often in ac circuits. The reason is given in the sections that follow.

We also note that the positive peaks occur at

TABLE 7A   Values of e, i, p used to plot Figure 7-1

angle ϕ degrees	voltage 162 sin ϕ volts	current 7.5 sin (ϕ − 50°) amperes	power p watts
0	0	−5.74	0
25	68.46	−3.17	−218
50	124.1	0	0
75	156.5	3.17	497
115	146.8	6.8	1000
155	68.5	7.25	497
180	0	5.75	0
205	−68.46	3.17	−218
230	−124.1	0	0

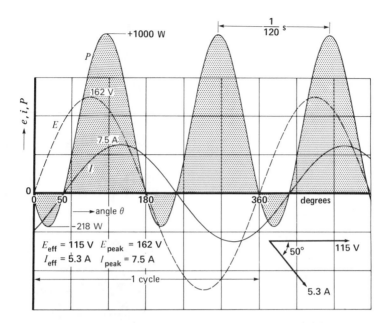

Figure 7-1
Instantaneous voltage, current and power in an ac circuit. (See Example 7-1).

intervals of 1/120s. This means that the frequency of the power cycle is 120 Hz, which is twice the frequency of the voltage and current that produce the power. Again, this phenomenon is quite normal: the frequency of ac power flow is always twice the line frequency.

7.2 Active power*

The simple ac circuit of Fig. 7-2a consists of a resistor connected to an ac generator. The effective voltage and current are designated E and I respectively and, as we would expect in a resistive circuit, phasors E and I are in phase (Fig. 7-2b). If we connect a wattmeter into the line, it will give a reading $P = EI$ watts (Fig. 7-2c).

To get a better picture of what goes on in such a circuit, we have drawn the sinusoidal curves of E and I (Fig. 7-2d). The peak values are respectively $\sqrt{2}E$ volts and $\sqrt{2}I$ amperes because, as stated previously, E and I are *effective* values. By multiplying the instantaneous values of voltage and current as we did before, we obtain the instantaneous power in watts.

The power wave consists of a series of positive pulses that vary from zero to a maximum of $(\sqrt{2}E) \times (\sqrt{2}I) = 2EI = 2P$ watts. The fact that power is always positive reveals that it always flows from the generator to the resistor. This is one of the basic properties of what is called *active power*: although it pulsates between zero and maximum, it *never* changes direction. The direction of power flow is shown by an arrow P (Fig. 7-2c).

The average power is clearly midway between $2P$ and zero, and so its value is $P = 2EI/2 = EI$ watts. This is the power indicated by the wattmeter.

Many persons refer to active power as "real power" or "true power" considering it to be more descriptive. In this book, we use the term "active power", because it conforms to the IEEE designation.

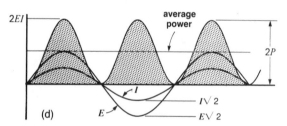

(a)

(b)

(c)

(d)

Figure 7-2
a. An ac voltage E produces an ac current I in this resistive circuit.
b. Phasors E and I are in phase.
c. A wattmeter indicates EI watts.
d. The active power is composed of a series of positive power pulses.

The two conductors leading to the resistor in Fig. 7-2a carry the active power. However, unlike current flow, power does not flow down one conductor and return by the other. Power flows over *both* conductors and, consequently, as far as power is concerned, we can replace the conductors by a single line, as shown in Fig. 7-2c.

In general, the line represents any transmission line connecting two devices, irrespective of the number of conductors it may have.

The generator is an *active source* and the resistor is an *active load*. The symbol for active power is P and the unit is the watt (W). The kilowatt (kW) and megawatt (MW) are frequently-used multiples of the watt.

Figure 7-3
Example of a high-precision wattmeter rated 50 V, 100 V, 200 V; 1 A, 5 A. The scale ranges from 0-50 W to 0-1000 W. (*Weston Instruments*)

7.3 Definition of active sources and active loads (sign notation)

We are often interested in determining whether a device is an active source or an active load. The reason is that many devices, besides resistors, can function as active loads. To enable us to positively identify an active source or active load, consider Fig. 7-4 in which a device A carries a current I. The voltage between the terminals is E, and one of the terminals bears a (+) sign.

This diagram, together with the arbitrary polarity and arbitrary current flow, permit us to state whether the device is an active load or active source. The following rule applies:

Figure 7-4
Device A may be an active source or active load depending upon the phasor relationship between E and I.

Figure 7-5
See Example 7-2.

1. **A device is an** *active load* **when:**

   a. **voltage E and current I are in phase** *and*

   b. **current I is shown as entering the (+) terminal.**

The following additional rules follow logically from the above basic rule:

2. A device is an *active source* when E and I are in phase and I is shown as leaving the (+) terminal.

3. A device is an active source when E and I are 180° out of phase and I is shown as entering the (+) terminal.

4. A device is an active load when E and I are 180° out of phase and I is shown as leaving the (+) terminal.

Example 7-2:
Two devices A and B are connected by a transmission line and an arbitrary polarity is assigned to the line voltage E and an arbitrary direction for line current I (Fig. 7-5). After solving the circuit, the equations show that phasors E and I are 180° out of phase. Which device is the active load?

Solution:
We could refer to the four rules listed above and select the one that applies to our problem. How-

ever, it is much easier to remember only rule 1 and reason as follows: If E and I *were* in phase, B would obviously be the load. However, because they are 180° out of phase, B must be the active source. It follows that device A is the active load, and the *actual* direction of active power flow is shown by arrow P.

Note that the phasor diagram applies simultaneously to device A, to device B and to the transmission line connecting them. In other words, a single phasor diagram suffices for all three elements of the circuit.

7.4 Definition of active sources and loads (double subscript notation)

We can also tell whether a device is an active source or an active load when the double subscript notation is used. Consider Figure 7-6 in which a device carries a current I flowing in the direction shown. The voltage between terminals **a, b** is E_{ab}. The following rule applies:

1. **The device is an** *active load* **when:**

   a. **voltage E_{ab} and current I are in phase** *and*

   b. **current I is shown as entering terminal a.**

The following additional rules follow naturally from the above basic rule:

Figure 7-6
Same circuit as in Fig. 7-4 except that double-subscript notation is used.

2. A device is an active source when E_{ab} and I are in phase and I is shown as flowing out of terminal **a**.

3. A device is an active source when E_{ab} and I are 180° out of phase and I is shown as entering terminal **a**.

4. A device is an active load when E_{ab} and I are 180° out of phase and I is shown as flowing out of terminal **a**.

7.5 Reactive power

The circuit of Fig. 7-7a is identical to the original resistive circuit except that the resistor is now replaced by an inductor. After writing the equations for the circuit, we discover that I lags 90° behind voltage E (Fig. 7-7b).

To see what really goes on in such a circuit, we have drawn the waveforms for E and I and, by again multiplying the instantaneous values, we obtain the curve of instantaneous power (Fig. 7-7c). This power p consists of a series of identical positive and negative pulses. The positive pulses correspond to instantaneous power delivered by the generator to the inductor and the negative pulses represent instantaneous power delivered from the inductor to the generator. The duration of each pulse corresponds to one quarter of a cycle of the line frequency. The fre-

quency of the power wave is therefore twice the line frequency.

Power which oscillates back and forth in this manner is called *reactive power* (symbol Q), to distinguish it from the unidirectional active power mentioned before. The reactive power in Fig. 7-7 is also given by the product EI. However, to distinguish this power from active power, another unit is used – the **var.** Frequent multiples are the kilovar (kvar) and megavar (Mvar).

Special instruments, called *varmeters*, are available to measure the reactive power in a circuit (Fig. 7-8). A varmeter registers the product of the effective line voltage E times the effective line current I times sin θ (where θ is the phase angle between E and I). A reading is only obtained when E and I are out of phase; if they are exactly in phase, the varmeter reads zero.

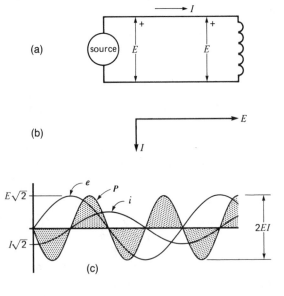

Figure 7-7
a. An ac voltage E produces an ac current I in this inductive circuit.
b. Phasor I lags 90° behind E.
c. Reactive power consists of a series of positive and negative power pulses.

Figure 7-8
Varmeter with a zero-center scale. It indicates positive or negative reactive power flow up to 100 Mvars.

Returning to Fig. 7-7, the dotted area under each pulse is the energy, in joules, transported in one direction or the other. Clearly, the energy is delivered in a continuous series of pulses of very short duration, every positive pulse being followed by a negative pulse. The energy flows back and forth between the generator and the inductor without ever being used up.

What is the reason for these positive and negative energy surges? The energy flows back and forth because magnetic energy is alternately being stored up and released by the inductor. Thus, when the power is positive, the magnetic field is building up inside the coil. A moment later when the power is negative, the energy in the magnetic field is decreasing (see Sec. 2-13).

We now have an explanation for the brief negative power pulses in Fig. 7-1. In effect, they represent magnetic energy previously stored up in the motor windings that is being returned to the source.

7.6 Definition of reactive sources and reactive loads (sign notation)

Reactive power involves energy that oscillates

back and forth over a transmission line; consequently, it is really impossible to say whether it originates at one end of the line or the other. Nevertheless, it is very useful to *assume* that some devices generate reactive power while others absorb it. In other words, some devices behave like *reactive sources* and others like *reactive loads*. In order to distinguish between the two, we use the same reasoning as in the case of active loads. Thus, suppose a device X is part of an ac circuit and that the voltage *E* and

Figure 7-9
Device X may be a reactive source or reactive load depending on the phasor relationship between *E* and *I*.

current *I* have respectively been assigned an arbitrary polarity and an arbitrary direction as shown in Fig. 7-9. After solving the circuit equations, suppose we find that the voltage and current are 90° out of phase. The following rule then applies*:

1. **A device is a** *reactive load* **when;**

   a. **current *I* lags 90° behind voltage *E* and**

   b. **current *I* is shown as entering the (+) terminal.**

The following additional rules follow logically from the above basic rule:

2. A device is a reactive source when *I* lags behind *E* by 90° and *I* is shown as leaving the (+) terminal.

---

*   These rules conform to definitions adopted by the IEEE.

3. A device is a reactive source when I leads E by 90° and I is shown as entering the (+) terminal.

4. A device is a reactive load when I leads E by 90° and I is shown as leaving the (+) terminal.

Example 7-3:

Two devices X and Y are connected by a transmission line and an arbitrary polarity is assigned to the line voltage E and an arbitrary direction to line current I. (Fig. 7-10). After solving the circuit, the equations show that phasor E lags 90° behind phasor I. Which device is the reactive source?

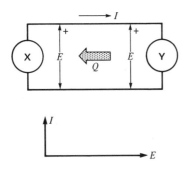

Figure 7-10
See Example 7-3.

Solution:
We could scan through the four rules given above and select the one that applies to our problem. However, it is much easier to remember only rule 1 and reason as follows: If the current lagged behind the voltage in Fig. 7-10, then Y would be the reactive load. But since the current is actually leading the voltage, Y must be a reactive source. It follows that X is the reactive load. The direction of reactive power flow is therefore from Y to X, as indicated by arrow Q.

The reader should note that the same

phasor diagram applies to devices X and Y as well as to the transmission line connecting them.

7.7 Definition of reactive sources and reactive loads (double subscript notation)

If the double-subscript notation is used, we can identify a reactive load and reactive source as follows. Consider Figure 7-11 in which a device

Figure 7-11
Same circuit as in Fig. 7-9 except that double-subscript notation is used.

carries a current I flowing into terminal **a**. The voltage between terminals **a, b** is E_{ab}. The following rule applies:

1. **The device is a reactive load when**

   a. current I lags 90° behind voltage E_{ab}
   and

   b. current I is shown as entering by terminal a.

The following rules are a corollary to the above basic rule:

2. A device is a reactive source when I lags 90° behind E_{ab} and I is shown as flowing out of terminal a.

3. A device is a reactive source when I leads E_{ab} by 90° and I is shown as flowing into terminal a.

4. A device is a reactive load when I leads E_{ab}

by 90° and I is shown as flowing out of terminal a.

7.8 The capacitor as a reactive source

Let us now study the behavior of a capacitor connected to an ac generator (Fig. 7-12a). Upon solving the circuit, we discover that the current leads the voltage E_{ab} by 90° (Fig. 7-12b). It follows from our definitions that the capacitor is a reactive source. Conse-

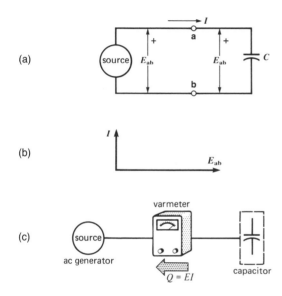

(a)

(b)

(c)

Figure 7-12
a. Capacitor connected to an ac source.
b. Phasor I leads E_{ab} by 90°.
c. Reactive power flows from the capacitor to the source.

quently, it delivers reactive power to the very generator to which it is connected! For most people, this takes a little time to accept because how, we may ask, can a passive device like a capacitor possibly produce any power? The answer is that reactive power really represents energy which, like a pendulum, swings back and forth without ever doing any useful work. The capacitor acts as a temporary energy-storing device repeatedly accepting energy for

brief periods and releasing it again. However, instead of storing magnetic energy as an inductor does, a capacitor stores electrostatic energy (see Sec. 2-14).

If we connect a varmeter into the circuit (Fig. 7-12c), it will give a negative reading of EI vars, showing that reactive power is indeed flowing from the capacitor to the generator. The generator is actually a reactive load, but we sometimes prefer to call it a *receiver* of reactive power, which, of course, amounts to the same thing.

7.9 Distinction between active and reactive power

There is a basic difference between active and reactive power and perhaps the most important thing to remember is that the one cannot be converted into the other. Active and reactive power function independently of each other and, consequently, they can be treated as separate quantities in electric circuits.

Both place a burden on the transmission line that carries them, but, whereas active power eventually produces a tangible result (heat, mechanical power, light, etc.), reactive power only represents energy that oscillates back and forth.

All ac inductive devices such as magnets, transformers, ballasts and induction motors, absorb reactive power because one component of the current they draw lags 90° behind the voltage. The reactive power plays a very important role because it produces the ac magnetic field in these devices.

A building, shopping center, or city may be considered to be a huge active/reactive load connected to an electric utility system. Such load centers contain thousands of induction motors and other electromagnetic devices which draw both reactive power (to sustain their magnetic fields) and active power (to do the useful work).

This leads us to the study of loads that absorb both active and reactive power.

7.10 Combined active and reactive loads: apparent power

Loads that absorb both active power P and reactive power Q may be considered to be made up of a resistance and an inductive reactance. Consider, for example, the circuit of Fig. 7-13a in which a resistor and inductor are connected to a source E. The resistor draws a current I_p, while the inductor draws a current I_q.

According to our definitions, the resistor is an active load while the inductor is a reactive load. Consequently, I_p is in phase with E_{ab} while I_q lags 90° behind. The phasor diagram (Fig. 7-13b) shows that the resultant line current I lags behind E_{ab} by an angle θ. Furthermore, the magnitude of I is given by

$$I = \sqrt{I_p^2 + I_q^2}$$

The active and reactive power components P and Q both flow in the same direction, as shown by the arrows in Fig. 7-13c. If we connect a wattmeter and a varmeter into the circuit, the readings will both be positive, indicating $P = EI_p$ watts and $Q = EI_q$ vars, respectively.

Furthermore, if we connect an ammeter into the line it will indicate a current of I amperes. As a result, we are inclined to believe that the power supplied to the load is equal to EI watts. But this is obviously incorrect because the power is composed of an active component (watts) and a reactive component (vars). For this reason, the product EI is called *apparent power*. The symbol for apparent power is S.

Apparent power is expressed neither in watts nor in vars, but in *voltamperes*. Multiples are the kilovoltampere (kVA) and megavoltampere (MVA).

7.11 Relationship between P, Q, S

Consider the single-phase circuit of Fig. 7-14a composed of a source, a load and appropriate meters. Let us assume that:

- the voltmeter indicates E volts;

- the ammeter indicates I amperes;

Figure 7-13
a. Circuit consisting of a source feeding an active and reactive load.
b. Phasor diagram of the voltage and currents.
c. Active and reactive power flow from source to load.

- the wattmeter indicates $+P$ watts;

- the varmeter indicates $+Q$ vars;

Knowing that P and Q are positive, we know that the load absorbs both active and reactive power. Consequently, the line current I lags behind E_{ab} by an angle θ.

Current I can be decomposed into two components I_p and I_q, respectively in phase, and in quadrature, with phasor E (Fig. 7-14b). The numerical values of I_p and I_q can be found directly from the instrument readings:

$$I_p = P/E \qquad (7-1)$$

$$I_q = Q/E \qquad (7-2)$$

Furthermore, the apparent power S transmitted

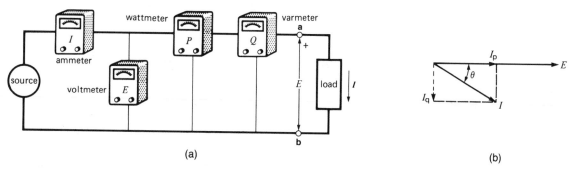

Figure 7-14
a. Instruments used to measure E, I, P, Q in a circuit.
b. The phasor diagram can be deduced from the instrument readings.

over the line is given by $S = EI$, from which:

$$I = S/E \qquad (7\text{-}3)$$

Referring to the phasor diagram (Fig. 7-14b), it is obvious that:

$$I^2 = I_p^2 + I_q^2$$

Consequently:

$$\left[\frac{S}{E}\right]^2 = \left[\frac{P}{E}\right]^2 + \left[\frac{Q}{E}\right]^2$$

that is,

$$S^2 = P^2 + Q^2 \qquad (7\text{-}4)^*$$

in which

S = apparent power [VA]
P = active power [W]
Q = reactive power [var]

We can also calculate the value of the angle θ because the tangent of θ is obviously equal to I_q/I_p. Thus, we have

$$\theta = \arctan I_q/I_p = \arctan Q/P \qquad (7\text{-}5)$$

---

* Equation 7-4 may be represented by a right-angle «power triangle» having sides P, Q, and hypothenuse S The relationship between the three sides is the well-known Pythagorean equation $S^2 = P^2 + Q^2$. The concept of the power triangle is useful when solving ac circuits that contain several active and reactive loads.

Example 7-4:
An alternating current motor absorbs 40 kW of active power and 30 kvar of reactive power. Calculate the apparent power supplied to the motor.

Solution:

$$S = \sqrt{P^2 + Q^2} \qquad \text{Eq. 7-4}$$
$$= \sqrt{40^2 + 30^2}$$
$$= 50 \text{ kVA}$$

Example 7-5:
A wattmeter and varmeter are connected into a 120 V single-phase line that feeds an ac motor. They respectively indicate 1800 W and 960 var. Calculate:

a. the in-phase and quadrature components I_p and I_q;

b. the line current I;

c. the apparent power supplied by the source.

d. the phase angle between the line voltage and line current

Solution:
Referring to Fig. 7-14b where the load is now a motor, we have:

a. $I_p = P/E = 1800/120 = 15\,\text{A}$ Eq. 7-1
   $I_q = Q/E = 960/120 = 8\,\text{A}$ Eq. 7-2

b. From the phasor diagram:

$$I = \sqrt{I_p^2 + I_q^2} = \sqrt{15^2 + 8^2}$$
$$= 17 \text{ A}$$

c. The apparent power is:

$$S = EI = 120 \times 17 = 2040 \text{ VA}$$

d. The phase angle θ between E and I is:

$$\theta = \arctan Q/P = \arctan 960/1800$$
$$= 28.07°$$

Example 7-6:
A voltmeter and ammeter connected into the circuit of Fig. 7-7a give readings of 140 V and 20 A, respectively. Calculate:

a. the apparent power of the circuit;

b. the reactive power of the circuit;

c. the active power of the circuit.

Solution:
a. The apparent power is:

$$S = EI = 140 \times 20$$
$$= 2800 \text{ VA} = 2.8 \text{ kVA}$$

b. The reactive power is:

$$Q = EI = 140 \times 20$$
$$= 2800 \text{ var} = 2.8 \text{ kvar.}$$

If a varmeter were connected into the circuit, it would give a reading of 2800 var.

c. The active power is zero.

If a wattmeter were connected into the circuit it would read zero.

To recapitulate, the apparent power is 2800 VA but because the current is 90° out of phase with the voltage, it is also equal to 2800 var.

7.12 Power factor

The power factor of an alternating current circuit is the ratio of the active power P to the apparent power S. It is given by the equation:

$$\text{power factor} = P/S \qquad (7\text{-}6)$$

where

$P =$ active power delivered or absorbed by the circuit [W]

$S =$ apparent power of the circuit [VA]

power factor is expressed as a simple number, or as a percentage.

Because the active power P can never exceed the apparent power S, it follows that the power factor can never be greater than unity (or 100 percent). The power factor of a resistor is 100 percent because the apparent power it draws is equal to the active power. On the other hand, the power factor of an ideal coil having no resistance is zero because it does not consume any active power.

To sum up, the power factor of a circuit or device is simply a way of stating what fraction of its apparent power is real, or active, power.

In a single-phase circuit the power factor is also a measure of the phase angle θ between the voltage and current. Thus, referring to Fig. 7-14:

$$\text{power factor} = P/S$$
$$= EI_p / EI$$
$$= I_p / I$$
$$= \cos \theta$$

consequently,

$$\text{power factor} = \cos \theta = P/S \qquad (7\text{-}7)$$

where

power factor = power factor of a single-phase circuit

$\theta =$ phase angle between the voltage and current

If we know the power factor, we know the cosine of the angle between E and I and, hence, we can calculate the angle. The power factor is said to be *lagging* if the current lags behind the voltage. Conversely, the power factor is said to be *leading* if the current leads the voltage.

Example 7-7:
Calculate the power factor of the motor in Example 7-5 and the phase angle between the line voltage and line current.

Solution:

$$\text{power factor} = P/S$$
$$= 1800/2040$$
$$= 0.882 \text{ or } 88.2 \text{ percent lagging}$$
$$\cos \theta = 0.882$$
$$\text{therefore } \theta = 28°$$

Example 7-8:
A single-phase motor draws a current of 5 A from a 120 V, 60 Hz line. The power factor of the motor is 65 percent. Calculate:
a. the active power absorbed by the motor;
b. the reactive power supplied by the line.

Solution:

a. The apparent power drawn by the motor is:
$$S_m = EI = 120 \times 5 = 600 \text{ VA}$$

The active power absorbed by the motor is:
$$P_m = S_m \cos \theta \qquad \text{Eq. 7-7}$$
$$= 600 \times 0.65 = 390 \text{ W}$$

b. The reactive power absorbed by the motor is:
$$Q_m = \sqrt{S_m^2 - P_m^2} \qquad \text{Eq. 7-4}$$
$$= \sqrt{600^2 - 390^2}$$
$$= 456 \text{ var}$$

Note that the motor draws even more reactive power from the line than active power. This is unfortunate, because it burdens the line with a lot of non-productive power.

The power triangle for the circuit is given in Fig. 7-15. Active power consumed (+) is drawn horizontally to the right. The reactive power absorbed (+) is drawn vertically upwards.

7.13 Some further aspects of sources and loads

Let us consider Fig. 7-16a in which a resistor and capacitor are connected to a source. The

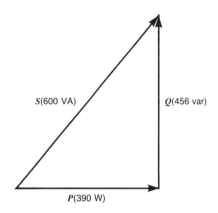

Figure 7-15
Power triangle of a motor. See Example 7-8.

circuit is similar to Fig. 7-13 except that the capacitor is a reactive source. As a result, reactive power flows from the capacitor to the source G while active power flows from the source G to the resistor. The active and reactive power components therefore flow in opposite directions over the transmission line. A wattmeter connected into the circuit will give a positive reading $P = EI_p$ watts but a varmeter will give a negative reading $Q = EI_q$. The so-called "source" G delivers active power P, but *receives* reactive power Q. Thus, G is both an active source and a reactive load.

It may seem unusual to have two powers flowing in opposite directions over the same transmission line, but we must again remember that active power P is not the same as a reactive power Q and that each flows independently of the other.

Speaking of sources and loads, the deceptively simple electrical outlet also deserves our attention. Such an outlet may be the 120 V receptacle in a home, the 3-phase 480 V service entrance to a factory, or the terminals of a high-power 345 kV transmission line. All such outlets are ultimately connected to the huge alternators that power the electrical transmis-

(a)

(b)

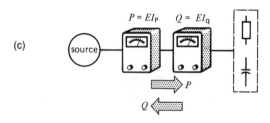

(c)

Figure 7-16
a. Source feeding an active and reactive (capacitive) load.
b. Phasor diagram of the circuit.
c. The active and reactive powers flow in opposite directions.

sion and distribution systems. Odd as it may seem, an electrical outlet can act not only as an active or reactive source (as we would expect), but it may also behave as an active or reactive load. What factors determine whether it will behave in one way or the other? It all depends upon the type of device (or devices) connected to the receptacle. If the device requires active power, the receptacle will provide it; if the device delivers reactive power, the receptacle will receive it. In other words, a receptacle outlet is at all times ready to deliver – or accept – either active power P or reactive power Q to meet the requirements of the devices connected to it.

Example 7-9:
A 50 μF paper capacitor is placed across the motor terminals in Example 7-8. Calculate:
a. the reactive power generated by the capacitor;
b. the active power absorbed by the motor;
c. the reactive power absorbed from the line;
d. the new line current.

Solution:
a. The impedance of the capacitor is:
$$X_C = 1/2\pi fC \qquad \text{Eq. 2-11}$$
$$= 1/2\pi \times 60 \times 50 \times 10^{-6}$$
$$= 53 \ \Omega$$

The current in the capacitor is:
$$I = E/X_c = 120/53$$
$$= 2.26 \ \text{A}$$

The reactive power generated by the capacitor is:
$$Q_C = EI = 120 \times 2.26$$
$$= 271 \ \text{var}$$

b. The motor continues to draw the same active power because it is still fully loaded. Consequently:
$$P_m = 390 \ \text{W}$$

The motor also draws the same reactive power as before because nothing has taken place to change its magnetic field. Consequently:
$$Q_m = 456 \ \text{var}$$

c. The motor draws 456 var from the line, but the capacitor furnishes 271 var to the same line. The net reactive power drawn from the line is therefore:
$$Q_L = Q_m - Q_C$$
$$= 456 - 271$$
$$= 185 \ \text{var}$$

The active power drawn from the line is:
$$P_L = P_m = 390 \ \text{W}$$

d. The apparent power drawn from the line is:

Figure 7-17
Power triangle of a motor and capacitor connected to an ac line. See Example 7-9.

$$S_L = \sqrt{P_L^2 + Q_L^2}$$
$$= \sqrt{390^2 + 185^2}$$
$$= 432 \text{ VA}$$

The new line current is:
$$I_L = S_L/E = 432/120$$
$$= 3.6 \text{ A}$$

Thus, the line current drops from 5 A to 3.6 A by placing the capacitor in parallel with the motor. This represents a big improvement because the line current is smaller and the operation of the motor has not been changed in the least.

The new power factor of the line is:
$$\cos \phi_L = P_L/S_L = 390/432$$
$$= 0.903 \text{ or } 90.3 \%$$
$$\phi_L = \text{arcos } 0.903 = 25.5°$$

The power triangle is shown in Fig. 7-17. The reactive power Q_C generated (–) by the capacitor is drawn vertically downwards. By comparing this power triangle with that in Fig. 7-15, we can visually observe the effect of the capacitor on the apparent power supplied by the line.

7.14 Systems comprising several loads

The concept of active and reactive power enables us to simplify the solution of some rather complex circuits. Consider, for example, a group of loads connected in a very unusual way to a 380 V source (Fig. 7-18a). We wish to calculate the apparent power of the system as well as the current supplied by the source.

Using the power approach, we do not have

(a)

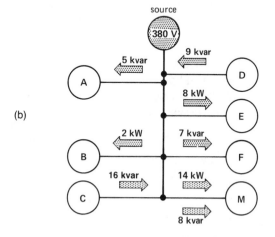

(b)

Figure 7-18.
a. Example of active and reactive loads connected to a 380 V source.
b. All loads are assumed to be directly connected to the 380 V receptacle.

to worry about the way the loads are interconnected. We simply draw a block diagram of the individual loads, indicating the direction (as far as the source is concerned) of active and reactive power flow (Fig. 7-18b). Thus, because load A is inductive, it absorbs reactive power; consequently, the 5 kvar arrow flows from the source to the load. On the other hand, because load C represents a capacitor, it delivers reactive power to the system. The 16 kvar arrow is directed therefore toward the source.

The distinct (and independent) nature of active and reactive power enables us to add all the active powers in a circuit to obtain the total active power P. In the same way, we can add the reactive powers to obtain the total reactive power Q. The resulting total apparent power S is then found by:

$$S = \sqrt{P^2 + Q^2} \qquad \text{Eq. 7-4}$$

When adding reactive powers, we assign a positive value to those that are absorbed by the system and a negative value to those that are generated (such as by a capacitor). In the same way, we assign a positive value to active powers that are absorbed and a negative value to those that are generated (such as by an alternator).

Note that usually we *cannot* add the apparent powers in various parts of a circuit to obtain the total apparent power S. We can only add them if their power factors are identical.

Let us now analyze the circuit of Fig. 7-18:

1. active power absorbed by the system:
   $$P = (2 + 8 + 14) = + 24 \text{ kW}$$

2. reactive power absorbed by the system:
   $$Q_1 = (5 + 7 + 8) = + 20 \text{ kvar}$$

3. reactive power supplied by the capacitors:
   $$Q_2 = (- 9 - 16) = - 25 \text{ kvar}$$

4. net reactive power Q absorbed by the system:
   $$Q = (+ 20 - 25) = - 5 \text{ kvar}$$

5. apparent power of the system:
   $$S = \sqrt{P^2 + Q^2} = \sqrt{24^2 + (-5)^2}$$
   $$= 24.5 \text{ kVA}$$

6. because the source furnishes the apparent power, the line current is:
   $$I = S/E = 24\,500/380 = 64.5 \text{ A}$$

7. the power factor of the system is:
   $$\cos \phi_L = P/S = 24/24.5 = 0.979 \text{ leading}$$

The 380 V source delivers 24 kW of active power, but it *receives* 5 kvar of reactive power. This reactive power flows into the local distribution system of the electrical utility company where it becomes available to create magnetic fields. The magnetic fields may be associated with distribution transformers, transmission lines or even relays and electromagnets of other customers connected to the same distribution system.

The power triangle for the system is shown in Fig. 7-18c. It does not provide any new information, but helps us visualize the power flows.

Figure 7-18c
Power triangle of the system.

7.15 Reactive power without magnetic fields

We sometimes encounter situations where loads absorb reactive power without creating any magnetic field at all. This can happen in electronic power circuits when the current flow is delayed by means of a rapid switching device, such as a thyristor.

Consider, for example, the circuit of Fig. 7-19 in which a 100 V, 60 Hz source is connected to a resistive load of 10 Ω by means of a

(a)

(b)

Figure 7-19
a. Active and reactive power flow in a switched resistive load.
b. The delayed current flow is the cause of the reactive power absorbed by the system.

synchronous switch. The switch opens and closes its contacts so that current only flows during the latter part of each half cycle. We can see, almost by intuition, that this forced delay causes the current to lag behind the voltage.

Indeed, if we connected a wattmeter and varmeter between the source and the switch, they would respectively read + 500 W and +318 var. This corresponds to a lagging power factor (sometimes called *displacement power factor*) of 84 percent. The reactive power is associated with the rapidly operating switch rather than with the resistor itself. Nevertheless, reactive power is consumed just as surely as if an inductor were present in the circuit.

7.16 Solving AC circuits using the power triangle method

We have seen that active and reactive powers can be added algebraically. This enables us to solve some rather complex ac circuits without ever having to draw a phasor diagram or resorting to complex (j) notation. We calculate the active and reactive powers associated with each circuit element and deduce the corresponding voltages and currents. The following example demonstrates the usefulness of this "power triangle" approach.

Example 7-10:
In Fig. 7-20a, the voltage between terminals 1 and 3 is 60 V. Calculate:
a. the current in each circuit element
b. the voltage between terminals 1 and 2.
c. the impedance between terminals 1 and 2.

Solution:
We know the impedances of the elements and that 60 V exists between terminals 3 and 1 (Fig. 7-20b). We now proceed in logical steps, as follows:

a. The current in the capacitor is:
   $$I_c = 60/5 = 12 \text{ A, from which the reactive power generated is:}$$
   $$Q_c = 12 \times 60 = -720 \text{ var.}$$

   The current in the resistor is:
   $$I_R = 60/12 = 5 \text{ A, from which the active power absorbed is:}$$
   $$P = 5 \times 60 = 300 \text{ W.}$$

(a)

(b)

Figure 7-20
a. Solving ac circuits by the power triangle method.
b. Voltages and currents in the circuit. See Example 7-10.

The apparent power supplied to terminals 1-3 is:

$$S = \sqrt{P^2 + Q^2} = \sqrt{300^2 + (-720)^2}$$
$$= 780 \text{ VA}$$

The current I_L must therefore be:

$$I_L = S/E_{31} = 780/60 = 13 \text{ A.}$$

The voltage across the inductive reactance is:

$$E_{23} = IX_L = 13 \times 8 = 104 \text{ V.}$$

The reactive power absorbed by the inductive reactance is:

$$Q_L = E_{23} \times I_L = 104 \times 13$$
$$= + 1352 \text{ var.}$$

The total reactive power absorbed the circuit is:

$$Q = Q_L + Q_C = 1352 - 720$$
$$= + 632 \text{ var.}$$

The total active power absorbed by the circuit is:

$$P = 300 \text{ W.}$$

The apparent power absorbed by the circuit is:

$$S = \sqrt{P^2 + Q^2} = \sqrt{300^2 + 632^2}$$
$$= 700 \text{ VA}$$

b. The voltage of the line is therefore:
$$E_{21} = S/I_L = 700/13 = 53.8 \text{ V.}$$

c. The impedance between terminals 2-1 is:
$$Z = E_{21}/I_L = 53.8/13 = 4.14 \text{ }\Omega.$$

7.17 Power and complex notation

If complex notation is used to solve an ac circuit, we can easily determine the active and reactive power associated with any component, including the sources. We simply multiply the phasor voltage E across the component by the conjugate $(I*)$ of the current that flows through it. The complex product $EI*$ gives the apparent power S in terms of $P + jQ$, where P is the active power and Q the reactive power absorbed (or delivered) by the component.

A positive value for P or Q means that the component absorbs active or reactive power. Negative values mean that the component produces active or reactive power.

An example will illustrate the procedure to be followed.

Example 7-11:

The circuit in Fig. 7-21 is composed of a 45 Ω resistor connected in series with a 28 Ω inductive reactance. The source is described by the phasor $E_{ab} = 159\angle65°$. Calculate:

a. the magnitude and phase of the current I.

b. the magnitude and phase of the voltage across the resistor and the reactance.

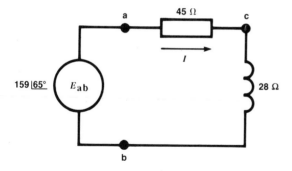

Figure 7-21
Solving an ac circuit using complex notation.

c. the active and reactive power associated
with the resistor, the reactance and the
source.

Solution:
a. Applying Kirchhoff's voltage law we obtain:

$$E_{ba} + E_{ac} + E_{cb} = 0$$
$$-E_{ab} + 45I + j28I = 0$$
$$-159\angle 65° + I(45 + j28) = 0$$
$$I = \frac{159\angle 65°}{45 + j28}$$

Transforming the denominator into polar
coordinates, we obtain:

$$\text{amplitude} = \sqrt{45^2 + 28^2} = 53$$

$$\text{phase angle} = \arctan 28/45 = 31.89°$$

hence $45 + j28 = 53\angle 31.89°$

and so $I = \dfrac{159\angle 65°}{53\angle 31.89°} = 3\angle(65° - 31.89°)$

$$= 3\angle 33.11°$$

b. Voltage across the resistor is:

$$E_{ac} = 45I$$
$$= 45 \times 3\angle 33.11°$$
$$= 135\angle 33.11°$$

Voltage across the reactance is:

$$E_{cb} = j28I$$
$$= j28 \times 3\angle 33.11°$$
$$= 84\angle(33.11° + 90°)$$
$$= 84\angle 123.11°$$

c. The conjugate I^* of the current I is:

$$I^* = 3\angle -33.11°$$

The complex power associated with the
resistor is:

$$P_r = E_{ac}I^*$$
$$= (135\angle 33.11°)(3\angle -33.11°)$$
$$= 405\angle 0°$$
$$= 405(\cos 0° + j\sin 0°)$$
$$= 405(1 + j0)$$
$$= 405$$

Thus, the resistor absorbs only real power
because there is no j component in P_r.

The complex power associated with the
reactance is:

$$P_x = E_{cb}I^*$$
$$= (84\angle 123.11°)(3\angle -33.11°)$$
$$= 252\angle 90°$$
$$= 252(\cos 90° + j\sin 90°)$$
$$= 252(0 + j1)$$
$$= j252$$

Thus, the reactance absorbs only reactive
power.

The power associated with the source is:

$$P_s = E_{ba}I^* = -E_{ab}I^*$$
$$= -(159\angle 65°)(3\angle -33.11°)$$
$$= -477\angle(65° - 33.11°)$$
$$= -477\angle 31.89°$$
$$= -477(\cos 31.89° + j\sin 31.89°)$$
$$= -477(0.849 + j0.528)$$
$$= -405 - j252$$

Because the active and reactive powers are
both negative proves that the source *delivers*
an active power of 405 W and a reactive power
of 252 var.

QUESTIONS AND PROBLEMS

Practical level

7-1   What is the unit of active power? reactive power? apparent power?

7-2   A capacitor of 500 kvar is placed in parallel with an inductor of 400 kvar. Calculate the apparent power of the group.

7-3   Name a static device which can generate reactive power.

7-4   Name a static device which absorbs reactive power.

7-5   What is the approximate power factor, in percent, of a capacitor? of a coil? of an incandescent lamp?

7-6   The current in a single-phase motor lags 50° behind the voltage. What is the power factor of the motor?

Intermediate level

7-7   A large motor absorbs 600 kW at a power factor of 90 percent. Calculate the apparent power and reactive power absorbed by the machine.

7-8   A 200 μF capacitor is connected to a 240 V, 60 Hz source. Calculate the reactive power it generates.

7-9   A 10 Ω resistor is connected across a 120 V, 60 Hz source. Calculate:
a. the active power absorbed by the resistor;
b. the apparent power absorbed by the resistor;
c. the peak power absorbed by the resistor;
d. the duration of each positive power pulse.

7-10   A 10 Ω reactance is connected to a 120 V, 60 Hz line. Calculate:
a. the reactive power absorbed by the reactor;
b. the apparent power absorbed by the reactor;

c. the peak power input to the reactor;
d. the peak power output of the reactor;
e. the duration of each positive power pulse.

7-11   Using the definitions in Sections 7-3, 7-4, 7-6 and 7-7, determine which of the devices in Fig. 7-22a through 7-22f acts as an active (or reactive) power source.

7-12   A single-phase motor draws a current of 12 A at a power factor of 60 percent. Calculate the in-phase and quadrature components of current I_p and I_q with respect to the line voltage.

7-13   A single-phase motor draws a current of 16 A from a 240 V, 60 Hz line. A wattmeter connected into the line gives a reading of 2765 W. Calculate the power factor of the motor and the reactive power it absorbs.

7-14   If a capacitor having a reactance of 30 Ω is connected in parallel with the motor of Problem 7-13, calculate:
a. the active power reading of the wattmeter;
b. the total reactive power absorbed by the capacitor and motor;
c. the apparent power of the ac line;
d. the line current;
e. the power factor of the motor/capacitor combination.

7-15   Using only power triangle concepts (Sec. 7-16) and without drawing any phasor diagrams, find the impedance of the circuits in Fig. 7-23.

7-16   An induction motor absorbs an apparent power of 400 kVA at a power factor of 80 percent. Calculate:
a. the active power absorbed by the motor;
b. the reactive power absorbed by the motor;
c. what purpose does the reactive power serve?

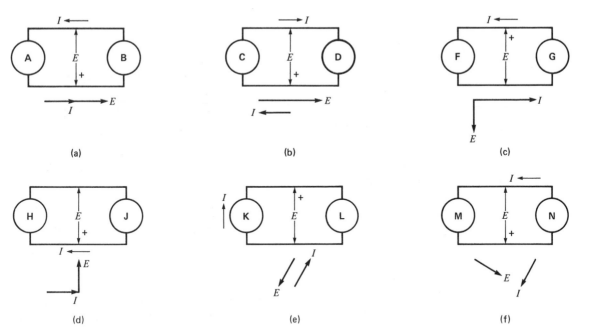

Figure 7-22
See Problem 7-11.

Figure 7-23
See Problem 7-15.

7-17 A circuit composed of a 12 Ω resistor in series with an inductive reactance of 5 Ω carries an ac current of 10 A. Calculate:
a. the active power absorbed by the resistor;
b. the reactive power absorbed by the inductor;
c. the apparent power of the circuit;
d. the power factor of the circuit.

7-18 A coil having a resistance of 5 Ω and an inductance of 2 H carries a direct current of 20 A. Calculate:
a. the active power absorbed;
b. the reactive power absorbed.

Advanced level

7-19 A motor having a power factor of 0.8 absorbs an active power of 1200 W. Calculate the reactive power drawn from the line.

7-20 In Problem 7-13, if we place a capacitor of 500 var in parallel with the motor, calculate:
a. the total active power absorbed by the system;
b. the apparent power of the system;
c. the power factor of the system.

7-21 A coil having a reactance of 10 Ω and a resistance of 2 Ω is connected in parallel with a capacitive reactance of 10 Ω. If the supply voltage is 200 V, calculate:
a. the reactive power absorbed by the coil;
b. the reactive power generated by the capacitor;
c. the active power dissipated by the coil;
d. the apparent power of the circuit;

7-22 The power factor at the terminals of a 120 V source is 0.6 lagging (Fig.7-24). Without using phasor diagrams, calculate:
a. the value of E;
b. the impedance of the load Z.

7-23 In Figs. 7-25a and 7-25b, indicate the magnitude and direction of the active and reactive power flow. Hint: decompose I into I_p and I_q and treat them independently.

Figure 7-24
See Problem 7-22.

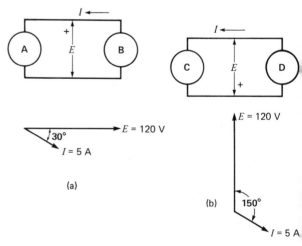

Figure 7-25
See Problem 7-23.

8

THREE-PHASE CIRCUITS

Electric power is generated, transmitted and distributed in the form of three-phase power. Homes and small establishments are wired for single-phase power, but this merely represents a tap-off from the basic three-phase system. Three-phase power is preferred over single-phase power for several important reasons:

a. 3-phase motors, generators and transformers are simpler, cheaper and more efficient;
b. 3-phase transmission lines can deliver more power for a given weight and cost;
c. the voltage regulation of 3-phase transmission lines is inherently better.

A knowledge of 3-phase power and 3-phase circuits is therefore essential to an understanding of power technology. Fortunately, the basic circuit techniques used to solve single-phase circuits can be directly applied to 3-phase circuits. Furthermore, we shall see that most 3-phase circuits can be reduced to elementary single-phase diagrams. In this regard, we assume the reader is familiar with the previous chapters dealing with ac circuits and power.

8.1 Polyphase systems

We can gain an immediate preliminary understanding of polyphase systems by referring to the common gasoline engine. A single-cylinder engine having one piston is comparable to a single-phase machine. On the other hand, a two-cylinder engine is comparable to a 2-phase machine. The more common six-cylinder engine could be called a six-phase machine. In a six-cylinder engine, identical pistons move up and down inside identical cylinders, but they do not move in unison. They are staggered in such a way so as to deliver power to the shaft in successive pulses rather than at the same time. As the reader may know from personal experience, this produces a smoother running engine and a much smoother output torque.

Similarly, in a 3-phase electrical system, the three phases are identical, but they deliver power at different times. As a result, the total power flow is very smooth. Furthermore, because the phases are identical, one phase may be used to represent the behavior of all three.

Although we must beware of carrying analogies too far, the above description reveals that a three-phase system is basically composed of three single-phase systems which operate in sequence. Once this basic fact is realized, much of the mystery surrounding 3-phase systems disappears.

8.2 Single-phase generator

Consider a permanent magnet NS revolving at constant speed inside a stationary iron ring (Fig. 8-1). The magnet is driven by an external mechanical source, such as a turbine. The ring (or stator) reduces the reluctance of the magnetic circuit; consequently, the flux density in the air gap is greater than if the ring were absent. A rectangular coil having terminals **a, 1** is mounted inside the ring, but insulated from it.

As the magnet turns, it sweeps across the conductors of the coil, inducing a voltage in them according to the equation:

$$E_{a1} = Blv \qquad \text{Eq. 2-25}$$

The sum of the voltages induced in all the conductors appears across the terminals. The terminal voltage E_{a1} is maximum when the poles are in the position of Fig. 8-1 because the flux cuts the conductors. On the other hand, the voltage is zero when the poles are in the position of Fig. 8-2 because flux does not cut the conductors at this moment.

If we plot E_{a1} as a function of the angle of rotation, we obtain the sinusoidal voltage shown in Fig. 8-3. This is an alternating voltage having a peak value of 20 V. Machines that produce such voltages are called alternating current generators or *synchronous generators*. The particular machine shown in Fig. 8-1 is called a *single-phase* generator.

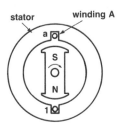

Figure 8-1
A single-phase generator with a multi-turn coil embedded in two slots. At this instant E_{a1} is maximum (+).

Figure 8-2
At this instant $E_{a1} = 0$ because the flux does not cut the conductors of winding A.

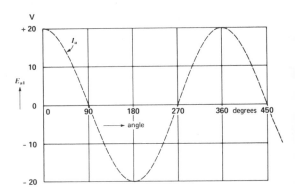

Figure 8-3
Voltage induced in winding A.

8.3 Power output of a single-phase generator

If a resistor is connected across terminals **a, 1** a current will flow and the resistor will heat up (Fig. 8-4). The current I_a is in phase with the voltage, and consequently the instantaneous power is composed of a series of positive pulses, as shown in Fig. 8-5. The average power is one-half the peak power. This electrical power is derived from the mechanical power provided by the turbine driving the generator. As a result, the turbine must deliver its mechanical energy in pulses, to match the pulsed electrical output. This sets up mechanical vibrations whose frequency is twice the electrical frequency. Consequently, the generator will vibrate and tend to be noisy.

8.4 Two-phase generator

Using the same single-phase generator, let us mount a second winding (B) on the stator, identical to winding A, but displaced from it by a mechanical angle of 90° (Fig. 8-6a).

As the magnet rotates, sinusoidal voltages are induced in each winding. They obviously have the same magnitude and frequency, but they do not reach their maximum value at the same time. In effect, at the moment when the magnet occupies the position shown in Fig. 8-6a, voltage E_{a1} passes through its maximum positive value, whereas voltage E_{b2} is zero. This is because the flux only cuts across the conductors in slots **1** and **a** at this instant. However, after the rotor has made one quarter turn (or 90°), voltage E_{a1} becomes zero and voltage E_{b2} attains its maximum positive value. The two voltages are therefore out of phase by 90°. They are represented by curves in Fig. 8-6b and by phasors in Fig. 8-6c. Note that E_{a1} leads E_{b2} because it reaches its peak positive value before E_{b2} does.

This machine is called a *two-phase generator*, and the stator windings are respectively called phase A and phase B.

Figure 8-4
Single-phase generator delivering power to a resistor.

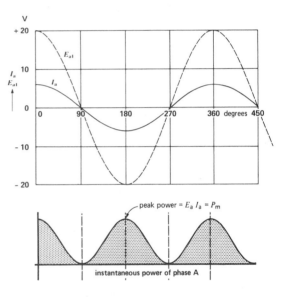

Figure 8-5
Graph of the voltage, current and power when the generator is under load.

Example 8-1:
The generator shown in Fig. 8-6a rotates at 6000 r/min and generates an effective sinusoidal voltage of 170 V per winding. Calculate:
a. the peak voltage across each phase;
b. the output frequency;
c. the time interval corresponding to a phase angle of 90°.

(a)

(b)

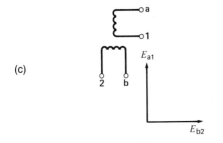

(c)

Figure 8-6
a. Schematic diagram of a two-phase generator.
b. Voltages induced in a 2-phase generator.
c. Phasor diagram of the induced voltages

Solution:

a. The peak voltage per phase is:
$$E_m = \sqrt{2}E = 1.414 \times 170 \qquad \text{Eq. 2-6}$$
$$= 240 \text{ V}$$

b. One cycle is completed every time the magnet makes one turn. The period of one cycle is:
$$T = 1/6000 \text{ min}$$
$$= 60/6000 \text{ s} = 0.01 \text{ s}$$
$$= 10 \text{ ms}$$

The frequency is:
$$f = 1/T = 1/0.01 = 100 \text{ Hz}$$

c. A phase angle of 90° corresponds to a time interval of one quarter revolution or 10 ms/4 = 2.5 ms. Consequently, phasor E_{b2} lags 2.5 ms behind phasor E_{a1}.

8.5 Power output of a 2-phase generator

Let us now connect two identical resistive loads across phases A and B (Fig. 8-7a). Currents I_a and I_b will flow in each resistor. They are respectively in phase with E_{a1} and E_{b2}. The currents are therefore 90° out of phase with each other (Fig. 8-7b). This means that I_a reaches its maximum value one quarter period before I_b does. Furthermore, the generator now produces a two-phase power output.

The instantaneous power supplied to each resistor is equal to the instantaneous voltage times the instantaneous current. This yields the two power waves shown in Fig. 8-8. Note that when the power of phase A is maximum, that of phase B is zero, and vice-versa. If we add the instantaneous powers of both phases, we discover that the resultant power is constant, and equal to the peak power P_m of one phase*. In other words, the total power output of the 2-phase generator is the same at every instant. As a result, the mechanical power needed to drive the generator is also constant. A two-phase generator does not vibrate and so it is less noisy. And as an important added benefit, it produces twice the power output without any increase in size.

8.6 Three-phase generator

A three-phase generator is similar to a two-phase generator, except that the stator has three identical windings instead of two. The three windings **a-1**, **b-2** and **c-3** are placed at 120° to each other, as shown in Fig. 8-9a.

When the magnet is rotated at constant speed, the voltages induced in the three windings have the same effective values, but

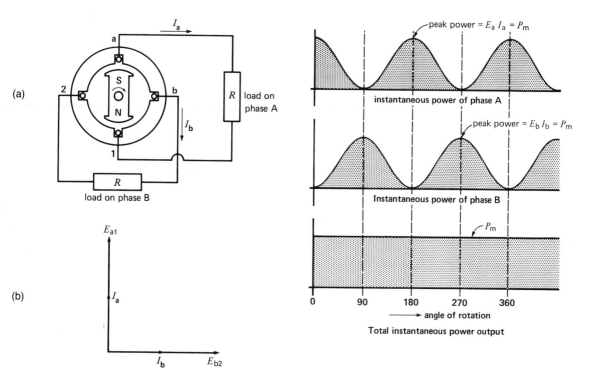

Figure 8-7
a. Two-phase generator under load.
b. Phasor diagram of the voltages and currents.

Figure 8-8
Power produced by a 2-phase generator.

---

* The term "phase" is used to designate different things. Consequently, it has to be read in context to be understood. The following examples show some of the ways in which the word "phase" is used.

1. the current is *out of phase* with the voltage (refers to phasor diagram);
2. the three *phases* of a transmission line (meaning the three conductors of the line);
3. the *phase-to-phase* voltage (meaning line voltage);
4. the phase sequence (the order in which the phasors follow each other);
5. the burned-out phase (the burned-out winding of a 3-phase machine);
6. the three-phase voltage (the line voltage of a 3-phase system);
7. the 3-phase currents are unbalanced (the currents in a 3-phase line or machine are unequal and not displaced at 120°);
8. phase-shift transformer (a device which can change the phase angle of the output voltage with respect to the input voltage);
9. phase-to-phase fault (a short-circuit between two line conductors);
10. phase-to-ground fault (a short-circuit between a line or winding, and ground);
11. the phases are unbalanced (the line voltages, or the line currents, are unequal or not displaced at 120° to each other).

(a)

(b)

(c)

Figure 8-9
a. Three-phase generator.
b. Voltages induced in a 3-phase generator
c. Phasor diagram of the induced voltages.

the peaks occur at different times. In effect, at the moment when the magnet is in the position shown in Fig. 8-9a, only voltage E_{a1} is at its maximum positive value.

Voltage E_{b2} will reach its positive peak after the rotor has turned through an angle of 120° (or one third of a turn). Similarly, voltage E_{c3} will attain *its* positive peak after the rotor has turned through 240° (or two-thirds of a turn) from its initial position.

Consequently, the three stator voltages, E_{a1}, E_{a2} and E_{a3} are respectively out of phase by 120°. They are shown as sine waves in Fig. 8-9b, and as phasors in Fig. 8-9c.

8.7 Power output of a 3-phase generator

Let us connect the three windings of the generator to three identical resistors. This arrangement requires six wires to deliver power to the individual single-phase loads (Fig. 8-10a). The resulting currents I_a, I_b and I_c are respectively in phase with voltages E_{a1}, E_{b2} and E_{c3}. Because the resistors are identical, the currents have the same effective values, but they are mutually out of phase by 120° (Fig. 8-10b). The fact that they are out of phase simply means that they reach their positive peaks at different times.

The instantaneous power supplied to each resistor is again composed of a wave which surges between zero and a maximum value P_m. However, the power peaks in the three resistors do not occur at the same time, owing to the phase angle between the voltages. If we add the instantaneous powers of all three resistors, we discover that the resulting power is constant, as in the case of a 2-phase generator. However, the output of a 3-phase generator has a magnitude of 1.5 P_m. Consequently, the mechanical power required to drive the rotor is also constant, and so a 3-phase generator does not vibrate. Furthermore, the power flow over the transmission line, connecting the generator to the load, is constant.

Example 8-2:
The 3-phase generator shown in Fig. 8-10a is connected to three 20 Ω load resistors. If the voltage induced in each phase is 120 V, calculate:
a. the power dissipated in each resistor;
b. the power dissipated in the 3-phase load;
c. the peak power P_m dissipated in each resistor;
d. the total 3-phase power compared to P_m.

Solution:
a. Each resistor behaves as a single-phase load subjected to an effective voltage of 120 V. The power dissipated in each resistor is therefore:

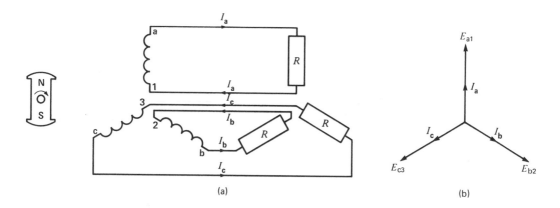

Figure 8-10
a. Three-phase, 6-wire system.

b. Corresponding phasor diagram.

$$P = E^2/R = 120^2/20$$
$$= 720 \text{ W}$$

b. The total power dissipated in the 3-phase load (all three resistors) is:
$$P_T = 3P = 3 \times 720$$
$$= 2160 \text{ W}$$

This power is constant from instant to instant.

c. The peak voltage across one resistor is
$$E_m = \sqrt{2}E = \sqrt{2} \times 120$$
$$= 169.7 \text{ V}$$

The peak current in each resistor is:
$$I_m = E_m/R = 169.7/20$$
$$= 8.485 \text{ A}$$

The peak power in each resistor is:
$$P_m = E_m I_m = 169.7 \times 8.485$$
$$= 1440 \text{ W}$$

d. The ratio of P_T to P_m is:
$$P_T/P_m = 2160/1440$$
$$= 1.5$$

Thus, whereas the power in each resistor pulsates between zero and a maximum of 1440 W, the total power for all three resistors is unvarying and equal to 2160 W.

8.8 Wye connection

The three single-phase circuits of Fig. 8-10 are electrically independent. Consequently, we can connect the three return conductors together to form a single return conductor (Fig. 8-11a). This reduces the number of trans-mission line conductors from 6 to 4. The return conductor, called *neutral* conductor (or simply, neutral), carries the sum of the three currents $(I_a + I_b + I_c)$. At first, it seems that the cross section of this conductor should be three times that of lines a, b, and c. However, the diagram of Fig. 8-11b clearly shows that the *sum of the three return currents is zero at every instant*. For example, at the instant corresponding to 240°, $I_c = +I_{max}$ and $I_b = I_a = -0.5I_{max}$, making $I_a + I_b + I_c = 0$. We arrive at the same result (and much more simply) by taking the sum of the phasors $(I_a + I_b + I_c)$ in Fig. 8-10b. The sum is clearly zero.

We can therefore remove the neutral wire altogether without in any way affecting the vol-tages or currents in the circuit (Fig. 8-12). In one stroke, we accomplish a great saving because the number of line conductors drops from six to three! However, the loads in Fig. 8-11a must be identical in order to remove

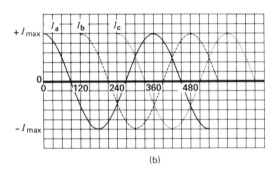

(a)

(b)

Figure 8-11
a. Three-phase, 4-wire system.

b. Line currents in a 3-phase, 4-wire system.

the neutral wire. If the loads are not identical, the absence of the neutral conductor produces unequal voltages across the three loads.

The circuit of Fig. 8-12 composed of the generator, transmission line and load, is called a *3-phase, 3-wire system*. The generator, as well as the load, are said to be connected in *wye*, because the three branches resemble the letter Y. For equally obvious reasons, some people prefer to use the term "connected in star".

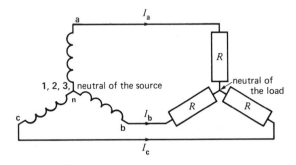

Figure 8-12
Three-phase, 3-wire system showing source and load.

The circuit of Fig. 8-11a is called a *3-phase, 4-wire system*. The neutral conductor in such a system is usually the same size or slightly smaller than the line conductors. Three-phase, 4-wire systems are widely used to supply electric power to commercial and industrial users.

The line conductors are often called *phases*, which is the same term applied to the generator windings.

8.9 Voltage relationships

Consider the wye-connected armature windings of a 3-phase generator (Fig. 8-13a). The induced voltage in each winding has an effective value E_{LN} and the corresponding phasor diagram is shown in Fig. 8-13b. Knowing that the line-to-neutral voltages are E_{an}, E_{bn} and E_{cn} the question is, what are the line-to-line voltages E_{ab}, E_{bc} and E_{ca}? We can write the following equations, based on Kirchhoff's voltage law:

$$E_{ab} = E_{an} + E_{nb} \qquad (8\text{-}1)$$
$$\quad = E_{an} - E_{bn} \qquad (8\text{-}1)$$

$$E_{bc} = E_{bn} + E_{nc} \qquad (8\text{-}2)$$
$$\quad = E_{bn} - E_{cn} \qquad (8\text{-}2)$$

$$E_{ca} = E_{cn} + E_{na} \qquad (8\text{-}3)$$
$$\quad = E_{cn} - E_{an} \qquad (8\text{-}3)$$

Referring first to Eq. 8-1, we draw phasor E_{ab} exactly as the equation indicates:

$$E_{ab} = E_{an} - E_{bn} = E_{an} + (-E_{bn})$$

The resulting phasor diagram shows that line voltage E_{ab} leads E_{an} by 30° (Fig. 8-13c). Using simple trigonometry, and based upon the fact that the length of the line-to-neutral

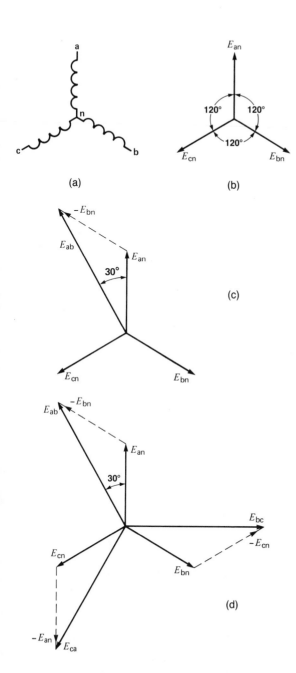

Figure 8-13
a. Wye-connected stator windings of a 3-phase generator.
b. Line-to-neutral voltages of the generator.
c. Method to determine line voltage E_{ab}.
d. Line voltages E_{ab}, E_{bc}, E_{ca} are equal and displaced at 120°.

phasors is E_{LN}, we have:

length E_L of phasor E_{ab} = 2 × E_{LN} cos 30°

$$E_L = 2 \times E_{LN} \sqrt{3}/2$$
$$= \sqrt{3} \, E_{LN}$$

The line-to-line voltage (called *line voltage*) is therefore $\sqrt{3}$ times the line-to-neutral voltage:

$$E_L = \sqrt{3} \, E_{LN} \qquad (8-4)$$

where

E_L = effective value of the line voltage [V]

E_{LN} = effective value of the line-to-neutral voltage [V]

$\sqrt{3}$ = a constant [approximate value = 1.73]

Owing to the symmetry of a 3-phase system, we conclude that the line voltage across *any* two generator terminals is equal to $\sqrt{3} \, E_{LN}$. The truth of this can be seen by referring to Fig. 8-13d, which shows phasors E_{bc} and E_{ca}. The phasors are drawn according to Eqs. 8-2 and 8-3, respectively. The line voltages are equal in magnitude and mutually displaced by 120°.

To further clarify these results, Fig. 8-14 shows the voltages between the terminals of a 3-phase generator whose line-to-neutral voltage is 100 V. The line voltages are all equal to 100 $\sqrt{3}$ or 173 V. The voltages between lines a, b, c constitute a 3-phase system, but the voltage between any two lines (a and b, b and c, b

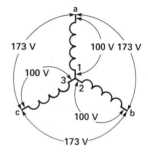

Figure 8-14
Voltages induced in a wye-connected generator.

and n, etc.) is nevertheless an ordinary single-phase voltage.

Example 8-3:

A 3-phase 60 Hz generator, connected in wye, generates a line voltage (meaning line-to-line) of 23 900 V. Calculate:

a. the line-to-neutral voltage;
b. the voltage induced in the individual windings;
c. the time interval between the positive peak voltage of phase A and the positive peak of phase B;
d. the peak value of the line voltage.

Solution:

a. The line to neutral voltage is:
$$E_{LN} = E_L/\sqrt{3} = 23\ 900/1.73$$
$$= 13\ 800\ V$$

b. The windings are connected in wye; consequently, the voltage induced in each winding is 13 800 V.

c. One complete cycle corresponds to 1/60s, or 360°. Consequently, a phase angle of 120° corresponds to an interval of
$$T = \frac{120}{360} \times \frac{1}{60} = 1/180\ s$$

$$= 5.55\ ms$$

The positive peaks are therefore separated by intervals of 5.55 ms.

d. The peak line voltage is:
$$E_m = \sqrt{2}\ E_L \qquad \text{Eq. 2-6}$$
$$= 1.414 \times 23\ 900$$
$$= 33800\ V$$

The same voltage relationships exist in a wye-connected load, such as that shown in Figs. 8-11 and 8-12. In other words, the line voltage is $\sqrt{3}$ times the line-to-neutral voltage.

Example 8-4:

The generator in Fig. 8-12 generates a line vol-tage of 865 V, and each load resistor has an impedance of 50 Ω. Calculate:

a. the voltage across each resistor;
b. the current in each resistor;
c. the total power output of the generator.

Solution:

a. The voltage across each resistor is:
$$E_{LN} = E_L/\sqrt{3} = 865/1.73 \qquad \text{Eq. 8-4}$$
$$= 500\ V$$

b. The current in each resistor is:
$$I = E_{LN}/R = 500/50$$
$$= 10\ A$$

All the line currents are therefore equal to 10 A.

c. Power absorbed by each resistor is:
$$P = E_{LN}I = 500 \times 10$$
$$= 5000\ W$$

The power delivered by the generator to all three resistors is:
$$P = 3 \times 5000 = 15\ kW$$

8.10 Delta connection

A 3-phase load is said to be *balanced* when the line voltages are equal and the line currents are equal. This corresponds to three identical impedances connected across the 3-phase line, a condition that is usually encountered in 3-phase circuits.

The three impedances may be connected in wye (as we already have seen) or in *delta* (Fig. 8-15a).

Let us determine the voltage and current relationships in such a delta connection,[*] assuming a resistive load. The resistors are connected across the line; consequently, resistor currents I_1, I_2 and I_3 are in phase with the respective line voltages E_{ab}, E_{bc} and E_{ca}. The line voltages are produced by an external generator (not shown). Furthermore, according to Kirchhoff's law, the line currents are given by:

---

[*] The connection is so named because it resembles the Greek letter Δ.

$$I_a = I_1 - I_3 \qquad (8\text{-}5)$$

$$I_b = I_2 - I_1 \qquad (8\text{-}6)$$

$$I_c = I_3 - I_2 \qquad (8\text{-}7)$$

Referring first to Eq. 8-5, we draw phasor I_a exactly as the equation indicates. The resulting phasor diagram shows that I_a leads I_1 by 30° (Fig. 8-15b). Using simple trigonometry and letting I_z be the effective resistor current, and I_L the line current, we have:

$$
\begin{aligned}
I_L &= 2 \times I_z \cos 30° \\
&= 2 \times I_z \sqrt{3}/2; \\
&= \sqrt{3} I_z
\end{aligned}
$$

The line current is therefore $\sqrt{3}$ times greater than the current in each delta-connected load:

$$I_L = \sqrt{3} I_z \qquad (8\text{-}8)$$

where
  I_L = effective value of the line current [A]
  I_z = effective value of the current in a delta-connected load [A]
  $\sqrt{3}$ = a constant [approximate value = 1.73]

The reader can easily determine the magnitude and position of phasors I_b and I_c, and observe that the three line currents are equal and displaced by 120°.

  Table 8A summarizes the basic relationships between the voltages and currents in wye-connected and delta-connected loads. The relationships are valid for any type of circuit element (resistor, capacitor, inductor, motor winding, generator winding, etc.) as long as the elements in the three phases are identical. In other words, the relationships apply to any balanced 3-phase load.

(a)

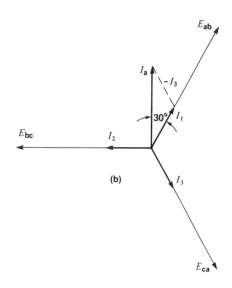

(b)

Figure 8-15
a. Impedances connected in delta.
b. Phasor relationships with a resistive load

Example 8-5:
Three identical impedances are connected in delta across a 3-phase 550 V line (Fig. 8-15c). If the line current is 10 A, calculate:
a. the current in each impedance;
b. the value of each impedance [Ω].

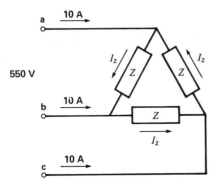

Figure 8-15c
See Example 8-5

Solution:
a. The current in each impedance is:
$$I_z = 10/\sqrt{3} = 5.78 \text{ A}$$

b. The voltage across each impedance is 550 V. Consequently,
$$Z = E/I_z = 550/5.78$$
$$= 95 \ \Omega.$$

8.11 Power transmitted by a 3-phase line

The apparent power supplied by a single-phase line is equal to the product of the line voltage E times the line current I. The question now arises: What is the apparent power supplied by a 3-phase line having a line voltage E and a line current I?

If we refer to the wye-connected load of Fig. 8-16a, the apparent power supplied to each impedance is:
$$S_z = \frac{E}{\sqrt{3}} \times I$$

The apparent power supplied to all three impedances is obviously three times as great.* Consequently, the total apparent power is:
$$S = \frac{E}{\sqrt{3}} \times I \times 3 = \sqrt{3} \ EI$$

In the case of a delta-connected load (Fig. 8-16b) the apparent power supplied to each impedance is:
$$S_z = E \times \frac{I}{\sqrt{3}}$$

which is the same as for a wye-connected load. Consequently, the total apparent power is also the same.

We therefore have:
$$S = \sqrt{3} \ EI \qquad (8\text{-}9)$$

where
$S =$ total apparent power delivered by a 3-phase line [VA]
$E =$ effective line voltage [V]
$I =$ effective line current [A]
$\sqrt{3} =$ a constant [approximative value = 1.73]

8.12 Active, reactive and apparent power in 3-phase circuits

The relationship between active power P, reactive power Q, and apparent power S is the same for balanced 3-phase circuits as for single-phase circuits. We therefore have:
$$S = \sqrt{P^2 + Q^2} \qquad (8\text{-}10)$$
and
$$\cos \theta = P/S \qquad (8\text{-}11)$$

___
*   In 3-phase balanced circuits, we can add the apparent powers of the three phases because they have identical power factors. If the power factors are not identical, the apparent powers cannot be added.

TABLE 8A   VOLTAGE AND CURRENT RELATIONSHIPS IN 3-PHASE CIRCUITS

Wye connection	**Delta connection**
Figure 8-16a Impedances connected in wye.	Figure 8-16b Impedances connected in delta.

- The current in each element is equal to the line current I,

- The voltage across each element is equal to the line voltage E divided by $\sqrt{3}$

- The voltages across the elements are 120° out of phase,

- The currents in the elements are 120° out of phase.

- The current in each element is equal to the line current I divided by $\sqrt{3}$

- The voltage across each element is equal to the line voltage E,

- The voltages across the elements are 120° out of phase,

- The currents in the elements are 120° out of phase.

where
$\quad S$ = total 3-phase apparent power [VA]
$\quad P$ = total 3-phase active power [W]
$\quad Q$ = total 3-phase reactive power [var]
$\cos\theta$ = power factor of the 3-phase load
$\quad\theta$ = phase angle between the line current and the line-to-neutral voltage [°]

Example 8-6:
A 3-phase motor, connected to a 440 V line, draws a line current of 5 A. If the power factor of the motor is 80 percent, calculate:

a. the apparent power;
b. the active power,
c. the reactive power absorbed by the machine.

Solution:
a. The total apparent power is:
$$S = \sqrt{3}EI = 1.73 \times 440 \times 5$$
$$= 3810 \text{ VA}$$
$$= 3.81 \text{ kVA}$$

b. The total active power is:
$$P = S \cos\theta = 3.81 \times 0.80$$
$$= 3.05 \text{ kW}$$

c. The total reactive power is:
$$Q = \sqrt{S^2 - P^2} = \sqrt{3.81^2 - 3.05^2}$$
$$= 2.28 \text{ kvar}$$

8.13 Solving 3-phase circuits

A *balanced* 3-phase load may be considered to be composed of three identical single-phase loads. Consequently, the easiest way to solve such a circuit is to consider only one phase. The following examples illustrate the method to be employed.

Example 8-7:
Three identical resistors are connected in wye across a 3-phase 550 V line, dissipating a total power of 3000 W (Fig. 8-17). Calculate:
a. the current in each line;
b. the value of each resistance.

Figure 8-17
See Example 8-7.

Solution:
a. The power dissipated by each resistor is:
   $$P = 3000 \text{ W}/3 = 1000 \text{ W}$$

   The voltage across the terminals of each resistor is:
   $$E = 550 \text{ V}/\sqrt{3} = 318 \text{ V}$$

   The current in each resistor is:
   $$I = P/E = 1000 \text{ W}/318 \text{ V} = 3.15 \text{ A}$$

   The current in each line is also 3.15 A.

b. The resistance of each element is:
   $$R = E/I = 318/3.15 = 101 \ \Omega$$

Example 8-8:
In the circuit of Fig. 8-18, calculate:
a. the current in each line;
b. the voltage across the inductor terminals.

Figure 8-18
See Example 8-8.

Solution:
a. Each phase is composed of an inductive reactance $X_L = 4 \ \Omega$ in series with a resistance $R = 3 \ \Omega$. Consequently, the impedance of each phase is:
   $$Z = \sqrt{4^2 + 3^2} = 5 \ \Omega \qquad \text{Eq. 2-12}$$

   The voltage across each phase is:
   $$E_{LN} = E_L/\sqrt{3} = 440 \text{ V}/\sqrt{3} = 254 \text{ V}$$

   The current in each circuit element is:
   $$I = E_{LN}/Z = 254/5 = 50.8 \text{ A}$$

   (This is also the line current.)

b. The voltage across each inductor is:
   $$E = IX_L = 50.8 \times 4$$
   $$= 203.2 \text{ V}$$

Example 8-9:
A 3-phase 550 V, 60 Hz line is connected to three identical capacitors connected in delta (Fig. 8-19). If the line current is 22 A, calculate the capacitance of each capacitor.

Solution:
The current in each capacitor is:
$$I = I_L/\sqrt{3} = 22 \text{ A}/\sqrt{3} = 12.7 \text{ A}$$

Voltage across each capacitor = 550 V

Capacitive reactance X_c of each capacitor is:
$$X_c = E_L/I = 550/12.7 = 43.3 \ \Omega$$

Figure 8-19
See Example 8-9.

The capacitance of each capacitor is:

$C = 1/2\pi f X_c$
$\quad = 1/(2\pi \times 60 \times 43.3)$ Eq. 2-11
$\quad = 61.3 \ \mu F$

8.14 Industrial loads

In many cases, we do not know whether a 3-phase load is connected in delta or in wye. For example, 3-phase motors, generators, transformers, capacitors, etc., often have only three external terminals, and there is no way to tell how the internal connections are made. Under these circumstances, we simply *assume* that the connection is in wye. (A wye connection is slightly easier to handle than a delta connection.)

In a wye connection, the impedance per phase is understood to be the line-to-neutral impedance. The voltage per phase is simply the line voltage divided by $\sqrt{3}$. Finally, the current per phase is equal to the line current.

The assumption of a wye connection can be made not only for individual loads, but for entire load centers such as a factory containing motors, lamps, heaters, furnaces, and so forth. We simply assume that the load center is connected in wye, and proceed with the usual calculations.

Example 8-10:
A manufacturing plant draws 415 kVA from a 2400 V (line-to-line) 3-phase line (Fig. 8-20a). If the plant power factor is 87.5 percent lagging, calculate:

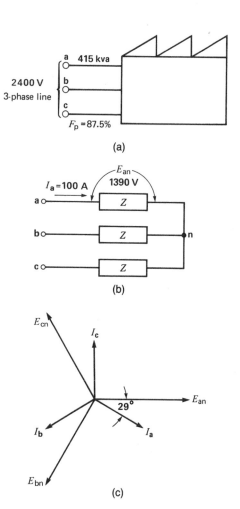

Figure 8-20
a. Power input to a factory. See Example 8-10.
b. Equivalent wye connection of the factory load.
c. Phasor diagram of the voltages and currents.

a. the impedance of the plant, per phase;
b. the phase angle between the line-to-neutral voltage and the line current;
c. the complete phasor diagram for the plant.

Solution:
a. We assume a wye connection composed of three identical impedances Z (Fig. 8-20b).

The voltage per phase is:
$$E = 2400/\sqrt{3}$$
$$= 1390 \text{ V}$$

The current per phase is:
$$I = S/(E\sqrt{3})$$
$$= 415\ 000/(2400\sqrt{3})$$
$$= 100 \text{ A} \qquad \text{Eq. 8-9}$$

The impedance per phase is:
$$Z = E/I = 1390/100$$
$$= 13.9 \ \Omega$$

b. The phase angle θ between the line-to-neutral voltage (1390 V) and the corresponding line current (100 A) is given by:
$$\cos \theta = 0.875 \qquad \text{Eq. 8-11}$$
$$\theta = 29°$$

The current in each phase lags 29° behind the phase voltage.

c. The complete phasor diagram is shown in Fig. 8-20c. In practice, we would show only one phase; for example, E_{an}, I_a and the phase angle between them.

Example 8-11:
A 5000 hp wye-connected motor is connected to a 4000 V (line-to-line), 3-phase, 60 Hz line (Fig. 8-21). A delta-connected capacitor bank rated at 1800 kvar is also connected to the line. If the motor produces an output of 3594 hp at

an efficiency of 93 percent and a power factor of 90 percent (lagging), calculate:
a. the active power absorbed by the motor;
b. the reactive power absorbed by the motor;
c. the reactive power supplied by the transmission line;
d. the apparent power supplied by the transmission line;
e. the transmission line current;
f. the motor line current.
g. Draw the complete phasor diagram for one phase.

Solution:
a. Power output of 3594 hp is equivalent to:
$$P_2 = 3594 \times 0.746 = 2681 \text{ kW}$$

Active power input to motor:
$$P_m = P_2/\eta = 2681/0.93 \qquad \text{Eq. 3-6}$$
$$= 2883 \text{ kW}$$

b. Apparent power absorbed by the motor:
$$S_m = P_m/\cos \theta = 2883/0.90$$
$$= 3203 \text{ kVA}$$

Reactive power absorbed by the motor:
$$Q_m = \sqrt{S_m^2 - P_m^2} = \sqrt{3203^2 - 2883^2}$$
$$= 1395 \text{ kvar}$$

c. Reactive power supplied by the capacitor bank:
$$Q_c = 1800 \text{ kvar}$$

Reactive power *returned* to the transmission line:
$$Q_L = Q_c - Q_m = 1800 - 1395$$
$$= 405 \text{ kvar}$$

This is an unusual situation. In most cases, the capacitor bank furnishes no more than Q_m kilovars of reactive power.

d. Active power supplied by the line is:
$$P_L = P_m = 2883 \text{ kW}$$

Apparent power supplied by the line is:
$$S_L = \sqrt{P_L^2 + Q_L^2} = \sqrt{2883^2 + 405^2}$$
$$= 2911 \text{ kVA}$$

Figure 8-21
Industrial motor and capacitor. See Example 8-11.

e. Transmission line current is:

$$I_L = S_L/(E_L\sqrt{3}) \qquad \text{Eq. 8-9}$$
$$= 2\,911\,000/(\sqrt{3} \times 4000)$$
$$= 420 \text{ A}$$

f. Motor line current is:

$$I_m = S_m/(E_L\sqrt{3})$$
$$= 3\,203\,000/(\sqrt{3} \times 4000)$$
$$= 462 \text{ A}$$

g. The line-to-neutral voltage is:

$$E_{LN} = 4000/\sqrt{3} = 2312 \text{ V}$$

Phase angle θ between the motor current and the line-to-neutral voltage is:

$$\cos\theta = 0.9$$
$$\theta = 25.8°$$

(The motor current lags 25.8° behind the voltage as shown in Fig. 8-22a.)

g. Line current drawn by the capacitor bank is:

$$I_c = Q_c/(E_L\sqrt{3})$$
$$= 1\,800\,000/(\sqrt{3} \times 4000)$$
$$= 260 \text{ A}$$

Where should phasor current I_c be located on the phasor diagram? The question is important because the capacitors are connected in delta, and we assumed a wye connection for the motor. This can create phase-angle problems if we try to follow the actual currents in the capacitor bank. The solution is to recognize that *if* the capacitors were connected in wye (while generating the same reactive power) the line current of 260 A would lead E_{LN} by 90°. Consequently, we draw I_c 90° ahead of E_{LN}. This is the correct position for phasor I_c no matter how the capacitors are connected.

Phase angle θ_L between the transmission line current and E_{LN} is:

$$\cos\theta = P_L/S_L = 2883/2911$$
$$= 0.99$$
$$\theta = 8°$$

(a)

(b)

Figure 8-22
a. Phasor relationships for one phase. See Example 8-11.
b. Line currents. Note that the motor currents exceed the currents of the source.

The line current (420 A) leads E_{LN} by 8° because the kvars supplied by the capacitor bank exceed the kvars absorbed by the motor.

The phasor diagram for one phase is shown in Fig. 8-22a.

The circuit diagram and current flows are shown in Fig. 8-22b.

We want to emphasize the importance of assuming a wye connection, irrespective of what the actual connection may be. By assuming a wye connection for all circuit elements, we simplify the calculations and eliminate confusion.

As a final remark, the reader has no doubt

noticed that the solution of a 3-phase problem involves active, reactive, and apparent power. Impedances of such devices as resistors, motors, and capacitors do not often appear. This is to be expected because most industrial loads involve electric motors, furnaces, lights, etc. which are seldom described in terms of resistance and reactance. They are usually represented as devices that draw a given amount of power at a given power factor.

The situation is somewhat different in the case of 3-phase transmission lines. Here, we can define resistances and reactances because the parameters are fixed. The same remarks apply to equivalent circuits describing the behavior of individual machines such as induction motors and synchronous machines.

In conclusion, the solution of 3-phase circuits may involve either active and reactive power or R, L, C elements - and sometimes both.

8.15 Phase sequence

In addition to line voltage and frequency, a three-phase system has an important property called *phase sequence*. Phase sequence is important because it determines the direction of rotation of 3-phase motors and whether one 3-phase system can be connected in parallel with another. Consequently, in 3-phase systems, phase sequence is as important as the frequency and voltage are.

Phase sequence means the order in which the three line voltages become successively positive. We can get a quick intuitive understanding of phase sequence by considering the following analogy. Suppose the letters **a, b, c** are printed at 120° intervals on a slowly revolving disc (Fig. 8-23).

If the disc turns counterclockwise, the letters appear in the sequence **a - b - c - a - b - c**. Let us call this the positive sequence. It can be described by any one of the three forms: **abc**, **bca** or **cab**.

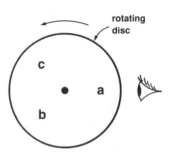

Figure 8-23
The letters are observed in the sequence a-b-c.

If the same disc is made to turn clockwise, the sequence becomes **a - c - b - a - c - b** . . . (Fig. 8-24). We call this the negative sequence and it can be described by any one of the three forms: **acb, cba** or **bac**. Clearly, there is a difference between a positive sequence and a negative sequence.

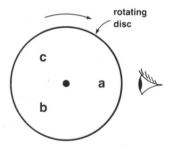

Figure 8-24
The letters are observed in the sequence a-c-b.

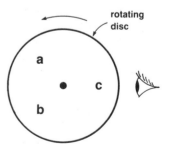

Figure 8-25
The letters are observed in the sequence a-c-b.

Suppose we interchange any two letters on the disc in Figure 8-23, while retaining the same counterclockwise rotation. If the letters **a** and **c** are interchanged, the result is shown in Fig. 8-25. The sequence now becomes **c - b - a - c - b - a** . . . which is the same as the negative sequence generated by the disc in Fig. 8-24.

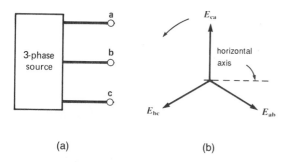

(a)   (b)

Figure 8-26
a. Determining the phase sequence of a 3-phase source.
b. Phase sequence depends upon the order in which the line voltages reach their positive peaks.

We conclude that for a given direction of rotation a positive sequence can be converted into a negative sequence by simply interchanging two letters. Similarly, a negative sequence can be converted into a positive sequence by interchanging any two letters.

Let us now consider a 3-phase source having terminals **a, b, c** (Fig. 8-26a). Suppose the line voltages E_{ab}, E_{bc}, E_{ca} are correctly represented by the revolving vectors shown in Fig. 8-26b. As they sweep past the horizontal axis in the conventional counterclockwise direction, they follow the sequence E_{ab} - E_{bc} - E_{ca} - E_{ab} - E_{bc}. . .

If we direct our attention to the first letter in each subscript, we find that the sequence is **a - b - c - a - b - c** . . . Therefore, the source shown in Fig. 8-26a is said to possess the sequence **a - b - c**. We can therefore state the following

rule: when using the double subscript notation, the sequence of the first subscripts corresponds to the phase sequence of the source.

Example 8-12:
In Fig. 8-17, the phase sequence of the source is known to be A - C - B. Draw the phasor diagram of the line voltages.

Solution:
The voltages follow the sequence A - C - B, which is the same as the sequence AC - CB - BA - AC . . . Consequently, the line voltage sequence is E_{AC} - E_{CB} - E_{BA} and the corresponding phasor diagram is shown in Fig. 8-27.

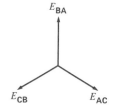

Figure 8-27
See Example 8-12.

We can reverse the phase sequence of a 3-phase line by interchanging any two conductors. Although this may appear to be a trivial change, it can become a major problem when large bus bars or high-voltage transmission lines have to be interchanged. In practice, measures are taken so that such drastic mechanical changes do not have to be made at the last minute. The phase sequence of all major distribution systems is known in advance, and any future connections are planned accordingly.

8.16 Determining the phase sequence

Special instruments are available to indicate the phase sequence, but we can also determine it by using two incandescent lamps and a capacitor. The devices are connected in wye. If we connect the circuit to a 3-phase line (without connecting the neutral), one lamp will always burn brighter than the other. The phase sequence is in the order: *bright lamp - dim lamp - capacitor.*

Suppose, for example, that a capacitor/lamp circuit is connected to a 3-phase line as shown in Fig. 8-28a. If the lamp connected to phase C burns more brightly, the phase sequence is C-B-A. The line voltages follow each other in the sequence CB-BA-AC, which is to say in the sequence E_{CB}, E_{BA}, E_{AC}. The corresponding phasor diagram is shown in Fig. 8-28b.

Figure 8-28
a. Determining phase sequence using two lamps and a capacitor.
b. Resulting phasor diagram.

8.17 Power measurement in ac circuits

Wattmeters are used to measure active power in single-phase and three-phase circuits.

Owing to its external connections and the way it is built, a wattmeter may be considered to be a voltmeter and ammeter combined in the same box. Consequently, it has 2 potential terminals and 2 current terminals. One of the potential terminals and one of the current terminals bears a ± sign.

The maximum voltage and current the instrument can tolerate are shown on the nameplate.

In single-phase circuits, the pointer moves upscale when the connections between source and load are made as shown in Fig. 8-29. Note that the ± current terminal is connected to the ± potential terminal. When the wattmeter is connected this way, an upscale reading means that power is flowing from terminals 1, 2 to terminals 3, 4.

Figure 8-29
Method of connecting a single-phase wattmeter.

8.18 Power measurement in 3-phase, 3-wire circuits

The active power supplied to a 3-phase load may be measured by two single-phase wattmeters connected as shown in Fig. 8-30. The total power is equal to the sum of the two wattmeter readings. If the power factor of the load is less than 100 percent, the instruments will give dif-

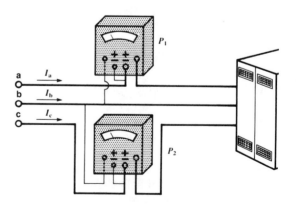

Figure 8-30
Measuring power in a 3-phase, 3-wire circuit using the two-wattmeter method.

ferent readings. Indeed, if the power factor is less than 50 percent, one of the wattmeters will give a negative reading. We must then reverse the connections of the potential coil, so as to obtain a reading of this negative quantity. In this case, the power of the three-phase circuit is equal to the *difference* between the two watt-meter readings.

The two-wattmeter method gives the active power absorbed whether the load is balanced or unbalanced.

Example 8-13:
A test on a three-phase motor yields the following results: $P_1 = +5950$ W, $P_2 = 2380$ W, the current in each of the three lines is 10 A and the line voltage is 600 V. Calculate the power factor of the motor.

Solution:
Apparent power supplied to the motor is:
$$S = \sqrt{3}E\,I = 1.73 \times 600 \times 10$$
$$= 10\ 380 \text{ VA}$$

Active power supplied to the motor is:
$$P = 5950 + 2380$$
$$= 8330 \text{ W}$$
$$\cos \theta = P/S = 8330/10\ 380$$
$$= 0.80 \text{ or } 80 \text{ percent.}$$

8.19 Power measurement in 3-phase, 4-wire circuits

In 3-phase, 4-wire circuits, three single-phase wattmeters are needed to measure the total power. The connections are made as shown in Fig. 8-31. Note that the ± current terminal is again connected to the ± potential terminal. When the wattmeters are connected this way, an upscale reading means that active power is flowing from source A, B, C, N to the load.

The total power supplied to the load is equal to the sum of the three wattmeter readings. The three-wattmeter method gives the active power for both balanced and unbalanced loads.

Some wattmeters, such as those used on switchboards, are specially designed to give a direct readout of the 3-phase power.

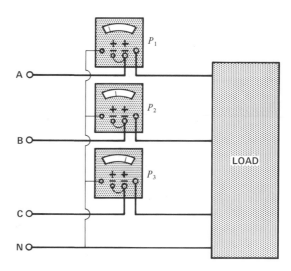

Figure 8-31
Measuring power in a 3-phase, 4-wire circuit.

Figure 8-32 shows a megawatt-range wattmeter circuit that measures the power in a generating station. The current transformers (CT) and potential transformers (PT) step down the line currents and voltages to values compatible with the instrument rating.

8.20 Varmeter

A varmeter indicates the reactive power in a circuit. It is built the same way as a wattmeter is, but an internal circuit shifts the line voltage by 90° before it is applied to the potential coil. Varmeters are mainly employed in the control rooms of generating stations and the substations of electrical utilities and large industrial consumers.

In 3-phase, 3-wire *balanced* circuits, we can calculate the reactive power from the two watt-meter readings (Fig. 8-30). We simply multiply the difference of the two readings by $\sqrt{3}$. For example, if the two wattmeters indicate +5950 W and +2380 W respectively, the reactive power is $(5950 - 2380) \times \sqrt{3} = 6176$ vars. Note that this method of var measurement is only valid for *balanced* three-phase circuits.

Figure 8-32
Measuring active power in a high-power circuit.

QUESTIONS AND PROBLEMS

Practical level

8-1 A 3-phase wye-connected generator generates 2400 V in each of its windings. Calculate the line voltage.

8-2 The generator in Fig. 8-9 generates a peak voltage of 100 V per phase.
   a. Calculate the instantaneous voltage between terminals **1, a** at 0°, 90°, 120°, 240° and 330°;
   b. What is the polarity of terminal **a** with respect to terminal **1** at each of these instants?
   c. What is the instantaneous value of the voltage across terminals **2, b** at each of these same instants?

8-3 Referring to Fig. 8-9c, voltage E_{b2} is 120° behind voltage E_{a1}. Could we also say that E_{b2} is 240° ahead of E_{a1}?

8-4 The voltage between lines a - b - c of Fig. 8-12 is 620 V.
   a. What is the voltage across each resistor?
   b. If $R = 15\,\Omega$, what is the current in each line?
   c. Calculate the power supplied to the 3-phase load.

8-5 Three resistors are connected in delta. If the line voltage is 13.2 kV and the line current is 1202 A, calculate:
   a. the current in each resistor;
   b. the voltage across each resistor;
   c. the power supplied to each resistor;
   d. the power supplied to the 3-phase load;
   e. the ohmic value of each resistor.

8-6 a. What is the phase sequence in Fig. 8-10?

b. Could we reverse it by changing the direction of rotation of the magnet?

8-7 A 3-phase motor connected to a 600 V line draws a line current of 25 A. Calculate the apparent power supplied to the motor.

8-8 Three incandescent lamps rated 60 W, 120 V are connected in delta. What line voltage is needed so that the lamps burn normally?

8-9 Three 10 Ω resistors are connected in delta on a 208 V 3-phase line.

a. What is the power supplied to the 3-phase load?

b. If the fuse in one line burns out, calculate the new power supplied to the load.

8-10 If one line conductor of a 3-phase line is cut, is the load then supplied by a single-phase or a 2-phase voltage?

8-11 A 3-phase heater dissipates 15 kW when connected to a 208 V, 3-phase line.

a. What is the line current if the resistors are connected in wye?

b. What is the line current if the resistors are connected in delta?

c. if the resistors are known to be connected in wye, calculate the resistance of each.

8-12 We wish to apply full load to a 100 kVA, 4 kV, 3-phase generator using a resistive load. Calculate the value of each resistance if the elements are connected;
a. in wye,   b. in delta.

8-13 The windings of a 3-phase motor are connected in delta. If the resistance

between two terminals is 0.6 Ω, what is the resistance of each winding?

8-14 Three 24 Ω resistors are connected in delta across a 600 V, 3-phase line. Calculate the resistance of three elements connected in wye which would dissipate the same power.

8-15 A 60 hp 3-phase motor absorbs 50 kW from a 600 V, 3-phase line. If the line current is 60 A, calculate:
a. the efficiency of the motor;

b. the apparent power absorbed by the motor;

c. the reactive power absorbed by the motor;

d. the power factor of the motor.

8-16 Three 15 Ω resistors and three 8 Ω reactors are connected as shown in Fig. 8-18. If the line voltage is 530 V, calculate:

a. the active, reactive and apparent power supplied to the 3-phase load;
b. the voltage across each resistor.

8-17 Two 60 W lamps and a 10 μF capacitor are connected in wye. The circuit is connected to the terminals X-Y-Z of a 3-phase 120 V outlet. The capacitor is connected to terminal Y, and the lamp which burns brighter is connected to terminal X.

a. What is the phase sequence?

b. Draw the phasor diagram for the line voltages.

Advanced level

8-18 Three 10 μF capacitors are connected in wye across a 2300 V, 60 Hz line. Calculate:

a. the line current;

b. the reactive power generated.

8-19 In Problem 8-17, if the capacitor is connected to terminal X, which lamp will be brighter?

8-20 Three delta-connected resistors absorb 60 kW when connected to a 3-phase line. If they are reconnected in wye, calculate the new power absorbed.

8-21 Three 15 Ω resistors (R) and three 8 Ω reactors (X) are connected in different ways across a 530 V, 3-phase line. Without drawing a phasor diagram, calculate the line current for each of the following connections:

   a. R and X in series, connected in wye;
   b. R and X in parallel, connected in delta;
   c. R connected in delta, X connected in wye.

8-22 In Fig. 8-19, calculate the line current if the frequency is 50 Hz instead of 60 Hz.

8-23 In Problem 8-15, assume that the motor is connected in wye, and that each branch can be represented by a resistance R in series with an inductive reactance X.

   a. Calculate the value of R and X.

   b. What is the phase angle between the line current and the corresponding line-to-neutral voltage?

8-24 An industrial plant draws 600 kVA from a 2.4 kV line at a power factor of 80 percent lagging.

   a. What is the equivalent line-to-neutral impedance of the plant?

   b. Assuming that the plant can be represented by an equivalent circuit similar to Fig. 8-18, determine the values of the resistance and reactance.

8-25 Two wattmeters connected into a 3-phase, 3-wire 220 V line indicate 3.5 kW and 1.5 kW, respectively. If the line current is 16 A, calculate:
   a. the apparent power;
   b. the power factor of the load.

8-26 An electric motor having a cos θ of 82 percent draws a current of 25 A from a 600 V 3-phase line.
   a. Calculate the active power supplied to the motor;
   b. If the motor has an efficiency of 85 percent, calculate the mechanical power output;
   c. How much energy does the motor consume in 3 h?

8-27 The wattmeters in Fig. 8-30 register +35 kW and −20 kW, respectively. If the load is balanced, calculate:
   a. the load power factor;
   b. the line current if the line voltage is 630 V.

9

THE IDEAL
TRANSFORMER

The transformer is probably one of the most useful electrical devices ever invented. It can raise or lower the voltage or current in an ac circuit, it can isolate circuits from each other, and it can increase or decrease the apparent value of a capacitor, an inductor, or a resistor. Furthermore, the transformer enables us to transmit electrical energy over great distances and to distribute it safely in factories and homes.

We will study some of the basic properties of transformers in this chapter. It will help us understand not only the commercial transformers covered in later chapters but, also the basic operating principle of induction motors, alternators, and synchronous motors. All these devices are based upon the laws of electromagnetic induction. Consequently, we encourage the reader to pay particular attention to the subject matter covered here.

9.1 Voltage induced in a coil

Consider the coil of Fig. 9-1a which surrounds (or "links") a variable flux Φ. The flux alternates

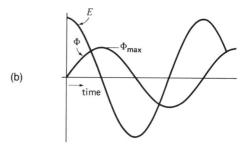

Figure 9-1
a. A voltage is induced in a coil when it links a variable flux.
b. A sinusoidal flux induces a sinusoidal voltage.

sinusoidally at a frequency f, periodically reaching positive and negative peaks Φ_{max}.

The alternating flux induces a sinusoidal ac voltage in the coil, whose value is given by:

$$E = 4.44\, fN\Phi_{max} \qquad (9\text{-}1)$$

where

$E =$ effective voltage induced [V]
$f =$ frequency of the flux [Hz]
$N =$ number of turns on the coil
$\Phi_{max} =$ peak value of the flux [Wb]
$4.44 =$ a constant [exact value $= 2\pi/\sqrt{2}$]

It does not matter where the ac flux comes from: it may be created by a moving magnet, a nearby ac coil, or even by an ac current which flows in the coil itself.

Equation 9-1 is deduced from Faraday's law $E = N\,\Delta\phi/\Delta t$ in which $\Delta\phi/\Delta t$ is the rate of change of flux. Thus, in Fig. 9-1b, when $\Delta\phi/\Delta t$ is positive the flux is increasing with time, and the voltage is positive. Then, when $\Delta\phi/\Delta t$ is negative, the induced voltage E is negative. Finally, when $\Delta\phi/\Delta t$ is momentarily zero, the voltage is also zero.

Example 9-1:
The coil in Fig. 9-1 possesses 4000 turns and links an ac flux having a peak value of 2 mWb. If the frequency is 60 Hz, calculate the value and frequency of the induced voltage E.

Solution:
$$\begin{aligned}
E &= 4.44\, fN\Phi_{max} \qquad \text{Eq. 9-1}\\
&= 4.44 \times 60 \times 4000 \times 0.002\\
&= 2131\ \text{V}
\end{aligned}$$

The induced voltage has an effective value of 2131 V and a frequency of 60 Hz. The peak voltage is $2131\sqrt{2} = 3013$ V.

9.2 Applied voltage and induced voltage

Figure 9-2a shows a coil of N turns connected to a sinusoidal ac source E_g. The coil has a reactance X_m and draws a current I_m. If the resistance of the coil is negligible, the effective current is given by:

$$I_m = E_g/X_m$$

As in any inductive circuit, I_m lags 90° behind E_g and Φ is in phase with the current (Fig. 9-2b).

The detailed behavior of the circuit can be explained as follows:

The sinusoidal current I_m produces a sinusoidal mmf NI_m which, in turn, creates a sinusoidal flux Φ. The peak value of this ac flux is Φ_{max}. The flux induces a voltage E across the terminals of the coil, whose value is given by Eq. 9-1. On the other hand, the applied voltage E_g and the induced voltage E *must* be identical because they appear between the same pair of conductors. Because $E_g = E$, we can write:

$$E_g = 4.44\, fN\Phi_{max}$$

from which

$$\Phi_{max} = \frac{E_g}{4.44fN} \qquad (9\text{-}2)$$

This equation shows that for a given frequency and a given number of turns, Φ_{max} varies in proportion to the applied voltage E_g. This means that if E_g is kept constant, the peak flux *must* remain constant.

(a)

(b)

Figure 9-2
a. The voltage E induced in a coil is equal to the applied voltage E_g.
b. Phasor relationships between E_g, E, I_m and Φ.

For example, suppose we gradually insert an iron core into the coil while keeping E_g fixed (Fig. 9-3). The peak value of the ac flux will remain absolutely constant during this operation, retaining its original value Φ_{max} even when the core is completely inside the coil. In effect, if the flux increased (as we would expect), the induced voltage E would also increase. But this is impossible because $E = E_g$ at every instant and as we said, E_g is fixed.

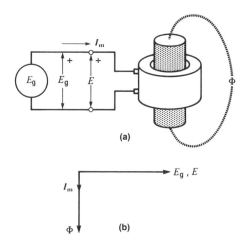

Figure 9-3
a. The flux in the coil remains constant so long as E_g is constant.
b. Phasor relationships.

For a given supply voltage E_g, the ac flux in Figs. 9-2 and 9-3 is therefore the same. However, the so-called *magnetizing current* I_m is much smaller when the iron core is inside the coil. In effect, to produce the same flux, a smaller magnetomotive force is needed with an iron core than with an air core. Consequently, the magnetizing current in Fig. 9-3 is much smaller than in Fig. 9-2.

Example 9-2:
A coil having 90 turns is connected to a 120 V, 60 Hz source. If the effective value of the magnetizing current is 4 A, calculate:
a. the peak value of flux;
b. the peak value of the mmf;
c. the inductive reactance of the coil;
d. the inductance of the coil.

Solution:
a. $\Phi_{max} = E_g/(4.44fN)$ Eq. 9-2

$\qquad\qquad = 120/(4.44 \times 60 \times 90)$

$\qquad\qquad = 0.005 = 5\ mWb$

b. The peak current is:
$$I_{m(peak)} = \sqrt{2}\,I = 1.41 \times 4$$
$$= 5.64\ A$$

The peak mmf U is:
$$U = NI_m = 90 \times 5.64$$
$$= 507.6\ A$$

The flux is equal to 5 mWb when the coil mmf is 507.6 ampere-turns.

c. The inductive reactance is:
$$X_m = E_g/I_m = 120/4$$
$$= 30\ \Omega$$

d. The inductance is:
$$L = X_m / 2\pi f \qquad\text{Eq. 2-10}$$
$$= 30/(2\pi \times 60)$$
$$= 0.0796$$
$$= 79.6\ mH$$

9.3 Elementary transformer

In Fig. 9-4, a coil having an air core is excited by an ac source E_g. The resulting current I_m produces a total flux Φ which is dispersed in the space around the coil. If we bring a second coil close to the first, it will surround a portion Φ_{m1} of the total flux. An ac voltage E_2 is therefore induced in the second coil and its value can actually be measured with a voltmeter. The combination of the two coils is called a *transformer*. The coil connected to the source is called the *primary winding* (or "primary") and the other one is called the *secondary winding* (or "secondary").

A voltage exists only between primary terminals 1-2 and secondary terminals 3-4. No voltage exists between the primary and secondary terminals. The secondary is therefore electrically isolated from the primary.

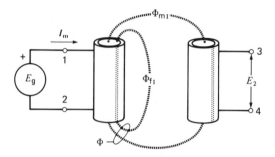

Figure 9-4
Voltage induced in a secondary winding. Mutual flux is Φ_{m1}; leakage flux is Φ_{f1}.

The flux Φ created by the primary can be broken up into two parts: 1) a *mutual flux* Φ_{m1}, which links the turns of both coils, and 2) a *leakage flux* Φ_{f1} which links only the turns of the primary. If the coils are far apart, the mutual flux is very small compared to the total flux Φ; we then say that the *coupling* between the two coils is weak. We can obtain a better coupling (and a higher secondary voltage E_2) by bringing the two coils closer together. However, even if we bring the secondary right up to the primary so that the two coils touch, the mutual flux will still be small compared to the total flux Φ. When the coupling is weak, voltage E_2 is relatively small and, worse still, it collapses almost completely when a load is connected across the secondary terminals. In most industrial transformers, the primary and secondary windings are wound on top of each other to improve the coupling between them.

9.4 Polarity of a transformer

In Fig. 9-4 , fluxes Φ_{f1} and Φ_{m1} are both produced by magnetizing current I_m. Conse-

quently, the fluxes are in phase, both reaching their peak values at the same instant. They also pass through zero at the same instant. It follows that voltage E_2 will reach *its* peak value at the same instant as E_g does. Suppose, during one

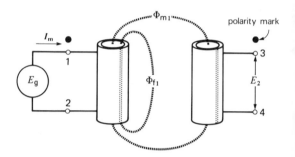

Figure 9-5
Terminals having the same instantaneous polarity are marked with a dot.

of these peak moments, that primary terminal 1 is positive with respect to primary terminal 2, and that secondary terminal 3 is positive with respect to secondary terminal 4 (Fig. 9-5). Terminals 1 and 3 are then said to possess the same *polarity*. This "sameness" can be shown by placing a large dot beside primary terminal 1 and another large dot beside secondary terminal 3. The dots are called polarity marks.

The polarity marks in Fig. 9-5 could equally well be placed beside terminals 2 and 4 because, as the voltage alternates, they, too, become simultaneously positive. Consequently, the polarity marks may be shown beside terminals 1 and 3 *or* beside terminals 2 and 4.

9.5 Properties of polarity marks

A transformer is usually enclosed in a metal container and so only the primary and secondary terminals are accessible. But although the transformer may not be visible, the following rules always apply to polarity marks:

1. A current entering a polarity-marked terminal produces a mmf that acts in a "positive" direction. As a result, it produces a flux in the "positive" direction* (Fig. 9-6). Conversely, a current flowing out of a polarity-marked terminal produces a mmf and flux in the "negative" direction. Thus, currents that flow into and out of polarity-marked terminals produce magnetomotive forces that buck each other.

is infinitely permeable. Furthermore, any flux produced by the primary is completely linked by the secondary, and vice versa. Consequently, an ideal transformer has no leakage flux of any kind. Furthermore, it cannot store magnetic energy and it cannot dissipate energy.

Practical transformers have properties which approach those of an ideal transformer. Consequently, our study of the ideal transformer will help us understand the proper-

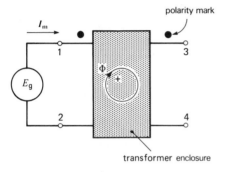

Figure 9-6
A current entering a polarity-marked terminal produces a flux in a "positive" direction.

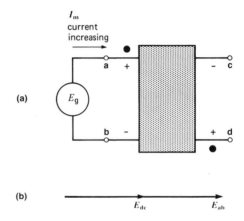

Figure 9-7
a. Instantaneous polarities when the magnetizing current is increasing.
b. Phasor relationship.

2. If one polarity-marked terminal is momentarily positive, then the other polarity-marked terminal is momentarily positive (with respect to its other terminal). This rule enables us to relate the phasor voltage on the primary side with the phasor voltage on the secondary side. For example, in Fig. 9-7, phasor E_{dc} is in phase with phasor E_{ab}.

9.6 Ideal transformer at no-load; voltage ratio

Before undertaking the study of practical, commercial transformers, we shall examine the properties of the so-called *ideal transformer*. An ideal transformer has no losses and its core

---

* "Positive" and "negative" are shown in quotation marks because we can rarely look inside a transformer to see in which direction the flux is actually circulating.

ties of transformers in general.

Figure 9-8a shows an ideal transformer in which the primary and secondary respectively possess N_1 and N_2 turns. The primary is connected to a sinusoidal source E_g and the magnetizing current I_m creates a flux Φ_m. The flux is completely linked by the primary and secondary windings and, consequently, it is a *mutual flux*. The flux varies sinusoidally, and reaches a peak value Φ_{max}. According to Eq. 9-1, we can therefore write:

$$E_1 = 4.44\, fN_1\Phi_{max} \qquad (9\text{-}3)$$

and

$$E_2 = 4.44\, fN_2\Phi_{max} \qquad (9\text{-}4)$$

From these equations, we deduce the expression for the *voltage ratio* of an ideal transformer:

$$\frac{E_1}{E_2} = \frac{N_1}{N_2} \qquad (9\text{-}5)$$

where

E_1 = voltage induced in the primary [V]
E_2 = voltage induced in the secondary [V]
N_1 = number of turns on the primary
N_2 = number of turns on the secondary

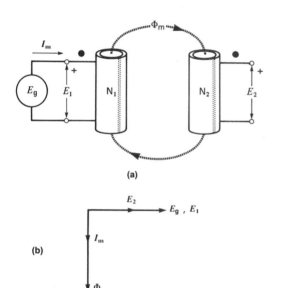

(a)

(b)

Figure 9-8
a. The ideal transformer at no-load. Primary and secondary are linked by a mutual flux.
b. Phasor relationships at no-load.

This equation shows that the ratio of the primary and secondary voltages is equal to the ratio of the number of turns. Furthermore, because the primary and secondary voltages are induced by the same mutual Φ_m, they are necessarily in phase.

The phasor diagram at no load is given in Fig. 9-8b. Phasor E_2 is in phase with phasor E_1

(and not 180° out of phase) on account of the polarity marks. If the transformer has fewer turns on the secondary than on the primary, phasor E_2 is shorter than phasor E_1. As in any inductor, current I_m lags 90° behind applied voltage E_g. The phasor representing flux Φ_m is obviously in phase with magnetizing current I_m which produces it.

However, because this is an *ideal* transformer, the magnetic circuit is infinitely permeable and so no magnetizing current is required to produce the flux Φ_m. Thus, under no-load conditions, the phasor diagram of such a transformer is identical to Fig. 9-8b except that phasor I_m is infinitesimally small.

Example 9-3:
A not quite ideal transformer having 90 turns on the primary and 2250 turns on the secondary is connected to a 120 V, 60 Hz source. The coupling between the primary and secondary is perfect, but the magnetizing current is 4 A. Calculate:

a. the effective voltage across the secondary terminals;
b. the peak voltage across the secondary terminals;
c. the instantaneous voltage across the secondary when the instantaneous voltage across the primary is 37 V.

Solution:
a. The turns ratio is:
$$N_2/N_1 = 2250/90 \qquad \text{Eq. 9-5}$$
$$= 25$$

The secondary voltage is therefore 25 times greater than the primary voltage because the secondary has 25 times more turns. Consequently:
$$E_2 = 25 \times E_1 = 25 \times 120$$
$$= 3000 \text{ V}$$

Instead of reasoning as above, we can apply Eq. 9-5:
$$E_1/E_2 = N_1/N_2$$

$120 / E_2 = 90 / 2250$ V

which again yields $E_2 = 3000$ V

b. The voltage varies sinusoidally; consequently, the peak secondary voltage is:
$$E_{2(peak)} = \sqrt{2} E = 1.414 \times 3000$$
$$= 4242 \text{ V}$$

c. The secondary voltage is 25 times greater than E_1 at every *instant*. Consequently, when $e_1 = 37$ V
$$e_2 = 25 \times 37 = 925 \text{ V}$$

9.7 Ideal transformer under load; current ratio

Pursuing our analysis, let us connect a load Z across the secondary of the ideal transformer (Fig. 9-9). A secondary current I_2 will immediately flow, given by:

$$I_2 = E_2 / Z$$

(a)

(b)

Figure 9-9
a. Ideal transformer under load. The mutual flux remains unchanged.
b. Phasor relationships under load.

Does E_2 change when we connect the load? To answer this question, we must remember two facts. First, in an ideal transformer the primary and secondary windings are linked by a mutual flux Φ_m, and by no other flux. In other words, an ideal transformer, by definition, has no leakage flux. Consequently, the voltage ratio under load is the same as at no-load, namely:

$$E_1 / E_2 = N_1 / N_2$$

Second, if the supply voltage E_g is kept fixed, then the primary induced voltage E_1 remains fixed. Consequently, mutual flux Φ_m also remains fixed. It follows that E_2 also remains fixed. We conclude that E_2 remains fixed whether a load is connected or not.

Let us now examine the magnetomotive forces created by the primary and secondary windings. First, current I_2 produces a secondary mmf $N_2 I_2$. If it acted alone, this mmf would produce a profound change in the mutual flux Φ_m. But we just saw that Φ_m does not change under load. We conclude that flux Φ_m can only remain fixed if the primary develops a mmf which *exactly* counterbalances $N_2 I_2$ at every instant. Thus, a primary current I_1 must flow so that:

$$N_1 I_1 = N_2 I_2 \qquad (9\text{-}6)$$

To obtain the required instant-to-instant bucking effect, currents I_1 and I_2 must increase and decrease at the same time. Thus, when I_2 goes through zero, I_1, goes through zero, and when I_2 is maximum ($+$) I_1 is maximum ($+$). In other words, they must be in phase. Furthermore, in order to produce the bucking effect, when I_1 *flows into* a polarity-mark on the primary side, I_2 must flow out of the polarity-mark on the secondary side (see Fig. 9-9a).

Using these facts, we can now draw the phasor diagram of an ideal transformer under load (Fig. 9-9b). Assuming a resistive-inductive load, current I_2 lags behind E_2 by an angle θ. Flux Φ_m lags $90°$ behind E_g, but no magnetizing

Figure 9-10
a. See Example 9-4.
b. Phasor relationships.

current I_m is needed to produce this flux because this is an ideal transformer. Finally, the primary and secondary currents are in phase. According to Eq. 9-6, they are related by the equation:

$$I_1/I_2 = N_2/N_1 \qquad (9\text{-}7)$$

where

I_1 = primary current [A]
I_2 = secondary current [A]
N_1 = number of turns on the primary
N_2 = number of turns on the secondary

Comparing Eq. 9-5 and Eq. 9-7, we see that the transformer current ratio is the inverse of the voltage ratio. In effect, what we gain in voltage, we lose in current and vice versa. This is consistent with the requirement that the instantaneous power input $E_1 I_1$ to the primary must equal the instantaneous power output $E_2 I_2$ of the secondary. If the power inputs and outputs were not identical, it would mean that the transformer itself absorbs power. By definition, this is impossible in an ideal transformer.

Example 9-4:
An ideal transformer having 90 turns on the primary and 2250 turns on the secondary is connected to a 200 V, 50 Hz source. The load across the secondary draws a current of 2 A at a power factor of 80 percent lagging

(Fig. 9-10a). Calculate:
a.   the effective value of the primary current;
b.   the instantaneous current in the primary when the instantaneous current in the secondary is 100 mA;
c.   the peak flux linked by the secondary winding;
d.   draw the phasor diagram.

Solution:
a.   The turns ratio is:
$$a = N_1/N_2 = 90/2250$$
$$= 1/25$$

The current ratio is therefore 25 and because the primary has fewer turns, the primary current is 25 times greater than the secondary current. Consequently:
$$I_1 = 25 \times 2 = 50 \text{ A}$$

Instead of reasoning as above, we can calculate the current by means of Eq. 9-6.
$$N_1 I_1 = N_2 I_2$$
$$90 I_1 = 2250 \times 2$$
$$I_1 = 50 \text{ A}$$

b.   The instantaneous current in the primary is *always* 25 times greater than the instantaneous current in the secondary. Thus, when $I_2 = 100$ mA, I_1 is:
$$I_{1 \text{ instantaneous}} = 25 \, I_{2 \text{ instantaneous}}$$
$$= 25 \times 0.1$$
$$= 2.5 \text{ A}$$

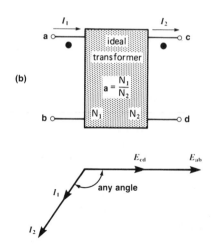

Figure 9-11
a. Symbol for an ideal transformer and phasor diagram using sign notation.
b. Symbol for an ideal transformer and phasor diagram using double-subscript notation.

c. In an ideal transformer, the flux linking the secondary is the same as that linking the primary. The peak flux in the secondary is:

$$\Phi_{max} = E_g/(4.44fN_1)$$
$$= 200/(4.44 \times 50 \times 90)$$
$$= 0.01$$
$$= 10 \text{ mWb}$$

d. To draw the phasor diagram, we reason as follows.
Secondary voltage is:

$$E_2 = 25 \times E_1 = 25 \times 200$$
$$= 5000 \text{ V}$$

E_2 is in phase with E_1 on account of the polarity marks. For the same reason, I_1 is in phase with I_2.

Phase angle between E_2 and I_2 is:

power factor $= \cos \theta$
$0.8 = \cos \theta$
$\theta = 36.9°$

The phase angle between E_1 and I_1 is also 36.9°.

The mutual flux lags 90° behind E_g (Fig. 9-10b).

9.8 Circuit symbol for an ideal transformer

To highlight the bare essentials of an ideal transformer, it is best to draw it in symbolic form. Thus, instead of drawing the primary and secondary windings and the mutual flux Φ_m, we simply show a box having primary and secondary terminals (Fig. 9-11). Polarity marks are added, enabling us to indicate the direction of current flow as well as the polarities of voltages E_1 and E_2. For example, a current I_1 flowing into one polarity-marked terminal is *always* accompanied by a current I_2 flowing out of the other polarity-marked terminal. Consequently, I_1 and I_2 are always in phase.

Furthermore, if we let the ratio of transformation $N_1/N_2 = a$, we obtain:

$$E_1 = aE_2$$

and

$$I_1 = I_2/a$$

In an ideal transformer, and specifically

referring to Fig. 9-11a, E_1 and E_2 are *always* in phase, and so are I_1 and I_2.*

If the double-subscript notation is used (Fig. 9-11b), E_{ab} and E_{cd} are *always* in phase and so are I_1 and I_2.

It is sometimes useful to make analogies between electrical and mechanical devices. For example, it is well known that a resistor is comparable to a mechanical brake, that a capacitor is similar to a mechanical spring and that an inductor behaves like a flywheel. Similarly, an ideal transformer may be compared to a gearbox having an input shaft and an output shaft. Thus, in the same way that a gearbox can increase or decrease the speed of rotation, a transformer can increase or decrease the current. The analogy between the two devices is best shown in tabular form:

Ideal transformer	Ideal gearbox
Primary terminals	Input shaft
Secondary terminals	Output shaft
Primary winding	Driving gear wheel
Secondary winding	Driven gear wheel
Primary turns	Teeth on the driving gear wheel
Secondary turns	Teeth on the driven gear wheel
Turns ratio	Gear ratio
Primary voltage	Input torque
Secondary voltage	Output torque
Primary current	Input r/min
Secondary current	Output r/min
Primary and secondary polarity marks	Arrows showing the input and output direction of rotation

---

*   Some texts show the respective voltages and currents as being 180° out of phase. This situation can arise depending upon how the behavior of the transformer is described, or how the voltage polarities and current directions are assigned. By using the methodology we have adopted, there is never any doubt as to how the phasors should be drawn.

9.9 Impedance ratio

Although a transformer is generally used to transform a voltage or current, it also has the important ability to transform an impedance. Consider, for example, Fig. 9-12a in which an ideal transformer T is connected between a

(a)

(b)

Figure 9-12
a. Impedance transformation using a transformer.
b. The impedance seen by the source differs from Z.

source E_g and a load Z. The ratio of transformation is **a**, and so we can write:

$$E_1/E_2 = a$$

and

$$I_1/I_2 = 1/a$$

As far as the source is concerned, it "sees" an impedance Z_x between the primary terminals given by:

$$Z_x = E_1/I_1$$

On the other hand, the secondary "sees" an impedance Z given by:

$$Z = E_2/I_2$$

However,

$$Z_x = \frac{E_1}{I_1} = \frac{aE_2}{I_2/a} = \frac{a^2E_2}{I_2} = a^2Z$$

Consequently,

$$Z_x = a^2Z \qquad (9\text{-}8)$$

This means that the impedance "seen" by the source is a^2 times the actual impedance (Fig. 9-12b). Thus, an ideal transformer has the amazing ability to increase or decrease the value of an impedance. In effect, the impedance seen across the primary terminals is identical to the actual impedance across the secondary terminals multiplied by the square of the turns ratio.

The impedance transformation is real, and not illusory like the image produced by a magnifying glass. An ideal transformer can modify the value of any component, be it a resistor, capacitor, or inductor. For example, if a 1000 Ω resistor is placed across the secondary of a transformer having a primary to secondary turns ratio of 1:5, it will appear across the primary as if it had a resistance of 1000 x $(1/5)^2$ = 40 Ω. Similarly, if a capacitor having a reactance of 1000 Ω is connected to the secondary, it appears as a 40 Ω capacitor across the primary. However, because the reactance of a capacitor is inversely proportional to its capacitance $(X_C = 1/2\pi fC)$, the apparent capacitance between the primary terminals is 25 times greater than its actual value. We can therefore artificially increase (or decrease) the microfarad value by means of a transformer.

9.10 Shifting impedances from secondary to primary and vice versa

As a further illustration of the impedance-changing properties of an ideal transformer, consider the circuit of Fig. 9-13a. It is composed of a source E_g, a transformer T and four impedances Z_1 to Z_4. The transformer has a turns ratio **a**.

We can progressively shift the secondary impedances to the primary side, as shown in Figs. 9-13b to 9-13e. As the impedances are shifted in this way, the circuit arrangement remains the same, but the shifted impedance values are multiplied by a^2.

If all the impedances are transferred to the primary side, the ideal transformer ends up at the extreme right-hand side of the circuit (Fig. 9-13d). In this position, the secondary of the transformer is on open-circuit. Consequently, both the primary and secondary currents are zero. We can therefore remove the ideal transformer altogether, yielding the equivalent circuit shown in Fig. 9-13e.

In comparing Figs. 9-13a and 9-13e, we may wonder how a circuit which contains a real transformer T can be reduced to a circuit which has no transformer at all. In effect, is there any meaningful relationship between the two circuits? The answer is yes - there is a useful relationship between the real circuit of Fig. 9-13a and the equivalent circuit of Fig. 9-13e. The reason is that the voltage E across each element in the secondary side becomes **a**E when the element is shifted to the primary side. Similarly, the current I in each element in the secondary side become I/a when the element is shifted to the primary side.

On account of this relationship, it is easy to solve a real circuit such as the one shown in Fig. 9-13a. We simply reduce it to the equivalent form shown in Fig. 9-13e and solve for all the voltages and currents. These values are then respectively multiplied by 1/**a** and by **a**, which yields the actual voltages and currents of each element in the secondary side.

To illustrate, suppose that the real voltage across Z_4 in Fig. 9-14 is E_4 volts, and that the real current through it is I_4 amperes. Then, in the equivalent circuit, the voltage across the equivalent a^2Z_4 impedance is equal to

(a)

(b)

(c)

(d)

(e)

Figure 9-13
a. The actual circuit showing the actual voltages and currents.
b. Impedance Z_2 is shifted to the primary side. Note the corresponding changes in E_2 and I_2.
c. Impedance Z_3 is shifted to the primary side. Note the corresponding change in E_3 and I_3.
d. Impedance Z_4 is shifted to the primary side. Note the corresponding change in E_4 and I_4. The currents in T are now zero.
e. All the impedances are now transferred to the primary side and the transformer is no longer needed.

$E_4 \times$ **a** volts. On the other hand, the current through the impedance is equal to $I_4 \div$ **a** amperes (Fig. 9-15). In other words, whenever an impedance is transferred to the primary side, the real voltage across the impedance increases by a factor **a**, while the real current decreases by the factor **a**.

In general, whenever an impedance is transferred from one side of a transformer to the other, the real voltage across it changes in proportion to the turns ratio. If the impedance is transferred to the side where the transformer voltage is higher, the voltage across the transferred impedance will also be higher. Conversely, if the impedance is transferred to the side where the transformer voltage is lower, the voltage across the transferred impedance is lower than the real voltage — again, of course, in the ratio of the number of turns.

In some cases, it is useful to shift impedances in the opposite way, that is from the primary side to the secondary side (Fig. 9-16a). The procedure is the same, but all impedances so transferred are now *divided* by a^2

Figure 9-14
Actual voltage and current in impedance Z_4.

Figure 9-15
Equivalent voltage and current in Z_4.

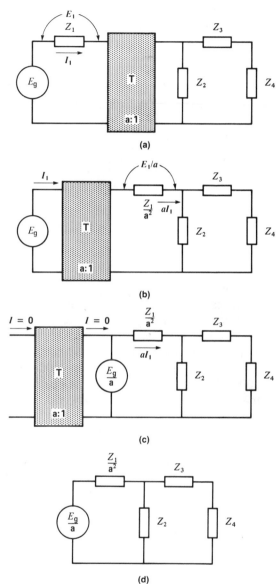

(a)

(b)

(c)

(d)

Figure 9-16
a. The actual circuit, showing the actual voltages and currents.
b. Impedance Z_1 is transferred to the secondary side. Note the corresponding change in E_1 and I_1.
c. The source is transferred to the secondary side. Note the corresponding change in E_g. Note also that the currents in T are zero.
d. All the impedances and even the source are now on the secondary side. The transformer is no longer needed because its currents are zero.

(Fig. 9-16b). We can even shift the source E_g to the secondary side, where it becomes a source having a voltage E_g/a. The ideal transformer is now located at the extreme left-hand side of the circuit (Fig. 9-16c). In this position, the primary of the transformer is on open-circuit. Consequently, both the primary and secondary currents are zero. As before, we can remove the transformer completely, leaving the equivalent circuit of Fig. 9-16d.

Example 9-5:
Calculate voltage E and current I in the circuit of Fig. 9-17, knowing that ideal transformer T has a primary to secondary turns ratio of 1:100.

Figure 9-17
See Example 9-5.

Solution:
The easiest way to solve this problem is to shift all the impedances to the primary side of the transformer. Because the primary has 100 times fewer turns than the secondary, the impedance values are divided by 100^2, or 10 000. Voltage E become $E/100$, but current I remains unchanged because it is already on the primary side (Fig. 9-18).

Figure 9-18
Equivalent circuit of Fig. 9-17.

a.   The impedance of the circuit in Fig. 9-18 is:
$$Z = \sqrt{R^2 + (X_L - X_C)^2} \qquad \text{Eq. 2-17}$$
$$= \sqrt{4^2 + (5 - 2)^2}$$
$$= \sqrt{16 + 9}$$
$$= 5 \ \Omega$$

The current in the circuit is:
$$I = E/Z = 10/5 = 2 \text{ A}$$

The voltage across the resistor is:
$$E/100 = IR = 2 \times 4 = 8 \text{ V}$$

The actual voltage E is therefore
$$E = 8 \times 100 = 800 \text{ V}$$

QUESTIONS AND PROBLEMS

9-1   The coil in Fig. 9-2a has 500 turns and a reactance of 60 Ω, but negligible resistance. If it is connected to a 120 V, 60 Hz source E_g, calculate:
   a. the effective value of the magnetizing current I_m;
   b. the peak value of I_m;
   c. the peak mmf produced by the coil;
   d. the peak flux Φ_{max}.

9-2   In Problem 9-1, if the voltage E_g is reduced to 40 V, calculate the new mmf developed by the coil, and the peak flux Φ_{max}.

9-3   What is meant by mutual flux? by leakage flux?

9-4   The ideal transformer in Fig. 9-9 has 500 turns on the primary and 300 turns on the secondary. The source produces a voltage E_g of 600 V, and the load Z is a resistance of 12 Ω. Calculate:
   a. the voltage E_2;
   b. the current I_2;
   c. the current I_1;
   d. the power delivered to the primary [W];
   e. the power output from the secondary [W];

9-5   In Problem 9-4, what is the impedance "seen" by the source E_g?

9-6   In Fig. 9-17, calculate the voltage across the capacitor, and the current flowing through it.

10

PRACTICAL TRANSFORMERS

In Chapter 9 we studied the ideal transformer and discovered its basic properties. However, in the real world transformers are not ideal and so our simple analysis must be modified to take this into account. Thus, the windings of practical transformers have resistance and the cores are not infinitely permeable. Furthermore, the flux produced by the primary is not completely captured by the secondary. Consequently, the leakage flux must be taken into account. And finally, the iron cores produce eddy-current and hysteresis losses which contribute to the temperature rise of the transformer.

In this chapter we discover that the properties of a practical transformer can be described by an equivalent circuit comprising an ideal transformer and resistances and reactances. The equivalent circuit is developed from fundamental concepts. This enables us to calculate such characteristics as voltage regulation and the behavior of transformers that are connected in parallel. The per-unit method is also used to illustrate its mode of application.

10.1 Ideal transformer with an imperfect core

The ideal transformer studied in the previous chapter had an infinitely permeable core. What happens if such a perfect core is replaced by an iron core having hysteresis and eddy-current losses, and whose permeability is rather low? We can represent these imperfections by two circuit elements R_m and X_m in parallel with the primary terminals of the ideal transformer (Fig. 10-1a).

The resistance R_m represents the iron losses and the heat they produce. To furnish these losses, a small current I_f is drawn from the line.

The so-called magnetizing reactance X_m is a measure of the permeability of the transformer core. Thus, if the permeability is low, X_m is relatively low. The current I_m flowing through X_m represents the magnetizing current needed to create the flux Φ_m in the core.

The values of the impedances R_m and X_m

Figure 10-1a
An imperfect core represented by a reactance X_m and a resistance R_m.

Figure 10-1b
Phasor diagram of a practical transformer at no-load.

may be deduced from the following equations:

$$R_m = E_1^2/P_m \qquad (10\text{-}1)$$

$$X_m = E_1^2/Q_m \qquad (10\text{-}2)$$

where

R_m = resistance representing the iron losses [Ω]

X_m = magnetizing reactance of the primary winding [Ω]

E_1 = Primary induced voltage [V]

P_m = iron losses [W]

Q_m = reactive power needed to set up the mutual flux Φ_m [var]

The total current needed to produce the flux Φ_m in an imperfect core is equal to the sum of I_f and I_m. It is called the *exciting current* I_o. The phasor diagram at no-load for this less-than-ideal transformer is given in Fig. 10-1b. The mutual flux Φ_m is again given by Eq. 9-2:

$$\Phi_m = E_1/(4.44\, fN_1) \qquad \text{Eq. 9-2}$$

Example 10-1:
A large transformer operating at no-load draws an exciting current I_o of 5 A when the primary is connected to a 120 V, 60 Hz source (Fig. 10-2). From a wattmeter test, it is known that the iron losses are equal to 180 W.*

* Iron losses are discussed in Sections 2-26 to 2-29.

Calculate:
a. the reactive power absorbed by the core;
b. the value of R_m and X_m;
c. the value of I_f, I_m and I_o

Solution:
a. The apparent power supplied to the core is:
$$S_m = E_1 I_o = 120 \times 5$$
$$= 600 \text{ VA}$$

The iron losses are:
$$P_m = 180 \text{ W}$$

The reactive power absorbed by the core is:
$$Q_m = \sqrt{S_m^2 - P_m^2} = \sqrt{600^2 - 180^2}$$
$$= 572 \text{ var}$$

b. The impedance corresponding to the iron losses is:
$$R_m = E_1^2/P_m = 120^2/180$$
$$= 80 \ \Omega$$

The magnetizing reactance is:
$$X_m = E_1^2/Q_m = 120^2/572$$
$$= 25.2 \ \Omega$$

c. The current needed to supply the iron losses is:
$$I_f = E_1/R_m = 120/80$$
$$= 1.5 \text{ A}$$

The magnetizing current is:
$$I_m = E_1/X_m = 120/25.2$$
$$= 4.8 \text{ A}$$

Figure 10-2a
See Example 10-1.

Figure 10-2b
Phasor diagram.

The exciting current I_o is:
$$I_o = \sqrt{I_f^2 + I_m^2} = \sqrt{1.5^2 + 4.8^2}$$
$$= 5 \text{ A}$$

The phasor diagram is given in Fig. 10-2b.

10.2 Ideal transformer with imperfect coupling

We have just seen how an ideal transformer behaves when it has an imperfect core. We now assume a transformer having a perfect core but imperfect coupling between its primary and secondary windings. We also assume that the primary and secondary windings have negligible resistance and the turns are N_1, N_2.

Consider the transformer in Fig. 10-3 connected to a source E_g and operating at no-load.

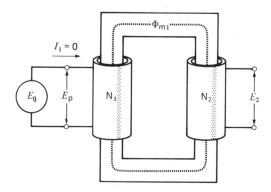

Figure 10-3
Transformer with infinitely permeable core at no-load.

Because the core is infinitely permeable, and because it has no losses, the no-load current $I_1 = 0$. The secondary voltage E_2 is given by $E_2 = (N_2/N_1) E_p$.

Let us now connect a load Z across the secondary, keeping the source voltage E_g fixed (Fig. 10-4). This simple operation sets off a train of events which we list as follows:

1. Currents I_1 and I_2 immediately begin to flow in the primary and secondary windings. They are related by the ideal-transformer equation $I_1/I_2 = N_2/N_1$.

2. I_2 produces a mmf N_2I_2 while I_1, produces a mmf N_1I_1. These magnetomotive forces are in direct opposition because when I_1 flows into the polarity-marked terminal 1, I_2 flows out of polarity-marked terminal 3.

3. The mmf N_2I_2 produces a total ac flux Φ_2. A portion of Φ_2 (Φ_{m2}) links with the primary winding while another portion (Φ_{f2}) does not. Flux Φ_{f2} is called the secondary leakage flux.

4. Similarly, the mmf N_1I_1 produces a total ac flux Φ_1. A portion of Φ_1 (Φ_{m1}) links with the secondary winding, while another portion (Φ_{f1}) does not. Flux Φ_{f1} is called the primary leakage flux.

The magnetomotive forces due to I_1 and I_2 upset the magnetic field that existed before the

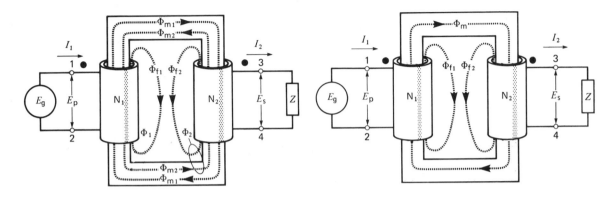

Figure 10-4
Mutual fluxes and leakage fluxes produced by a transformer under load. The leakage fluxes are due to the imperfect coupling between the coils.

Figure 10-5
A transformer possesses two leakage fluxes and a mutual flux.

load was connected. The question is, how can we analyze this new situation?

Referring to Fig. 10-4, we reason as follows:

First, the total flux produced by I_1 is composed of two parts: a new mutual flux Φ_{m1} and a leakage flux Φ_{f1}. (The mutual flux Φ_{m1} in Fig. 10-4 is not the same as that in Fig. 10-3).

Second, the total flux produced by I_2 is composed of a mutual flux Φ_{m2} and a leakage flux Φ_{f2}.

Third, we combine Φ_{m1} and Φ_{m2} into a single mutual flux Φ_m (Fig. 10-5). This mutual flux is created by the joint action of the primary and secondary mmfs.

Fourth, we note that the primary leakage flux Φ_{f1} is created by N_1I_1, while the secondary leakage flux is created by N_2I_2. Consequently, leakage flux Φ_{f1} is in phase with I_1 and leakage flux Φ_{f2} is in phase with I_2.

Fifth, the voltage E_s induced in the secondary is actually composed of two parts:

1. a voltage E_{f2} induced by leakage flux Φ_{f2} and given by:
$$E_{f2} = 4.44fN_2\Phi_{f2} \tag{10-3}$$

2. a voltage E_2 induced by mutual flux Φ_m and given by:
$$E_2 = 4.44fN_2\Phi_m \tag{10-4}$$

In general, E_{f2} and E_2 are not in phase.

Similarly, the voltage E_p induced in the primary is composed of two parts:

1. a voltage E_{f1} induced by leakage flux Φ_{f1} and given by:
$$E_{f1} = 4.44fN_1\Phi_{f1} \tag{10-5}$$

2. a voltage E_1 induced by mutual flux Φ_m and given by:
$$E_1 = 4.44fN_1\Phi_m \tag{10-6}$$

Sixth, induced voltage E_p = applied voltage E_g.

Using these six basic facts, we now proceed to develop the equivalent circuit of the transformer.

10.3 Primary and secondary leakage reactance

We can better identify the four induced voltages E_1, E_2, E_{f1} and E_{f2} by rearranging the transformer circuit as shown in Fig. 10-6. Thus, the secondary winding is drawn twice to show

Figure 10-6
Separating the various induced voltages due to the mutual flux and the leakage fluxes.

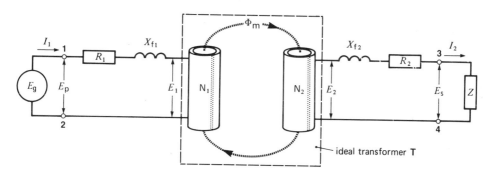

Figure 10-7
Resistance and leakage reactance of the primary and secondary windings.

even more clearly that the N_2 turns are linked by both Φ_{f2} and Φ_m. This rearrangement does not change the value of the induced voltages, but it does make each voltage stand out by itself. Thus, it becomes clear that E_{f2} is really a voltage drop across a reactance. This so-called *secondary leakage reactance* X_{f2} is given by:

$$X_{f2} = E_{f2}/I_2 \qquad (10\text{-}7)$$

The primary winding is also shown twice, to separate E_1 from E_{f1}. Again, it is clear that E_{f1} is simply a voltage drop across a reactance. This so-called *primary leakage reactance* X_{f1} is

given by:

$$X_{f1} = E_{f1}/I_1 \qquad (10\text{-}8)$$

The primary and secondary leakage reactances are shown in Fig. 10-7. We have also added the primary and secondary winding resistances R_1 and R_2 which, of course, act in series with the respective windings.

10.4 Equivalent circuit of a practical transformer

The circuit of Fig. 10-7 is composed of resistive and inductive elements (R_1, R_2, X_{f1}, X_{f2}, Z) cou-

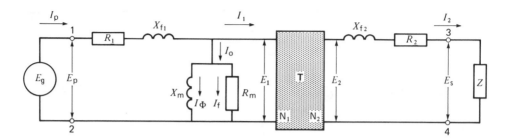

Figure 10-8
Complete equivalent circuit of a practical transformer.

pled together by a mutual flux Φ_m which links the primary and secondary windings. The leakage-free magnetic coupling enclosed in the dotted square is actually an ideal transformer. It possesses the same properties and obeys the same rules as the ideal transformer discussed in Chapter 9. For example, we can shift impedances to the primary side by multiplying their values by $(N_1/N_2)^2$, as we did before.

If we add circuit elements X_m and R_m to represent a practical core, we obtain the complete equivalent circuit of a practical transformer (Fig. 10-8). In this circuit, only the primary and secondary terminals 1-2 and 3-4 are accessible; all other components are "buried" inside the transformer itself. However, by appropriate tests, we can find the values of all the circuit elements that make up a practical transformer.

Example 10-2:
The secondary winding of a transformer possesses 180 turns. When the transformer is under load, the secondary current has an effective value of 18 A, 60 Hz. Furthermore, the mutual flux Φ_m has a peak value of 20 mWb. The secondary leakage flux Φ_{f2} has a peak value of 3 mWb. Calculate:
a. the voltage induced in the secondary winding by its leakage flux;
b. the value of the secondary leakage reactance;
c. the value of E_2 induced by the mutual flux Φ_m.

Solution:
a. The voltage induced by the secondary leakage flux is:
$$E_{f2} = 4.44 f N_2 \Phi_{f2} \qquad \text{Eq. 10-3}$$
$$= 4.44 \times 60 \times 180 \times 0.003$$
$$= 143.9 \text{ V}$$

b. The secondary reactance is:
$$X_{f2} = E_{f2}/I_2 \qquad \text{Eq. 10-7}$$
$$= 143.9/18$$
$$= 8 \ \Omega$$

c. The voltage induced by the mutual flux is:
$$E_2 = 4.44 f N_2 \Phi_m \qquad \text{Eq. 10-4}$$
$$= 4.44 \times 60 \times 180 \times 0.02$$
$$= 959 \text{ V}$$

10.5 Construction of a power transformer

Power transformers are usually designed so that they approach the characteristics of an ideal transformer. Thus, to attain high permeability, the core is made of iron (Fig. 10-9a). The resulting magnetizing current I_m is at least 5000 times smaller than it would be if an air core were used. Furthermore, to keep the iron losses down, the core is laminated, and high resistivity, high-grade silicon steel is used. Consequently, the current I_f needed to supply the iron losses is usually 2 to 10 times smaller than I_m.

Leakage reactances X_{f1} and X_{f2} are made as small as possible by winding the primary and secondary coils on top of each other, and by

Figure 10-9a
Construction of a simple transformer.

Figure 10-9b
Stacking laminations inside a coil.

spacing them as closely together as insulation considerations will permit. The coils are carefully insulated from each other and from the core. Such tight coupling between the coils means that the secondary voltage at no-load is almost exactly equal to N_2/N_1 times the primary voltage. It also guarantees good voltage regulation when a load is connected to the secondary terminals.

Winding resistances R_1 and R_2 are kept low both to reduce the I^2R loss and resulting heat and to ensure high efficiency. Figure 10-9a is a simplified version of a power transformer in which the primary and secondary are wound on one leg. In practice, the primary and secondary coils are distributed over both core legs in order to reduce the amount of copper. For the same reason, in larger transformers the cross section of the laminated iron core is not square (as shown) but is built up so as to be nearly round.

Figure 10-9b shows how the laminations of a small transformer are stacked to build up the core. Figure 10-9c shows the primary winding of a very big transformer.

Figure 10-9c
Primary winding of a large transformer; rating 128 kV, 290 A. *(courtesy ABB)*

The number of turns on the primary and secondary windings depends upon their respective voltages. A high-voltage winding has far more turns than a low-voltage winding does. On the other hand, the current in a HV winding is much smaller, enabling us to use a smaller size conductor. As a result, the amount of copper in the primary and secondary windings is about the same. In practice, the outer coil (coil 2, in Fig. 10-9a) weighs more because the length per turn is greater. Aluminum or copper conductors are used.

A transformer is reversible in the sense that either winding can be used as the primary winding.

10.6 Standard terminal markings

We saw in Sec. 9-4 that the polarity of a transformer can be shown by means of dots on the primary and secondary terminals. This type of marking is used on instrument transformers. On power transformers, however, the terminals are designated by the symbols H_1 and H_2 for the high-voltage (HV) winding and by X_1 and X_2 for the low-voltage (LV) winding. By convention, H_1 and X_1 have the same polarity.

Although the polarity is known when the symbols H_1, H_2, X_1 and X_2 are given, it is common practice to mount the four terminals on the transformer tank in a standard way so that the transformer has either *additive or subtractive* polarity. A transformer is said to have *additive* polarity when terminal H_1 is diagonally opposite terminal X_1. Similarly, a transformer has *subtractive* polarity when terminal H_1 is adjacent to terminal X_1 (Fig. 10-10). If we know that a transformer has additive (or subtractive) polarity, we do not have to identify the terminals by symbols.

Subtractive polarity is standard for all single-phase transformers above 200 kVA, provided the high-voltage winding is rated above 8660 V. All other transformers have additive polarity.

Figure 10-10
Additive and subtractive polarity depends upon the location of the H_1-X_1 terminals.

10.7 Polarity tests

To determine whether a transformer possesses additive or subtractive polarity, we proceed as follows (Fig. 10-11):

1. connect the high-voltage winding to an ac source E_g.

2. connect a jumper J between any two adjacent HV and LV terminals.

3. connect a voltmeter E_x between the other two adjacent HV and LV terminals.

4. Connect another voltmeter E_p across the HV winding. If E_x gives a higher reading than E_p, the polarity is additive. This tells us that H_1 and X_1 are diagonally opposite. On the other hand, if E_x gives a lower reading than E_p, the polarity is subtractive, and terminals H_1 and X_1 are adjacent.

Figure 10-11
Determining the polarity of a transformer using an ac source.

In this polarity text, jumper J effectively connects the secondary voltage E_s in series with the primary voltage E_p. Consequently, E_s either adds to or subtracts from E_p. In other words, $E_x = E_p + E_s$ or $E_x = E_p - E_s$, depending on the polarity. We can now see how the terms "additive" and "subtractive" originated.

In making the polarity test, an ordinary 120 V, 60 Hz source can be connected to the HV winding, even though its nominal voltage may be several hundred kilovolts.

Example 10-3:
During a polarity test on a 500 kVA, 69 kV/600 V transformer (Fig. 10-11), the following readings were obtained: $E_p = 118$ V, $E_x = 119$ V. Determine the polarity markings of the terminals.

Solution:
The polarity is additive because E_x is greater than E_p. Consequently, the HV and LV terminals connected by the jumper must respectively be labelled H_1 and X_2 (or H_2 and X_1).

Figure 10-12 shows another circuit which may be used to determine the polarity of a transformer. A dc source, in series with an open switch, is connected to the LV winding of the transformer. The transformer terminal connected to the positive side of the source is marked X_1. A dc voltmeter is connected across the HV terminals. When the switch is closed, a voltage is momentarily induced in the HV winding. If, at this moment, the pointer of the voltmeter moves upscale, the transformer terminal connected to the (+) terminal of the voltmeter is marked H_1 and the other is marked H_2.

10.8 Transformer taps

Owing to voltage drops in transmission lines, the voltage in a particular section of a distribution system may consistently be lower than normal. Thus, a transformer having a ratio of 2400 V to 120 V may be connected to a transmission line where the voltage is never higher than 2000 V. Under these conditions, the vol-

Figure 10-12
Determining the polarity of a transformer using a dc source.

tage across the secondary is considerably less than 120 V. Incandescent lamps are dim, electric stoves take longer to cook food, and electric motors may stall under moderate overloads.

To correct this problem, *taps* are provided on the primary windings of distribution transformers (Fig. 10-13). Taps enable us to change the turns ratio so as to raise the secondary voltage by 4 ½, 9 or 13 ½ percent. We can therefore maintain a satisfactory secondary voltage, even though the primary voltage may be 4 ½, 9 or 13 ½ percent below normal. Thus, referring to the transformer of Fig. 10-13, if the line voltage is only 2076 V (instead of the normal 2400 V) we would use taps 1 and 5 to obtain 120 V on the secondary side.

HV tap	percent	primary voltage	secondary voltage
1 - 2	0	2400 V	120 V
1 - 3	4½	2292 V	120 V
1 - 4	9	2184 V	120 V
1 - 5	13½	2076 V	120 V

Figure 10-13
Taps on a transformer enable an output of 120 V whether the primary voltage is high or low.

Some transformers are designed to change the taps automatically whenever the secondary voltage is above or below a preset level. Such *tap-changing transformers* help maintain the secondary voltage within ±2 percent of its rated value throughout the day.

10.9 Losses and transformer rating

Like any electrical machine, a transformer has losses. They are composed of:

1. I^2R losses in the windings;

2. hysteresis and eddy-current losses in the core;

3. stray losses due to currents induced in the tank and metallic supports, by the primary and secondary leakage fluxes.

The losses appear in the form of heat and produce 1) an increase in temperature and 2) a drop in efficiency. Under normal operating conditions, the efficiency of transformers is very high; it may reach 99.5 percent for large power transformers.

The heat produced by the iron losses depends upon the peak value of the mutual flux Φ_m, which, in turn, depends upon the applied voltage. On the other hand, the heat dissipated in the windings depends upon the current they carry. Consequently, to keep the transformer temperature at an acceptable level, we must set limits to both the applied voltage and the current drawn by the load. These two limits determine the *nominal voltage* and *nominal current* of the transformer.

The *power rating* of a transformer is equal to the product of the nominal voltage times the nominal current. However, the result is not expressed in watts, because the phase angle between the voltage and current may have any value at all, depending on the nature of the load. Consequently, the power-handling capacity of a transformer is expressed in voltamperes (VA), in kilovoltamperes (KVA) or in megavoltamperes (MVA), depending on the

size of the transformer. In effect, the temperature rise of a transformer is directly related to the *apparent* power which flows through it. This means that a 500 kVA transformer will get just as hot feeding a 500 kvar inductive load as a 500 kW resistive load.

The rated kVA, frequency and voltage are always shown on the nameplate. In large transformers, the corresponding rated currents are also shown.

Example 10-4:
The nameplate of a distribution transformer indicates 250 kVA, 60 Hz, primary 4160 V, secondary 480 V.
a. Calculate the nominal primary and secondary currents.
b. If we apply 2000 V to the 4160 V primary, can we still draw 250 kVA from the transformer?

Solution:

a. Nominal current of the 4160 V winding is:

$$I_p = \frac{\text{nominal } S}{\text{nominal } E_p} = \frac{250 \times 1000}{4160}$$

$$= 60 \text{ A}$$

Nominal current of the 480 V winding is:

$$I_p = \frac{\text{nominal } S}{\text{nominal } E_s} = \frac{250 \times 1000}{480}$$

$$= 521 \text{ A}$$

b. If we apply 2000 V to the primary, the flux and the iron losses will be lower than normal and the core will be cooler. However, the load current should not exceed its nominal value, otherwise the windings will overheat. Consequently, the maximum power output using this lower voltage is:

$$S = 2000 \text{ V} \times 60 \text{ A} = 120 \text{ kVA}$$

10.10 No-load saturation curve

Let us gradually increase the voltage E_p on the primary of a transformer, with the secondary open-circuited. As the voltage rises, the mutual flux Φ_m increases in direct proportion, in accordance with Eq. 9-2. Exciting current I_o will therefore increase but, when the iron begins to saturate, the magnetizing current I_m has to increase very steeply to produce the required flux. If we draw a graph of E_p versus I_o, we see the dramatic increase in current as we pass the normal operating point (Fig. 10-14).

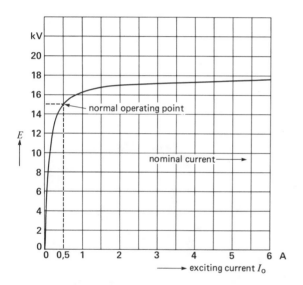

Figure 10-14
No-load saturation curve of a 167 kVA, 14.4 kV/480 V, 60 Hz transformer.

Transformers are usually designed to operate at a peak flux density of about 1.5 T, which corresponds roughly to the knee of the saturation curve. Thus, when nominal voltage is applied to a transformer, the corresponding flux density is about 1.5 T. We can exceed the nominal voltage by perhaps 20 percent, but if we were to apply twice the nominal voltage, the exciting current could become greater than the nominal full-load current.

The non-linear relationship between E_p and I_o shows that the exciting branch (composed of R_m and X_m in Fig. 10-1a) is not as constant as it appears. In effect, although R_m is reasonably constant, X_m decreases rapidly with increasing saturation. However, transformers usually operate at close to rated voltage, and so R_m and X_m remain essentially constant.

10.11 Cooling methods

To prevent rapid deterioration of the oil and insulating materials inside a transformer, adequate cooling of the windings and core must be provided.

Indoor transformers below 200 kVA can be directly cooled by the natural flow of the surrounding air. The metallic housing is fitted with ventilating louvres so that convection currents may flow over the windings and around the core (Fig. 10-15). Larger trans-

Figure 10-15
Single-phase dry-type transformer, type AA, rated 15 kVA, 600 V/240 V, 60 Hz, insulation class 150°C for indoor use. Height: 600 mm; width: 434 mm; depth: 230 mm; weight 79.5 kg. (*Hammond*)

Figure 10-16
Two single-phase transformers, type OA, rated 75 kVA, 14.4 kV/240 V, 60 Hz, 55°C temperature rise, impedance 4.2 %. The small radiators at the side increase the effective cooling area.

For transformers in the megawatt range, cooling may be effected by an oil-water heat exchanger. Hot oil drawn from the transformer tank is pumped to a heat exchanger where it flows through pipes that are in contact with cool water. Such a heat exchanger is very effective, but also very costly, because the water itself has to be continuously cooled and recirculated.

Some big transformers are designed to have a variable rating, depending on the method of cooling used. Thus, a transformer may have a triple rating of 18 000/24 000/32 000 kVA depending on whether it is cooled

a)   by the natural circulation of air (AO) (18 000 kVA) or

Figure 10-17
Three-phase, type OA grounding transformer, rated 1900 kVA, 26.4 kV, 60 Hz. The power of this transformer is 25 times greater than that of the transformers shown in Fig. 10-16, but it is still self-cooled. Note, however, that the radiators occupy as much room as the transformer itself.

formers can be built the same way, but forced circulation of clean air must be provided. Such *dry-type* transformers are used inside buildings, away from hostile atmospheres.

Distribution transformers below 200 kVA are usually immersed in mineral oil and enclosed in a steel tank. Oil carries the heat away to the tank, where it is dissipated by radiation and convection to the outside air (Fig. 10-16). Oil is a much better insulator than air is; consequently, it is invariably used on high-voltage transformers.

As the power rating increases, external radiators are added to increase the cooling surface of the oil-filled tank (Fig. 10-17). Oil circulates around the transformer windings and moves through the radiators where the heat is again released to surrounding air. For still higher ratings, cooling fans blow air over the radiators (Fig. 10-18).

OA - oil-immersed, self cooled
OA/FA - oil-immersed, self-cooled /
forced-air cooled
AO/FA/FOA - oil-immersed self-cooled /
forced-air cooled / forced-air,
forced-oil-cooled

The temperature rise by resistance of oil-immersed transformers is either 55°C or 65°C. The temperature must be kept low to preserve the quality of the oil. By contrast, the temperature rise of a dry-type transformer may be as high as 180°C, depending on the type of insulation used.

Figure 10-18
Three-phase, type FOA, transformer rated 1300 MVA, 24.5 kV/ 345 kV, 60 Hz, 65°C temperature rise, impedance: 11.5 %. This step-up transformer, installed at a nuclear power generating station, is one of the largest units ever built. The forced-oil circulating pumps can be seen just below the cooling fans. (*Westinghouse*)

b) by forced-air cooling with fans (FA) (24 000 kVA) or

c) by the forced circulation of oil accompanied by forced air cooling (FOA) (32 000 kVA).

These elaborate cooling systems are nevertheless economical because they enable a much bigger output from a transformer of a given size (Fig. 10-19).

The type of transformer cooling is designated by the following symbols:*

AA - dry-type self-cooled
AFA - dry-type forced-air cooled

* Drawn from IEEE Std. 462-1973

Figure 10-19
Three-phase, type OA/FA/FOA transformer rated 36/48/60 MVA, 225 kV/26.4 kV, 60 Hz, impedance 7.4 %. The circular tank enables the oil to expand as the temperature rises and reduces the surface of the oil in contact with air. other details:
weight of core and coils: 37.7 t
weight of tank and accessories: 28.6 t
weight of coil (44.8 m³): 38.2 t
Total weight: 104.5 t

10.12 Transformers and the per-unit system

The per-unit system of measurement (Sec. 1-9) is particularly useful when dealing with transformers. The nominal power rating S_n of the transformer is selected as the base power and the nominal voltage E_n is selected as the base voltage. It follows that the base current I_n is equal to S_n/E_n which is the same as the nominal current. The base impedance Z_n is therefore equal to E_n/I_n. Note that E_n and I_n are the nominal voltage and nominal current on either the primary or secondary side. The base impedance may also be calculated by the equation:

$$Z_n = E_n^2/S_n \qquad (10\text{-}9)$$

where

Z_n = base impedance of the transformer [Ω]

S_n = power rating of the transformer [VA]

E_n = nominal voltage of the transformer on either the primary or secondary side [V]

Because the primary and secondary voltages are usually different, it follows that a transformer has two base impedances, one for the primary and one for the secondary. Depending upon the respective voltages, the two impedances may be vastly different.

Example 10-5:
Calculate the base impedance of a 250 kVA, 4160 V/480 V, 60 Hz transformer, on the primary and secondary side.

a. Base primary impedance is:

$Z_{np} = E_p^2/S_n = 4160^2/250\ 000$
$= 69\ \Omega$

a. Base secondary impedance is:

$Z_{ns} = E_s^2/S_n = 480^2/250\ 000$
$= 0.92\ \Omega$

10.13 Per-unit impedances

We can get a better idea of the relative magnitude of the winding resistance, leakage reactance, etc., of a transformer by comparing these impedances with the base impedance of the transformer. In making the comparison, circuit elements located on the primary side are compared with the primary base impedance. Similarly, circuit elements on the secondary side are compared with the secondary base impedance. The comparison can be made either on a percent or on a per-unit basis; we shall use the latter. Typical per-unit values are listed in Table 10A, for transformers ranging from 3 kVA to 100 MVA.

Example 10-6:
Using the information given in Table 10A, calculate the approximate real values of the impedances of a 250 kVA, 4160 V/480 V, 60 Hz distribution transformer.

Solution:
We first determine the base impedances on the primary and secondary side. From the results of Example 10-5, we have:

$Z_{np} = 69\Omega$
$Z_{ns} = 0.92\ \Omega.$

We now calculate the various impedances by multiplying Z_{np} and Z_{ns} by the per-unit values given in Table 10A. Thus, referring to Fig. 10-20, we find:

$R_1 = 0.005 \times 69\ \Omega = 0.35\ \Omega$
$R_2 = 0.005 \times 0.92\ \Omega = 4.6\ m\Omega$
$X_{f1} = 0.025 \times 69\ \Omega = 1.7\ \Omega$
$X_{f2} = 0.025 \times 0.92\ \Omega = 23\ m\Omega$
$X_m = 30 \times 69\ \Omega = 2070\ \Omega = 2\ k\Omega$
$R_m = 50 \times 69\ \Omega = 3450\ \Omega = 3.5\ k\Omega$

This example shows the usefulness of the per-unit method of estimating impedances. The equivalent circuit of the 250 kVA transformer is shown in Fig. 10-21. The true values may be 20

TABLE 10 A   TYPICAL PER-UNIT VALUES OF TRANSFORMERS

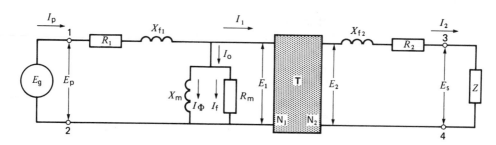

Figure 10-20
Equivalent circuit of a transformer

Circuit Element (see Fig. 11-30)	Typical per-unit values	
	3 kVA to 250 kVA	1 MVA to 100 MVA
R_1 or R_2	0.009 - 0.005	0.005 - 0.002
X_{f1} or X_{f2}	0.008 - 0.025	0.03 - 0.06
X_m	20 - 30	50 - 200
R_m	20 - 50	100 - 500
I_0	0.05 - 0.03	0.02 - 0.005

percent higher or lower than estimated because the per-unit values given in Table 10A are broad estimates covering a wide range of transformers.

$$a = \frac{4160}{480} = 8.67$$

Figure 10-21
See Example 10-6.

10.14 Simplifying the equivalent circuit

The equivalent circuit of a transformer (Figs. 10-8 and 10-20) is relatively simple, but it gives far more detail than is needed in most practical problems. Consequently, we can simplify it to make the calculations easier. Let us try to simplify the circuit when the transformer operates (a) at no-load and (b) at full-load.

a) *At no-load,* (Fig. 10-22) I_2 is zero and so is I_1 because T is an ideal transformer. Consequently, only the exciting current I_0 flows in R_1 and X_{f1}. These impedances are so small that the voltage drop across them is negligible. Furthermore, the current in R_2 and X_{f2} is zero. We can therefore neglect these four impedances, giving us the much simpler circuit of Fig. 10-23. The turns ratio, a = N_1/N_2, is obviously equal to the ratio of the primary to secondary voltages E_p/E_s measured across the terminals.

b) *At full-load*, I_p (Fig. 10-22) is at least 20 times larger than I_o. Consequently, we can neglect I_o and the corresponding magnetizing branch. The resulting circuit is shown in Fig. 10-24. This simplified circuit may be used whenever the load is 10 percent or more of the rated capacity of the transformer.

   We can further simplify the circuit by shifting everything to the primary side, thus eliminating transformer T (Fig. 10-25). This technique is explained in Section 9-10. Then, by summing the respective resistances and reactances, we obtain the circuit of Fig. 10-26. In this circuit,

$$R_p = R_1 + a^2R_2 \qquad (10\text{-}10)$$

$$X_p = X_{f1} + a^2X_{f2} \qquad (10\text{-}11)$$

where

R_p = total transformer resistance referred to the primary side

X_p = total transformer leakage reactance referred to the primary side

The combination of R_p and X_p constitutes the total transformer impedance Z_p referred to the primary side. From Eq. 2-12, we have:

$$Z_p = \sqrt{R_p^2 + X_p^2} \qquad (10\text{-}12)$$

Impedance Z_p is usually expressed as either a percent or per-unit of the base primary impedance Z_{np}, mentioned before. The per-unit impedance is always given on the nameplate of the transformer.

   Transformers above 500 kVA possess a leakage reactance X_p at least five times greater than R_p. In such transformers, we can neglect R_p, as far as voltages and currents are concerned.* The equivalent circuit is thus reduced to a simple reactance X_p between the source

---

* From the standpoint of temperature rise and efficiency R_p can never be neglected.

Figure 10-22
Complete equivalent circuit of a transformer at no-load.

Figure 10-23
Simplified circuit at no-load.

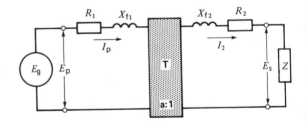

Figure 10-24
Simplified equivalent circuit of a transformer at full load.

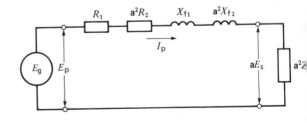

Figure 10-25
Equivalent circuit with impedances shifted to the primary side.

Figure 10-27
The internal impedance of a large transformer is mainly reactive.

Figure 10-27
The internal impedance of a large transformers is mainly reactive.

and the load (Fig. 10-27). It is quite remarkable that the relatively complex circuit of Fig. 10-20 can be reduced to a simple reactance in series with the load.

Example 10-7:
A single-phase transformer rated 3000 kVA, 69 kV/4.16 kV, 60 Hz has an impedance of 8 percent. Calculate:
a. the total impedance of the transformer referred to the primary side;
b. the voltage regulation from no-load to full-load for a 2000 kW resistive load;
c. the primary and secondary currents if the secondary is accidentally short-circuited.

Solution:
a. The base impedance on the primary side is:

$$Z_{np} = Z_{np} = E_n^2/S_n$$
$$= 69\ 000^2/3\ 000\ 000$$
$$= 1587\ \Omega$$

The transformer impedance referred to the primary side is:

$$Z_p = Z(pu) \times Z_{np} = 0.08 \times 1587$$
$$= 127\ \Omega$$

Because the transformer exceeds 500 kVA, the windings have negligible resistance compared to their leakage reactance; we can therefore write:

$$Z_p = X_p = 127\ \Omega \text{ (Fig. 10-28a)}$$

b. Approximate impedance Z of the 2000 kW load:

$$Z = E_s^2/P = 4160^2/2\ 000\ 000$$
$$= 8.65\ \Omega$$

Impedance referred to primary side:

$$a^2Z = (69/4.16)^2 \times 8.65 = 2380\ \Omega$$

Referring to Fig. 10-28b, we have:

$$I_p = 69\ 000\ /\sqrt{127^2 + 2380^2}$$
$$= 28.95\ A$$

Figure 10-28a
See Example 10-7.

Figure 10-28b
See Example 10-7.

$$aE_s = (a^2Z)\,I_p = 2380 \times 28.95$$
$$= 68\,902 \text{ V}$$
$$E_s = 68\,902 \times (4.16/69) = 4154 \text{ V}$$

Secondary no-load voltage = 4160 V

Voltage regulation in percent is:

$$\frac{\text{no-load voltage} - \text{full-load voltage}}{\text{no-load voltage}} \times 100$$

$$= \frac{4160 - 4154}{4160} \times 100 = 0.14 \text{ percent}$$

The voltage regulation is excellent.

c. Referring again to Fig. 10-28b, if the secondary is accidentally short-circuited, we find:
$$I_p = E_p/X_p = 69\,000/127$$
$$= 543 \text{ A}$$

The corresponding current I_s on the secondary side:
$$I_s = aI_p = (69/4.16) \times 543$$
$$= 9006 \text{ A}$$

The short-circuit currents in the primary and secondary windings are 12.5 times greater than the rated values. The I^2R losses are therefore $12.5^2 = 156$ times greater than normal. The circuit-breaker or fuse protecting the transformer must open immediately to prevent overheating. Very powerful electromagnetic forces are also set up. They, too, are 156 times greater than normal and unless the windings are firmly braced and supported, they may be damaged or torn apart.

10.15 Measuring transformer impedances

We can determine all the transformer impedances shown in Fig. 10-20 by means of an open-circuit test and a short-circuit test on the transformer.

During the *open-circuit test*, rated voltage is applied to the primary winding and current I_o, voltage E_p, and active power P_m are measured (Fig. 10-29). The secondary open-circuit vol-

Figure 10-29
Open-circuit test and determination of R_m, X_m, and turns ratio.

tage E_s is also measured. These test results give us the following information:

active power absorbed by core = P_m
apparent power absorbed by core = $S_m = E_p I_o$
reactive power absorbed by core = Q_m
$$= \sqrt{S_m^2 - P_m^2}$$

Resistance R_m corresponding to the core loss is:

$$R_m = E_p^2/P_m \qquad \text{Eq. 10-1}$$

Magnetizing reactance X_m is:

$$X_m = E_p^2/Q_m \qquad \text{Eq. 10-2}$$

Turns ratio **a** is:

$$\mathbf{a} = N_1/N_2 = E_p/E_s$$

During the *short-circuit test*, the secondary winding is short-circuited and a voltage E_g much lower than normal (usually less than 5 percent of rated voltage) is applied to the primary. The primary current I_{sc} should be less than its nominal value to prevent overheating and a consequent rapid change in winding resistance while the test is being made.

Voltage E_{sc}, current I_{sc} and power P_{sc} are measured as before (Fig. 10-30) and the following calculations made:

Transformer impedance referred to the primary is:

Figure 10-30
Short-circuit test to determine leakage reactance and winding resistance.

$$Z_p = E_{sc}/I_{sc} \qquad (10\text{-}13)$$

Transformer resistance referred to the primary is:

$$R_p = P_{sc}/I_{sc}^2 \qquad (10\text{-}14)$$

Transformer leakage reactance referred to the primary is:

$$X_p = \sqrt{Z_p^2 - R_p^2} \qquad \text{Eq. 10-12}$$

Example 10-8:
During a short-circuit test on a 500 kVA, 69 kV/4.16 kV, 60 Hz transformer, the following voltage, current and power measurements were made. Terminals X_1, X_2 were in short-circuit. (see Fig. 10-30):

$$E_{sc} = 2600 \text{ V};$$
$$I_{sc} = 4 \text{ A};$$
$$P_{sc} = 2400 \text{ W}$$

Calculate the value of the reactance and resistance of the transformer, referred to the HV side, as well as the percent impedances.

Solution:
Referring to the equivalent circuit of the transformer under short-circuit conditions (Fig. 10-31a), we find the following values:

Transformer impedance referred to the primary is:

$$Z_p = E_{sc}/I_{sc} = 2600/4$$
$$= 650 \ \Omega$$

Resistance referred to the primary is:
$$R_p = P_{sc}/I_{sc}^2 = 2400/16$$
$$= 150 \ \Omega$$

Leakage reactance referred to the primary is:
$$X_p = \sqrt{650^2 - 150^2}$$
$$= 632 \ \Omega$$

Nominal primary voltage is:
$$E_p = 69 \text{ kV}$$

Nominal primary current is:
$$I_p = 500 \text{ kVA}/69 \text{ kV} = 7.25 \text{ A}$$

Base impedance referred to the primary is:
$$Z_{np} = E_p/I_p = 69\ 000/7.25$$
$$= 9517 \ \Omega$$

Figure 10-31a
See Example 10-8.

Per-unit impedance $= Z_p/Z_{np}$
$= 650/9517$
$= 0.0682$
(Percent impedance $= 6.82$)

Per-unit reactance $= X_p/Z_{np}$
$= 632/9517$
$= 0.0664$
(Percent reactance $= 6.64$)

Per-unit resistance $= R_p/Z_{np}$
$= 150/9517$
$= 0.0158$
(Percent resistance $= 1.58$)

Example 10-9:
An open-circuit test was conducted on the transformer given in Example 10-8. The following results were obtained when the *low voltage*

winding was excited. (In some cases, such as in a repair shop, a 69 kV voltage may not be available and the open-circuit test has to be done by exciting the LV winding.)

$$E_s = 4160 \text{ V} \quad I_o = 2 \text{ A} \quad P_m = 5000 \text{ W}$$

Using this information and the transformer characteristics found in Example 10-8, calculate the voltage regulation and efficiency of the transformer when it supplies a 250 kVA, 80 percent power factor (lagging) load.

Solution:
Industrial loads and voltages fluctuate all the time. Thus, when we state that a load is 250 kVA, with $\cos \theta = 0.8$, it is understood that the load is *about* 250 kVA and that the power factor is *about* 0.8. Furthermore, the primary voltage is *about* 69 kV.

Consequently, in calculating voltage regulation and efficiency, there is no point in arriving at a precise mathematical answer, even if we were able to give it. Knowing this, we can make certain assumptions that make it much easier to arrive at a solution.

The equivalent circuit is represented by Fig. 10-31b. The values of R_p and X_p are already known, and so we only have to add the magnetizing branch. Let us assume that the voltage across the load is 4160 V. We now calculate the efficiency of the transformer.

a. The load current is:
$$I_2 = S/E_s = 250\,000/4160$$
$$= 60 \text{ A}$$

The current in the primary is:
$$I_1 = I_2/a = 60 \times 4.16/69$$
$$= 3.62 \text{ A}$$

The total copper loss is:
$$P_{copper} = I_1{}^2 R_p = 3.62^2 \times 150$$
$$= 1966 \text{ W}$$

The iron loss is the same as that measured on the LV side of the transformer:
$$P_{iron} = 5000 \text{ W}$$

Figure 10-31b
See Example 10-9.

Total losses are:
$$P_{losses} = 5000 + 1966$$
$$= 6966 \text{ W} = 7 \text{ kW}$$

The active power delivered by the transformer to the load is:
$$P_o = S \cos \theta = 250 \times 0.8$$
$$= 200 \text{ kW}$$

The active power received by the transformer is:
$$P_i = P_o + P_{losses} = 200 + 7$$
$$= 207 \text{ kW}$$

The efficiency is:
$$\eta = P_o/P_i = 200/207$$
$$= 0.966 \text{ or } 96.6 \text{ percent}$$

Note that in making the above calculations, we only consider the active power. The reactive power does not enter into efficiency calculations. Let us now determine the voltage regulation.

b. In examining Fig. 10-31b, it is obvious that the presence of the magnetizing branch does not affect the voltage drop across R_p and X_p. Consequently, the magnetizing branch does not affect the voltage regulation.

To determine the voltage regulation we shall use the per-unit method explained in Sec. 1-11, together with complex notation. We shall refer all voltages, impedances and currents to the HV (69 kV) side. We assume the voltage between terminals 1, 2 is 69 kV, and that it remains fixed.

The base power P_B is 500 kVA
The base voltage E_B is 69 kV

Consequently, the base current is:
$$I_B = P_B/E_B = 500\ 000/69\ 000$$
$$= 7.246\ A$$

and the base impedance is:
$$Z_B = E_B/I_B = 69\ 000/7.246 = 9522\ \Omega$$

The per-unit value of R_p is
$$R_p(pu) = 150/9522 = 0.0158$$

The per-unit value of X_p is:
$$X_p(pu) = 632/9522 = 0.0664$$

The per-unit value of voltage E_{12} is
$$E_{12}(pu) = 69\ 000/69\ kV = 1.0$$

The per-unit value of the apparent power absorbed by the load is:
$$S(pu) = 250\ kVA/500\ kVA = 0.5$$

The per-unit value of the active power absorbed by the load is:
$$P(pu) = S(pu) \cos\theta = 0.5 \times 0.8 = 0.4$$

The per-unit value of the reactive power absorbed by the load is:
$$Q(pu) = \sqrt{S^2(pu) - P^2(pu)}$$
$$= \sqrt{0.5^2 - 0.4^2}$$
$$= 0.3$$

The per-unit load resistance R_L corresponding to P is:
$$R_L(pu) = \frac{E^2(pu)}{P(pu)} = \frac{1.0^2}{0.4} = 2.50$$

The per-unit load reactance X_L corresponding to Q is:
$$X_L(pu) = \frac{E^2(pu)}{Q(pu)} = \frac{1.0^2}{0.3} = 3.333$$

We now draw the equivalent per-unit circuit shown in Figure 10-31c. The magnetizing

Figure 10-31c
Per-unit equivalent circuit of a 500 kVA transformer feeding a 250 kVA load.

branch is not shown because it does not enter into the calculations. Note that the load appears across the primary terminals 3, 4 of the circuit shown in Fig. 10-31b. (These "terminals" are not accessible; they exist only in the equivalent circuit diagram)

The per-unit impedance between terminals 3, 4 is:

$$Z_{34}(pu) = \frac{2.50 \times j\ 3.33}{2.50 + j\ 3.33}$$
$$= 2\angle 36.87°$$
$$= 1.6 + j\ 1.2$$

The per-unit impedance between terminals 1, 2 is:
$$Z_{12}(pu) = 0.0158 + 1.6$$
$$+ j(1.2 + 0.0664)$$
$$= 1.616 + j\ 1.266$$
$$= 2.053\angle 38.07°$$

The per-unit current I_1 is:
$$I_1(pu) = \frac{E_{12}(pu)}{Z_{12}(pu)} = \frac{1.0}{2.053\angle 38.07°}$$
$$= 0.4872\angle -38.07°$$

The per-unit voltage E_{34} across the load is:

$E_{34}(\text{pu}) = I_1(\text{pu}) \times Z_{34}(\text{pu})$
$= (0.4872\angle-38.07°)(2\angle36.87°)$
$= 0.9744\angle-1.20°$

The per-unit voltage regulation is:

$$\frac{E_{34}(\text{pu}) \text{ at no-load} - E_{34}(\text{pu}) \text{ at full-load}}{E_{34}(\text{pu}) \text{ at no-load}}$$

$$= \frac{1.0 - 0.9744}{1.0}$$

$$= 0.0256$$

The voltage regulation is 2.56 percent.

If we wish, we can calculate the actual values of the voltage and current as follows:

a) Voltage across "terminals" 3, 4 is:
$E_{34} = E_{34}(\text{pu}) \times E_B$
$= 0.9744 \times 69\ 000$
$= 67.23$ kV

b) Actual voltage across the load is:
$E_{56} = E_{34} \times (4160/69\ 000) = 4053$
$= 4053$ V

c) Actual line current is:
$I_1 = I_1(\text{pu}) \times 1_B$
$= 0.4872 \times 7.246$
$= 3.53$ A.

10.16 Transformers in parallel

When a growing load eventually exceeds the power rating of an installed transformer, we sometimes connect a second transformer in parallel with it. To ensure proper load-sharing between the two transformers, they must possess:

a. the same primary and secondary voltages;

b. the same per-unit impedance.

Particular attention must be paid to the polarity of each transformer, so that only terminals having the same polarity are connected together (Fig. 10-32). An error in polarity produces a short-circuit as soon as the transformers are excited.

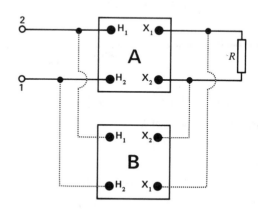

Figure 10-32
Connecting transformers in parallel to share a load.

In order to calculate the currents flowing in each transformer when they are connected in parallel, we must first determine the equivalent circuit of the system.

Consider first the equivalent circuit when a single transformer feeds a load Z_L (Fig. 10-33a). The primary voltage is E_p and the impedance of the transformer referred to the primary side is Z_{p1}. If the ratio of transformation is **a**, the circuit can be simplified to that shown in Fig. 10-33b, a procedure we are already familiar with.

If a second transformer having an impedance Z_{p2} is connected in parallel with the first, the equivalent circuit becomes that shown in Fig. 10-33c. In effect, the impedances of the transformers are in parallel. The primary currents in the transformers are respectively I_1 and I_2. Because the voltage drop E_{13} across the impedances is the same, we can write

$$I_1 Z_{p1} = I_2 Z_{p2} \qquad (10\text{-}15)$$

Figure 10-33a
Equivalent circuit of a transformer feeding a load Z_L.

Figure 10-33b
Equivalent circuit with all impedances referred to the primary side.

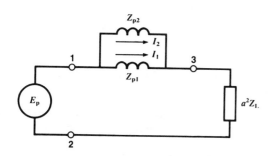

Figure 10-33c
Equivalent circuit of two transformers in parallel feeding a load Z_L. All impedances referred to the primary side.

that is

$$\frac{I_1}{I_2} = \frac{Z_{p2}}{Z_{p1}} \qquad (10\text{-}16)$$

The ratio of the primary currents is therefore determined by the magnitude of the respective primary impedances – and not by the ratings of the two transformers. But in order that the temperature rise be the same for both transformers, the currents must be proportional to the respective kVA ratings. Consequently, we want to fulfill the following condition:

$$\frac{I_1}{I_2} = \frac{S_1}{S_2} \qquad (10\text{-}17)$$

From Eqs. (10-16) and (10-17) it can readily be proved that the desired condition is met if the transformers have the same per-unit impedances. The following example shows what happens when the per-unit impedances are different.

Example 10-10:
A 100 kVA transformer is connected in parallel with an existing 250 kVA transformer to supply a load of 330 kVA. Both transformers are rated 7200 V/240 V, but the 100 kVA unit has an impedance of 4 percent while the 250 kVA transformer has an impedance of 6 percent (Fig. 10-34a). Calculate:
a. the nominal primary current of each transformer;
b. the impedance of the load referred to the primary side;
c. the impedance of each transformer referred to the primary side;
d. the actual primary current in each transformer.

Solution:
a. Nominal primary current of the 250 kVA transformer is:
$$I_{n1} = 250\ 000/7200 = 34.7 \text{ A}$$

Nominal primary current of the 100 kVA transformer is:
$$I_{n2} = 100\ 000/7200 = 13.9 \text{ A}$$

Figure 10-34a
Actual transformer connections.

b. The equivalent circuit of the two transformers and the load, referred to the primary side, is given in Fig. 10-33c. Note that transformer impedances Z_{p1} and Z_{p2} are considered to be entirely reactive. This assumption is justified because the transformers are fairly big.

Load impedance referred to the primary side is:
$$Z = E_p^2/S_{load} = 7200^2/330\,000$$
$$= 157\ \Omega$$

The approximate load current is:
$$I_L = S_{load}/E_p = 330\,000/7200 = 46\text{ A}$$

c. The base impedance of the 250 kVA unit is:
$$Z_{np1} = 7200^2/250\,000 = 207\ \Omega$$

Figure 10-34b
Equivalent circuit. Calculations show that the 100 kVA transformer is seriously overloaded.

Transformer impedance referred to the primary side is:
$$Z_{p1} = 0.06 \times 207 = 12.4\ \Omega$$

Base impedance of the 100 kVA unit is:
$$Z_{np2} = 7200^2/100\,000 = 518\ \Omega$$

Transformer impedance referred to the primary side is:
$$Z_{p2} = 0.04 \times 518 = 20.7\ \Omega$$

d. Referring to Fig. 10-34b, we find that the 46 A load current divides in the following way:
$$I_1 = 46 \times 20.7/(12.4 + 20.7) = 28.8\text{ A}$$
$$I_2 = 46 - 28.8 = 17.2\text{ A}$$

The 100 kVA transformer is seriously overloaded because it carries a primary current of 17.2 A, which is 25 percent above its rated value of 13.9 A. The 250 kVA unit is not overloaded because it only carries a current of 28.8 A versus its rated value of 34.7 A. Clearly, the two transformers are not carrying their proportional share of the load.

The 100 kVA transformer is overloaded because of its low impedance (4 percent), compared to the impedance of the 250 kVA transformer (6 percent). A low-impedance transformer always tends to carry more than its proportionate share of the load. If the percent impedances were equal, the load would be shared between the transformers in proportion to their respective power ratings.

QUESTIONS AND PROBLEMS

Practical level

10-1 Name the principal parts of a transformer.

10-2 Explain how a voltage is induced in the secondary winding of a transformer.

10-3 The secondary winding of a transformer has twice as many turns as the primary. Is the secondary voltage higher or lower than the primary voltage?

10-4 Which winding is connected to the load: the primary or secondary?

10-5 State the voltage and current relationships between the primary and secondary windings of a transformer under load. The primary and secondary windings have N_1 and N_2 turns, respectively.

10-6 Name the losses produced in a transformer.

10-7 What purpose does the no-load current of a transformer serve?

10-8 Name 3 conditions that must be met in order to connect two transformers in parallel?

10-9 What is the purpose of taps on a transformer?

10-10 Name three methods used to cool transformers.

10-11 The primary of a transformer is connected to a 600 V, 60 Hz source. If the primary has 1200 turns and the secondary 240, calculate the secondary voltage.

10-12 The windings of a transformer respectively have 300 and 7500 turns. If the low-voltage winding is excited by a 2400 V source, calculate the voltage

across the HV winding.

10-13 A 6.9 kV transmission line is connected to a transformer having 1500 turns on the primary and 24 turns on the secondary. If the load across the secondary has an impedance of 5 Ω, calculate:
a. the secondary voltage;
b. the primary and secondary currents.

10-14 The primary of a transformer has twice as many turns as the secondary. The primary voltage is 220 V and a 5 Ω load is connected across the secondary. Calculate the power delivered by the transformer, as well as the primary and secondary currents.

10-15 A 3000 kVA transformer has a ratio of 60 kV to 2.4 kV. Calculate the nominal current of each winding.

Intermediate level

10-16 In Problem 10-11, calculate the peak value of the flux in the core.

10-17 Explain why the flux in a 60 Hz transformer remains fixed as long as the ac supply voltage is fixed.

10-18 The transformer in Fig. 10-35 is excited by a 120 V, 60 Hz source, and draws a no-load current I_o of 3 A. The primary and secondary windings respectively

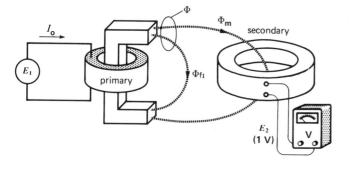

Figure 10-35
See Problem 10-18.

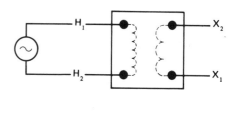

Figure 10-36
See Problem 10-19.

possess 200 and 600 turns. If 40 percent of the primary flux is linked by the secondary, calculate:

a. the voltage indicated by the voltmeter;
b. the peak value of flux Φ;
c. the peak value of Φ_m.
d. Draw the phasor diagram showing E_1, E_2, I_o, Φ_m and Φ_{f1}.

10-19 In Fig. 10-36, when 600 V is applied to terminals H_1 and H_2, 80 V is measured across terminals X_1, X_2.

a. What is the voltage between terminals H_1 and X_2?
b. If terminals H_1, X_1 are connected together, calculate the voltage across terminals H_2, X_2.
c. Does the transformer have additive or subtractive polarity?

10-20 a. Referring to Fig. 10-32, what would happen if we reversed terminals H_1, and H_2 of transformer B?

b. Would the operation of the transformer bank be affected if terminals H_1, H_2 and X_1, X_2 of transformer B were reversed? Explain.

10-21 Explain why the secondary voltage of a practical transformer decreases with increasing resistive load.

10-22 What is meant by:
a. transformer impedance;
b. percent impedance of a transformer?

10-23 The transformer in Problem 10-15 has an impedance of 6 percent. Calculate the impedance [Ω] referred to:
a. the 60 kV primary;
b. the 2.4 kV secondary.

10-24 A 2300 V line is connected to terminals 1 and 4 in Fig. 10-13. Calculate:
a. the voltage between terminals X_1 and X_2;

b. the current in each winding, if a 12 kVA load is connected across the secondary.

10-25 A 66.7 MVA transformer has an efficiency of 99.3 percent when it delivers full power to a load having a power factor of 100 percent.
a. Calculate the losses in the transformer under these conditions;
b. Calculate the losses and efficiency when the transformer delivers 66.7 MVA to a load having a power factor of 80 percent.

11-26 If the transformer shown in Fig. 10-15 were placed in a tank of oil, the temperature rise would have to be reduced to 65°. Explain.

Advanced level

10-27 Referring to Fig. 10-37, calculate the peak value of flux in the core if the transformer is supplied by a 50 Hz source.

10-28 The impedance of a transformer increases as the coupling is reduced between the primary and secondary windings. Explain.

10-29 The following information is given for the transformer circuit of Fig. 10-24.

$R_1 = 18\ \Omega$ $E_p = 14.4$ kV (nominal)
$R_2 = 0.005\ \Omega$ $E_s = 240$ V (nominal)
$X_{f1} = 40\ \Omega$ $X_{f2} = 0.01\ \Omega$

If the transformer has a nominal rating of 75 kVA, calculate:

a. the transformer impedance [Ω] referred to the primary side;
b. the percent impedance of the transformer;
c. the impedance [Ω] referred to the secondary side;
d. the percent impedance referred to the secondary side;
e. the total copper losses at full load;

f. the percent resistance and percent reactance of the transformer.

10-30 During a short-circuit test on a 10 MVA, 66 kV/7.2 kV transformer (see Fig. 10-30), the following results were obtained:

$E_g = 2640$ V
$I_{sc} = 72$ A
$P_{sc} = 9.85$ kW

Calculate:

a. the total resistance and the total leakage reactance referred to the 66 kV primary side;
b. the nominal impedance of the transformer referred to the primary side;
c. the percent impedance of the transformer.

10-31 In Problem 10-30, if the iron losses at rated voltage are 35 kW, calculate the full-load efficiency of the transformer if the power factor of the load is 85 percent.

Figure 10-37
See Problem 10-33. The primary is wound on one leg and the secondary on the other.

10-32 a. The windings of a transformer operate at a current density of 3.5 A/mm². If they are made of copper, and operate at a temperature of 75°C, calculate the copper loss per kilogram.

b. If aluminum windings were used, calculate the loss per kilogram under the same conditions.

10-33 If a transformer were actually built according to Fig. 10-37, it would have very poor voltage regulation. Explain why and propose a method of improving it.

11

SPECIAL TRANSFORMERS

Many transformers are designed to meet specific industrial applications. In this chapter, we study some of the special transformers that are used in distribution systems, neon signs, laboratories and induction furnaces. Although they are special, they still possess the basic properties of the standard transformers discussed in Chapter 10. As a result, the following approximations can be made when the transformers are under load:

1. the voltage induced in a winding is directly proportional to the number of turns, and to the flux in the core;

2. the ampere-turns of the primary are equal and opposite to the ampere-turns of the secondary;

3. the apparent power input to the transformer is equal to the apparent power output;

4. the exciting current in the primary winding may be neglected;

11.1 Dual-voltage distribution transformer

Transformers which supply electric power to residential areas generally have two secondary winding, each rated at 120 V. The windings are connected in series, and so the total voltage between the lines is 240 V while that between the lines and the middle conductor is 120 V (Fig. 11-1a). The middle conductor, called *neutral*, is always connected to ground.

In special cases, the two secondaries are connected in parallel by connecting the terminals of the same polarity together (Fig. 11-1b). The resulting output voltage is only 120 V, but the power rating of the transformer remains unchanged. The nominal rating of these so-called *distribution transformers* ranges from 3 kVA to 500 kVA. They are mounted on poles of the electrical utility company (Fig. 11-2) to supply power to anywhere from 1 to 20 customers.

The load on distribution transformers varies greatly throughout the day, depending on cus-

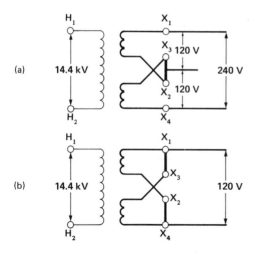

Figure 11-1
a. Distribution transformer with 120 V/240 V second-ary. The central conductor is the neutral.
b. Same distribution transformer reconnected to give only 120 V.

Figure 11-2
Single-phase pole-mounted distribution transformer rated: 100 kVA, 14.4 kV/240 V/120 V, 60 Hz.

tomer demand. In residential districts, a peak occurs in the morning and another in the late afternoon. The power peaks never last for more than one or two hours, with the result that dur-ing most of the 24-hour day the transformers operate far below their normal rating. Because thousands of such transformers are connected to the public utility system, every effort is made to keep the no-load losses small. This is achieved by using special low-loss silicon-steel in the core.

11.2 Autotransformer

Consider a single transformer winding com-posed of N_1 turns, mounted on a iron core (Fig. 11-3). The winding is connected to a fixed ac source E_1, and the resulting exciting current I_o creates an ac flux Φ_m in the core. As in any transformer, the flux is fixed so long as E_1 is fixed (Sec. 9.1).

Suppose a tap C is taken off the winding, so that there are N_2 turns between terminals A and

C. Because the induced voltage between these terminals is proportional to the number of turns, E_2 is given by:

$$E_2 = (N_2/N_1) \times E_1 \qquad (11\text{-}1)$$

Clearly, this simple coil resembles a transformer having a primary voltage E_1 and a secondary voltage E_2. However, the primary terminals B, A and the secondary terminals C, A are no longer isolated from each other, because of the common terminal A.

If we connect a load to secondary terminals CA, the resulting current I_2 immediately causes a primary current I_1 to flow (Fig. 11-4).

The BC portion of the winding obviously car-ries current I_1. Therefore, according to Kirchhoff's current law, the CA portion carries a current $(I_2 - I_1)$. Furthermore, the mmf due to I_1 must be equal and opposite to the mmf pro-

Figure 11-3
Autotransformer having N_1 turns on the primary and N_2 turns on the secondary.

duced by $(I_2 - I_1)$. As a result, we have:

$$I_1 (N_1 - N_2) = (I_2 - I_1) N_2$$

which reduces to:

$$I_1 N_1 = I_2 N_2 \qquad (11\text{-}2)$$

Finally, assuming that both the transformer losses and exciting current are negligible, the apparent power drawn by the load must equal the apparent power supplied by the source. Consequently:

$$E_1 I_1 = E_2 I_2 \qquad (11\text{-}3)$$

Equations (11-1), (11-2) and (11-3) are identical to those of a standard transformer having a turns ratio N_1/N_2. However, in this so-called *autotransformer*, the secondary winding is actually part of the primary winding. In effect, an

autotransformer eliminates the need for a separate secondary winding. As a result, autotransformers are always smaller, lighter and cheaper than standard transformers of equal power output. The difference in size becomes particularly important when the ratio of transformation E_1/E_2 lies between 0.5 and 2. On the other hand, the absence of electrical isolation between the primary and secondary windings is a serious drawback in some applications.

Autotransformers are used to start induction motors, to regulate the voltage of transmission lines and, in general, to transform voltages when the primary to secondary ratio is close to 1.

Example 11-1:
The autotransformer in Fig. 11-4 has an 80 percent tap and the supply voltage is 300 V. If a 3.6 kW load is connected across the secondary, calculate:

a. the secondary voltage and current;

b. the currents which flow in the winding;

c. the relative size of the conductors on windings BC and CA.

Solution:
a. The secondary voltage is:
$$E_2 = 80\% \times 300 = 240 \text{ V}$$

The secondary current is:
$$I_2 = P/E_2 = 3600/240 = 15 \text{ A} \quad \text{(Fig. 11-5)}$$

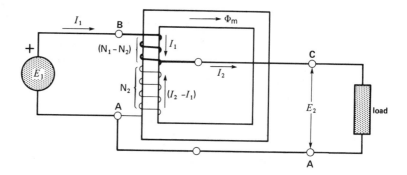

Figure 11-4
Autotransformer under load. The currents flow in opposite directions in the upper and lower windings.

Figure 11-5
Autotransformer of Example 11-1.

b. The current supplied by the source is:

$I_1 = P/E_1 = 3600/300 = 12$ A;

the current in winding BC = 12 A;

the current in winding CA = 15 − 12 = 3 A.

c. The conductors in winding CA can be one quarter the size of those in winding BC because the current is 4 times smaller (see Fig. 11-5). But winding CA has 4 times as many turns. Consequently, the two windings require essentially the same amount of copper.

11.3 Conventional transformer connected as an autotransformer

Any two-winding transformer can be changed into an autotransformer by connecting the primary and secondary windings in series. Depending upon how the connection is made, the secondary voltage may add to or substract from the primary voltage. The basic operation and behavior of a transformer is unaffected by a mere change in external connections. Consequently, the following rules apply whenever a standard transformer is connected as an autotransformer:

1. The current in any winding should not exceed its nominal current rating.

2. The voltage across any winding should not exceed its nominal voltage rating.

3. If rated current flows in one winding, rated current will *automatically* flow in the other winding (reason: the ampere-turns of the windings are always equal).

4. If rated voltage exists across one winding, rated voltage *automatically* exists across the other winding (reason: the same mutual flux links both windings).

5. If a current I_1 flows into terminal H_1, a current I_2 must flow *out* of terminal X_1 and vice versa (reason: the ampere-turns always oppose each other).

Example 11-2:
The standard single-phase transformer shown in Fig. 11-6 has a rating of 15 kVA, 600 V/120 V, 60 Hz. We wish to reconnect it as an autotransformer in three different ways to obtain three different voltage ratios:

a. 600 V primary to 480 V secondary;
b. 600 V primary to 720 V secondary;
c. 120 V primary to 480 V secondary.

Calculate the maximum load the transformer can carry in each case.

Solution:
Nominal current of the 600 V winding is:

$I_1 = S/E_1 = 15\ 000/600 = 25$ A

Nominal current of the 120 V winding is:

$I_2 = S/E_2 = 15\ 000/120 = 125$ A

a. To obtain 480 V, the secondary voltage (120 V) between terminals X_1, X_2 must sub-

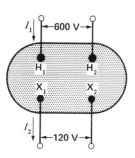

Figure 11-6
Standard 15 kVA, 600 V/120 V transformer.

Figure 11-7
Transformer reconnected as an autotransformer to give a ratio of 600 V/480 V.

tract from the primary voltage (600 V). Consequently, we connect terminals having the same polarity together, as shown in Fig. 11-7. The corresponding schematic diagram is given in Fig. 11-8.

Note that the 120 V winding is connected in series with the load. Because this winding has a nominal current rating of 125 A, the load can draw a maximum power:

$$S_a = 125 \text{ A} \times 480 \text{ V} = 60 \text{ kVA}$$

The currents flowing in the circuit at full load are shown in Fig. 11-8. Note that:

1. If we assume that the current of 125 A flows *into* terminal X_1, a current of 25 A must flow *out* of terminal H_1. All other currents are then found by applying Kirchhoff's current law.

2. The apparent power supplied by the source is equal to that absorbed by the load:

$$S = 100 \text{ A} \times 600 \text{ V} = 60 \text{ kVA}.$$

b. To obtain a ratio of 600 V/720 V, the secondary voltage must *add* to the primary voltage: 600 + 120 = 720 V. Consequently, terminals of opposite polarity must be connected together, as shown in Fig. 11-9.

Figure 11-8
Schematic diagram of Fig. 11-7 showing voltages and current flows.

Figure 11-9
Transformer reconnected to give a ratio of 600 V/720 V.

The secondary winding is again in series with the load, and the maximum load current is again 125 A. The maximum load is therefore:

$$S_b = 125 \text{ A} \times 720 \text{ V} = 90 \text{ kVA}$$

The previous examples show that when a conventional transformer is connected as an autotransformer, it can supply a load far greater than the rated capacity of the transformer. As mentioned earlier, this is one of the advantages of using an autotransformer instead of a conventional transformer. However, this is not always true, as the next example shows.

Figure 11-10
Transformer reconnected to give a ratio of 120 V/480 V.

c. To obtain a ratio of 120 V to 480 V, we again connect H_1 and X_1 (as in solution a.), but the source is now connected to terminals $X_1 X_2$ (Fig. 11-10).

This time, the 600 V winding is in series with the load; consequently, the maximum load current cannot exceed 25 A. The corresponding maximum load is therefore:

$$S_c = 25 \text{ A} \times 480 \text{ V} = 12 \text{ kVA}$$

which is *less* than the nominal rating (15 kVA) of the standard transformer.

We want to make one final remark concerning the three autotransformer connections. The temperature rise of the transformer is the same in each case, even though the loads are respectively 60 kVA, 90 kVA and 12 kVA. The reason is that the currents in the windings and the flux in the core are identical in each case and so the losses are the same.

11.4 Voltage transformers

Voltage transformers (formerly called potential transformers) are high-precision transformers in which the ratio of primary voltage to secondary voltage is a known constant, which changes very little with load.* Furthermore, the secondary voltage is almost exactly in phase with the primary voltage. The nominal secondary voltage is usually 115 V, irrespective of what the rated primary voltage may be. This permits standard instruments and relays to be used on the secondary side. Voltage transformers are used to measure or monitor the voltage on transmission lines and to isolate the metering equipment from these lines (Fig. 11-11).

The construction of voltage transformers is similar to that of conventional transformers. However, the insulation between the primary and secondary windings must be particularly great to withstand the full line voltage on the HV side.

In this regard, one terminal of the secondary winding is always connected to ground to eliminate the danger of a fatal shock when touching one of the secondary leads. Although the secondary *appears* to be isolated from the primary, the distributed capacitance between the two windings makes an invisible connection which can produce a very high voltage between the secondary winding and ground. By grounding one of the secondary terminals, the highest voltage between the secondary lines and ground is limited to 115 V.

The nominal rating of voltage transformers is usually less than 500 VA. As a result, the volume of insulation is often far greater than the volume of copper or steel.

---

* In the case of voltage transformers and current transformers, the «load» is called burden.

Figure 11-11
Potential transformer installed on a 69 kV line. Note the distributed capacitance between the windings.

Figure 11-12
7000 VA, 80.5 kV, 50/60 Hz potential transformer having an accuracy of 0.3 % and a BIL of 650 kV. The primary terminal at the top of the bushing is connected to the HV line while the other is connected to ground. The secondary is composed of two 115 V windings each tapped at 66.4 V. Other details: total height: 2565 mm; height of porcelain bushing: 1880 mm; oil: 250 L; weight: 740 kg. (*Ferranti-Packard*)

Voltage transformers installed on HV lines always measure the line-to-neutral voltage. This eliminates the need for two HV bushings because one side of the primary is connected to ground. For example, the 7000 VA, 80.5 kV transformer shown in Fig. 11-12 has one large porcelain bushing to isolate the HV line from the grounded case. The latter houses the actual transformer.

11.5 Current transformers

Current transformers are high-precision transformers in which the ratio of primary to secondary current is a known constant that changes very little with the burden. The phase angle between the primary and secondary current is very small, usually much less than one degree. The highly accurate current ratio and small phase angle are achieved by keeping the exciting current small.

Current transformers are used to measure or monitor the current in a line and to isolate the metering and relay equipment connected to the secondary side. The primary is connected in series with the line, as shown in Fig. 11-13. The nominal secondary current is usually 5 A, irrespective of the primary current rating.

Because current transformers (CT's) are only used for measurement and system protection, their power rating is small – generally between 15 VA and 200 VA. As in the case of

Figure 11-13
Current transformer installed on a 69 kV line.

Figure 11-14
500 VA, 100 A/5 A, 60 Hz current transformer, insulated for a 230 kV line and having an accuracy of 0.6 %. (*Westinghouse*)

Figure 11-15
Current transformer in the final process of construction. (*Ferranti-Packard*)

conventional transformers, the current ratio is inversely proportional to the number of turns on the primary and secondary windings. A current transformer having a ratio of 150 A/5 A has therefore 30 times more turns on the secondary than on the primary.

For safety reasons, current transformers must always be used when measuring currents in HV transmission lines. The insulation between the primary and secondary windings must be great enough to withstand the full line-to-neutral voltage, including line surges. The maximum voltage the CT can withstand is always shown on the nameplate.

As in the case of voltage transformers (and for the same reasons) one of the secondary terminals is always connected to ground.

Figure 11-14 shows a 500 VA, 100 A/5 A current transformer designed for a 230 kV line.

The large bushing serves to isolate the HV line from the ground. The CT is housed in the grounded steel case at the lower end of the bushing. The upper end of the bushing has two terminals that are connected in series with the HV line. The line current flows into one terminal, down the bushing, through the primary of the transformer, then up the bushing and out by the other terminal. The internal construction of a CT is shown in Fig. 11-15 and a typical installation is shown in Fig. 11-16.

By way of comparison, the 50 VA current transformer shown in Fig 11-17 is much smaller, mainly because it is insulated for only 36 kV.

Example 11-3:
The current transformer is Fig. 11-17 has a rating of 50 VA, 400 A/5 A, 36 kV, 60 Hz. It is con-

Figure 11-16
Current transformer in series with one phase of a 220 kV, 3-phase line inside a substation.

Figure 11-17
Epoxy-encapsulated current transformer rated 50 VA, 400 A/5 A, 60 Hz and insulated for 36 kV. (*Montel, Sprecher & Schuh*)

nected to an ac line, having a line-to-neutral voltage of 24.9 kV, in a manner similar to that shown in Fig. 11-13. The ammeters, relays and connecting wires on the secondary side possess a total impedance (burden) of 1.2 Ω. If the transmission-line current is 280 A, calculate:

a. the secondary current;
b. the voltage across the secondary terminals;
c. the voltage drop across the primary.

Solution:

a. The current ratio is:
$$I_1/I_2 = 400/5 = 80$$

The turns ratio is:
$$N_1/N_2 = 1/80$$

The secondary current is:
$$I_2 = 280/80 = 3.5 \text{ A}$$

b. The voltage across the burden is:
$$E_2 = IR = 3.5 \times 1.2 = 4.2 \text{ V}$$

The secondary voltage is therefore 4.2 V.

c. The primary voltage is:
$$E_1 = 4.2/80 = 0.0525 = 52.5 \text{ mV}$$

This is a negligible drop, compared to the 24.9 kV line voltage.

11.6 Opening the secondary of a CT is dangerous

Every precaution must be taken to *never* open the secondary circuit of a current transformer while current is flowing in the primary circuit. If the secondary is accidentally opened, the primary current I_1 continues to flow unchanged because the impedance of the primary is negligible compared to that of the electrical load. The line current thus becomes the *exciting* current of the transformer because there is no fur-

Figure 11-18
Primary current, flux and secondary voltage when a CT is open-circuited.

ther bucking effect due to the secondary ampere-turns. Because the line current may be 100 to 200 times greater than the normal exciting current, the flux in the core reaches peaks much higher than normal. The flux is so large that the core is totally saturated for the greater part of every half cycle. Referring to Fig. 11-18, as the primary current I_1 rises and falls during the first half cycle, flux Φ in the core also rises and falls, but it remains at a fixed, saturation level Φ_s for most of the time. The same thing happens during the second half cycle. During these saturated intervals, the induced voltage is quite low because the flux changes very little.

However, during the unsaturated intervals, the flux changes at an extremely high rate, inducing voltage peaks of several thousand volts across the open-circuited secondary. This is a dangerous situation because an unsuspecting operator could easily receive a bad shock. The voltage is particularly high in current transformers having a rating above 50 VA.

In view of the above, if a meter or relay in the secondary circuit of a CT has to be disconnected, we must first short-circuit the secondary winding *and then* remove the component . Short-circuiting a current transformer does no harm because the primary current remains unchanged and the secondary current can be no greater than that determined by the turns ratio. The short-circuit across the winding may be removed *after* the secondary circuit is again closed.

11.7 Toroidal current transformers

When the line current exceeds 100 A, we can sometimes use a *toroidal* current transformer. It consists of a laminated toroidal (ring-shaped) core which carries the secondary winding. The primary is composed of a single conductor that simply passes through the center of the ring (Fig. 11-19). The position of the primary conductor is unimportant as long as it is more or less centered. If the secondary possesses N turns, the ratio of transformation is N. Thus, a toroidal CT having a ratio of 1000 A/5 A has 200 turns on the secondary winding.

Figure 11-19
Toroidal transformer having a ratio of 1000 A/5 A, connected to measure the current in a line.

Toroidal CT's are simple and inexpensive and are widely used in low-voltage (LV) and medium-voltage (MV) indoor installations. They are also incorporated in circuit-breaker bushings to monitor the line current (Fig. 11-20). If the current exceeds a predetermined limit, the CT causes the circuit-breaker to trip.

toroidal iron core. A movable carbon brush in sliding contact with the winding serves as a variable tap. The brush can be set in any position between zero and 330°. Manual or motorized positioning may be used (Figs. 11-21 and 11-23).

As the brush slides over the bared portion of the winding, the secondary voltage E_2

Figure 11-20
Toroidal transformer surrounding a conductor inside a bushing.

Figure 11-21
Cutaway view of a manually operated 0-140 V, 15 A variable autotransformer showing (1) the laminated toroidal core; (2) the single-layer winding; (3) the movable brush. (*American Superior Electric*)

Example 11-4:
A potential transformer rated 14 400 V/115 V and a current transformer rated 75/5 A are used to measure the voltage and current in a transmission line. If the voltmeter indicates 111 V and the ammeter reads 3 A, what is the voltage and current in the line?

Solution:
The voltage on the line is:
$$E = 111 \times (14\ 400/115) = 13\ 900 \text{ V}$$

The current in the line is:
$$I = 3 \times (75/5) = 45 \text{ A}$$

11.8 Variable autotransformer

A variable autotransformer is often used when we wish to obtain a variable ac voltage from a fixed-voltage ac source. It is composed of a single-layer winding wound uniformly on a

Figure 11-22
Schematic diagram of a variable autotransformer having a fixed 90% tap.

increases in proportion to the number of turns swept out (Fig. 11-22). The input voltage E_1 is usually connected to a fixed 90 percent tap on the winding. This enables E_2 to vary from zero to 110 percent of the input voltage.

Variable autotransformers are efficient and provide good voltage regulation under variable loads. The secondary line should always be

Figure 11-23
Variable autotransformer rated at 200 A, 0-240 V, 50/60 Hz. It is composed of eight 50 A, 120 V units, connected in series-parallel. This motorized unit can vary the output voltage from zero to 240 V in 5 s. Dimensions: 400 mm × 1500 mm. (*American Superior Electric*)

protected by a fuse or circuit-breaker so that the output current I_2 never exceeds the current rating of the autotransformer.

11.9 High-impedance transformers

The transformers we have studied so far are all designed to have a relatively low leakage reactance, ranging perhaps from 0.03 to 0.1 per unit (Sec. 10.13). However, some industrial and commercial applications require much higher reactances, sometimes reaching values as high as 0.9 p.u. Such high-impedance transformers are used in the following typical applications:

electric toys	arc welders
fluorescent lamps	electric arc furnaces
neon signs	reactive power regulators
oil-burners	

Let us briefly examine these special applications.

1. A toy transformer is often accidentally short-circuited, but being used by children, it is neither practical nor safe to protect it with a fuse. Consequently, the transformer is designed so that its leakage reactance is so high that even a permanent short-circuit will not cause overheating.

   The same remarks apply to some bell transformers that distribute low-voltage signalling power throughout a home. If a short-circuit occurs on the secondary side, the current is automatically limited by the high reactance so as not to burn out the transformer or damage the fragile annunciator wiring.

2. Electric arc furnaces and discharges in gases possess a negative E/I characteristic, meaning that once the arc is struck, the current increases as the voltage falls. To maintain a steady arc, or a uniform dis-

charge, we must add an impedance in series with the load. The series impedance may be either a resistor or reactor, but we prefer the latter because it consumes very little active power.

However, if a transformer is used to supply the load, it is usually more economical to incorporate the reactance in the transformer itself, by designing it to have a high leakage reactance. A typical example is the neon-sign transformer shown in Fig. 11-24.

The primary winding P is connected to a 240 V ac source, and the two secondary windings S are connected in series across the long neon tube. Owing to the large leakage fluxes Φ_a and Φ_b, the secondary voltage E_2 falls rapidly with increasing current, as seen in the regulation curve of the transformer (Fig. 11-24c). The high open-circuit voltage (20 kV) initiates the discharge, but as soon as the neon tube lights up, the secondary current is automatically limited to 15 mA. The corresponding voltage across the neon tube falls to 15 kV. The power of these transformers ranges from 50 VA to 1500 VA. The secondary voltages range from 2 kV to 20 kV, depending mainly upon the length of the tube.

Returning to Fig. 11-24a, we note that the center of the secondary winding is grounded. This ensures that the secondary line-to-ground voltage is only one-half the voltage across the neon tube. As a result, less insulation is needed to isolate the winding from ground.

Fluorescent lamp transformers (called ballasts) have properties similar to neon-sign transformers. Capacitors are usually added to improve the power factor of the total circuit.

Oil-burner transformers possess essentially the same characteristics as

(a)

(b)

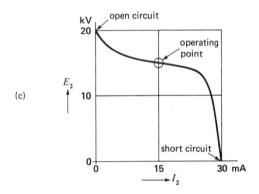

(c)

Figure 11-24
a. Schematic diagram of a neon-sign transformer.
b. Construction of the transformer.
c. Typical E-I characteristic of the transformer.

neon-sign transformers do. A secondary open-circuit voltage of about 10 kV creates an arc between two closely-spaced electrodes situated immediately above the oil jet. The arc continually ignites the vaporized oil while the burner is in operation.

3. Some electric furnaces generate heat by maintaining an intense arc between two carbon electrodes. A relatively low sec-

ondary voltage is used and the large secondary current is limited by the leakage reactance of the transformer. Such transformers have ratings between 100 kVA and 500 MVA. In very big furnaces, the leakage reactance of the secondary together with the reactance of the conductors is usually sufficient to provide the necessary limiting impedance.

4. Arc-welding transformers are also designed to have a high leakage react-

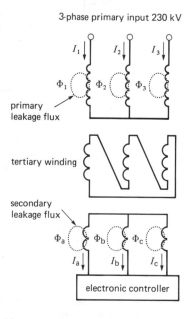

Figure 11-25
Three-phase static var compensator having high leakage reactance.

ance so as to stabilize the arc during the welding process. The open-circuit voltage is about 70 V, which facilitates striking the arc when the electrode touches the work. However, as soon as the arc is established, the secondary voltage falls to about 15 V, a value which depends upon the length of the arc and the intensity of the welding current.

5. As a final example of high-impedance transformers, we mention the enormous 3-phase units that absorb reactive power from a 3-phase transmission line. These transformers are intentionally designed to produce leakage flux and, consequently, the primary and secondary windings are very loosely coupled. The three primary windings are connected to the HV line (typically between 230 kV and 765 kV) while the three secondary windings (typically 6 kV) are connected to an electronic controller (Fig. 11-25). The controller permits more or less secondary current to flow, causing the leakage flux to vary accordingly. A change in the leakage flux produces a corresponding change in the reactive power absorbed by the transformer. The transformer, incorporated in a static var compensator, is further discussed in Chapter 25.

Figure 11-26
Coreless induction furnace. The flux Φ produces eddy currents in the molten metal. The capacitor furnishes the reactive power absorbed by the coil.

11.10 Induction heating transformers

High-power induction furnaces also use the transformer principle to produce high-quality steel and other alloys. The induction principle can be understood by referring to Fig. 11-26. A relatively high-frequency 500 Hz ac source is connected to a coil that surrounds a large crucible containing molten iron. The coil is the primary and the molten iron acts like a single secondary turn, short-circuited upon itself. Consequently, it carries a very large secondary current. This current provides the energy that keeps the iron in a liquid state, melting other scrap metal as it is added to the pool.

Such induction furnaces have ratings between 15 kVA and 40 000 kVA. The operating frequency becomes progressively lower as the power rating increases. Thus, a frequency of 60 Hz is used when the power exceeds about 3000 kVA.

The power factor is very low (typically 20 percent) because a large magnetizing current is required to drive the flux through the molten iron and through the air. In this regard, we must remember that the temperature of molten iron is far above the Curie point, and so it behaves like air as far as permeability is concerned. That is why these furnaces are often called coreless induction furnaces.

Capacitors are installed close to the coil to supply the reactive power it absorbs.

In another type of furnace, known as a *channel furnace*, a transformer having a laminated iron core is made to link with a channel of molten iron, as shown in Fig. 11-27. The channel is a ceramic pipe that is fitted to the bottom of the crucible. The primary coil is excited by a 60 Hz source, and the secondary current I_2 flows in the liquid channel and through the molten iron in the crucible. In effect, the channel is equivalent to a single turn short-circuited on itself.

The magnetizing current is low because the flux is confined to a highly permeable iron core. On the other hand, the leakage flux is large because the secondary "turn" is obviously not tightly coupled to the primary coil. Nevertheless, the power factor is higher than that in Fig. 11-26, being typically between 60 and 80 percent. As a result, a smaller capacitor bank is required to furnish the reactive power.

Owing to the very high ambient temperature, the primary windings of induction furnace transformers are always made of hollow, water-

Figure 11-27
Channel induction furnace and its water-cooled transformer.

cooled copper conductors. Induction furnaces are used for melting aluminum, copper, and other metals, as well as iron.

Figure 11-28 shows a very special application of the induction heating principle.

QUESTIONS AND PROBLEMS

Practical level

11-1 What is the difference between an autotransformer and a conventional transformer?

11-2 What is the purpose of a potential transformer? Of a current transformer?

11-3 Why must we never open the secondary of a current transformer?

11-4 Explain why the secondary winding of a CT or PT must be grounded.

Figure 11-28
Special application of the transformer effect. This picture shows one stage in the construction of the rotor of a steam-turbine generator. It consists of expanding the diameter of a 5 t coil-retaining ring. A coil of asbestos-insulated wire is wound around the ring and connected to a 35 kW, 2000 Hz source (left foreground). The coil creates a 2000 Hz magnetic field, which induces large eddy currents in the ring, bringing its temperature up to 280°C in about 3 h. The resulting expansion enables the ring to be slipped over the coil-ends, where it cools and contracts. This method of induction heating is clean and produces a very uniform temperature rise of the large mass. *(courtesy ABB)*

11-5   A toroidal current transformer has a ratio of 1500 A/5 A. How many turns does it have?

11-6   A current transformer has a rating of 10 VA, 50 A/5 A, 60 Hz, 2.4 kV. Calculate the nominal voltage across the primary winding.

Intermediate level

11-7   A single-phase transformer has a rating of 100 kVA, 7200 V/600 V, 60 Hz. If it is reconnected as an autotransformer having a ratio of 7800 V/7200 V, calculate the load it can carry.

11-8   In Problem 11-7, how should the transformer terminals (H_1, H_2, X_1, X_2) be connected?

11-9   The transformer (in problem 11-8) is reconnected again as an autotransformer having a ratio of 6.6 kV/600 V. What load can it carry and how should the connections be made?

Advanced level

11-10 A current transformer has a rating of 100 VA, 2000 A/5 A, 60 Hz, 138 kV. It has a primary to secondary capacitance of 250 pF. If it is installed on a transmission line where the line-to-neutral voltage is 138 kV, calculate the capacitive leakage current that flows to ground (see Fig. 11-13).

11-11 The toroidal current transformer of Fig. 11-19 has a ratio of 1000 A/5 A. The line conductor carries a current of 600 A.

  a. Calculate the voltage across the secondary winding if the ammeter has an impedance of 0.15 Ω.

  b. Calculate the voltage drop the transformer produces on the line conductor.

  c. If the primary conductor is looped four times through the toroidal opening, calculate the new current ratio.

12

THREE-PHASE TRANSFORMERS

Power is distributed throughout North America by means of three-phase transmission lines. In order to transmit this power efficiently and economically the voltages must be at appropriate levels. These levels depend upon the amount of power that has to be transmitted and the distance it has to be carried. Another aspect is the appropriate voltage levels used in factories and homes. These are fairly uniform, ranging from 120/240 V single-phase systems to 600 V, 3-phase systems. Clearly, this requires the use of 3-phase transformers to transform the voltages from one level to another.

The transformers may be inherently 3-phase, having three primary windings and three secondary windings mounted on a 3-legged core. However, the same result can be achieved by using three single-phase transformers connected together to form a so-called 3-phase transformer bank.

12.1 Basic properties of 3-phase transformer banks

When three single-phase transformers are used to transform a 3-phase voltage, the windings can be connected in several ways. Thus, the primaries may be connected in delta and the secondaries in wye, or vice versa. As a result, the ratio of the 3-phase input voltage to the 3-phase output voltage depends not only upon the turns ratio of the transformers, but also upon how they are connected.

A 3-phase transformer bank can also produce a *phase shift* between the 3-phase input voltage and the 3-phase output voltage. The amount of phase shift depends again upon the turns ratio of the transformers, and on how the primaries and secondaries are interconnected. Furthermore, the phase-shift feature enables us to change the *number* of phases. Thus, a 3-phase system can be converted into a 2-phase, a 6-phase, or a 12-phase system. Indeed, if there were a practical application for it, we could even convert a 3-phase system into a 5-phase system by an appropriate choice of single-phase transformers and interconnections.

In making the various connections, it is

important to observe transformer polarities. An error in polarity may produce a short-circuit or unbalance the line voltages and currents.

The basic behavior of balanced 3-phase transformer banks can be understood by making the following simplifying assumptions:

1. the exciting currents are negligible;

2. the transformer impedances are negligible;

3. the total apparent input power to the transformer bank is equal to the total apparent output power;

Furthermore, when single-phase transformers are connected into a 3-phase system, they retain all their basic single-phase properties, such as current ratio, voltage ratio and flux in the core.Note that the phase shift between primary and secondary is either zero or 180°.

12.2 Delta-delta connection

The three single-phase transformers P, Q, and R of Fig. 12-1 transform the voltage of the incoming (source) transmission line A, B, C to a level appropriate for the outgoing (load) transmission line 1, 2, 3. The transformers are connected in *delta-delta*. Terminal H_1 of each transformer is connected to terminal H_2 of the next transformer. Similarly, terminals X_1 and X_2 of successive transformers are connected together. The actual physical layout of the transformers is shown in Fig. 12-1. The corresponding schematic diagram is given in Fig. 12-2.

The schematic diagram is drawn in such a way to show not only the connections, but also the phasor relationship between the primary and secondary voltages. Thus, each secondary winding is drawn parallel to the corresponding primary winding to which it is coupled. Furthermore, if source G produces voltages E_{AB}, E_{BC}, E_{CA} according to the indicated phasor diagram, the primary windings are oriented the

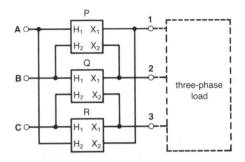

Figure 12-1
Delta-delta connection of three single-phase transformers. The incoming lines (source) are A, B, C and the outgoing lines (load) are 1, 2, 3.

same way, phase by phase. For example, the primary of transformer P is oriented horizontally in the same direction as phasor E_{AB}.

Because the primary and secondary voltages of a given transformer must be in phase, it follows that E_{12} (secondary voltage of transformer P) must be in phase with E_{AB} (primary of the same transformer). Similarly, E_{23} is in phase with E_{BC}, and E_{31} with E_{CA}.

In such a delta-delta connection, the voltages between the incoming and outgoing transmission lines are in phase.

If a balanced load is connected to lines 1-2-3 the resulting line currents are equal in magnitude. This produces balanced line currents in the incoming lines A-B-C. As in any delta connection, the line currents are $\sqrt{3}$ times greater than the respective currents I_p and I_s flowing in the primary and secondary windings (Fig. 12-2). The power rating of the transformer bank is 3 times the rating of a single transformer.

Note that although the transformer bank constitutes a 3-phase arrangement, each transformer, considered alone, acts as if it were placed in a single-phase circuit. Thus, a current I_p flowing into a primary polarity mark H_1 is associated with a current I_s flowing out of a secondary polarity mark X_1.

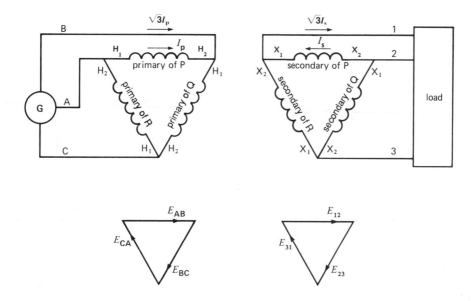

Figure 12-2
Schematic diagram of a delta-delta connection and associated phasor diagram.

Example 12-1:

Three single-phase transformers are connected in delta-delta to step down a line voltage
of 138 kV to 4160 V to supply power to a manufacturing plant. The plant draws 21 MW at a
lagging power factor of 86 percent. Calculate:
a. the apparent power drawn by the plant;
b. the apparent power furnished by the HV
    line;
c. the current in the HV lines;
d. the current in the LV lines;
e. the currents in the primary and secondary
    windings of each transformer;
f. the load carried by each transformer.

Solution:

a. The apparent power drawn by the plant is:

$$S = P/\cos\theta \qquad \text{Eq. 7-7}$$
$$= 21/0.86$$
$$= 24.4 \text{ MVA}$$

b. The transformer bank itself absorbs a neg-

ligible amount of active and reactive power
because the I^2R losses and the reactive
power associated with the mutual flux and
leakage flux are small. It follows that the
apparent power furnished by the HV line is
also 24.4 MVA.

c. The current in each HV line is:
$$I_1 = S/(\sqrt{3}E) \qquad \text{Eq. 8-9}$$
$$= (24.4 \times 10^6)/(\sqrt{3} \times 138\,000)$$
$$= 102 \text{ A.}$$

d. The current in the LV lines is:
$$I_2 = S/(\sqrt{3}\,E)$$
$$= (24.4 \times 10^6)/(\sqrt{3} \times 4160)$$
$$= 3384 \text{ A}$$

e. Referring to Fig. 12-2, the current in the pri-
    mary winding is:
$$I_p = 102/\sqrt{3} = 58.9 \text{ A}$$

The current in the secondary winding is:
$$I_s = 3384/\sqrt{3} = 1954 \text{ A}$$

f. Because the plant load is balanced, each transformer carries one third of the total load, or 24.4/3 = 8.13 MVA.

The transformer load is also obtained by the product of the primary voltage times the primary current:

$$S = E_p I_p = 138\ 000 \times 58.9$$
$$= 8.13\ \text{MVA}$$

Note that we can calculate the line currents and the currents in the transformer windings even though we do not know how the 3-phase load is connected. In effect, the plant load (shown as a box in Fig. 12-2) is composed of hundreds of individual loads, some of which are connected in delta, others in wye. Furthermore, some are single-phase loads operating at much lower voltages than 4160 V, powered by smaller transformers located inside the plant. The sum total of these loads usually results in a reasonably well-balanced 3-phase load, represented by the box.

12.3 Delta-wye connection

When the transformers are connected in *delta-wye*, the three primary windings are connected the same way as in Fig. 12-1. However, the secondary windings are connected so that all the X_2 terminals are joined together, creating a common neutral N (Fig. 12-3). In such a delta-wye connection, the voltage across each primary winding is equal to the incoming line voltage. However, the outgoing line voltage is $\sqrt{3}$ times the secondary voltage across each transformer.

The relative values of the currents in the transformer windings and transmission lines are given in Fig. 12-4. Thus, the line currents in phases A, B, and C are $\sqrt{3}$ times the currents in the primary windings. The line currents in phases 1, 2, 3 are the same as the currents in the secondary windings.

A delta-wye connection produces a 30° phase shift between the line voltages of the incoming and outgoing transmission lines.

Figure 12-3
Delta-wye connection of three single-phase transformers

Thus, outgoing line voltage E_{12} is 30° ahead of incoming line voltage E_{AB}, as can be seen from the phasor diagram. If the outgoing line feeds an isolated group of loads, the phase shift creates no problem. But, if the outgoing line has to be connected in parallel with a line coming from another source, the 30° shift may make such a parallel connection impossible, even if the line voltages are the same.

One of the important advantages of the wye connection is that it reduces the amount of insulation needed inside the transformer. In effect, the HV winding has to be insulated for only $1/\sqrt{3}$ or 58 percent of the line voltage.

Example 12-2:
Three single-phase step-up transformers rated 40 MVA, 13.2 kV/80 kV are connected in delta-wye on a 13.2 kV transmission line (Fig. 12-5). If they feed a 90 MVA load, calculate:
a. the secondary line voltage;
b. the currents in the transformer windings;
c. the incoming and outgoing transmission line currents.

Solution:
The easiest way to solve this problem is to consider the windings of only one transformer, say transformer P.

a. The voltage across the primary winding is obviously 13.2 kV.

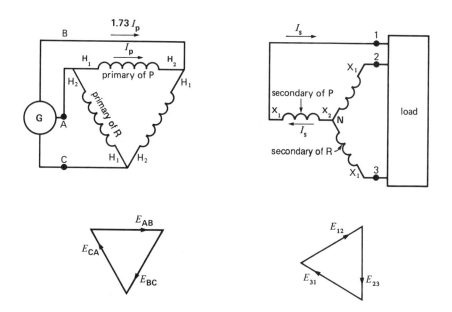

Figure 12-4
Schematic diagram of a delta-wye connection and associated phasor diagram.

The voltage across the secondary is therefore 80 kV.

The voltage between the outgoing lines 1, 2, and 3 is:
$$E_s = 80 \sqrt{3} = 138 \text{ kV}$$

b. The load carried by each transformer is:
$$S = 90/3 = 30 \text{ MVA}$$

The current in the primary winding is:
$$I_p = 30 \text{ MVA}/13.2 \text{ kV} = 2272 \text{ A}$$

Figure 12-5
See Example 13-2.

The current in the secondary winding is:
$$I_s = 30 \text{ MVA}/80 \text{ kV} = 375 \text{ A}$$

c. The current in each incoming line A, B, C is:
$$I = 2272 \sqrt{3} = 3932 \text{ A}$$

The current in each outgoing line 1, 2, 3 is:
$$I = 375 \text{ A}$$

12.4 Wye-delta connection

The currents and voltages in a *wye-delta* connection are identical to those in the delta-wye connection of Sec. 12-3. The primary and secondary connections are simply interchanged. In other words, the H_2 terminals are connected together to create a neutral, and the X_1, X_2 terminals are connected in delta. Again, there results a 30° phase shift between the voltages of the incoming and outgoing lines.

12.5 Wye-wye connection

When transformers are connected in wye-wye, special precautions have to be taken to prevent severe distortion of the line-to-neutral voltages. One way to prevent the distortion is to connect the neutral of the primary to the neutral of the source, usually by way of the ground (Fig. 12-6). Another way is to provide each transformer with a third winding, called *tertiary* winding. The tertiary windings of the 3 transformers are connected in delta (Fig. 12-7). They often provide the substation service voltage where the transformers are installed.

Note that there is no phase shift between the incoming and outgoing transmission line voltages of a wye-wye connected transformer.

12.6 Open-delta connection

It is possible to transform the voltage of a 3-phase system by using only 2 transformers, connected in open-delta. The open-delta arrangement is identical to a delta-delta connection, except that one transformer is absent (Fig. 12-8). However, the open-delta connection is seldom used because the load capacity of the transformer bank is only 86.6 percent of the installed transformer capacity. For example, if two 50 kVA transformers are connected in open-delta, the installed capacity of the transformer bank is obviously $2 \times 50 = 100$ kVA. But strange as it may seem, it can only deliver 86.6 kVA before the transformers begin to overheat.

The open-delta connection is mainly used in emergency situations. Thus, if three transformers are connected in delta-delta and one of them becomes defective and has to be removed, it is possible to feed the load on a temporary basis with the two remaining transformers.

Example 12-3:
Two single-phase 150 kVA, 7200 V/600 V transformers are connected in open-delta. Calculate the maximum 3-phase load they can carry.

Figure 12-6
Wye-wye connection with neutral of the primary connected to the neutral of the source.

Figure 12-7
Wye-wye connection using a tertiary winding.

Solution:
Although each transformer has a rating of 150 kVA, the two together *cannot* carry a load of 300 kVA. The following calculations show why:

The nominal secondary current of each transformer is:

$$I_s = 150 \text{ kVA}/600 \text{ V} = 250 \text{ A}$$

The current I_s in lines 1, 2, 3 cannot, therefore, exceed 250 A (Fig. 12-8). Consequently, the maximum load which the transformers can carry is:

$$S = \sqrt{3}EI \qquad\qquad \text{Eq. 8-9}$$
$$= 1.73 \times 600 \times 250 = 259\,500 \text{ VA}$$
$$= 260 \text{ kVA}$$

Thus, the ratio

Figure 12-8a
Open-delta connection

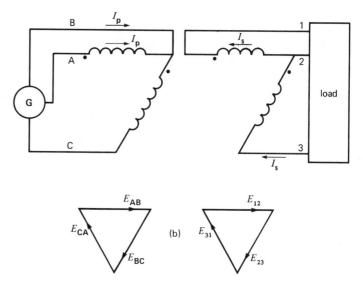

Figure 12-8b
Associated schematic and phasor diagram.

$$\frac{\text{maximum load}}{\text{installed transformer rating}} = \frac{260 \text{ kVA}}{300 \text{ kVA}}$$

$$= 0.866 \text{ or } 86.6 \text{ percent}$$

12.7 Three-phase transformers

A transformer bank composed of three single-phase transformers may be replaced by one 3-phase transformer (Fig. 12-9). The magnetic core of such a transformer has 3 legs which carry the primary and secondary windings of each phase. The windings are connected inter-

Figure 12-9
Three-phase transformer for an electric arc furnace, rated 36 MVA, 13.8 kV/160 V to 320 V, 60 Hz. The secondary voltage is adjustable from 160 V to 320 V by means of 32 taps on the primary winding (not shown). The three large bus bars in the foreground deliver a current of 65 000 A. Other characteristics: impedance: 3.14%; diameter of each leg of the core: 711 mm; overall height of core: 3500 mm; center line distance between adjacent core legs: 1220 mm. (*Ferranti-Packard*)

nally, either in wye or in delta, with the result that only six terminals have to be brought outside the tank. For a given capacity, a 3-phase transformer is always smaller and cheaper than three single-phase transformers having the same capacity. Nevertheless, single-phase transformers are sometimes preferred, particularly when a replacement unit is essential. For example, suppose a manufacturing plant absorbs 5000 kVA. To guarantee continued service we can install one 3-phase 5000 kVA transformer and keep a second one as a spare. Alternatively, we can install three single-phase transformers each rated at 1667 kVA, plus one spare. The 3-phase transformer option is more expensive (total capacity: 2 × 5000 = 10 000 kVA) than the single-phase option (total capacity: 4 × 1667 = 6667 kVA).

Figure 12-10 shows successive stages of construction of a 3-phase 110 MVA, 222.5 kV/34.5 kV tap-changing transformer.* Note that in addition to the three main legs, the magnetic core has two additional lateral legs. They enable the designer to reduce the overall height of the transformer, which simplifies the problem of shipping. In effect, whenever large equipment has to be shipped, the designer is faced with the problem of overhead clearances on highways and rail lines.

The 34.5 kV windings (connected in delta) are mounted next to the core. The 222.5 kV windings (connected in wye) are mounted on top of the 34.5 kV windings. A space of several centimeters separates the two windings to ensure good isolation and to allow cool oil to flow freely between them. The HV bushings that protrude from the oil-filled tank are connected to a 220 kV line. The medium voltage (MV) bushings are much smaller, and cannot be seen in the photograph.

---

* A tap-changing transformer regulates the secondary voltage by automatically switching from one tap to another on the primary winding. The tap-changer is a motorized device under the control of a sensor that continually monitors the voltage that has to be held constant.

Figure 12-10a
Core of a 110 MVA, 222.5 kV/34.5 kV, 60 Hz, 3-phase transformer. By staggering laminations of different widths, the core legs can be made almost circular. This reduces the coil diameter to a minimum, resulting in less copper and also lower I^2R losses. The legs are tightly bound to reduce vibration. Mass of core: 53 560 kg. (*ASEA*)

Figure 12-10b
Same transformer with coils in place. The primary winding are connected in wye, and the secondaries in delta. Each primary has 8 taps to change the voltage in steps of ±2.5%. The motorized tap-changer can be seen in the right upper corner of the transformer. Mass of copper: 15 230 kg.

Figure 12-10c
Same transformer ready for shipping. It has been subjected to a 1050 kV impulse test on the HV side and a similar 250 kV test on the LV side. Other details: power rating: 110 MVA/146.7 MVA (OA/FA); total mass including oil: 158.7 t; overall height: 9 m; width: 8.2 m, length: 9.2 m. (*courtesy* ABB)

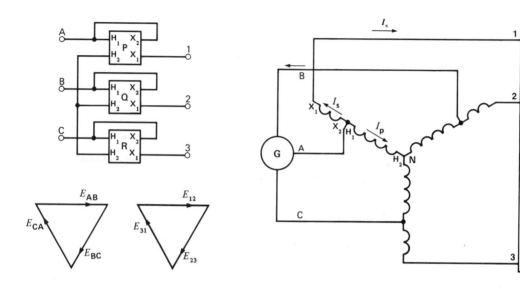

Figure 12-11a
Wye-connected autotransformer.

Figure 12-11b
Associated schematic diagram.

12.8 Step-up and step-down autotransformer

When the voltage of a 3-phase line has to be stepped up or stepped down by a moderate amount, it is economically advantageous to use three single-phase transformers connected as a wye-connected *autotransformer*. The actual physical connections are shown in Fig. 12-11a, and the corresponding schematic diagram is given in Fig. 12-11b. It is obvious that the incoming and outgoing transmission line voltages are in phase. The neutral is connected to the system neutral, otherwise a tertiary winding must be added to prevent the voltage distorsion mentioned previously.

For a given power output, an autotransformer is smaller and cheaper than a conventional transformer (see Sec. 11-2). This is particularly true if the ratio of the incoming line voltage to outgoing line voltage lies between 0.5 and 2.

Figure 12-11c shows a large single-phase autotransformer rated 404 kV/173 kV with a tertiary winding rated 11.9 kV. It is part of a 3-phase transformer bank used to connect a 700 kV transmission line to an existing 300 kV system.

Example 12-4:
The voltage of a 3-phase, 230 kV line has to be stepped up to 345 kV to supply a load of 200 MVA. Three single-phase transformers connected as autotransformers are to be used. Calculate the basic power and voltage rating of each transformer, assuming they are connected as shown in Fig. 12-11b.

Solution:
To simplify the calculations, let us consider only one phase (phase A, say).

The line-to-neutral voltage between X_1 and H_2 is:

$$E_{1N} = 345/\sqrt{3} = 200 \text{ kV}$$

The line-to-neutral voltage between H_1 and H_2 is:

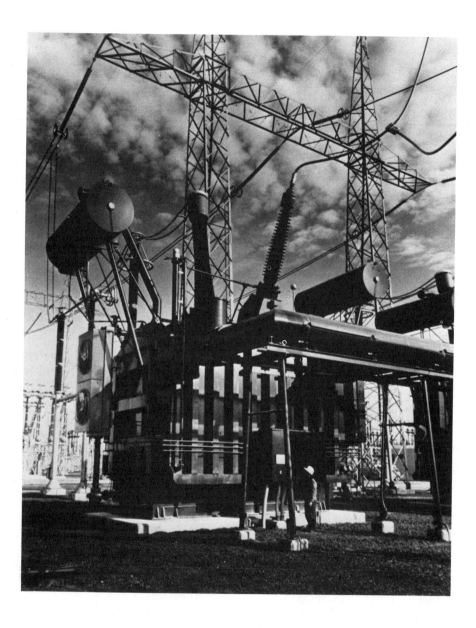

Figure 12-11c
Single-phase autotransformer (one of a group of three) connecting a 700 kV, 3-phase, 60 Hz transmission line to an existing 300 kV system. The transformer ratio is 404 kV/173 kV, to give an output of 200/267/333 MVA per transformer, at a temperature rise of 55°C. Cooling is OA/FA/FOA. A tertiary winding rated 35 MVA, 11.9 kV maintains balanced and distorsion-free line-to-neutral voltages, while providing power for the substation. Other properties of this transformer: weight of core and windings: 132 t; tank and accessories: 46 t; oil: 87 t; total weight: 265 t. BIL rating is 1950 kV and 1050 kV on the HV and LV side, respectively. Note the individual 700 kV (right) and 600 kV(left)bushings protruding from the tank. (*Hydro-Québec*)

$E_{AN} = 230/\sqrt{3} = 133$ kV

The voltage of winding X_1X_2 is:
$E_{1A} = 200 - 133 = 67$ kV

This means that each transformer has an effective primary to secondary voltage rating of 133 kV to 67 kV.

The current in each phase of the outgoing line is:

$I_s = S/\sqrt{3}\,E$ Eq. 8-9
$= (200 \times 10^6) / (1.73 \times 345\,000)$
$= 335$ A

The power associated with winding X_1X_2 is:
$S_a = 67\,000 \times 335 = 22.4$ MVA

Winding H_1H_2 has the same power rating. The basic rating of each single-phase transformer is therefore 22.4 MVA.

The basic rating of the 3-phase transformer bank = $22.4 \times 3 = 67.2$ MVA

The basic transformer rating (as far as size is concerned) is considerably less than its load-carrying capacity of 200 MVA. This is in keeping with the fact that the ratio of transformation (345/230 = 1.5) lies between 0.5 and 2.0.

12.9 Phase-shift principle

A 3-phase system enables us to shift the phase-angle of a voltage very simply. Such phase-shifting enables us to create 2-phase, 6-phase and 12-phase systems from an ordinary 3-phase line. Such multi-phase systems are used in large electronic converter stations and in special electric controls. Phase shifting is also used to control power flow over transmission lines that are part of a power grid.

To understand the phase shifting principle, consider a potentiometer connected between phases B and C of a 3-phase line (Fig. 12-12). As we slide contact P from phase B towards phase C, voltage E_{AP} changes both in amplitude and phase. We obtain a 60° phase shift in moving from one end of the potentiometer to the other. Thus, as we move from B to C, vol-

Figure 12-12
Voltage E_{AP} can be phase-shifted with respect to E_{AC} by means of a potentiometer.

tage E_{AP} gradually advances in phase with respect to E_{AB}. At the same time, the magnitude of E_{AP} varies slightly, from E (voltage between the lines) to $0.866\,E$ when the contact is in the middle of the potentiometer.

Such a simple phase-shifter can only be used in circuits where the load between terminals A and P draws a few milliamperes. If a heavier load is applied, the resulting IR drop in the potentiometer completely changes the voltage and phase angle from what it was on open-circuit.

To get around this problem, we connect a multitap autotransformer between phases B and C (Fig. 12-13). By moving contact P, we obtain the same open-circuit voltages and phase shifts as before, but this time they remain essentially unchanged when a load is connected between terminals A and P. Why is this so? The reason is that the flux in the autotrans-

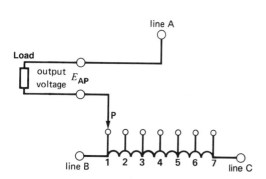

Figure 12-13
Autotransformer used as a phase-shifter.

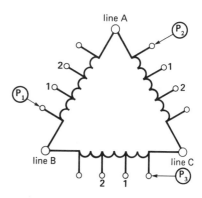

Figure 12-14
Three-phase phase shifter.

former is fixed because E_{BC} is fixed. As a result, the voltage across each turn remains fixed (both in magnitude and phase) whether the autotransformer delivers a current to the load or not.

Figure 12-14 shows 3 tapped autotransformers connected between phases A, B and C. Contacts P_1, P_2, P_3 move in tandem as we switch from one set of taps to the next. This arrangement enables us to create a 3-phase source P_1, P_2, P_3 whose phase angle changes step-wise with respect to source ABC. We obtain a maximum phase shift of 60° as we move from one end of the autotransformers to the other. We now discuss some practical applications of the phase-shift principle.

12.10 Three-phase to 2-phase transformation

The voltages in a 2-phase system are equal, but displaced from each other by 90°. There are several ways we can create a 2-phase system from a 3-phase source. One of the simplest and cheapest is to use a single-phase autotransformer having taps at 50 percent and 86.6 percent. We connect it between any two phases of a 3-phase line, as shown in Fig. 12-15. If the voltage between phases A, B, C is 100 V, voltages E_{AT} and E_{CN} are both equal to 86.6 V.

Furthermore, they are displaced by 90°. This relationship can be seen by referring to the phasor diagram (Fig. 12-15c), and reasoning as follows:

1. Phasors E_{AB}, E_{BC} and E_{CA} are fixed by the source;

2. Phasor E_{AN} is in phase with phasor E_{AB} because the same ac flux links the turns of the autotransformer;

3. Phasor E_{AT} is in phase with phasor E_{AB} for the same reason;

4. From Kirchhoff's voltage law, $E_{AN} + E_{NC} + E_{CA} = 0$. Consequently, phasor E_{NC} must have the value and direction shown in the figure.

Loads 1 and 2 must be isolated from each other, such as the isolated windings of a 2-phase induction motor. The ratio of transformation (3-phase voltage to 2-phase voltage) is fixed and given by $E_{AB}/E_{AT} = 100/86.6 = 1.15$.

Another way to produce a 2-phase system is to use the so-called *Scott connection*. It consists of two identical single-phase transformers, the one having a 50% tap and the other an 86.6% tap on the primary winding. The transformers are connected as shown in Fig. 12-16. The 3-phase source is connected to

(a)

(b)

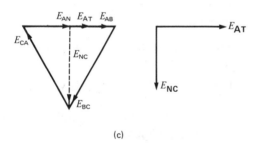

(c)

Figure 12-15
a. Simple method to obtain a 2-phase system from a 3-phase line, using a single transformer winding.
b. Schematic diagram of the connections.
c. Phasor diagram of the voltages.

terminals A, B, C and the 2-phase load is connected to the secondary windings. The ratio of transformation (3-phase line voltage to 2-phase line voltage) is given by E_{AB}/E_{12}. The

Scott connection has the advantage of isolating the 3-phase and 2-phase systems and providing any desired voltage ratio between them.

Except for servomotor applications, two-phase systems are seldom encountered today.

Example 12-5:
A 2-phase, 7.5 kW (10 hp), 240 V, 60 Hz motor has an efficiency of 0.83 and a power factor of 0.80. It is to be fed from a 600 V, 3-phase line using a Scott-connected transformer bank (Fig. 12-16c). Calculate:

a. the apparent power drawn by the motor;
b. the current in each 2-phase line;
c. the current in each 3-phase line.

Solution:
a.   The active power drawn by the motor is:

$$P_1 = P_o/\eta = 7500/0.83$$
$$= 9036 \text{ W}$$

The apparent power drawn by the motor is:
$$S = P/\cos\phi = 9036/0.8$$
$$= 11\ 295 \text{ VA}$$

The apparent power per phase is:
$$S = 11\ 295/2 = 5648 \text{ VA}$$

b.   The current in each 2-phase line is:
$$I = S/E = 5648/240$$
$$= 23.5 \text{ A}$$

c.   The transformer bank itself consumes very little active and reactive power; consequently, the 3-phase line supplies only the active and reactive power absorbed by the motor. The total apparent power furnished by the 3-phase line is therefore 11 295 VA.

The 3-phase line current is:
$$I = S/(\sqrt{3}E) = 11\ 295/(\sqrt{3} \times 600)$$
$$= 10.9 \text{ A}$$

Figure 12-16c shows the power circuit and the line voltages and currents.

Figure 12-16c
See Example 12-5.

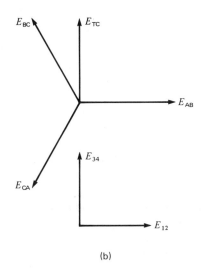

(a)

(b)

Figure 12-16
a. Scott connection
b. Phasor diagram of the Scott connection.

12.11 Phase-shift transformer

A phase-shift transformer is a special type of 3-phase autotransformer that shifts the phase angle between the incoming and outgoing lines without changing the voltage ratio.

Consider a 3-phase transmission line connected to the terminals A, B, C of a phase-shift transformer (Fig. 12-17). The transformer twists all the incoming line voltages through an angle α without, however, changing their magnitude. The result is that all the voltages of the outgoing transmission line 1, 2, 3 are shifted with respect to the voltages of the incoming line ABC. The angle may be leading or lagging, and is usually variable between zero and ±20°.

The phase angle is sometimes varied in discrete steps by means of a motorized tap-changer.

The basic power rating of the transformer

Figure 12-17a
Phase-shift transformer.

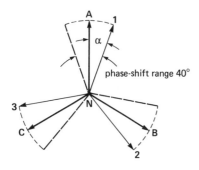

Figure 12-17b
Phasor diagram showing the range over which the phase angle of the outgoing line can be varied.

(which determines its size) depends upon the apparent power carried by the transmission line, and upon the phase shift. For angles less than 20°, it is given by the approximate equation:

$$S_T = 0.025 \, S_L \, \alpha_{max} \qquad (12\text{-}1)$$

where

S_T = basic power rating of the 3-phase transformer bank [VA]

S_L = apparent power carried by the transmission line [VA]

α_{max} = maximum transformer phase shift [°]

0.025 = an approximate coefficient

Example 12-6:
A phase-shift transformer is designed to control 150 MVA on a 230 kV, 3-phase line. The phase angle is variable between zero and ± 15°.
a. Calculate the approximate basic power rating of the transformer;
b. Calculate the line currents in the incoming and outgoing transmission lines.

Solution:
a. The basic power rating is:
$$S_T = 0.025 \, S_L \, \alpha_{max} \qquad \text{Eq. 12-1}$$
$$= 0.025 \times 150 \times 15$$
$$= 56 \text{ MVA}$$

Note that the power rating is much less than the power which the transformer carries. This is a feature of all autotransformers.

b. The line currents are the same in both lines, because the voltages are the same. The line current is:
$$I = S_L/\sqrt{3} \, E \qquad \text{Eq. 8-9}$$
$$= (150 \times 10^6)/(\sqrt{3} \times 230\,000)$$
$$= 377 \text{ A}$$

Figure 12-17c is an example of a 3-phase transformer that could be used to obtain a phase shift of 20 degrees. The transformer has two windings on each leg. Thus, the leg associated with phase A has one winding PN with a tap brought out at terminal A and a second winding having terminals a, 3. The windings of the 3 phases are interconnected as shown. The incoming line is connected to terminals A, B, C and the outgoing line to terminals 1, 2, 3.

The result is that E_{1N} lags 20° behind E_{AN}. Similarly, E_{2N} lags 20° behind E_{BN} and E_{3N} lags 20° behind E_{CN} (Fig. 12-17d).

The basic principle of obtaining a phase shift is to connect two voltages in series that are generated by two different phases. Thus, voltage E_{1b} generated by phase B is connected in series with E_{PN} generated by phase A. The values of E_{PN} and E_{1b} are selected so that the output voltage is equal to the input voltage

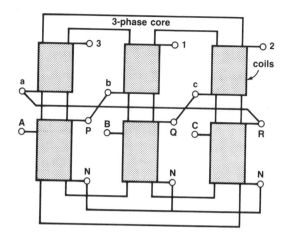

Figure 12-17c
Construction of a 3-phase-shift transformer. The incoming terminals are A, B, C; the outgoing terminals are 1, 2 3.

while obtaining the desired phase angle between them. In our particular example, if E is the line-to-neutral voltage of the incoming line, the respective voltages across the windings of phase A are:

$$E_{AN} = E;$$
$$E_{PN} = 1.14E;$$
$$E_{3a} = 0.40E$$

The internal circuit of a tap-changing, phase-shift transformer is much more complex. However, it rests upon the basic principles we have just discussed. The purpose of such transformers will be covered in Chapter 25.

12.12 Voltage regulation

The voltage regulation of a 3-phase transformer bank is calculated the same way as for a single-phase transformer. In making the calculations, we proceed as follows:

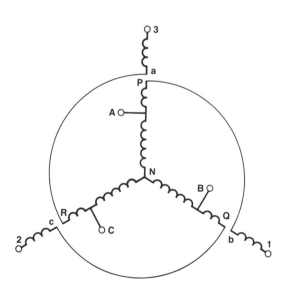

Figure 12-17d
Schematic diagram of the transformer in Fig. 12-17c.

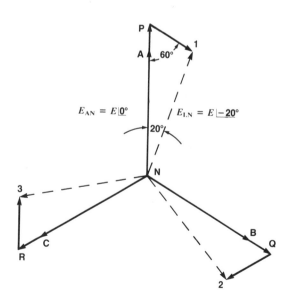

Figure 12-17e
Phasor diagram of a transformer that gives a phase-shift of 20°.

1. We assume that the primary and secondary windings are both connected in wye, *even if they are not* (see Sec. 8-14). This eliminates the problem of having to deal with delta-wye and delta-delta voltages and currents.

2. We consider only one transformer (single phase) of this assumed wye-wye transformer bank.

3. The primary voltage of this transformer is the line-to-neutral voltage of the incoming line.

4. The secondary voltage of this transformer is the line-to-neutral voltage of the outgoing line.

5. The nominal power rating of this transformer is one third the rating of the 3-phase transformer bank.

6. The load on this transformer is one third the load on the transformer bank.

Example 12-7:
The 3-phase step-up transformer shown in Fig. 10-18 (Chap 10) is rated 1300 MVA, 24.5 kV/345 kV, 60 Hz, impedance 11.5%.
a. Determine the equivalent circuit of this transformer, per phase.
b. Calculate the voltage across the generator terminals when the HV line delivers 810 MVA at 370 kV with a lagging power factor of 0.90.

Solution:
a. First, we note that the primary and secondary winding connections are not specified. We don't need this information. However, we assume that both windings are connected in wye.

We shall use the per-unit method to solve this problem. We select the nominal voltage of the secondary winding as our base voltage, E_B.

The base voltage is:

$E_B = 345/\sqrt{3} = 199.2$ kV

Ratio of transformation is:
$a = 345/24.5 = 14.08$

The nominal power rating of the transformer will be used as the base power, S_B. Thus:
$S_B = 1300/3 = 433.3$ MVA

This is a very large transformer; consequently, the transformer impedance is almost entirely reactive. The per-unit impedance is therefore:
$Z_T(pu) = j\,0.115$

The equivalent circuit is shown in Fig. 12-18.

b. The power of the load per phase is:
$S_L = 810/3 = 270$ kVA

The voltage across the load is:
$E_L = 370$ kV/$\sqrt{3} = 213.6$ kV

The per-unit power of the load is:
$S_L(pu) = 270$ kVA/433.3 kVA $= 0.6231$

By selecting E_L as the reference phasor, the per-unit voltage across the load is:
$$E_L(pu) = 213.6 \text{ kV}/199.2 \text{ kV}$$
$$= 1.0723\angle 0°$$

The per-unit current in the load is:
$$I_L(pu) = \frac{S_L(pu)}{E_L(pu)} = \frac{0.6231}{1.0723} = 0.5811$$

The power factor of the load is 0.9. Consequently, I_L lags behind E_L by an angle of $\cos^{-1}0.90 = 25.84°$.

Consequently, the amplitude and phase of the per-unit load current is given by:
$$I_L(pu) = 0.5811\angle{-25.84°}$$

The per-unit voltage E_s is:
$$E_s(pu) = E_L(pu) + I_L(pu) \times Z_T(pu)$$
$$= 1.0723\angle 0° +$$
$$(0.5811\angle{-25.84°}) \times (.115\angle 90°)$$
$$= 1.0723 + 0.0668\angle 64.16°$$
$$= 1.0723 +$$
$$.0668(\cos 64.16° + j\sin 64.16°)$$
$$= 1.1014 + j0.0601$$
$$= 1.103\angle 3.12°$$

therefore:
$$E_s = 1.103 \times 345 \text{ kV} = 380 \text{ kV}$$

The per-unit voltage on the primary side is also
$$E_p = 1.103\angle3.12°$$

The voltage across the terminals of the generator is therefore:
$$\begin{aligned}E_g &= E_p(\text{pu}) \times E_B(\text{primary}) \\ &= 1.103 \times 24.5 \text{ kV} \\ &= 27.02 \text{ kV}\end{aligned}$$

Figure 12-18
See Example 12-7.

12.13 Polarity marking of 3-phase transformers

The HV terminals of a 3-phase transformer are marked H_1, H_2, H_3 and the LV terminals are marked X_1, X_2, X_3. The following rules have been standardized:

1. If the primary windings and secondary windings are connected wye-wye or delta-delta, the voltages between similarly-marked terminals are in phase. Thus:
   $E_{H1,H2}$ is in phase with $E_{X1,X2}$;
   $E_{H2,H1}$ is in phase with $E_{X2,X1}$;
   $E_{H1,H3}$ is in phase with $E_{X1,X3}$
   ... *and so on* ...

2. If the primary and secondary windings are connected wye-delta or delta-wye, there results a 30° phase shift between the primary and secondary line voltages. The internal connections are made so that the

voltages on the HV side *always* lead the voltages of similarly-marked terminals on the LV side. Thus:
   $E_{H1,H2}$ leads $E_{X1,X2}$ by 30°;
   $E_{H2,H1}$ leads $E_{X2,X1}$ by 30°;
   $E_{H3,H2}$ leads $E_{X3,X2}$ by 30°;
   ... *and so on* ...

3. These rules are not affected by the phase sequence of the line voltage applied to the primary side.

QUESTIONS AND PROBLEMS

Practical level

12-1 Assuming that the transformer terminals have polarity marks H_1, H_2, X_1, X_2, make schematic drawings of the following connections:
   a. delta-wye;
   b. open-delta

12-2 Three single-phase transformers rated at 250 kVA, 7200 V/600 V, 60 Hz, are connected in wye-delta on a 12 470 V, 3-phase line. If the load is 450 kVA, calculate the currents:
   a. in the incoming and outgoing transmission lines;
   b. in the primary and secondary windings.

12-3 The transformer in Fig. 12-9 has a rating of 36 MVA, 13.8 kV/320 V. Calculate the nominal currents in the primary and secondary lines.

12-4 Calculate the nominal currents in the primary and secondary windings of the transformer shown in Fig. 10-18, knowing that the windings are connected delta-wye.

Intermediate level

12-5 The transformer shown in Fig. 10-19 operates in the forced-air mode during the morning peaks.

a. Calculate the currents in the second-ary lines if the primary line voltage is 225 kV and the primary line current is 150 A.

b. Is the transformer overloaded?

12-6 The transformers in Problem 12-2 are used to raise the voltage of a 3-phase 600 V line to 7.2 kV.

a. How must they be connected?

b. Calculate the line currents for a 600 kVA load.

c. Calculate the corresponding primary and secondary currents.

12-7 In order to meet an emergency, three single-phase transformers rated 100 kVA, 13.2 kV/2.4 kV are connected in wye-delta on a 3-phase 18 kV line.

a. What is the maximum load that can be connected to the transformer bank?

b. What is the outgoing line voltage?

12-8 Two transformers rated 250 kVA, 2.4 kV/600 V are connected in open-delta to supply a load of 400 kVA.

a. Are the transformers overloaded?

b. What is the maximum load the bank can carry on a continuous basis?

12-9 Referring to Figs. 12-3, and 12-4, the line voltage between phases A-B-C is 6.9 kV and the voltage between lines 1, 2 and 3 are balanced and equal to 600 V. Then, in a similar installation the secondary windings of transformer P are by mistake connected in reverse.

a. Determine the voltages measured between lines 1-2, 2-3, and 3-1.

b. Draw the new phasor diagram.

13

THREE-PHASE INDUCTION MOTORS

Three-phase induction motors are the motors most frequently encountered in industry. They are simple, rugged, low-priced, and easy to maintain. They run at essentially constant speed from zero to full load. However, the speed is frequency-dependent and, consequently, these motors are not easily adapted to speed control. We usually prefer direct-current motors when large speed variations are required. Nevertheless, variable frequency electronic sources are now available to drive commercial induction motors.

In this chapter, we cover the basic principles of the 3-phase induction motor and develop the fundamental equations describing its behavior. We then discuss the general construction and the way the windings are made.

Squirrel-cage, wound-rotor and linear induction motors ranging from a few horsepower to several thousand horsepower permit the reader to see that the same principles can be applied to all.

13.1 Principal components

A 3-phase induction motor has two main parts: a stationary stator and a revolving rotor. The rotor is separated from the stator by a small air gap which ranges from 0.4 mm to 4 mm, depending on the power of the motor.

The *stator* consists of a steel frame which supports a hollow, cylindrical core made up of stacked laminations. A number of evenly spaced slots, punched out of the internal circumference of the laminations, provide the space for the stator winding (Fig. 13-1).

The *rotor* is also composed of punched laminations. These are carefully stacked to create a series of rotor slots to provide space for the rotor winding. We use two types of rotor windings: (a) conventional three-phase windings made of insulated wire and (b) squirrel-cage windings. The type of winding gives rise to two main classes of motors: *squirrel-cage induction motors* and *wound-rotor induction motors*.

Figure 13-1
Stator of a 2 kW, 1725 r/min, 60 Hz, 3-phase induction motor. (*Brook Crompton Parkinson Ltd*)

Figure 13-2
Squirrel-cage rotor of a 2 kW, 1725 r/min, 60 Hz, 3-phase induction motor. (*Brook Crompton Parkinson Ltd*)

A **squirrel-cage rotor** (Fig. 13-2) is composed of bare copper bars, slightly longer than the rotor, which are pushed into the slots. The opposite ends are welded to two copper end-rings, so that all the bars are short-circuited together. The entire construction (bars and end-rings) resembles a squirrel-cage, from which the name is derived. In small and medium-size motors, the bars and end-rings are made of die-cast aluminum, molded to form an integral block (Fig. 13-3a). Figures 13-3b and 13-3c show progressive stages in the manufacture of a squirrel-cage motor.

A **wound rotor** has a three-phase winding, similar to the one on the stator. The winding is uniformly distributed in the slots and is usually connected in wye. The terminals are connected to three slip rings which turn with the rotor (Fig. 13-4). The revolving slip rings and associated stationary brushes enable us to connect external resistors in series with the rotor winding. The external resistors are mainly used during the start-up period; under normal running conditions, the three brushes are short-circuited.

Figure 13-3a
Die-cast aluminum squirrel-cage rotor with integral cooling fan (*Lab-Volt*)

13.2 Principle of the induction motor

The operation of a 3-phase induction motor is based upon the application of Faraday's Law and the Lorentz force on a conductor (Secs. 2-20, 2-21, and 2-22). The behavior can readily be understood by means of the following example.

Figure 13-3b
Progressive steps in the manufacture of stator and rotor laminations. Sheet steel is sheared to size (1), blanked (2), punched (3), blanked (4) and punched (5). (*Lab-Volt*)

(a)

(b)

(c)

Figure 13-3c
Progressive steps in the injection molding of a squirrel-cage rotor.

a) Molten aluminum is poured into a cylindrical cavity. The laminated rotor stacking is firmly held between two molds.

b) Compressed air rams the mold assembly into the cavity. Molten aluminum is forced upwards through the rotor bar holes and into the upper mold.

c) Compressed air withdraws the mold assembly, now completely filled with hot (but hardened) aluminum.

d) The upper and lower molds are pulled away, revealing the die-cast rotor. The cross section view shows that the upper and lower end rings are joined by the rotor bars. (*Lab-Volt*)

(d)

Figure 13-4a
Exploded view of a 5 hp, 1730 r/min wound-rotor induction motor.

Figure 13-4b
Close-up of the slip ring end of the rotor. (*Brook Crompton Parkinson Ltd*)

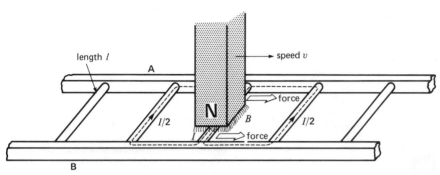

Figure 13-5a
Moving magnet cutting across a conducting ladder.

Consider a series of conductors of length l whose extremities are short-circuited by two bars A and B (Fig. 13-5a). A permanent magnet, placed above this conducting "ladder", moves rapidly to the right at a speed v, so that its magnetic field B sweeps across the conductors. The following sequence of events then takes place:

1. a voltage $E = Blv$ is induced in each conductor while it is being cut by the flux (Faraday's Law);

2. the induced voltage immediately produces a current I, which flows down the conductor underneath the pole-face, through the end-bars, and back through the other conductors;

3. because the current-carrying conductor lies in the magnetic field of the permanent magnet, it experiences a mechanical force (Lorentz force);

4. the force always acts in a direction to drag the conductor along with the magnetic field (Sec. 2-23).

If the conducting "ladder" is free to move, it will accelerate towards the right. However, as it picks up speed, the conductors will be cut less

Figure 13-5b
Ladder bent upon itself to form a squirrel-cage.

rapidly, with the result that the induced voltage E and the current I will diminish. Consequently, the force acting on the conductors will also decrease. If the ladder were to move at the same speed as the magnetic field, the induced voltage E, the current I, and the force dragging the ladder along would all become zero.

In an induction motor, the ladder is closed upon itself to form a squirrel-cage (Fig. 13-5b) and the moving magnet is replaced by a rotating field. The field is produced by the 3-phase currents which flow in the stator windings, as we shall now explain.

13.3 The rotating field

Consider a simple stator having six salient poles, connected as shown in Fig. 13-6. Three identical windings AN, BN, CN, each composed of two coils in series, are arranged to be mechanically spaced at 120° to each other. The respective coil ends, aa, bb, cc are joined but we have neglected to show the connections so as not to encumber the drawing. The three sets of coils are connected in wye, thus forming a common neutral N. Owing to the perfectly symmetrical arrangement, the line-to-neutral impedances are identical. In other words, with regard to terminals A, B, C, the windings constitute a balanced 3-phase system.

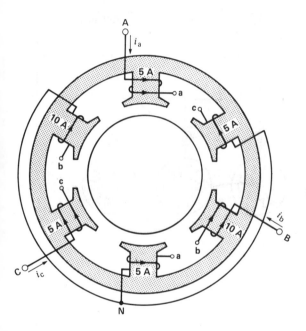

Figure 13-6
Elementary stator having terminals A, B, C, connected to a 3-phase source (not shown).

If we apply a 3-phase source to terminals A, B, C, alternating currents I_a, I_b and I_c will flow in the windings. The currents will have the same value but will be displaced in time by an angle of 120°. These currents produce magnetomotive forces which, in turn, create a magnetic flux. It is this flux we are interested in.

In order to follow the sequence of events, we assume that positive currents always flow from line to neutral. Conversely, negative currents flow from neutral to line. Furthermore, to enable us to work with numbers, suppose that each

coil has 5 turns and that the peak current per phase is 10 A. Thus, when $I_a = +7$ A, the two coils of phase A will together produce a mmf of 70 amperes-turns and a corresponding value of flux. Because the current is positive, the flux is directed vertically upwards, according to the right-hand rule.

As time goes by, we can determine the instantaneous value and direction of the current in each winding, and thereby establish the successive flux patterns. Thus, referring to Fig. 13-7 at instant 1, current I_a has a value of +10 A, whereas I_b and I_c both have a value of −5 A. The mmf of phase A is 10 A × 10 turns = 100 ampere-turns, while the mmf of phases B and C are each 50 ampere-turns. The direction of the mmf depends upon the instantaneous current flows and, using the right-hand rule, we find that the resulting magnetic field has the shape shown in Fig. 13-8a. Note that as far as the rotor is concerned, the six salient poles together produce a magnetic field having essentially one north and one south pole. The combined magnetic field points upwards.

At instant 2, one sixth cycle later, current I_c attains a peak of −10 A, while I_a and I_b both have a value of +5 A (Fig. 13-8b). We discover that the new field has the same shape as before, except that it has moved clockwise by an angle of 60°. In other words, the flux makes one sixth of a turn between instants 1 and 2.

Proceeding in this way for each of the successive instants 3, 4, 5, 6, and 7, separated by intervals of 1/6 cycle, we find that the magnetic field makes one complete turn during one cycle (see Figs. 13-8a to 13-8f).

The rotational speed of the field depends therefore, upon the duration of one cycle which, in turn, depends on the frequency of the source. If the frequency is 60 Hz, the resulting field makes one turn in 1/60 s, or 3600 revolutions per minute. On the other hand, if the frequency were 5 Hz, the field would make one turn in 1/5s, giving a speed of only 300 r/min. Because

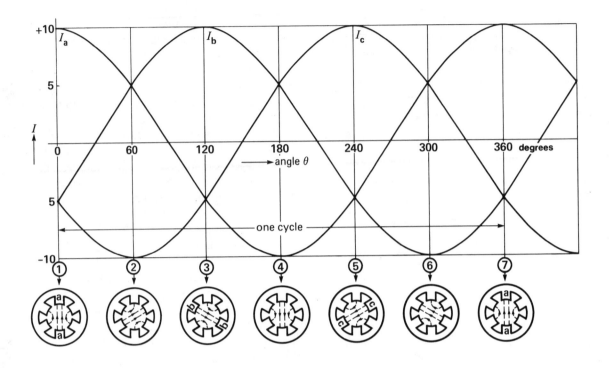

Figure 13-7
Instantaneous values of currents and position of the flux in Fig. 13-6.

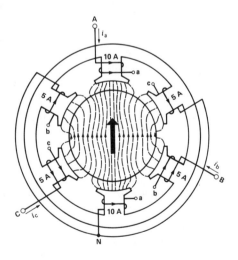

Figure 13-8a
Flux pattern at instant 1

Figure 13-8b
Flux pattern at instant 2.

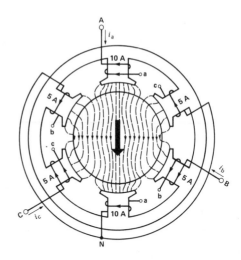

Figure 13-8c
Flux pattern at instant 3.

Figure 13-8d
Flux pattern at instant 4.

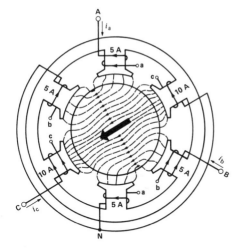

Figure 13-8e
Flux pattern at instant 5.

Figure 13-8f
Flux pattern at instant 6.

the speed of the rotating field is necessarily synchronized with the frequency of the source, it is called *synchronous speed.*

13.4 Direction of rotation

The positive crests of the currents in Fig. 13-7 follow each other in the order A-B-C. This phase sequence produces a field that rotates clockwise. If we interchange any two of the lines connected to the stator, the new phase sequence will be A-C-B. Following the same line of reasoning developed in Sec. 13-3, we find that the field now revolves at synchronous speed in the opposite, or counterclockwise direction. Interchanging any two lines of a 3-phase motor will therefore reverse its direction of rotation.

Although early machines were built with salient poles, the stators of modern motors are smooth. Thus, the two-pole stator of Fig. 13-6 is built as shown in Fig. 13-9a. In effect, the original two coils of phase A (Aa and aN) now become what is called phase group 1 and phase group 2 (Fig. 13-9a). A phase group is equivalent to one salient pole. Each phase group (or simply "group") is composed of 1, 2 or more coils. When there are 2 or more coils, they are distributed in successive stator slots and connected in series. For example, the upper group of phase A is composed of 5 coils. The five coils together produce one pole. When current I_a flows in the two groups, it produces the 2-pole flux pattern shown in Fig. 13-9a.

The groups (poles) of the other two phases are identical and, as seen in Fig. 13-9b, the groups are displaced at 120° to each other. The resulting field due to all three phases again consists of 2 poles.

13.5 Number of poles - synchronous speed

Soon after the invention of the induction motor it was found that the speed of the revolving flux could be reduced by increasing the number of poles.

To construct a 4-pole stator, the coils are distributed as shown in Fig. 13-10a. The 4 identical groups of phase A now span only 90° of the stator circumference. The groups are connected in series and in such a way that adjacent groups produce magnetomotive forces acting in opposite directions. In other words, when a current I_a flows only in the stator winding of phase A (Fig. 13-10a), it creates 4 alternate N-S poles.

The windings of the other two phases are identical but are displaced from each other (and from phase A) by a mechanical angle of 60°. When the wye-connected windings are connected to a 3-phase source, a revolving field having 4 poles is created (Fig. 13-10b). This field rotates at only half the speed of the 2-pole field shown in Fig. 13-9b. We will shortly explain why this is so.

phase group 1
phase A
(I_a = + 10 A)

A ○ → I_a
B ○ → I_b
C ○ → I_c

phase group 2
phase A

Figure 13-9a
Coil group 1 is composed of several coils distributed in the stator slots. The coils are connected in series. Coil group 2 is identical to coil group 1. The two coil groups produce a N and S pole when current flows in the coils. The resulting flux pattern is shown dotted.

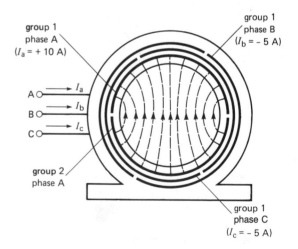

group 1
phase A
(I_a = + 10 A)

group 1
phase B
(I_b = – 5 A)

A ○ → I_a
B ○ → I_b
C ○ → I_c

group 2
phase A

group 1
phase C
(I_c = – 5 A)

Figure 13-9b
Two-pole, full-pitch lap wound stator and resulting magnetic field when the current in phase A = + 10 A, and I_b = I_c = − 5 A.

Figure 13-10a
The four phase groups of phase A produce a 4-pole magnetic field.

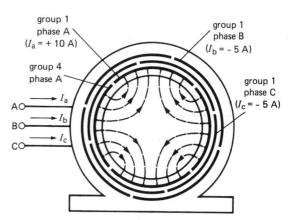

Figure 13-10b
Four-pole, full-pitch lap-wound stator and resulting magnetic field when $I_a = +10$ A, and $I_b = I_c = -5$ A.

We can increase the number of poles as much as we please provided there are enough slots. Thus, Fig. 13-11 shows a 3-phase, 8-pole stator. Each phase consists of 8 groups and the groups of all the phases together produce an 8-pole rotating field. When connected to a 60 Hz source, the poles turn, like the spokes of a wheel, at a synchronous speed of 900 r/min.

Figure 13-11
Eight-pole, full-pitch lap-wound stator and resulting magnetic field when $I_a = +10$ A, and $I_b = I_c + -5$ A.

How can we tell what the synchronous speed will be? Without going into all the details of current flow in the three phases, let us restrict our attention to phase A. In Fig. 13-11, each phase group covers a mechanical angle of $360/8 = 45°$. Suppose the current in phase A is at its maximum positive value. The magnetic flux is then centered on phase A, and the N-S poles are located as shown in Fig. 13-12a. A few moments later, the current in phase A will reverse and when it reaches its maximum negative value, the flux pattern will be the same as before, except that all the N poles will have become S poles and vice versa (Fig. 13-12b). In comparing the two figures, it is clear that the entire magnetic field has shifted by an angle of $45°$ – and this gives us the clue to finding the speed of rotation. The flux moves $45°$ or one pole pitch in one half cycle, and so it takes 8 half cycles ($= 4$ cycles) to make a complete turn. On a 60 Hz system the time to make one turn is therefore $4 \times 1/60 = 1/15$ s. Consequently, the flux turns at the rate of 15 r/s or 900 r/min.

We can generalize by stating that a rotating field advances by one pole pitch in one-half cycle. Thus, if a motor has p poles, the flux will make a complete turn in $p/2$ cycles.

Figure 13-12a
Flux pattern when the current in phase A is at its maximum positive value.

Figure 13-12b
Flux pattern when the current in phase A is at its maximum negative value. The pattern is the same as in Fig. 13-12a but it has advanced by 1 pole pitch.

The speed of a rotating field depends therefore upon the frequency of the source and the number of poles on the stator. Using the same reasoning as above, we can prove that the synchronous speed is always given by:

$$n_s = \frac{120f}{p} \qquad (13\text{-}1)$$

where

n_s = synchronous speed [r/min]
f = frequency of the source [Hz]
p = number of poles per phase

This equation shows that the synchronous speed increases with frequency and decreases with the number of poles.

Example 13-1:
Calculate the synchronous speed of a three-phase induction motor having 20 poles per phase when it is connected to a 50 Hz source.

Solution:
$$n_s = 120f/p = 120 \times 50/20$$
$$= 300 \text{ r/min}$$

13.6 Starting characteristics of a squirrel-cage motor

Let us connect the stator of an induction motor to a 3-phase source, with the rotor locked. The revolving field created by the stator cuts across the rotor bars and induces a voltage in all of them.

This is an ac voltage because each conductor is cut, in rapid succession, by a N pole followed by a S pole. The frequency of the voltage depends upon the number of N and S poles that sweep across a conductor per second; when the rotor is at rest, it is always equal to the frequency of the source.

Because the rotors bars are short-circuited by the end-rings, the induced voltage causes a large current to flow – usually several hundred amperes per bar in machines of medium power.

The current-carrying conductors are in the path of the flux created by the stator, consequently, they all experience a strong mechanical force. These forces tend to drag the rotor along with the revolving field.

In summary:

1. a revolving magnetic field is set up when a 3-phase voltage is applied to the stator of an induction motor;

2. the revolving field induces a voltage in the rotor bars;

3. the induced voltage creates large circulating currents which flow in the rotor bars and end-rings;

4. the current-carrying rotor bars are immersed in the magnetic field created by the stator; they are therefore subjected to a strong mechanical force;

5. the sum of the mechanical forces on all the rotor bars produces a torque which tends to drag the rotor along in the direction of the revolving field.

13.7 Acceleration of the rotor – slip

As soon as the rotor is released, it rapidly accelerates in the direction of the rotating field. As it picks up speed, the relative velocity of the field with respect to the rotor diminishes progressively. This causes both the value and the frequency of the induced voltage to decrease because the rotor bars are cut more slowly. The rotor current, very large at first, decreases rapidly as the motor picks up speed.

The speed will continue to increase, but it will never catch up with the revolving field. In effect, if the rotor *did* turn at the same speed as the field (synchronous speed), the flux would no longer cut the rotor bars and the induced voltage and current would fall to zero. Under these conditions, the force acting on the rotor bars would also become zero and the friction and windage would immediately cause the rotor to slow down.

The rotor speed is always slightly less than synchronous speed so as to produce a current in the rotor bars sufficiently large to overcome the braking torque. At no-load the difference in speed between the rotor and field (called slip), is very small: usually less than 0.1% of synchronous speed.

13.8 Motor under load

Suppose the motor is initially running at no-load. If we apply a mechanical load to the shaft, the motor will begin to slow down and the revolving field will cut the rotor bars at a higher and higher rate. The induced voltage and the resulting current in the bars will increase progressively, producing a greater and greater motor torque. The question is, for how long can this go on? Will the speed continue to drop until the motor comes to a halt?

No; the motor and the mechanical load will reach a state of equilibrium when the motor torque is exactly equal to the load torque. When this state is reached, the speed will cease to drop and the motor will turn at a constant rate. It is very important to understand that a motor only turns at constant speed when its torque is *exactly* equal to the torque exerted by the mechanical load. The moment this state of equilibrium is upset, the motor speed will start to change (Sec. 3-11).

Under normal loads, induction motors run very close to synchronous speed. Thus, at full load, the slip for large motors (1000 kW and more) rarely exceeds 0.5% of synchronous speed, and for small machines (10 kW and less), it seldom exceeds 3%. That is why induction motors are considered to be constant-speed machines. However, because they never actually turn at synchronous speed, they are sometimes called *asynchronous* machines.

13.9 Slip

The slip s of an induction motor is the difference between the synchronous speed and the rotor speed, expressed as a percent (or per-unit) of synchronous speed. The per-unit slip is given by the equation:

$$s = \frac{n_s - n}{n_s} \qquad (13\text{-}2)$$

where

s = slip

n_s = synchronous speed [r/min]

n = rotor speed [r/min]

The slip is practically zero at no-load and is equal to 1 (or 100%) when the rotor is locked.

Example 13-2:
A 6-pole induction motor is excited by a 3-phase, 60 Hz source. If the full-load speed is 1140 r/min, calculate the slip.

Solution:
The synchronous speed of the motor is:

$$n_s = 120f/p = 120 \times 60/6 \qquad \text{Eq. 13-1}$$
$$= 1200 \text{ r/min}$$

The difference between the synchronous speed of the revolving flux and rotor speed is the slip speed:

$$n_s - n = 1200 - 1140 = 60 \text{ r/min}$$

The slip is:

$$s = (n_s - n)/n_s = 60/1200 \qquad \text{Eq. 13-2}$$
$$= 0.05 \text{ or } 5\%$$

13.10 Voltage and frequency induced in the rotor

The voltage and frequency induced in the rotor both depend upon the slip. They are given by the following equations:

$$f_2 = sf_o \qquad (13\text{-}3)$$

$$E_2 = sE_{oc} \text{ (approx.)} \qquad (13\text{-}4)$$

where

f_2 = frequency of the voltage and current in the rotor [Hz]

f_o = frequency of the source connected to the stator [Hz]

s = slip

E_2 = voltage induced in the rotor at slip s

E_{oc} = open-circuit voltage induced in the rotor when at rest [V]

The open-circuit voltage E_{oc} is the voltage that would be induced in the rotor bars if the bars were disconnected from the end-rings. In the case of a wound-rotor motor, the open-circuit voltage is $1/\sqrt{3}$ times the voltage between the open-circuit slip rings.

It should be noted that Eq. 13-3 *always* holds true, but Eq. 13-4 is valid only if the revolving flux (in webers) remains absolutely constant. However, between zero and full load E_2 is only slightly smaller than the value given by the equation.

Example 13-3:
A 6-pole wound-rotor induction motor is excited by a 3-phase 60 Hz source. Calculate the frequency of the rotor current under the following conditions:
a.   at standstill;
b.   motor turning at 500 r/min in the same direction as the revolving field;
c.   motor turning at 500 r/min in the opposite direction to the revolving field;
d.   motor turning at 2000 r/min in the same direction as the revolving field.

Solution:
From Example 13-2, the synchronous speed of the motor is 1200 r/min.

a.   At standstill, the motor speed $n = 0$. Consequently, the slip is:
$$s = (n_s - n)/n_s = (1200 - 0)/1200 = 1$$

The frequency of the induced voltage (and of the induced current) is:
$$f_2 = sf_o = 1 \times 60 = 60 \text{ Hz}$$

b.   When the motor turns in the same direction as the field, the motor speed n is positive. The slip is:
$$s = (n_s - n)/n_s = (1200 - 500)/1200$$
$$= 7/12 = 0.583$$

The frequency of the induced voltage (and of the rotor current) is:

$$f_2 = sf_o = 0.583 \times 60 = 35 \text{ Hz}$$

c. When the motor turns in the opposite direction to the field, the motor speed is *negative*: thus, $n = -500$. The slip is:

$$
\begin{aligned}
s &= (n_s - n)/n_s \\
&= \{1200 - (-500)\}/1200 \\
&= (1200 + 500)/1200 = 17/12 \\
&= 1.417
\end{aligned}
$$

A slip greater than 1 implies that the motor is operating as a brake.

The frequency of the induced voltage and rotor current is:

$$f_2 = sf_o = 1.417 \times 60 = 85 \text{ Hz}$$

d. The motor speed is positive because the rotor turns in the same direction as the field: $n = +2000$.

The slip is:

$$
\begin{aligned}
s &= (n_s - n)/n_s \\
&= (1200 - 2000)/1200 \\
&= -2/3 = -0.667
\end{aligned}
$$

A negative slip implies that the motor is actually operating as a generator.

The frequency of the induced voltage and rotor current is:

$$f_2 = sf_o = -0.667 \times 60 = -40 \text{ Hz}$$

A negative frequency means that the phase sequence of the voltages induced in the rotor windings is reversed. Thus, if the phase sequence of the rotor voltages is A-B-C when the frequency is positive, the phase sequence is A-C-B when the frequency is negative. As far as a frequency meter is concerned, a "negative" frequency gives the same reading as a "positive" frequency. Consequently, we can say that the frequency is simply 40 Hz.

13.11 Characteristics of squirrel-cage induction motors

Table 13A lists the typical properties of squirrel-cage inductions motors in the power range between 1 kW and 20 000 kW. Note that the current and torque are expressed in per-unit values. The following explanations will help us understand the values and the percentages given in the table.

TABLE 13A   TYPICAL CHARACTERISTICS OF SQUIRREL - CAGE INDUCTION MOTORS *(per unit values)*

Loading	Current (per-unit)		Torque (per-unit)		Slip (per-unit)		Efficiency		Power factor	
motor size →	small*	big*	small	big	small	big	small	big	small	big
Full load	1	1	1	1	0.03	0.004	0.7 to 0.9	0.96 to 0.98	0.8 to 0.85	0.87 to 0.9
No-load	0.5	0.3	0	0	≈ 0	≈ 0	0	0	0.2	0.05
Locked rotor	5 to 6	4 to 6	1.5 to 3	0.5 to 1	1	1	0	0	0.4	0.1

* **small** means under 11 kW (15 hp)   **big** means over 1120 kW (1500 hp) and up to 25 000 hp.

1. **Motor at no load.** When the motor runs at no load, the stator current lies between 0.5 and 0.3 p.u. (of full-load current). The no-load current is similar to the exciting current in a transformer. Thus, it is composed of a magnetizing component which creates the revolving flux Φ_m and a small active component that supplies the windage and friction losses in the rotor plus the iron losses in the stator. The flux Φ_m links both the stator and the rotor; consequently it is similar to the mutual flux in a transformer (Fig. 13-13).

very similar to the opposing mmf's of the secondary and primary in a transformer. As a result, leakage flux Φ_{f1} and Φ_{f2} are created, in addition to the mutual flux Φ_m (Fig. 13-14). The total reactive power needed to produce these three fluxes is slightly greater than when the motor is operating at no-load. Nevertheless, both the exciting current and reactive power under load remain about the same as at no-load. However, the active power (kW) absorbed by the motor increases in proportion

Figure 13-13
At no-load, the flux in the motor is mainly the mutual flux Φ_m. To create this flux, considerable reactive power is needed.

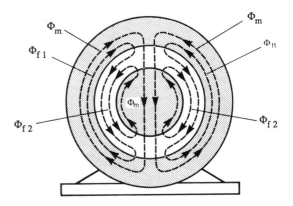

Figure 13-14
At full-load, the mutual flux decreases but stator and rotor leakage fluxes are created. The reactive power needed is slightly greater than in Fig. 13-13.

Considerable reactive power is needed to create the revolving field and, to keep it within acceptable limits, we use as short an air gap as mechanical tolerances will permit. The power factor at no-load is therefore low; it ranges from 0.2 (or 20%) for small machines to 0.05 for large machines. The efficiency is zero because the output power is zero.

2. **Motor under load.** When the motor is under load, the current in the rotor produces a mmf which tends to change the mutual flux Φ_m. This sets up an opposing current flow in the stator. The opposing mmf's of the rotor and stator are

to the mechanical load. It follows that the power factor of the motor improves dramatically as the mechanical load increases. At full load, it ranges from 0.70 for small machines to 0.90 for large machines. The efficiency at full load is particularly high; it can attain 98% for very large machines.

3. **Locked rotor characteristics.** The locked-rotor current is 5 to 6 times the full-load current, making the I^2R losses 25 to 36 times higher than normal. The rotor must therefore never remain locked for more than a few moments. Although the mechanical power is zero, the

motor develops a strong torque. The power factor is low because considerable reactive power is needed to produce the leakage flux in the rotor and stator windings. These leakage fluxes are much larger than in a transformer because the stator and the rotor windings are not as tightly coupled (see Sec. 10-2).

13.12 Estimating the currents in an induction motor

The full-load current of a three-phase induction motor may be calculated by means of the following approximate equation:

$$I = 600 \, P_h/E \qquad (13\text{-}5)$$

where

I = full-load current [A]
P_h = output power [horsepower]
E = rated line voltage (V)
600 = empirical constant

Recalling that the starting current is 5 to 6 times the full-load current and the no-load current is between 0.5 and 0.3 p.u., we can readily estimate the critical currents for any induction motor.

Example 13-4:
a. Calculate the approximate full-load current, locked-rotor current and no-load current of a 3-phase induction motor having a rating of 500 hp, 2300 V.
b. Estimate the apparent power drawn under locked rotor conditions.
c. State the nominal rating of this motor, expressed in kilowatts.

Solution:
a. The full-load current is:

$$\begin{aligned} I &= 600 \, P_h/E && \text{Eq. 13-5} \\ &= 600 \times 500/2300 \\ &= 130 \text{ A (approx.)} \end{aligned}$$

The no-load current is:

$$\begin{aligned} I_o &= 0.3I = 0.3 \times 130 \\ &= 39 \text{ A (approx.)} \end{aligned}$$

The starting current is:

$$\begin{aligned} I_{LR} &= 6I = 6 \times 130 \\ &= 780 \text{ A (approx.)} \end{aligned}$$

b. The apparent power under locked-rotor conditions is:

$$\begin{aligned} S &= \sqrt{3} \, EI \\ &= 1.73 \times 2300 \times 780 && \text{Eq. 8-9} \\ &= 3100 \text{ kVA (approx.)} \end{aligned}$$

c. When the power of a motor is expressed in kilowatts, it always relates to the mechanical output, and *not* to the electrical input. The nominal rating of this motor expressed in SI units is therefore:

$$\begin{aligned} P &= 500/1.34 \\ &= 373 \text{ kW (see Power conversion} \\ &\qquad \text{chart)} \end{aligned}$$

13.13 Active power flow

Voltages, currents, and phasor diagrams enable us to understand the detailed behavior of an induction motor. However, it is easier to see how electrical energy is converted into mechanical energy by following the active power as it flows through the machine. Thus, referring to Fig. 13-15, active power P_e flows from the line into the three-phase stator. Owing to the stator copper losses, a portion P_{js} is dissipated as heat in the windings. Another portion P_f is dissipated as heat in the stator core, owing to the iron losses. The remaining active power P_r is carried across the air gap and transferred to the rotor by electromagnetic induction.

Owing to the I^2R losses in the rotor, a third portion P_{jr} is dissipated as heat and the remainder is finally available in the form of mechanical power P_m. By subtracting a small fourth portion P_v, representing windage and bearing-friction losses, we finally obtain P_L, the mechanical power available at the shaft to drive the load.

The power flow diagram of Fig. 13-15 enables us to identify and to calculate three important properties of the induction motor: 1) its *efficiency,* 2) its *power* and 3) its *torque.*

Figure 13-15
Active power flow in a 3-phase induction motor.

1. **Efficiency.** By definition, the efficiency of a motor is the ratio of the output power to the input power:

$$\text{efficiency } (\eta) = P_L/P_e \qquad (13\text{-}6)$$

2. **I^2R losses in the rotor.** It can be shown* that the rotor I^2R losses P_{jr} are related to the rotor input power P_r by the equation:

$$P_{jr} = sP_r \qquad (13\text{-}7)$$

where

P_{jr} = rotor I^2r losses [W]
s = slip
P_r = power transmitted to the rotor [W]

Equation 13-7 shows that as the slip increases, the rotor I^2R losses consume a larger and larger proportion of the power P_r transmitted to the rotor. A rotor turning at half synchronous speed ($s = 0.5$) dissipates in the form of heat 50% of the active power it receives. When the rotor is locked ($s = 1$), all the power transmitted to the rotor is dissipated as heat.

3. **Mechanical power.** The mechanical power P_m developed by the motor is equal to the power transmitted to the rotor minus its I^2R losses. Thus,

$$P_m = P_r - P_{jr}$$
$$= P_r - sP_r \qquad \text{Eq. 13-7}$$

---

*
$$\begin{bmatrix} \text{mechanical} \\ \text{power output} \\ \text{of rotor} \end{bmatrix} = \begin{bmatrix} \text{electromagnetic} \\ \text{power transferred} \\ \text{to rotor} \end{bmatrix} - \begin{bmatrix} \text{electrical} \\ \text{losses} \\ \text{in rotor} \end{bmatrix}$$

$$P_m = P_r - P_{jr} \qquad (i)$$

but from Eq. 3-5

$$P_m = \frac{\text{rotor speed} \times \text{mechanical torque}}{9.55}$$

hence $$P_m = \frac{nT_m}{9.55} \qquad (ii)$$

also from Eq. 3-5 we can write :

$$P_r = \frac{\text{speed of flux} \times \text{electromagnetic torque}}{9.55}$$

$$P_r = \frac{n_s T_{mag}}{9.55} \qquad (iii)$$

but the mechanical torque T_m must equal the electromagnetic torque T_{mag}

thus $$T_m = T_{mag} \qquad (iv)$$

substituting (ii),(iii),(iv) in (i) we find

$$P_{jr} = sP_r$$

whence

$$P_m = (1 - s)P_r \qquad (13\text{-}8)$$

The actual mechanical power available to drive the load is slightly less than P_m, owing to the power needed to overcome the windage and friction losses. In most calculations we can neglect this small loss.

4. **Motor torque.** The torque T_m developed by the motor *at any speed* is given by:

$$T_m = \frac{9.55\ P_m}{n} = \frac{9.55\ (1 - s)P_r}{n_s\ (1 - s)}$$

$$= 9.55 P_r/n_s$$

therefore

$$T_m = 9.55 P_r/n_s \qquad (13\text{-}9)$$

where

T_m = torque developed by the motor *at any speed* [N·m]
P_r = power transmitted to the rotor [W]
n_s = synchronous speed [r/min]
9.55 = multiplier to take care of units [exact value: $60/2\pi$]

The actual torque T_L available at the shaft is slightly less than T_m, owing to the torque required to overcome the windage and friction losses. However, in most calculations we can neglect this small difference.

Equation 13-9 shows that the torque is directly proportional to the active power transmitted to the rotor. Thus, to develop a high locked-rotor torque, the rotor must absorb a large amount of active power. The latter is dissipated in the form of heat, consequently, the temperature of the rotor rises very rapidly.

Example 13-5:
A 3-phase induction motor having a synchronous speed of 1200 r/min draws 80 kW from a 3-phase feeder. The copper losses and iron losses in the stator amount to 5 kW. If the motor runs at 1152 r/min, calculate:

a. the active power transmitted to the rotor;
b. the rotor I^2R losses;
c. the mechanical power developed;
d. the mechanical power delivered to the load, knowing that the windage and friction losses are equal to 2 kW;
e. the efficiency of the motor.

Solution:

a. Active power to the rotor is:
$$P_r = P_e - P_{js} - P_f$$
$$= 80 - 5 = 75 \text{ kW}$$

b. The slip is:
$$s = (n_s - n)/n_s$$
$$= (1200 - 1152)/1200$$
$$= 48/1200 = 0.04$$

Rotor I^2R losses are:
$$P_{jr} = sP_r = 0.04 \times 75 = 3 \text{ kW}$$

c. The mechanical power developed is:
$$P_m = P_r - I^2R \text{ losses in rotor}$$
$$= 75 - 3 = 72 \text{ kW}$$

d. The mechanical power P_L delivered to the load is slightly less than P_m owing to the friction and windage losses.
$$P_L = P_m - P_v = 72 - 2 = 70 \text{ kW}$$

e. The efficiency is:
$$\nu = P_L/P_e = 70/80$$
$$= 0.875 \text{ or } 87.5\%$$

Example 13-6:
A 3-phase squirrel-cage induction motor, connected to a 60 Hz line, possesses a synchronous speed of 900 r/min. The motor absorbs 40 kW, and the stator copper and iron losses amount to 5 kW and 1 kW, respectively. Calculate the torque developed by the motor.

Solution:
The power transmitted across the air gap to the rotor is:

$$P_r = P_e - P_{js} - P_f$$
$$= 40 - 5 - 1 = 34 \text{ kW}$$
$$T_m = 9.55\ P_r/n_s \qquad \text{Eq. 13-9}$$
$$= 9.55 \times 34\ 000/900$$
$$= 361 \text{ N·m}$$

Note that the solution to this problem (the torque) is independent of the speed of rotation. The motor could be at a standstill or running at full speed, but as long as the power P_r transmitted to the rotor is equal to 34 kW, the motor develops a torque of 361 N·m.

Example 13-7:
A 3-phase induction motor having a nominal rating of 100 hp (~75 kW) and a synchronous speed of 1800 r/min is connected to a 600 V source (Fig. 13-16a). The two-wattmeter

$$R = 0.34/2 = 0.17\ \Omega;$$

Stator I^2R losses are:
$$P_{js} = 3I^2R = 3 \times 0.17 \times (78)^2$$
$$= 3.1\ kW;$$

Iron losses $P_f = 2$ kW;

Power supplied to the rotor:
$$P_r = P_e - P_{js} - P_f$$
$$= (70 - 3.1 - 2) = 64.9\ kW;$$

Figure 13-16a
See Example 13-7.

method shows a total power consumption of 70 kW, and an ammeter indicates a line current of 78 A. Precise measurements give a rotor speed of 1763 r/min. In addition, the following characteristics are known about the motor:

stator iron losses $P_f = 2$ kW
windage and friction losses $P_v = 1.2$ kW
resistance between two
stator terminals = 0.34 Ω

Calculate:
a. power supplied to the rotor;
b. rotor I^2R losses;
c. mechanical power supplied to the load, in horsepower;
d. efficiency;
e. torque developed at 1763 r/min.

Solution:
a. Power supplied to the stator is:
$$P_e = 70\ kW;$$

Stator resistance per phase (assume a wye connection) is:

b. The slip is:
$$s = (n_s - n)/n_s$$
$$= (1800 - 1763)/1800$$
$$= 0.0205;$$

Rotor I^2R losses:
$$P_{jr} = sP_r = 0.0205 \times 64.9 = 1.33\ kW;$$

c. Mechanical power developed is:
$$P_m = P_r - P_{jr} = 64.9 - 1.33 = 63.5\ kW$$

Mechanical power P_L to the load:
$$P_L = 63.5 - P_v = 63.5 - 1.2$$
$$= 62.3 = 62.3 \times 1.34\ (hp)$$
$$= 83.5\ hp$$

d. Efficiency of the motor is:
$$v = P_L/P_e = 62.3/70 = 0.89\ or\ 89\%$$

e. Torque at 1763 r/min:
$$T = 9.55\,P_r/n_s = 9.55 \times 64\,900/1800$$
$$= 344\ N·m$$

The above calculations are summarized in Fig. 13-16b.

3.1 kW   2 kW

P_{js} P_f 1.33 kW   1.2 kW

        P_{jr} P_v

70 kW → P_e 64.9 kW → P_r 62.3 kW → P_L

Figure 13-16b
Power flow in Example 13-7

13.14 Torque versus speed curve

The torque a motor develops depends upon its speed, but the relationship between the two cannot be expressed by a simple equation. Consequently, we prefer to show the relationship in the form of a curve. Figure 13-17 shows the torque-speed curve of a conventional 3-phase induction motor whose nominal full-load torque is T. The starting torque is 1.5 T and the maximum torque (called *breakdown torque*) is 2.5 T.

At full load, the motor runs at a speed n. If the mechanical load increases, the speed will drop until the motor torque is again equal to the load torque. As soon as the two torques are in balance, the motor will turn at a constant but lower speed. However, if the load torque exceeds 2.5 T, the motor will quickly stop.

Small motors (10 kW and less) develop their breakdown torque at a speed n_d of about 80% of synchronous speed. Big motors (1000 kW and more) attain their breakdown torque at about 98% of synchronous speed.

13.15 Effect of rotor resistance

The rotor resistance of a squirrel-cage rotor is essentially constant, except that it increases with temperature. Thus, the resistance increases with increasing load because the temperature rises.

In designing a squirrel-cage motor, the rotor resistance can be varied over wide limits by using copper, aluminum or other metals in the rotor bars and end-rings. The torque-speed curve is greatly affected by such a change in resistance. The only characteristic that remains unchanged is the breakdown torque. The following example illustrates the changes that occur.

Figure 13-18a shows the torque-speed curve of a 10 kW (13.4 hp), 50 Hz, 380 V motor having a synchronous speed of 1000 r/min and

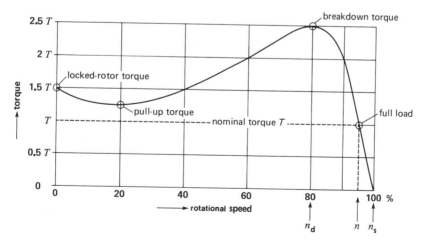

Figure 13-17
Typical torque-speed curve of a 3-phase squirrel-cage induction motor.

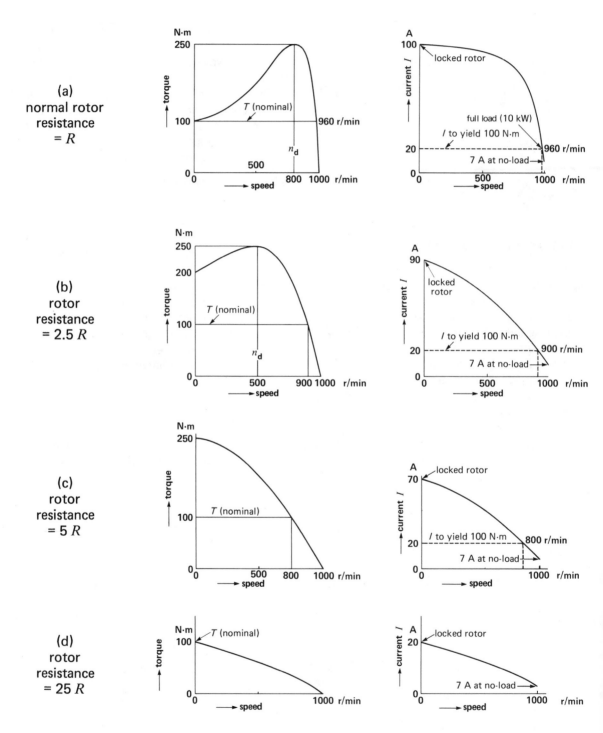

Figure 13-18
Rotor resistance affects the motor characteristics.

a full-load torque of 100 N·m ($\sim$ 73.7 ft·lbf). The full-load current is 20 A and the locked-rotor current, 100 A. The rotor has an arbitrary resistance R.

Let us increase the rotor resistance by a factor of 2.5. This can be achieved by using a material of higher resistivity, such as bronze, for the rotor bars and end-rings. The result is that the starting torque doubles and the locked-rotor current decreases from 100 A to 90 A (Fig. 13-18b). The motor develops its breakdown torque at a speed N_d of 500 r/min, compared to the original breakdown speed of 800 r/min.

If we again double the rotor resistance so that it becomes 5 R, the locked-rotor torque attains a maximum value of 250 N·m for a corresponding current of 70 A (Fig. 13-18c).

A further increase in rotor resistance decreases both the locked-rotor torque and locked-rotor current. For example, if the rotor resistance is increased 25 times (25 R), the locked-rotor current drops to 20 A, but the motor develops the same starting torque (100 N·m), as it did when the locked-rotor current was 100 A (Fig. 13-18d).

In summary, a high rotor resistance is desirable because it produces a high starting torque and a relatively low starting current (Fig. 13-18c). Unfortunately, it also produces a rapid fall-off in speed with increasing load. Furthermore, because the slip at rated torque is high, the motor I^2R losses are high. The efficiency is therefore low and the motor tends to get hot.

Under running conditions, it is preferable to use a low rotor resistance (Fig. 13-18a). The speed decreases much less with increasing load, and the slip at rated torque is small. Consequently, the efficiency is high and the motor tends to run cool.

We can obtain both a high starting resistance and a low running resistance by designing the rotor bars in a special way (see Fig. 14-5, Chapter 14). However, if the rotor resistance has to be varied over a wide range, it may be necessary to use a wound-rotor induction motor. Such a motor enables us to vary the rotor resistance at will by means of an external rheostat.

13.16 Wound-rotor motor

We explained the basic difference between a squirrel-cage motor and a wound-rotor motor in Sec. 13-1. Although a wound-rotor motor costs more than a squirrel-cage motor, it offers the following advantages;

1. The locked-rotor current can be drastically reduced by inserting three external resistors in series with the rotor. Nevertheless, the locked-rotor torque will still be as high as that of a squirrel-cage motor;

2. The speed can be varied by varying the external rotor resistors;

3. The motor is ideally suited to accelerate high-inertia loads, which require a long time to come up to speed.

Figure 13-19 is a diagram of the circuit used to start a wound-rotor motor. The rotor windings are connected to three wye-connected external resistors by means of a set of slip rings and brushes. Under locked-rotor (LR) conditions, the variable resistors are set to their highest value. As the motor speeds up, the resistance is gradually reduced until full-load speed is reached, whereupon the brushes are short-circuited. By properly selecting the resistance values, we can produce a high accelerating torque with a stator current that never exceeds twice full-load current.

To start large motors, we often use liquid rheostats because they are easy to control and have a large thermal capacity. A liquid rheostat is composed of three electrodes immersed in a suitable electrolyte. To vary its resistance, we simply vary the level of the electrolyte surrounding the electrodes. The large thermal capacity of the electrolyte limits the tempera-

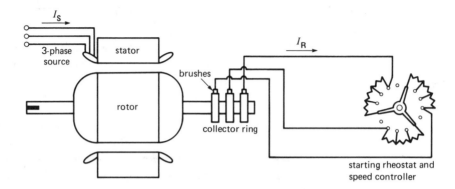

Figure 13-19
External resistors connected to the 3 slip rings of a wound-rotor induction motor.

ture rise. For example, a liquid rheostat is typically used with a 1260 kW wound-rotor motor to bring a large synchronous motor up to speed.

We can also regulate the speed of a wound-rotor motor by varying the resistance of the rheostat. As we increase the resistance, the speed will drop. This method of speed control has the disadvantage that a lot of heat is dissipated in the resistors; the efficiency is therefore low. Furthermore, for a given rheostat setting, the speed varies considerably if the mechanical load varies.

The power rating of a self-cooled wound-rotor motor depends upon the speed at which it operates. Thus, for the same temperature rise, a motor that can develop 100 kW at 1800 r/min can deliver only about 40 kW at 900 r/min. However, if the motor is cooled by a separate fan, it can deliver 50 kW at 900 r/min.

13.17 Three-phase windings

In 1883, a 27-year old Yougoslav scientist named Nikola Tesla invented the three-phase induction motor. His first model had a salient-pole stator winding similar to the one shown in Fig. 13-6. Since then, the design of induction motors has evolved considerably; modern machines are built with so-called *lap windings* distributed in slots around the stator.

A lap winding consists of a set of phase groups evenly distributed around the stator circumference. The number of groups is given by the equation:

$$\text{groups} = \text{poles} \times \text{phases}$$

Thus, a 4-pole 3-phase stator must have $4 \times 3 = 12$ phase groups. Because a group must have at least one coil, it follows that the minimum number of coils is equal to the number of groups. A 4-pole, 3-phase stator must therefore have at least 12 coils. Furthermore, in a lap winding, the stator has as many slots as it has coils. Consequently, a 4-pole, 3-phase stator must have at least 12 slots. However, motor designers have discovered that it is preferable to use 2, 3, or more coils per group rather than only one. The number of coils and slots increases in proportion. For example, a 4-pole, 3-phase stator having 5 coils per group must have a total of $(4 \times 3 \times 5) = 60$ coils, lodged in 60 slots. The coils in each group are connected in series and are staggered at one-slot intervals (Fig. 13-20). The coils are identical and may possess one or more turns. The width of each coil is called the *coil pitch*.

Such a *distributed* winding is obviously more costly to build than a concentrated winding having only 1 coil per group. However, it improves the starting torque and reduces the noise under running conditions.

Figure 13-20
The 5 coils are connected in series to create one phase group.

When the stator windings are excited from a 3-phase source, a multipole revolving field is produced. The distance between adjacent poles is called the *pole pitch*. It is equal to the internal circumference of the stator divided by the number of poles. For example, a 12-pole stator having a circumference of 600 mm has a pole pitch of 600/12 or 50 mm.

In practice, the coil pitch is between 80% and 100% of the pole pitch. The coil pitch is usually made less than the pole pitch in order to save copper and to improve the flux distribution in the air gap. The shorter coil width reduces the cost and weight of the windings, while the more sinusoidal flux distribution improves the torque during start-up, and often results in a quieter machine. In the case of two-pole machines, the shorter pitch also makes the coils much easier to insert in the slots.

To get an overall picture of a lap winding, let us consider a 24-slot stator laid out flat as shown in Fig. 13-21a. The 24 coils are held upright, with one coil-side standing in each slot. If the windings are now laid down so that all the other coil sides fall into the slots, we obtain the classical appearance of a 3-phase lap winding (Fig. 13-21b). The coils are connected together to create 3 identical windings, one for each phase. Each winding consists of a number of groups equal to the number of poles. Finally,

the groups of each phase are symmetrically distributed around the circumference of the stator. The following examples show how this is done.

Example 13-8:
The stator of a 3-phase, 10-pole induction motor possesses 120 slots. If a lap winding is used, calculate:
a. the total number of coils;
b. the number of coils per phase;
c. the number of coils per group;
d. the pole pitch;
e. the coil pitch (expressed as a percent of the pole pitch), if the coil width extends from slot 1 to slot 11.

Figure 13-21a
Coils held upright in 24 stator slots.

Figure 13-21b
Coils laid down to make a typical lap winding.

Solution:
a. A 120-slot stator requires 120 coils.

b. Coils per phase = 120 ÷ 3 = 40.

c. Number of groups per phase = number of poles = 10

   Coils per group = 40 ÷ 10 = 4.

d. The pole pitch corresponds to:

pole pitch = slots/poles = 120/10

= 12 slots

One pole pitch extends therefore from slot 1 (say) to slot 13.

e. The coil pitch covers 10 slots (slot 1 to slot 11). The percent coil pitch = 10/12 = 83.3%.

The next example shows in greater detail how the coils are interconnected in a typical 3-phase stator winding.

Example 13-9:
A stator having 24 slots has to be wound with a 3-phase, 4-pole winding. Determine:
1) the connections between the coils;
2) the connections between the phases.

Solution:
The 3-phase winding has 24 coils. Assume that they are standing upright, with one coil-side in each slot (Fig. 13-22). We shall first determine the coil distribution for phase A and then proceed with the connections for that phase. Similar connections will then be made for phases B and C. Here is the line of reasoning:

a. The revolving field creates 4 poles; the motor therefore has 4 groups per phase, or 4 × 3 = 12 phase groups in all. Each rectangle in Fig. 13-22a represents one group. Because the stator contains 24 coils, each group consists of 24/12 = 2 consecutive coils.

b. The groups (poles) of each phase must be uniformly spaced around the stator. The group distribution for phase A is shown in Fig. 13-22b. Each shaded rectangle represents 2 upright coils connected in series, producing the two terminals shown. Note that the mechanical distance between two successive groups always corresponds to an *electrical* phase angle of 180°.

c. Successive groups of phase A must have opposite magnetic polarities. Consequently, the 4 groups of phase A are connected in series to produce successive N-S-N-S poles (Fig. 13-22c). Phase A now has two terminals, a "starting" terminal A_1 and a "finishing" terminal A_2.

d. The phase groups of phases B and C are spaced the same way around the stator. However, the "starting" terminals B_1 and C_1 are respectively located at 120° and 240° (electrical) with respect to the starting terminal A_1 of phase A (Fig. 13-22d).

e. The groups in phases B and C are connected in series in the same way as those of phase A are (Fig. 13-22e). This yields six terminals: A_1A_2, B_1B_2 and C_1C_2. They may be connected either in wye or in delta inside the machine. The resulting 3 wires corresponding to the 3 phases are brought out to the terminal box of the machine (Fig. 13-22f). In practice, the connections are made, not while the coils are upright (as shown) but only after they have been laid down in the slots.

f. Because the pole pitch corresponds to a span of 24/4 = 6 slots, the coil pitch may be shortened to 5 slots (slot 1 to slot 6). Thus, the first coil of phase A is lodged in the first and sixth slots (Fig. 13-23). All the other coils and connections follow suit, according to Fig. 13-22e.

Figures 13-24a and 13-24b show the coil and stator of a 450 kW (600 hp) induction motor. Figure 13-25 illustrates the procedure used in winding a smaller 37.5 kW (50 hp) stator.

Figure 13-22a
The 24 coils are grouped two-by-two to make 12 groups.

Figure 13-22b
The 4 groups of phase A are selected so as to be evenly spaced from each other.

Figure 13-22c
The groups of phase A are connected in series to create alternate N-S poles.

Figure 13-22d
The start of phases B and C begins 120° and 240° after the start of phase A.

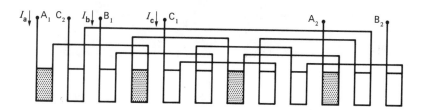

Figure 13-22e
When all phase groups are connected, only six leads remain.

Figure 13-22f
The phases may be connected in wye or in delta and 3 leads are brought out to the terminal box.

Figure 13-23
The pole pitch is from slot 1 slot 7; the coil pitch from slot 1 to slot 6.

Figure 13-24a
Stator of a 3-phase, 450 kW, 1180 r/min, 575 V, 60 Hz induction motor. The lap winding is composed of 108 preformed coils having a pitch from slots 1 to 15. One coil side falls into the bottom of a slot and the other at the top. Rotor diameter: 500 mm; axial length; 460 mm. (*H. Roberge*)

Figure 13-24b
Close-up view of the preformed coil in Fig. 13-24a.

(a)

(b)

(c)

(d)

Figure 13-25

Stator winding of a 3-phase, 50 hp, 575 V, 60 Hz, 1765 r/min induction motor. The stator possesses 48 slots carrying 48 coils.

a. Coil made of 27 turns # 15 copper wire, covered with a high-temperature polyimide insulation, ready to be placed into 2 slots.

b. One coil-side is threaded into slot 1 (say) and the other side goes into slot 12. The coil-pitch is therefore from 1 to 12.

c. Each coil-side fills half a slot, and is covered with a paper spacer so that it does not touch the second coil-side placed in the same slot. The photograph shows 3 empty and uninsulated slots and 4 empty slots insulated with a composition paper liner. The remaining 10 slots each carry one coil-side.

d. A varnished cambric cloth, cut in the shape of a triangle, provides extra insulation between adjacent phase groups. (*Electro-Mecanik*)

13.18 Sector motor

Consider a standard 3-phase, 4-pole, wye-connected motor having a synchronous speed of 1800 r/min. Let us cut the stator in half, so that half the winding is removed and only two complete N and S poles are left (per phase). Next, let us connect the three phases in wye, without making any other changes to the existing coil connections. Finally, we mount the original rotor above this sector *stator*, leaving a small air gap (Fig. 13-26).

Figure 13-26
Two-pole sector induction motor.

If we connect the stator terminals to a 3-phase, 60 Hz source, the rotor will again turn at close to 1800 r/min. To prevent saturation, the voltage should be reduced to half its original value because the stator winding now has only one-half the original number of turns. Under these conditions, this remarkable truncated *sector motor* still develops about 30 percent of its original rated power.

The sector motor produces a "revolving" field that moves at the same peripheral speed as the flux in the original 3-phase motor. However, instead of making a complete turn, the field simply travels continuously from one end of the stator to the other.

13.19 Linear induction motor

It is obvious that the sector stator could be laid out flat, without affecting the shape or speed of the magnetic field. Such a flat stator produces a field that moves at constant speed, in a straight line. Using the same reasoning as in Sec. 13-5, we can prove that the flux travels at a linear synchronous speed given by:

$$v_s = 2wf \qquad (13\text{-}10)$$

where

v_s = linear synchronous speed [m/s]
w = width of one pole-pitch [m]
f = frequency [Hz]

Note that the linear speed does not depend upon the number of poles but only on the pole-pitch. Thus, it is possible for a 2-pole linear stator to create a field moving at the same speed as that of a 6-pole linear stator (say), provided they have the same pole-pitch.

If a flat squirrel-cage winding is brought near the flat stator, the travelling field drags the squirrel cage along with it (Sec. 13-2). In practice, we generally use a simple aluminum or copper plate as a "rotor" (Fig. 13-27). Furthermore, to increase the power and to reduce the

Figure 13-27
Components of a 3-phase linear induction motor.

reluctance of the magnetic path, two flat stators are usually mounted, face-to-face, on opposite sides of the aluminum plate. The combination is called a *linear induction motor*. The direction of the motor can be reversed by interchanging any two stator leads.

In many practical applications, the rotor is stationary, while the stator moves. For example, in some high-speed trains, the rotor is composed of a thick aluminum plate fixed to the ground and extending over the full length of the track. The linear stator is bolted to the undercarriage of the train and straddles the plate. Train speed is varied by changing the frequency applied to the stator (Fig. 13-31).

Example 13-10:
The stator of a linear induction motor is excited from a 75 Hz electronic source. If the distance between consecutive phase groups of phase A is 300 mm, calculate the linear speed of the magnetic field.

Solution:
The pole pitch is 300 mm. Consequently,

$$v_s = 2wf \qquad \text{Eq. 13-10}$$
$$= 2 \times 0.3 \times 75$$
$$= 45 \text{ m/s or } 162 \text{ km/h}$$

13.20 Traveling waves

We are sometimes left with the impression that when the flux reaches the end of a linear stator, there must be a delay before it "returns" to restart once more at the beginning. This is not the case. The linear motor produces a traveling wave of flux which moves continuously and smoothly from one end of the stator to the other. Figure 13-28 shows how the flux moves from left to right in a 2-pole linear motor. The flux cuts off sharply at extremities A, B of the stator. However, as fast as a N pole "disappears" at the right, it builds up again at the left.

13.21 Properties of a linear induction motor

The properties of a linear induction motor are almost identical to those of a standard rotating

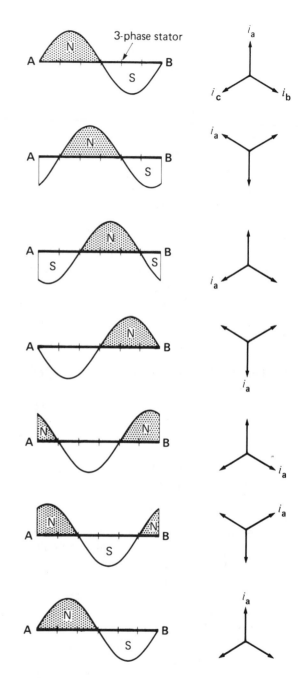

Figure 13-28
Shape of the magnetic field created by a 2-pole, 3-phase linear stator, over one complete cycle. The successive frames are separated by an interval of time equal to 1/6 cycle or 60°.

machine. Consequently, the equations for slip, thrust, power, etc., are also similar.

1. **Slip.** Slip is defined by:

$$s = (v_s - v)/v_s \qquad (13\text{-}11)$$

where

s = slip
v_s = synchronous linear speed [m/s]
v = speed of rotor (or stator) [m/s]

2. **Active power flow.** With reference to Fig. 13-15, active power flows through a linear motor in the same way it does through a rotating motor, except that the stator and rotor are flat. Consequently, Eqs. 13-6, 13-7 and 13-8 apply to both types of machines

$$\eta = P_L/P_e \qquad \text{Eq. 13-6}$$
$$P_{jr} = sP_r \qquad \text{Eq. 13-7}$$
$$P_m = (1 - s)P_r \qquad \text{Eq. 13-8}$$

3. **Thrust.** The thrust or force developed by a linear induction motor is given by:

$$F = P_r/v_s \qquad (13\text{-}12)$$

where

F = thrust [N]
P_r = power transmitted to the rotor [W]
v_s = linear synchronous speed [m/s]

Example 13-11:
An overhead crane in a factory is driven horizontally by means of two linear induction motors whose "rotors" are the two steel I-beams upon which the crane rolls. The 3-phase, 4-pole linear stators (mounted on opposite sides of the crane and facing the respective webs of the I-beams) have a pole pitch of 8 cm and are driven by a variable frequency electronic source. During a test on one of the motors, the following results were obtained:

Stator frequency:	15 Hz
Power to stator:	5 kW
Copper loss + iron loss in stator:	1 kW
Crane speed:	1.8 m/s

Calculate:
a. synchronous speed and slip;
b. power to the rotor;
c. I^2R loss in rotor;
d. mechanical power and thrust.

Solution:
a. Linear synchronous speed is:

$$\begin{aligned} v_s &= 2wf \qquad \text{Eq. 13-10} \\ &= 2 \times 0.08 \times 15 \\ &= 2.4 \text{ m/s} \end{aligned}$$

The slip is:

$$\begin{aligned} s &= (v_s - v)/v_s \qquad \text{Eq. 13-11} \\ &= (2.4 - 1.8)/2.4 \\ &= 0.25 \end{aligned}$$

b. Power to the rotor is:

$$\begin{aligned} P_r &= P_e - P_{js} - P_f \text{ (see Fig. 13-15)} \\ &= 5 - 1 \\ &= 4 \text{ kW} \end{aligned}$$

c. I^2R loss in the rotor is:

$$\begin{aligned} P_{jr} &= sP_r \qquad \text{Eq. 13-7} \\ &= 0.25 \times 4 \\ &= 1 \text{ kW} \end{aligned}$$

d. Mechanical power is:

$$\begin{aligned} P_m &= P_r - P_{jr} \qquad \text{(Fig. 13-15)} \\ &= 4 - 1 \\ &= 3 \text{ kW} \end{aligned}$$

The thrust is:

$$\begin{aligned} F &= P_r/v_s \qquad \text{Eq. 13-12} \\ &= 4000/2.4 \\ &= 1666 \text{ N} = 1.67 \text{ kN } (\sim375 \text{ lb}) \end{aligned}$$

13.22 Magnetic levitation

In Sec. 13-2, we saw that a moving permanent magnet, sweeping across a conducting ladder, tends to drag the ladder along with the magnet.

We will now show that this horizontal tractive force is also accompanied by a *vertical* force which tends to push the magnet away from the ladder.

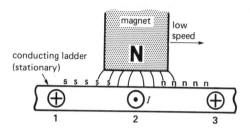

Figure 13-29
Currents and magnetic poles at low speed.

Figure 13-30
Currents and magnetic poles at high speed.

Referring to Fig. 13-29, suppose that conductors 1, 2, 3 are three conductors of the stationary ladder. The center of the N pole of the magnet is sweeping across the top of conductor 2. The voltage induced in this conductor is maximum because the flux density is greatest at the center of the pole. If the magnet moves very slowly, the resulting induced current reaches *its* maximum value at virtually the same time. This current, returning by conductors 1 and 3, creates magnetic poles nnn and sss as shown in Fig. 13-29. According to the

laws of attraction and repulsion, the front half of the magnet is repelled upwards while the rear half is attracted downwards. Because the distribution of the nnn and sss poles is symmetrical with respect to the center of the magnet, the vertical forces of attraction and repulsion are equal, and the resulting vertical force is nil. Consequently, there is only a horizontal tractive force.

But suppose now that the magnet moves very rapidly. Owing to its inductance, the current in conductor 2 reaches its maximum value a fraction of a second after the voltage has attained *its* maximum. Consequently, by the time the current in conductor 2 is maximum, the center of the magnet is already some distance ahead of the conductor (Fig. 13-30). The current returning by conductors 1 and 3 again creates nnn and sss poles; however, the N pole of the magnet is now directly above a nnn pole, with the result that a strong vertical force tends to push the magnet upwards.* This effect is called the principle of *magnetic levitation*.

Magnetic levitation is used in some ultra-high-speed trains that glide on a magnetic cushion rather than on wheels. A powerful electromagnet fixed underneath the train moves above a conducting rail inducing currents in the rail in the same way as in our ladder. The force of levitation is always accompanied by a small horizontal braking force which must, of course, be overcome by the linear motor that propels the train. See Figs. 13-31 and 13-32.

---

*   The current is always delayed (even at low speeds) by an interval of time Δt, which depends upon the L/R time constant of the rotor. This delay is so brief that, at low speeds, the current reaches its maximum at virtually the same time and place as the voltage does. On the other hand, at high speeds, the same delay Δt produces a significant shift in *space* between the points where the voltage and current reach their respective maximum values.

Figure 13-31
This 17 t electric train is driven by a linear motor. The motor consists of a stationary rotor and a flat stator fixed to the undercarriage of the train. The rotor is the vertical aluminum plate mounted in the center of the track. The 3-tonne stator is energized by a 4.7 MVA electronic dc to ac inverter whose frequency can be varied from zero to 105 Hz. The motor develops a maximum thrust of 35 kN (7800 lb) and the top speed is 200 km/h. Direct current power at 4 kV is fed into the inverter by means of a brush assembly in contact with 6 stationary dc bus bars mounted on the left-hand side of the track.

Electromagnetic levitation is obtained by means of superconducting electromagnet. The magnet is 1300 mm long, 600 mm wide and 400 mm deep and weighs 500 kg. The coils of the magnet are maintained at a temperature of 4 K by the forced circulation of liquid helium. The current density is 80 A/mm^2 and the resulting flux density is 3 T. The vertical force of repulsion attains a maximum of 60 kN and the vertical gap between the magnet and the reacting metallic track varies from 100 mm to 300 mm depending on the current. (*Siemens*)

Figure 13-32
Cross section view of the main components of the high-speed train shown in Fig. 13-31. (*Siemens*)

QUESTIONS AND PROBLEMS

Practical level

13-1 Name the principal components of an induction motor.

13-2 Explain how a revolving field is set up in 3-phase induction motor.

13-3 If we double the number of poles on the stator of an induction motor, will its synchronous speed also double?

13-4 The rotor of an induction should never be locked while full voltage is being applied to the stator. Explain.

13-5 Why does the rotor of an induction motor turn slower than the revolving field?

13-6 What happens to the rotor speed and rotor current when the mechanical load on an induction motor increases?

13-7 Would you recommend using a 50 hp induction motor to drive a 10 hp load? Explain.

13-8 Give two advantages of a wound-rotor motor over a squirrel-cage motor.

13-9 Both the voltage and frequency induced in the rotor of an induction motor decrease as the rotor speeds up. Explain.

13-10 A 3-phase, 20-pole induction motor is connected to a 600 V, 60 Hz source.
   a. What is the synchronous speed?
   b. If the voltage is reduced to 300 V, will the synchronous speed change?
   c. How many groups are there, per phase?

13-11 Describe the principle of operation of a linear induction motor.

13-12 Calculate the approximate values of starting current, full-load current and no-load current of a 150 horsepower, 575 V, 3-phase induction motor.

13-13 Make a drawing of the magnetic field created by a 3-phase, 12-pole induction motor.

13-14 How can we change the direction of rotation of a 3-phase induction motor?

Intermediate level

13-15 a. Calculate the synchronous speed of a 3-phase, 12-pole induction motor that is excited by a 60 Hz source.
   b. What is the nominal speed if the slip at full load is 6 percent?

13-16 A 3-phase 6-pole induction motor is connected to a 60 Hz supply. The voltage induced in the rotor bars is 4 V when the rotor is locked. If the motor turns in the same direction as the flux, calculate the approximate voltage induced and its frequency:
   a. at 300 r/min;
   b. at 1000 r/min
   c. at 1500 r/min.

13-17 a. Calculate the approximate values of full-load current, starting current and no-load current of a 75 kW, 4000 V, 3-phase, 900 r/min induction motor.
   b. Calculate the nominal full-load speed and torque knowing that the slip is 2 percent.

13-18 A 3-phase, 75 hp, 440 V induction motor has a full-load efficiency of 91% and a power factor of 83%. Calculate the nominal current per phase.

13-19 An open-circuit voltage of 240 V appears across the slip rings of a wound rotor induction motor when the rotor is locked. The stator has 6 poles and is excited by a 60 Hz source. If the rotor is driven by a variable-speed dc motor, calculate the open-circuit voltage and frequency across the slip-rings if the dc motor turns:
   a. at 600 r/min, in the same direction as the rotating field;
   b. at 900 r/min, in the same direction as the rotating field;
   c. at 3600 r/min, opposite to the rotating field.

13-20 a. Referring to Fig. 13-7, calculate the instantaneous values of I_a, I_b and I_c for an angle of 150°.
   b. Determine the actual direction of current flow in the three phases at this instant and calculate the mmf developed by the windings.
   c. Does the resulting mmf point in a direction intermediate between the mmf's corresponding to instants 3 and 4?

13-21 A 3-phase lap-wound stator possessing 72 slots, produces a synchronous speed of 900 r/min when connected to a 60 Hz source. Calculate the number of coils per phase group as well as the probable coil pitch. Draw the complete coil connection diagram, following steps (a) to (f) in Fig. 13-22.

13-22 The 3-phase 4-pole stator of Fig. 13-25 has an internal diameter of 250 mm and a stacking (axial length) of 200 mm. If the maximum flux density per pole is 0.7 T, calculate:
   a. the peripheral speed [m/s] of the revolving field when the stator is connected to a 60 Hz source;
   b. the peak voltage induced in the rotor bars;
   c. the pole pitch.

13-23 A large 3-phase, 4000 V, 60 Hz squirrel-cage induction motor draws a current of

385 A and a total active power of 2344 kW when operating at full load. The corresponding speed is accurately measured and is found to be 709.2 r/min. The stator is connected in wye and the resistance between two stator terminals is 0.10 Ω. The total iron losses are 23.4 kW and the windage and friction losses are 12 kW. Calculate:
a. the power factor at full load;
b. the active power supplied to the rotor;
c. the total I^2R losses in the rotor;
d. the load mechanical power [kW], torque [kN·m] and efficiency [%].

13-24 If we slightly increase the rotor resistance of an induction motor, what effect does this have (increase or decrease) upon:
a. starting torque
b. starting current
c. full load speed
d. efficiency
e. power factor
f. temperature rise of the motor at its rated power output

13-25 Explain the principle of magnetic levitation.

Advanced level;

13-26 In Fig. 13-5a, the permanent magnet has a width of 100 mm and moves at 30 m/s. The flux density in the air gap is 0.5 T and the effective resistance per rotor bar is 1 mΩ. Calculate the current I and the tractive force.

13-27 If the conducting ladder in Fig. 13-5a is pulled along with a force of 20 N, what is the braking force exerted on the magnet?

13-28 A 3-phase, 5000 hp, 6000 V, 60 Hz 12-pole wound rotor induction motor turns at 594 r/min. What are the approximate rotor I^2R losses?

13-29 The motor in Problem 13-28 has the following characteristics:
1. dc resistance between stator terminals = 0.112 Ω at 17°C
2. dc resistance between rotor slip rings = 0.0073 Ω at 17°C
3. open-circuit voltage induced between slip rings with rotor locked = 1600 V
4. line-to-line stator voltage = 6000 V
5. no-load stator current, per phase = 100 A
6. active power supplied to motor at no-load = 91 kW
7. windage and friction losses = 51 kW
8. iron losses in the stator = 39 kW
9. locked rotor current at 6000 V = 1800 A
10. active power to stator with rotor locked = 2207 kW

Calculate:
a. rotor and stator resistance per phase at 75°C (assume a wye connection);
b. voltage and frequency induced in the rotor when it turns at 200 r/min and at 594 r/min;
c. reactive power absorbed by the motor to create the revolving field, at no-load;
d. I^2R losses in the stator when the motor runs at no-load (winding temperature 75°C);
e. active power supplied to the rotor at no-load;

13-30 Referring to the motor described in Problem 13-29, calculate, under full-voltage LR (locked-rotor) conditions:
a. reactive power absorbed by the motor;
b. I^2R losses in the stator;
c. active power supplied to the rotor;
d. mechanical power output;
e. torque developed by the rotor.

13-31 We wish to control the speed of the motor given in Problem 13-29 by inserting resistors in series with the rotor (see Fig. 13-19). If the motor has to develop a torque of 20 kN·m at a speed of 450 r/min, calculate:
   a. voltage between the slip rings;
   b. rotor resistance (per phase) and the total power dissipated;
   c. approximate rotor current, per phase.

13-32 The train shown in Fig. 13-31 moves at 200 km/h when the stator frequency is 105 Hz. By supposing a negligible slip, calculate the length of the pole-pitch of the linear motor [mm].

13-33 A 3-phase, 300 kW, 2300 V, 60 Hz, 1780 r/min induction motor is used to drive a compressor. The motor has a full-load efficiency and power factor of 92% and 86%, respectively. If the terminal voltage rises to 2760 V while the motor operates at full load, determine the effect (increase or decrease) upon:
   a. mechanical power delivered by the motor;
   b. motor torque;
   c. rotational speed;
   d. full-load current;
   e. power factor and efficiency;
   f. starting torque;
   g. starting current;
   h. breakdown torque;
   i. motor temperature rise;
   j. flux per pole;
   k. exciting current;
   l. iron losses.

13-34 A 3-phase, 60 Hz linear induction motor has to reach a top no-load speed of 12 m/s and it must develop a standstill thrust of 10 kN. Calculate the required pole-pitch and the minimum I^2R loss in the rotor, at standstill.

14

SELECTION AND APPLICATION OF THREE-PHASE INDUCTION MOTORS

When purchasing a three-phase induction motor for a particular application, we often discover that several types can fill the need. Consequently, we have to make a choice. The selection is generally simplified because the manufacturer of the lathe, fan, pump, and so forth indicates the type of motor which is best suited to drive the load. Nevertheless, it is useful to know something about the basic construction and characteristics of the various types of three-phase induction motors that are available on the market.

In this chapter, we also cover some special applications of induction machines, such as asynchronous generators and frequency converters. These interesting devices will enable the reader to gain an even better understanding of induction motors in general.

14.1 Standardization and classification of induction motors*

The frames of all industrial motors under 500 hp have standardized dimensions. Thus, a 25 hp, 1725 r/min, 60 Hz motor of one manu-

facturer can be replaced by that of any other manufacturer, without having to change the mounting holes, the shaft height, or the type of coupling. The standardization covers not only frame sizes, but also establishes limiting values for electrical, mechanical and thermal characteristics. Thus, motors must satisfy minimum requirements as to starting torque, locked-rotor current, overload capacity, and temperature rise.

14.2 Classification according to environment and cooling methods

Motors are grouped into several categories, depending upon the environment in which they have to operate. We shall limit our discussion to five important classes.

---

* Standards in the United States are governed by National Electrical Manufacturers (NEMA) publication MG-1 titled **Motors and Generators**. Standards in Canada are similarly governed by Canadian Standards Association (CSA) publication C 154. The two standards are essentially identical.

1. **Drip-proof motors.** The frame in a drip-proof motor protects the windings against liquid drops and solid particles which fall at any angle between 0 and 15 degrees downward from the vertical. The motors are cooled by means of a fan directly coupled to the rotor. Cool air, drawn into the motor through vents in the frame, is blown over the windings and then expelled. The maximum allowable temperature rise (measured by the change in winding resistance) may be 60°C, 80°C, 105°C, or 125°C, depending on the type of insulation used in the windings. Drip-proof motors (Fig. 14-1)can be used in most locations.

2. **Splash-proof motor.** The frame in a splash-proof motor protects the windings against liquid drops and solid particles which fall at any angle between 0 and 100 degrees downward from the vertical. Cooling is similar to that in drip-proof motors and the maximum temperature rise is also the same. These motors are mainly used in wet locations.

3. **Totally enclosed, nonventilated motors.** These motors have closed frames which prevent the free exchange of air between the inside and the outside of the case. They are designed for very wet and dusty locations. Most are rated below 10 kW because it is difficult to get rid of the heat of larger machines. The motor losses are dissipated by natural convection and radiation from the frame. The permissible temperature rise is 65°C, 85°C, 110°C or 130°C, depending on the class of insulation (see Fig. 14-2).

4. **Totally enclosed, fan-cooled motors.** Medium and high-power motors that are totally enclosed are usually cooled by an external blast of air. An external fan, directly coupled to the shaft, blows air over the ribbed motor frame (Fig. 14-3). A concentric outer shield prevents physical contact with the fan and serves to channel the air-stream. The permissible tem-

Figure 14-1
Energy efficient drip-proof, 3-phase squirrel-cage induction motor rated 230 V/460 V, 3 hp, 1750 r/min, 60 Hz. (*Gould*)

Figure 14-2
Totally-enclosed induction motor for a centrifugal pump. (*Brook Crompton-Parkinson Ltd*)

Figure 14-3
Totally-enclosed fan-cooled induction motor rated 350 hp, 1760 r/min, 440 V, 3-phase, 60 Hz. (*Gould*)

perature rise is the same as for drip-proof motors.

5. **Explosion-proof motors.** Explosion-proof motors are used in highly inflammable or explosive surroundings such as are found in coal mines, oil refineries and grain elevators. They are totally enclosed (but not air-tight) and the frames are designed to withstand the enormous pressure which may build up inside the motor due to an internal explosion. Furthermore, the flanges on the end-bells are made extra-long in order to cool any escaping gases generated by such an explosion. The permissible temperature rise is the same as for totally enclosed motors (see Fig. 14-4).

14.3 Classification according to electrical and mechanical properties

In addition to the various enclosures just mentioned, 3-phase squirrel-cage motors can have special electrical and mechanical characteristics, as listed below.

1. **Motors with standard locked-rotor torque (NEMA Design B).** Most induction motors belong to this group. The locked-rotor torque

depends upon the size of the motor, and ranges from 130% to 70% of full-load torque, as the power increases from 20 hp to 200 hp (15 kW to 150 kW). The corresponding locked-rotor current should not exceed 6.4 times the rated full-load current. These general-purpose motors are used to drive fans, centrifugal pumps, machine-tools, and so forth.

2. **High starting-torque motors (NEMA Design C).** These motors are employed when starting conditions are difficult. Pumps and piston-type compressors that have to start under load are two typical applications. In the range from 20 hp to 200 hp, the locked-rotor torque is 200 % of full-load torque and the locked-rotor current should not exceed 6.4 times the rated full-load current.

In general, these motors are equipped with a double-cage rotor. The excellent performance of a double-cage rotor (Fig. 14-5) is based upon the fact that:

a. the frequency of the rotor current diminishes as the motor speeds up;
b. a conductor that lies close to the rotor surface (cage 1) has a lower inductive react-

Figure 14-4
Totally-enclosed, fan-cooled, explosion-proof motor. Note the particularly rugged construction of this type of motor. (*Brook Crompton-Parkinson Ltd*)

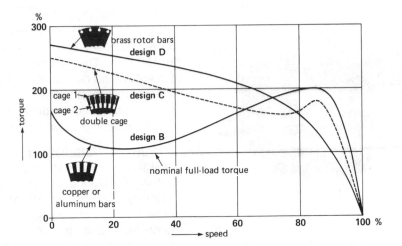

Figure 14-5
Typical torque-speed curves of NEMA design B, C and D motors. Each curve corresponds to the minimum NEMA values of locked-rotor torque, pull-up torque and breakdown torque of a 3-phase 1800 r/min, 10 hp, 60 Hz squirrel-cage induction motor. The cross section of the respective rotors indicates the type of rotor bars used.

ance than one buried deep inside the core (cage 2).

c. the conductors of cage 1 are much smaller than those of cage 2.

When the motor is connected to the line with the rotor at standstill, the frequency of the rotor current is equal to line frequency. Owing to the high inductive reactance of squirrel-cage 2, the rotor current flows mainly in the small bars of cage 1. The effective motor resistance is therefore high, being essentially equal to that of cage 1. Consequently, a high starting torque is developed.

As the motor speeds up, the rotor frequency falls, with the result that the inductive reactance of both squirrel-cage windings diminishes. At rated speed, the rotor frequency is so low that the reactance of both windings is negligible. The rotor current is then limited only by the *resistance* of cage 1 and cage 2, operating in parallel. Because the conductors of cage 2 are much larger than those of cage 1, the effective rotor resistance at rated speed is much lower

than at standstill. For this reason, the double-cage rotor develops both a high starting torque and a low slip at full load.

Despite their high torque, Design C motors are not recommended for the starting of high-inertia loads. The reason is that most of the rotor I^2R losses during start-up are concentrated in cage 1. Owing to its small size, it tends to overheat and the bars may melt.

3. **High-slip motors (NEMA Design D).** The nominal speed of high-slip, Design D motors is between 85% and 95% of synchronous speed. These motors drive high inertia loads (such as centrifugal dryers) which take a relatively long time to reach full speed. The high-resistance squirrel cage is made of brass, and the motors are usually designed for intermittent operation to prevent overheating.

The large drop in speed with increasing load is ideal to drive impact machine-tools such as shears and punch presses that are equipped with a flywheel. The flywheel gradually increases its speed and stores mechanical energy during the relatively long idling period and releases it in a fraction of a second during the high-power impulse load. For example, suppose a punch-press requires the release of 100 hp during 1/10 of a second and the time between impacts is 5 s. A design D motor having a rating of 100 hp (x 1/10) (÷5) = 2 hp would suffice to drive the punch-press. A design D motor can drop in speed without causing a sharp increase in the stator current.

The torque speed curves of Fig. 14-5 enable us to compare the characteristics of these motors. The rotor construction is also shown, and it can be seen that the distinguishing properties are obtained by changing the rotor design. For example, if we increase the rotor resistance (by using brass instead of copper or aluminum), the locked-rotor torque increases, but the speed at rated torque drops.

14.4 Choice of motor speed

The choice of motor speed is rather limited because the synchronous speed of induction motors changes by quantum jumps, depending upon the frequency and number of poles. For example, it is impossible to build a conventional 60 Hz induction motor having an acceptable efficiency and running at a speed, say, of 2000 r/min. Such a motor would require 2 poles and a corresponding synchronous speed of 3600 r/min. The slip of (3600 − 2000)/3600 = 0.444 means that 44.4% of the power supplied to the rotor would be dissipated as heat. (See Sec. 13.13)

The speed of a motor is obviously determined by the speed of the machine it has to drive. However, for low-speed machines, it is often preferable to use a high-speed motor and a gear box instead of directly coupling a low-speed motor to the load. There are several advantages to using a gear-box:

1. For a given output power, the size and cost of a high-speed motor is less than that of a low-speed motor, and its effi-

ciency and power factor are higher;
2. The locked-rotor torque of a high-speed motor is always greater (as a percent of full-load torque) than that of a similar low-speed motor of equal power.

By way of example, Table 14A compares the properties of two commercial 3-phase 60 Hz induction motors having the same power but different synchronous speeds. The difference in price alone would justify the use of a high-speed motor and a gear box to drive a load operating at, say, 900 r/min.

When equipment has to operate at very low speeds (100 r/min or less), a gear box is mandatory. The gears are often an integral part of the motor, making for a very compact unit (Fig. 14-6).

A gear box is also mandatory when equipment has to run above 3600 r/min. For example, in one industrial application a large gear unit is used to drive a 1200 hp, 5000 r/min centrifugal compressor from a 3560 r/min induction motor.

Figure 14-6
Gear motor rated at 2.25 kW, 1740 r/min, 60 Hz. The output torque and speed are respectively 172 N·m and 125 r/min. (*Reliance Electric*)

TABLE 14A   COMPARISON BETWEEN TWO MOTORS OF DIFFERENT SPEEDS

power		synchronous speed	power factor	efficiency	locked rotor torque	mass	cost (1990)
hp	kW	r/min	%	%	%	kg	$
10	7.5	3600	89	87	150	50	500
10	7.5	900	82	83	125	115	1400

14.5 Two-speed motors

The stator of a squirrel-cage induction motor can be designed so that the motor can operate at two different speeds. Such motors are often used on drill-presses, blowers and pumps. One way to obtain two speeds is to wind the stator with two separate windings having, say, 4 poles and 6 poles. The problem is that only one winding is in operation at a time and so only half the copper in the slots is being used.

However, special windings have been invented whereby the speed is changed by simply changing the external stator connections. The synchronous speeds obtained are usually in the ratio 2 to 1: 3600/1800 r/min; 1200/600 r/min; etc. The lower speed is produced by the creation of *consequent poles*.

Consider, for example, one phase of a two-pole, 3-phase motor (Fig. 14-7a). When the two poles are connected in series to a 60 Hz ac source, current I_1 flows into terminal 1 and current I_2 ($=I_1$) flows out of terminal 2. As a result, one N pole and one S pole are created and the flux has the pattern shown. The synchronous speed is:

$$n_s = 120f/p = 120 \times 60/2$$
$$= 3600 \text{ r/min} \qquad \text{Eq. 13-1}$$

Note that each pole covers only one quarter of the stator circumference. This is achieved by using a coil pitch equal to 50 percent of the pole pitch.

Let us now connect the two poles in parallel, as shown in Fig. 14-7b. In this case, at the instant current I_1 flows into terminal 1, current I_2 flows into terminal 2. As a result, two N poles are created by the windings.

Because every N pole must be accompanied by a S pole, it follows that two S poles will appear between the two N poles. The south poles created in this ingenious way are called *consequent* poles. The new connection produces 4 poles in all and the synchronous speed is 1800 r/min. Thus, we can double the number of poles by simply changing the stator connections. It is upon this principle that two-speed motors are based.

Figure 14-8 shows the stator connections for a two-speed, 4-pole/8-pole, 3-phase motor. Six leads, numbered 1 to 6, are brought out from the stator winding. For the high-speed connection power is applied to terminals 1-2-3 and terminals 4-5-6 are open. The resulting delta connection produces 4 poles per phase having alternate N and S polarities (Fig. 14-8a). Note

(a)

(b)

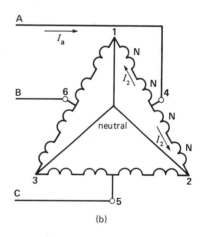

(b)

Figure 14-7
a. Two short-pitch coils connected in series produce a two-pole motor.
b. When the coils are connected in parallel, a 4-pole motor is produced. Two of the poles are consequent poles.

Figure 14-8
a. High-speed connection of a 3-phase stator, yielding 4 poles.
b. Low-speed connection of same motor yielding 8 poles.

that the poles are connected in series.

The low-speed connection is made by short-circuiting terminals 1-2-3 and applying power to terminals 4-5-6. The resulting double-wye connection again produces 4 poles per phase but now they all possess the same polarity (Fig. 14-8b). The coil-pitch of two-speed

motors is always one half the high-speed pole-pitch. The short coil-pitch provides room for the creation of the consequent poles at the lower speed.

Two-speed motors have a relatively lower efficiency and power factor than single-speed motors do. They can be designed to develop (at both speeds) either constant power, constant torque or variable torque, depending upon the load that has to be driven.

The two-speed motors described so far have pole ratios of 2 to 1. However, it is possible to design the 3-phase windings to obtain pole ratios such as 8/10, 14/16, 26/28, 10/14, and 38/46. These so-called *pole amplitude modulation* or PAM motors are particularly useful in driving large fans in the hundred horsepower range and more. The reason is that the power output of a fan varies as the cube of the speed. Consequently, if the speed is reduced by half, the power drops to 1/8, which is often too low to be of interest. The PAM motor permits a significant reduction in fan power by simply reconnecting the windings to give the lower speed.

14.6 Induction motor characteristics under load

The complete torque-speed characteristics shown in Fig. 14-5 are important, but it must be recognized that most of the time a motor runs at close to synchronous speed, supplying a torque that varies from zero to full-load torque T_n. It so happens that between these limits, the torque-speed curve is essentially a straight line (Fig. 14-9). The slope of the line depends mainly upon the rotor resistance: the lower the resistance, the sharper the slope.

This linearity enables us to establish a simple equation showing how the line voltage and rotor resistance affect the behavior of the motor under load. In effect, when we know the characteristics of a motor under a given load condition, we can calculate its speed, torque, power, etc.,

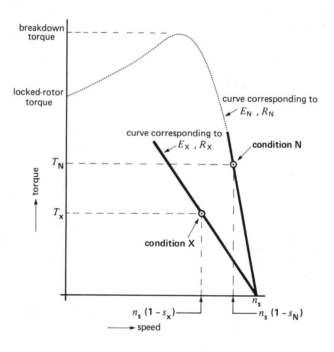

Figure 14-9
The torque-speed curve is essentially a straight line between the no-load and rated torque operating points.

under any other load condition.

These quantities are related by the equation:

$$s_x = s_n \left[\frac{T_x}{T_n}\right]\left[\frac{R_x}{R_n}\right]\left[\frac{E_n}{E_x}\right]^2 \qquad (14\text{-}1)$$

where

  n = subscript referring to the initial, or given load conditions (the given conditions may correspond to the nominal rating of the motor).

  x = subscript referring to the new load conditions.

  s = slip

  T = torque [N·m]

  R = rotor resistance [Ω]

  E = stator voltage [V]

In applying the equation, the only restriction is that the new torque T_x must not be greater than $T_n(E_x/E_n)^2$. Under these conditions, Eq. 14-1 yields an accuracy of better than 5%, which is sufficient for most practical problems.

Example 14-1:
A 3-phase, 208 V induction motor having a synchronous speed of 1200 r/min turns at a speed of 1140 r/min when connected to a 215 V line. Calculate the speed if the voltage increases to 240 V.

Solution:
a. The slip at 215 V is:

$$s = (n_s - n)/n_s$$
$$= (1200 - 1140)/1200$$
$$= 0.05$$

When the voltage rises to 240 V, the load torque and rotor resistance remain the same. Consequently, in applying Eq. 14-1, we can write:

$$s_x = s_n(E_n/E_x)^2 = 0.05(215/240)^2$$
$$= 0.04$$

The slip speed is therefore
$$0.04 \times 1200 = 48 \text{ r/min}$$

The new speed at 240 V is:
$$n_x = 1200 - 48 = 1152 \text{ r/min}$$

Example 14-2:
A 3-phase induction motor driving a compressor runs at a speed of 873 r/min immediately after it is connected to a fixed 460 V, 60 Hz line. The initial cold rotor temperature is 23°C. The speed drops to 864 r/min after the machine has run for several hours. Calculate:
a. the hot rotor resistance in terms of the cold resistance;
b. the approximate hot temperature of the rotor bars, knowing they are made of copper.

Solution:
a. The initial and final slips are:

$$s_n = (900 - 873)/900 = 0.03$$
$$s_x = (900 - 864)/900 = 0.04$$

The voltage, and torque, etc., are fixed; consequently, the speed change is entirely due to the change in rotor resistance. We can therefore write:

$$s_x = s_n(R_x/R_n)$$
$$0.04 = 0.03 \ (R_x/R_n)$$
$$R_x = 1.33 \ R_n$$

The hot rotor resistance is 33% greater than the cold rotor resistance.

b. The hot rotor temperature is:

$$t_2 = \frac{R_2}{R_1} \ (234 + T_1) - 234 \qquad \text{Eq. 6-3}$$
$$= 1.33 \ (234 + 23) - 234$$
$$= 108°C$$

Example 14-3:
A 3-phase wound-rotor induction motor has a rating of 110 kW, 1760 r/min, 2.3 kV, 60 Hz. Three external resistors of 2 Ω are connected in wye across the rotor slip rings. The motor develops a torque of 300 N·m at a speed of 1000 r/min.

a. Calculate the speed for a torque of 400 N·m;
b. Calculate the value of the external resistors so that the motor develops 10 kW at 200 r/min.

Solution:

a. The given conditions are:

$$T_n = 300 \text{ N·m}$$
$$s_n = (1800 - 1000)/1800 = 0.444$$

All other conditions being fixed, we have for a torque of 400 N·m:

$$s_x = s_n(T_x/T_n) = 0.444(400/300)$$
$$= 0.592$$

The slip speed = 0.592 × 1800 = 1067 r/min. Consequently, the motor speed is:

$$n = 1800 - 1067$$
$$= 733 \text{ r/min}$$

Note that the speed drops from 1000 r/min to 733 r/min with increasing load.

b. The torque corresponding to 10 kW at 200 r/min is:

$$T_x = 9.55 \ P/n \qquad \text{Eq. 3-5}$$
$$= 9.55 \times 10\,000/200$$
$$= 478 \text{ N·m}$$

The rated torque is:

$$T_{rated} = 9.55 \ P/n$$
$$= 9.55 \times 110\,000/1760$$
$$= 597 \text{ N·m}$$

because T_x is less than T_{rated}, we can apply Eq. 14-1.

The slip is:

$$s_x = (1800 - 200)/1800 = 0.89$$

All other conditions being fixed, we have:

$$s_x = s_n \ (T_x/T_n) \ (R_x/R_n)$$
$$0.89 = 0.44 \ (478/300) \ (R_x/2)$$

whence $R_x = 2.5 \ \Omega$

Three 2.5 Ω wye-connected resistors in the rotor circuit will enable the motor to develop 10 kW at 200 r/min.

14.7 Starting an induction motor

High-inertia loads put a strain on induction motors because they prolong the starting period. The starting current in both the stator and rotor is high during this interval so that overheating becomes a major problem. For motors of several thousand horsepower, a prolonged starting period may even put a strain on the transmission line feeding the plant where the motor is installed. The line voltage may fall below normal for many seconds, thus affecting other connected loads. To relieve the problem, induction motors are often started on reduced voltage. This limits the power drawn by the motor, and consequently reduces the line voltage drop as well as the heating rate of the windings. Reduced voltage lengthens the start-up time, but this is usually less important. However, whether the start-up time is long or short, it is worth remembering the following rule:

> **Rule 1 -** The heat dissipated in the rotor during the starting period (from zero to nominal speed) is equal to the final kinetic energy stored in all the revolving parts.

This rule holds true, irrespective of the stator voltage or the torque-speed curve of the motor. Thus, if a motor brings a heavy flywheel up to speed, and if the energy stored in the flywheel is 5000 joules, the rotor will dissipate 5000 joules in the form of heat. Depending upon the size of the rotor and its cooling system, this energy could easily produce overheating.

14.8 Plugging an induction motor

In some industrial applications, the induction motor and its load have to be brought to a quick stop. This can be done by interchanging two stator leads, so that the revolving field suddenly turns in the opposite direction to the rotor. During this so-called *plugging* period, the motor acts as a *brake*.

It absorbs kinetic energy from the still-

LEGEND

P_e — active power supplied to stator
P_{js} — stator copper losses
P_f — stator iron losses
P_r — active power transmitted to the rotor
P_m — mechanical power supplied to the rotor
P_{jr} — I^2R losses in the rotor
P_v — windage and friction losses

Figure 14-10
When a 3-phase induction motor is plugged, the rotor I^2R losses are very high.

revolving load, causing its speed to fall. The associated mechanical power P_m is entirely dissipated as heat in the rotor. Unfortunately, the rotor also continues to receive electromagnetic power P_r from the stator, which is also dissipated as heat (Fig. 14-10). Consequently, plugging produces I^2R losses in the rotor which even exceed those when the rotor is locked. Motors should not be plugged too frequently because high rotor temperatures may melt the rotor bars or overheat the stator winding. In this regard, it is worth remembering the following rule for plugging operations:

> **Rule 2 -**   The heat dissipated in the rotor during the plugging period (nominal-speed to zero) is *three times* the original kinetic energy of all the revolving parts.

Example 14-4:
A 100 kW, 60 Hz, 1175 r/min motor is coupled to a flywheel by means of a gear box. The kinetic energy of all the revolving parts is 300 kJ when the motor runs at nominal speed. The motor is plugged to a stop and allowed to run up to 1175 r/min in the reverse direction. Calculate the energy dissipated in the rotor if the flywheel is the only load.

Solution:
During the plugging period, the motor speed

drops from 1175 r/min to zero. The heat dissipated in the rotor is 3×300 kJ = 900 kJ. The motor then accelerates to nominal speed in the reverse direction. The energy dissipated in the rotor during this period is 300 kJ. The total heat dissipated in the rotor from start to finish is therefore 900 + 300 = 1200 kJ.

14.9 Braking with direct current

An induction motor and its high-inertia load can also be brought to a quick stop by circulating dc current in the stator winding. Any two terminals can be connected to the dc source.

The direct current produces stationary N, S poles in the stator. The number of poles created is equal to the number of poles which the motor develops normally. Thus, a 3-phase, 4-pole induction motor produces 4 dc poles, no matter how the motor terminals are connected to the dc source.

When the rotor sweeps past the stationary field, an ac voltage is induced in the rotor bars. The voltage produces an ac current and the resulting rotor I^2R losses are dissipated at the expense of the kinetic energy stored in the revolving parts. The motor finally comes to rest when all the kinetic energy has been dissipated as heat in the rotor.

The advantage of dc braking is that it produces far less heat than does plugging. In effect, the energy dissipated in the rotor is only equal to the original kinetic energy stored in the revolving masses, and not three times that energy. The energy dissipated in the rotor is independent of the magnitude of the dc current. However, a smaller dc current increases the braking time. The dc current can be two or three times the rated current of the motor. Even larger values can be used, provided that the stator does not become too hot. The braking torque is proportional to the square of the dc braking current.

Example 14-5:
A 50 hp, 1760 r/min, 440 V, 3-phase induction motor drives a load having a total moment of inertia of 25 kg·m². The dc resistance between two terminals is 0.32 Ω, and the rated motor current is 62 A. We want to stop the motor by connecting a 24 V battery across the stator. Calculate:

a. the dc current in the stator;
b. the energy dissipated in the rotor;
c. the average braking torque if the stopping time is 4 min.

Solution:
a. The dc current is:

$$I = E/R = 24/0.32 = 75 \text{ A}$$

This current is slightly higher than the rated current of the motor. However, the stator will not overheat, because the braking time is short.

b. The kinetic energy in the rotor and load at 1760 r/min is:

$$E_k = 5.48 \times 10^{-3} J n^2 \qquad \text{Eq. 3-8}$$
$$= 5.48 \times 10^{-3} \times 25 \times 1760^2$$
$$= 424 \text{ kJ}$$

Consequently, the rotor absorbs 424 kJ during the braking period.

c. The average braking torque can be calculated from the equation:

$$\Delta n = 9.55 \, T\Delta t/J \qquad \text{Eq. 3-14}$$

$$1760 = 9.55T \times (4 \times 60)/25$$
$$T = 19.2 \text{ N·m}$$

14.10 Abnormal conditions

Abnormal motor operation may be due to internal causes (short-circuit in the stator, overheating of the bearings, etc.), or to external conditions. External problems may be caused by:

1. mechanical overload;
2. supply voltage changes;
3. single phasing;
4. frequency changes.

We shall examine the nature of these problems in the sections that follow.

According to national standards, a motor shall operate satisfactorily on any voltage within ± 10% of the nominal voltage, and for any frequency within ± 5% of the nominal frequency. If the voltage and frequency *both* vary, the sum of the two percentage changes must not exceed 10%. Finally, all motors are designed to operate satisfactorily at altitudes up to 1000 m above sea level. At higher altitudes, the temperature may exceed the permissible limits owing to the poor cooling afforded by the thinner air.

14.11 Mechanical overload

Although standard induction motors can develop as much as twice their rated power for short periods, they should not be allowed to run continuously beyond their rated capacity. Overloads cause overheating, which deteriorates the insulation and reduces its useful life. In practice, the higher motor current causes the thermal overload relays to trip, bringing the motor to a stop before its temperature gets too high.

Some drip-proof motors are designed to carry a continuous overload of 15 percent. This overload capacity is shown on the nameplate by the so-called *service factor* 1.15. The allowable temperature rise is then 10°C higher than that permitted for drip-proof motors operating at normal load.

During emergencies, a drip-proof motor can be made to carry overloads of as much as 150% as long as strong external ventilation is provided. This is not recommended for long periods because even if the external frame is cool, the temperature of the windings may be excessive.

14.12 Line voltage changes

The most important consequence of a line voltage change is its effect upon the torque-speed curve of the motor. In effect, the torque at any speed is proportional to the *square* of the applied voltage. Thus, if the stator voltage decreases by 10%, the torque at every speed will drop by 20%. A line voltage drop is often produced during start-up, due to the heavy starting current drawn from the line. As a result, the starting torque may be much less than its rated value.

On the other hand, if the line voltage is too high when the motor is running, the flux per pole will be too high. For a motor running at full load, this increases both the iron losses and the magnetizing current, with the result that the temperature increases slightly and the power factor is slightly reduced.

An *unbalance* of the 3-phase voltages produces a serious unbalance of the three line currents. This condition increases the stator and rotor losses, yielding a higher temperature. A voltage unbalance of a little as 3.5% can cause the temperature to increase by 15°C. The utility company should be notified whenever the phase-to-phase line voltages differ by more than 1%.

14.13 Single-phasing

If one line of a 3-phase line is accidentally opened, or if a fuse blows while the motor is running, the machine will continue to run as a *single-phase motor*. The current drawn from the remaining two lines will almost double, and the motor will begin to overheat. The thermal relays protecting the motor will eventually trip the circuit-breaker, thereby disconnecting the motor from the line.

The torque-speed curve is seriously affected when a 3-phase motor operates on single phase. The breakdown torque decreases to

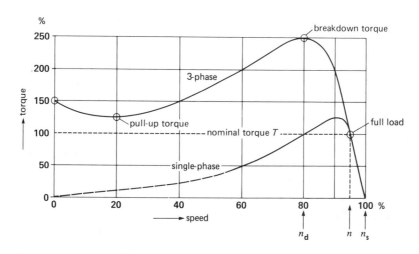

Figure 14-11
Typical torque-speed curves when a 3-phase squirrel-cage motor operates normally and when it operates on single-phase.

about 40% of its original value, and the motor develops no starting torque at all. Consequently, a fully-loaded 3-phase motor may simply stop if one of its lines is suddenly opened. The resulting locked-rotor current is about 90% of the normal 3-phase L.R. current. It is therefore large enough to trip the circuit breaker or to blow the fuses.

Figure 14-11 shows the typical torque-speed curves of a 3-phase motor when it runs normally and when it is single-phasing. Note that the curves follow each other closely until the torque approaches the single-phase breakdown torque.

14.14 Frequency variation

Important frequency changes never take place on a large distribution system, except during a major disturbance. However, the frequency may vary significantly on isolated, low-power systems where electrical energy is generated by diesel engines or gas turbines. The emergency supply in a hospital, the electrical system on a ship, the generators in a lumber camp, etc., are examples of this type of supply.

The most important consequence of a frequency change is the resulting change in motor speed: if the frequency drops by 20%, the motor speed drops by 20%.

Machine tools and other motor-driven equipment imported from countries where the frequency is 50 Hz, may cause problems when they are connected to a 60 Hz system. Everything runs 20% faster than normal, and this may not be acceptable in some applications. In such cases, we either have to gear down the motor speed, or supply an expensive auxiliary 50 Hz source.

A 50 Hz motor operates well on a 60 Hz line, but its terminal voltage should be raised to 6/5 (or 120%) of the nameplate rating. The new break-down torque is then equal to the original break-down torque and the starting torque is only slightly reduced. Power factor, efficiency and temperature rise remain satisfactory.

A 60 Hz motor can also operate on a 50 Hz line, but its terminal voltage should be reduced to 5/6 (or 83%) of its nameplate value. The breakdown torque and starting torque are then about the same as before, and the power factor, efficiency and temperature rise remain satisfactory.

14.15 Induction motor operating as a generator

Consider an electric train powered by a squirrel-cage induction motor which is directly coupled to the wheels. As the train climbs up a hill, the motor will run at slightly less than synchronous speed, developing a torque sufficient to overcome both friction and the force of gravity. At the top of the hill, on level ground, the force of gravity no longer comes into play and the motor has only to overcome the friction of the rails and the air. The motor runs essentially at no-load and very close to synchronous speed.

What happens when the train begins to move downhill? The force of gravity causes the train to accelerate and because the motor is coupled to the wheels, it begins to rotate *above* synchronous speed. However, as soon as this takes place, the motor develops a counter torque which *opposes* the increase in speed. This torque has the same effect as a brake. However, instead of being dissipated as heat, the mechanical braking power is returned to the line in the form of electrical energy. An induction motor which turns faster than synchronous speed acts therefore as a generator. It converts the mechanical energy it receives into electrical energy, and this energy is released by the stator. Such a machine is called an *asynchronous generator*.

Although induction motors are rarely used to drive trains (Fig. 14-12), there are several industrial applications which may cause a motor to run above synchronous speed. In cranes, for example, during the lowering cycle, the motor receives power from the mechanical

Figure 14-12
The electric train makes the round trip between Zermatt (1604 m) and Gornergrat (3089 m), in Switzerland. The drive is provided by four 3-phase wound-rotor induction motors, rated 78 kW, 1470 r/min, 700 V, 50 Hz. Two aerial conductors constitute phases A and B, and the rails provide phase C. A toothed gear-wheel 573 mm in diameter engages a stationary rack on the roadbed to drive the train up and down the steep slopes. The speed can be varied from zero to 14.4 km/h by means of variable resistors in the rotor circuit. The rated thrust is 78 kN. (ABB)

"load" and returns it to the line.

We can make an asynchronous generator by connecting an ordinary squirrel-cage motor to a 3-phase line and coupling it to a gasoline engine (Fig. 14-13). As soon as the engine speed exceeds the synchronous speed, the motor becomes a source, delivering active power P to the electrical system to which it is connected. However, to create its magnetic field, the motor has to absorb reactive power Q. This power can only come from the ac line, with the result that the reactive power Q flows in the opposite direction to the active power P (Fig. 14-13).

The active power delivered to the line is directly proportional to the slip above synchronous speed. Thus, a higher engine speed produces a greater electrical output. However, the rated output is reached at very small slips, usually less than 3%.

The reactive power may be supplied by a group of capacitors connected to the terminals of the motor. With this arrangement, we can supply a 3-phase load without using an external 3-phase source (Fig. 14-14). The frequency generated is slightly less than that corresponding to the speed of rotation. Thus, a 4-pole

motor driven at a speed of 2400 r/min produces a frequency slightly less than

$$f = pn/120 = 4 \times 2400/120 = 80 \text{ Hz}.$$

The terminal voltage of the generator increases with the capacitance, but its magnitude is limited by saturation in the iron. If the capacitance is insufficient, the generator voltage will not build up. The capacitor bank must be able to supply at least the reactive power the machine normally absorbs when operating as a motor.

Figure 14-13
Gasoline engine driving an asynchronous generator connected to a 3-phase line. Note that P and Q flow in opposite directions.

Figure 14-14
Capacitors can provide the reactive power for an asynchronous generator. This eliminates the need for a 3-phase external source.

Example 14-6:
We wish to use a 40 hp, 1760 r/min, 440 V, 3-phase squirrel-cage induction motor as an asynchronous generator. The rated current of the motor is 41 A, and the full-load power factor is 84%.
a. Calculate the capacitance required per phase if the capacitors are connected in delta.
b. At what speed should the driving engine run to generate a frequency of 60 Hz?

Solution:
a. The apparent power drawn by the machine when it operates as a motor is:
$$S = \sqrt{3}\, EI$$
$$= 1.73 \times 440 \times 41$$
$$= 31.2 \text{ kVA}$$

The corresponding active power absorbed is:
$$P = S \cos \theta$$
$$= 31.2 \times 0.84$$
$$= 26.2 \text{ kW}$$

The corresponding reactive power absorbed is:
$$Q = \sqrt{S^2 - P^2}$$
$$= \sqrt{31.2^2 - 26.2^2}$$
$$= 17 \text{ kvar}$$

When the machine operates as an asynchronous generator, the capacitor bank must supply *at least* $17 \div 3 = 5.7$ kvar per phase. The voltage per phase is 440 V because the capacitors are connected in delta. Consequently, the capacitive current per phase is:
$$I_c = Q/E = 5700/440$$
$$= 13 \text{ A}$$

The capacitance per phase is:
$$X_c = E/I = 440/13$$
$$= 34 \; \Omega$$

The capacitance must be at least:
$$C = 1/2\pi f X_c$$
$$= 1/(2\pi \times 60 \times 34)$$
$$= 78 \; \mu\text{F}$$

Figure 14-15 shows how the generating system is connected. Note that if the load itself absorbs reactive power, the capacitor bank must be increased to provide it.

b. The driving engine must turn at slightly more than synchronous speed. Typically, the slip should be equal to the full-load slip when the machine operates as a motor. Consequently,
$$\text{slip} = 1800 - 1760$$
$$= 40 \text{ r/min}$$

The engine should therefore run at an approximate speed of:
$$n = 1800 + 40$$
$$= 1840 \text{ r/min}$$

Figure 14-15
See Example 14-6.

T = torque developed
by the machine

n = speed of rotation

n_s = synchronous speed
of the revolving field

Figure 14-16
Complete torque-speed curve of a 3-phase induction machine.

14.16 Complete torque-speed characteristic of an induction machine

We have seen that a 3-phase squirrel-cage induction motor can also function as a generator or as a brake. These three modes of operation – motor, generator, brake – merge into each other, as can be seen from the torque speed curve of Fig. 14-16. This curve, together with the adjoining power flow diagrams illustrate the overall properties of a 3-phase squirrel-cage induction machine.

We see, for example, that when the shaft turns in the *same* direction as the revolving field, the induction machine operates in either the motor or the generator mode. But to operate in the generator mode, the shaft must turn faster than synchronous speed. Similarly, to operate as a motor, the shaft must turn at less than synchronous speed.

Finally, in order to operate as a brake, the shaft must turn in the opposite direction to the revolving flux.

14.17 Features of a wound-rotor induction motor

So far, we have directed our attention to the squirrel-cage induction motor and its related properties as a generator and brake. The reason is that this is the type of motor most frequently used in industry. However, the wound-rotor induction motor has certain features that make it attractive in special industrial applications. These may be listed as follows:

1. start-up of very high inertia loads
2. variable-speed drives
3. frequency converter.

We now examine these applications.

14.18 Start-up of high-inertia loads

We recall that whenever a load is brought up to speed, the energy dissipated in the rotor is equal to the kinetic energy imparted to the load. This means that a high inertia load will produce very high losses in the rotor, causing it to become excessively hot. The advantage of the wound-rotor motor is that the heat is dissipated in the external resistors connected to the slip rings. Thus, the rotor itself remains cool.

Another advantage is that the external resistors can be varied as the motor picks up speed. As a result, it is possible for the motor to develop its maximum torque during the entire acceleration period. Thus, the final speed can be attained as quickly as possible.

14.19 Variable-speed drives

We have already seen that an increase in rotor resistance will cause the speed of an induction motor to fall. Thus, by varying the external resistors of a wound-rotor motor we can obtain any speed we want, so long as it is below synchronous speed. The problem is that the power dissipated as heat in the resistors makes for a very inefficient system, which becomes too costly when the motors have ratings of several thousand horsepower. We get around this problem by connecting the slip rings to an electronic converter. The converter changes the power at low rotor frequency into power at line frequency and feeds this power back into the 3-phase systems. (Fig. 14-17). As a result, such variable-speed control becomes very efficient, in the sense that much less power is lost.

14.20 Frequency converter

A conventional wound-rotor motor may be used as a *frequency converter* to generate a frequency different from that of the utility company. The stator of the wound-rotor machine is connected to the utility line, and the rotor is driven at an appropriate speed by a motor M (Fig. 14-18). The wound-rotor machine then behaves as a rotating transformer in which the stator is the primary and the rotor the secondary. The rotor supplies power to the three-phase load at a voltage E_2 and frequency f_2 that depends upon the slip. Thus, according to Eqs. 13-3 and 13-4, we have:

$$f_2 = sf_o \qquad \text{Eq. 13-3}$$

$$E_2 = sE_{oc} \qquad \text{Eq. 13-4}$$

Figure 14-17
The water supply in the City of Stuttgart, Germany is provided by a pipeline that is 1.6 m in diameter and 110 km long. The water is pumped from Lake Constance in the Alps. The pump in the background is driven by a wound-rotor induction motor rated at 3300 kW, 425 to 595 r/min, 5 kV, 50 Hz. The variable speed enables the water supply to be varied according to the needs of the city.

The enclosed motor housing seen in the foreground contains an air/water heat exchanger which uses the 5°C water for cooling purposes. During start-up, liquid rheostats are connected to the slip rings but when the motor is up to speed the slip rings are connected to an electronic converter which feeds the rotor power back into the line. (*Siemens*)

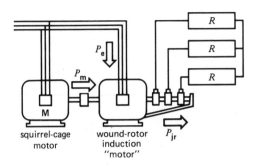

Figure 14-18
Wound-rotor motor used as a frequency converter.

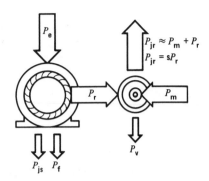

Figure 14-19
Power flow in a frequency converter when the output frequency is greater than the line frequency.

In general, the desired frequency is two or three times that of the utility company. To attain this frequency, the slip must be positive and greater than 1. As a result, the shaft must be rotated against the direction of the revolving flux.

The operation of the frequency converter is then identical to that of an induction motor operating as a brake. However, the power P_{jr}, usually dissipated as heat in the rotor, is now available to supply power to the load. The converter acts as a generator, and the active power flow is as shown in Fig. 14-19. Note how similar this is to the power flow when an induction motor runs as a brake (Fig. 14-16).

Example 14-7:
A 3-phase wound-rotor induction motor has a rating of 150 hp ($\sim$ 110 kW), 1760 r/min, 2.3 kV, 60 Hz. The open-circuit rotor voltage between the slip rings is 500 V. The rotor is driven by variable-speed dc motor. Calculate:
a. the turns ratio of the stator to rotor windings;
b. the rotor voltage and frequency when the rotor is driven at 720 r/min in the same direction as the revolving field;
c. the rotor voltage and frequency when the rotor is driven at 720 r/min opposite to the revolving field;

Solution:
a. The turns ratio is:
$$a = E_1/E_{oc} = 2300/500$$
$$= 4.6$$

b. The slip at 720 r/min is:
$$s = (n_s - n)/n_s = (1800 - 720)/1800$$
$$= 0.6$$

The rotor voltage at 720 r/min is:
$$E_2 = sE_{oc} = 0.6 \times 500$$
$$= 300 \text{ V}$$

The rotor frequency is:
$$f_2 = sf_o = 0.6 \times 60$$
$$= 36 \text{ Hz}$$

The motor speed is considered to be negative ($-$) when it turns opposite to the revolving field. The slip at -720 r/min is:
$$s = (n_s - n)/n_s$$
$$= (1800) - (-720))/1800$$
$$= (1800 + 720)/1800$$
$$= 1.4$$

The rotor voltage and frequency at -720 r/min are:
$$E_2 = sE_{oc} = 1.4 \times 500$$
$$= 700 \text{ V}$$
$$f_2 = sf_o \ 1.4 \times 60$$
$$= 84 \text{ Hz}$$

Example 14-8:

We wish to use a 30 kW, 900 r/min, 60 Hz wound-rotor motor to generate 60 kW at an approximate frequency of 180 Hz. If the supply-line frequency is 60 Hz, calculate:

a. the speed of the induction motor driving the frequency converter;
b. the active power input to the stator of the frequency converter;
c. the power of the induction motor;
d. Will the frequency converter overheat under these conditions?

Solution:

a. To generate 180 Hz the slip must be:

$$f_2 = sf_o \qquad \text{Eq. 13-3}$$
$$180 = s \times 60$$

from which

$$s = 3$$

The converter must be driven at a speed n given by:

$$s = (n_s - n)/n_s \qquad \text{Eq. 13-2}$$
$$3 = (900 - n)/900$$

from which

$$n = -1800 \text{ r/min}$$

The converter must therefore be driven at a speed of 1800 r/min. The negative sign indicates that the rotor must run opposite to the revolving field. The induction motor driving the converter must therefore have a synchronous speed of 1800 r/min.

b. The rotor delivers an output of 60 kW. This corresponds to P_{jr}, but instead of being "dissipated" in the rotor, P_{jr} is useful power delivered to a load (Fig. 14-20). The power P_r transferred from the stator to the rotor is:

$$P_r = P_{jr}/s = 60/3 \qquad \text{Eq. 13-7}$$
$$= 20 \text{ kW}$$

The power input to the stator of the frequency converter is equal to 20 kW plus the small copper and iron losses in the stator.

c. The remaining power input to the rotor $(60 - 20 = 40 \text{ kW})$ is derived from the mechanical input to the shaft. By referring to Fig. 14-19 and Fig. 14-20, we can see how the active power flows into (and out of) the converter.

In summary, the rotor receives 20 kW of electrical power from the stator and 40 kW of mechanical power from the driving motor M. The rotor converts this power into 60 kW of electrical power at a frequency of 180 Hz. Induction motor M must therefore have a rating of 40 kW, 60 Hz, 1800 r/min.

Figure 14-20
See Example 14-8.

d. The stator of the converter will not overheat because the 20 kW it absorbs is much less than its nominal rating of 30 kW. The rotor will not overheat either, even though it delivers 60 kW. The increased power arises from the fact that the voltage induced in the rotor is three times higher than at standstill. The iron losses in the rotor will be high because the frequency is 180 Hz; however, because the rotor turns at twice normal speed, the cooling is more effective, and the rotor will probably not overheat. The stator frequency is 60 Hz, consequently, the iron losses in the stator are normal.

QUESTIONS AND PROBLEMS

Practical level

14-1   What is the difference between a drip-proof motor and an explosion-proof motor?

14-2   What is the approximate life expectancy of a motor?

14-3   Explain why a NEMA design D motor is unsatisfactory for driving a pump.

14-4   Identify the motor components shown in Fig. 14-3.

14-5   Show the flow of active power in a 3-phase induction motor when it operates:
a. as a motor;
b. as a brake.

14-6   Will a 3-phase motor continue to rotate if one of the lines becomes open? Will the motor be able to start on such a line?

14-7   What type of ac motor would you recommend for the following application:
a. a saw in a lumber mill?
b. a variable speed pump?

14-8   Give some of the advantages of standardization as it relates to induction motors.

14-9   We can bring an induction motor to a quick stop either by plugging it or by exciting the stator from a dc source. Which method produces the least amount of heat in the motor? Explain.

14-10 A standard squirrel-cage induction motor rated at 50 hp, 440 V, 60 Hz, 1150 r/min is connected to a 208 V, 3-phase line. By how much are the breakdown torque and locked-rotor torque reduced?

Intermediate level

14-11 A 3-phase squirrel-cage induction motor having a rated voltage of 575 V, is connected to a 520 V line. Explain how the following parameters are affected:
a. locked-rotor current;
b. locked-rotor torque;
c. no-load current;
d. no-load speed;
e. full-load current;
f. full-load power factor;
g. full-load efficiency.

14-12 a. Referring to Fig. 14-6, if we eliminated the gear box and used another motor directly coupled to the load, what would its power output have to be [hp]?
b. How many poles would the motor have?

14-13 Draw the typical torque-speed curve of a NEMA Design C squirrel-cage induction motor, rated at 30 hp, 900 r/min (see Fig. 14-5). Give the values of the LR, pull-up and breakdown torques and the corresponding speeds [ft·lbf and r/min].

14-14 A 300 hp, 2300 V, 3-phase, 60 Hz squirrel-cage induction motor turns at a full load speed of 590 r/min.
a. Calculate the approximate value of the rotor I^2R losses. If the line voltage then drops to 1944 V, calculate:
b. the new speed, knowing that the load torque remains the same;
c. the new power output;
d. the new I^2R losses in the rotor.

14-15 We wish to make an asynchronous generator using a standard squirrel-cage induction motor rated at 40 hp, 208 V, 870 r/min, 60 Hz (Fig. 14-14). The generator is driven at 2100 r/min by a gasoline engine, and the load consists of three 5 Ω resistors connected in wye. The generator voltage builds up when three 100 μF capacitors are connected in wye across the terminals. If the line voltage is 520 V, calculate:

a. the approximate frequency generated;

b. the active power supplied to the load;

c. the reactive power supplied by the capacitor bank;

d. the stator current;

e. if the following gasoline engines are available: 30 hp, 100 hp, 150 hp, which one is best suited to drive the generator?

f. discuss the copper and iron losses when the machine operates as above, compared to when it functions as a standard motor at 60 Hz.

14-16 A 30 000 hp, 13.2 kV, 3-phase, 60 Hz air-to-water cooled induction motor drives a turbo compressor in a large oxygen-manufacturing plant. The motor runs at an exact full-load speed of 1792.8 r/min and by means of a gearbox, it drives the compressor at a speed of 4930 r/min. The motor has an efficiency of 98.1% and a power factor of 0.90. The LR torque and current are respectively 0.7 p.u. and 4.9 p.u. Calculate:

a. the full-load current;

b. the total losses at full load;

c. the exact rotor I^2R losses if the windage and friction losses amount to 62 kW;

d. the LR current and torque ;

e. the torque developed at the compressor shaft.

14-17 The motor in Problem 14-16 is cooled by circulating 350 gallons (U.S.) of water through the heat exchanger per minute.

Calculate the increase in water temperature as the water flows through the heat exchanger.

14-18 The motor and compressor in Problem 14-16 are started on reduced voltage and the average starting torque during the acceleration period is 0.25 p.u. The compressor has a moment of inertia of 130 000 lb·ft^2 referred to the motor shaft. The squirrel cage rotor alone has a J of 18 000 lb·ft^2.

a. How long will it take to bring the motor and compressor up to speed, at no-load?

b. what is the energy dissipated in the rotor during the starting period [Btu]?

14-19 A 3-phase induction motor rated at 10 kW, 1450 r/min, 380 V, 50 Hz has to be connected to a 60 Hz line.

a. What line voltage should be used, and what will be the approximate speed of the motor?

b. What power [hp] can the motor deliver without overheating?

Advanced level

14-20 A 1 hp, squirrel-cage, Design B induction motor accelerates an inertia load of 1.4 kg·m^2, from zero to 1800 r/min. Could this motor be replaced by a class D motor and if so:

a. which motor has the shortest acceleration time from zero to 1200 r/min?

b. which of the two rotors will be the hottest, after reaching the no-load speed?

14-21 A three-phase, wound-rotor induction motor having a rating of 150 hp, 1760 r/min, 2.3 kV, 60 Hz, drives a belt conveyor. The rotor is connected in wye and the nominal open-circuit voltage between the slip rings is 530 V. Calculate:

a. the rotor winding resistance per phase;

b. the resistance which must be placed in series with the rotor (per phase) so that the motor will deliver 40 hp at a speed of 600 r/min, knowing that the line voltage is 2.4 kV.

14-22 A 150 hp, 1165 r/min, 440 V, 60 Hz, 3-phase induction motor is running at no-load, close to its synchronous speed of 1200 r/min. The stator leads are suddenly reversed, and the stopping time is clocked at 1.3 s. Assuming that the torque exerted during the plugging interval is equal to the starting torque (1.2 p.u.), calculate:

a. the magnitude of the plugging torque;

b. the moment of inertia of the rotor.

14-23 In Problem 14-22 calculate the energy dissipated in the rotor during the plugging interval.

14-24 A 3-phase, 8-pole induction motor has a rating of 40 hp, 575 V, 60 Hz. It drives a steel flywheel having a diameter of 31.5 inches and a thickness of 7 7/8 in. The torque-speed curve corresponds to that of a design D motor given in Fig. 14-5.

a. calculate the mass of the flywheel and its moment of inertia [lb·ft²];

b. calculate the rated speed of the motor and the corresponding torque [ft·lbf];

c. calculate the locked-rotor torque [ft·lbf].

d. draw the torque-speed curve of the 40 hp motor and give the torques [N·m] at 0, 180, 360, 540, 720 and 810 r/min.

14-25 a. In Problem 14-24, calculate the average torque between zero and 180 r/min.

b. Using Eq. 3-14, calculate the time required to accelerate the flywheel from zero to 180 r/min, assuming no other load on the motor.

c. Using Eq. 3-8, calculate the kinetic energy in the flywheel at 180 r/min.

d. Calculate the time required to accelerate the flywheel from zero to 540 r/min, knowing that this time, the load exerts a fixed counter torque of 300 N·m, in addition to the flywheel load.

14-26 The train in Fig. 14-12 has a mass of 78 500 lb and can carry 240 passengers. Calculate:

a. the speed of rotation of the gear wheel when the train moves at 9 miles per hour.

b. the speed ratio between the motor and the gear wheel.

c. the approximate transmission line current when the motors are operating at full load;

d. the total mass if the average weight of a passenger is 60 kg;

e. the energy required to climb from Zermatt to Gornergrat [MJ];

f. the minimum time required to make the trip [min];

g. assuming that 80% of the electrical energy is converted into mechanical energy when the train is going uphill, and that 80% of the mechanical energy is reconverted to electrical energy when going downhill, calculate the total electrical energy consumed during a round trip [kW·h].

15

EQUIVALENT CIRCUIT OF THE INDUCTION MOTOR

The preceding three chapters have shown that we can describe the important properties of squirrel-cage and wound-rotor induction motors without using a circuit diagram. However, if we want to gain even a better understanding of the properties of the motor, an equivalent circuit diagram is indispensible. In this chapter,* we develop the equivalent circuit from basic principles. We then analyze the characteristics of a low-power and high-power motor and observe their basic differences.

Finally, we develop the equivalent circuit of an asynchronous generator and determine its properties under load.

15.1 The wound-rotor induction motor

A 3-phase wound-rotor induction motor is very similar in construction to a 3-phase transformer. Thus, the motor has 3 identical primary windings and 3 identical secondary windings - one set for each phase. On account

*This chapter can be skipped by the reader who does not have time to study the more theoretical aspects of induction motor behavior.

of the perfect symmetry, we can consider a single primary winding and a single secondary winding in analyzing the behavior of the motor.

When the motor is at standstill, it acts exactly like a conventional transformer, and so its equivalent circuit (Fig. 15-1) is the same as that previously developed in Chapter 10, Fig. 10-20.

We assume a wye connection and the circuit parameters, *per phase*, are identified as follows:

E_g = source voltage, line to neutral

r_1 = stator winding resistance

x_1 = stator leakage reactance

x_2 = rotor leakage reactance

r_2 = rotor winding resistance

R_x = external resistance connected between one slip-ring and the neutral of the rotor

X_m = magnetizing reactance

R_m = resistance corresponding to the iron losses and windage and friction losses

T = ideal transformer having a turns ratio of 1:1

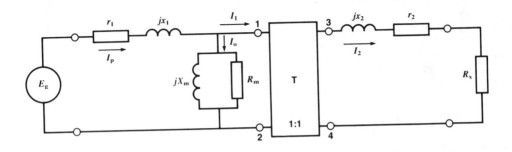

Figure 15-1
Equivalent circuit of a wound-rotor induction motor at standstill.

Figure 15-2
Approximation of the equivalent circuit is acceptable for motors above 5 hp.

In the case of a conventional 3-phase transformer, we would be justified in removing the magnetizing branch composed of jX_m and R_m because the exciting current I_o is negligible compared to the load current I_p. However, in a motor this is no longer true: I_o may be as high as 40 percent of I_p because of the air gap. Consequently, we cannot eliminate the magnetizing branch. However, for motors exceeding 2 hp, we can shift it to the input terminals as shown in Fig. 15-2. This greatly simplifies the equations that describe the behavior of the motor, without compromising accuracy.* Figure 15-2 is a true representation

of the motor when the rotor is locked. How is it affected when the motor starts turning?

Suppose the motor runs at a slip s, meaning that the rotor speed is $n_s(1 - s)$, where n_s is the synchronous speed. This will modify the values of E_1, I_1 and E_2, I_2 on the primary and secondary side of the ideal transformer T. Furthermore, the frequency in the secondary winding will become sf, where f is the frequency of the source. Figure 15-3 shows these new conditions.

Directing our attention to the secondary side, the amplitude of the induced voltage E_2 would be equal to E_1 (the turns ratio is 1:1) if the motor were stationary. But because the slip is s, the actual voltage induced is

$$E_2 = sE_1$$

---

* For motors under 1 hp the exact circuit should be used.

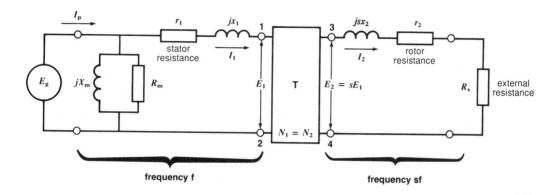

Figure 15-3
Equivalent circuit of a wound-rotor motor when it is running at a slip s. The frequency of the voltages and currents in the stator is f. But the frequency of the voltages and currents in the rotor is sf.

The frequency is sf and this changes the impedance of the secondary leakage reactance from jx_2 to jsx_2. Because resistors are not frequency-sensitive, the values of r_2 and R_x remain the same. Let us lump the two together to form a single secondary resistance R_2, given by:

$$R_2 = r_2 + R_x \qquad (15\text{-}1)$$

The details of the secondary circuit are shown in Fig. 15-4a and the resulting current I_2 is:

$$I_2 = \frac{sE_1}{R_2 + jsx_2} = \frac{sE_1 \angle -\theta}{\sqrt{R_2^2 + (sx_2)^2}} \qquad (15\text{-}2)$$

where $\theta = \arctan sx_2/R_2$ (15-3)

The corresponding phasor diagram is shown in Fig. 15-4b. It is important to realize that this phasor diagram relates to the frequency sf. Consequently, it cannot be integrated into the phasor diagram on the primary side, where the frequency is f. Nevertheless, there is a direct relationship between I_2 (frequency sf) in the rotor and I_1 (frequency f) in the stator. In effect, the absolute value and phase angle of I_1 is exactly

(a)

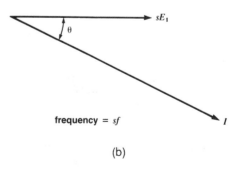

(b)

Figure 15-4
a. Equivalent circuit of the rotor; E_2 and I_2 have a frequency sf.
b. phasor diagram showing the current lagging behind the voltage by angle θ.

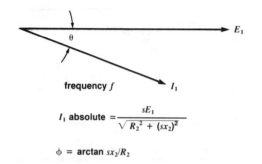

$$I_1 \text{ absolute} = \frac{sE_1}{\sqrt{R_2^2 + (sx_2)^2}}$$

$$\phi = \arctan sx_2/R_2$$

Figure 15-5
The voltage and current in the stator are separated by the same phase angle θ, even though the frequency is different.

the same as that of I_2. This enables us to draw the phasor diagram for E_1 and I_1 as shown in Fig. 15-5.

This means that:
1. The effective value of I_1 is equal to the effective value of I_2 even though their frequencies are different.

2. The effective value of E_1 is equal to the effective value of E_2 divide by the slip s.

3. The phase angle between E_1 and I_1 is the same as that between E_2 and I_2.

Thus, on the primary side we can write:

$$I_1 = I_2 = \frac{sE_1}{R_2 + jsx_2} \qquad (15\text{-}4)$$

therefore

$$I_1 = \frac{E_1}{\dfrac{R_2}{s} + jx_2} = \frac{E_1}{Z_2} \qquad (15\text{-}5)$$

The impedance Z_2 «seen» between the primary terminals 1,2 of the ideal transformer is therefore:

$$Z_2 = \frac{R_2}{s} + jx_2 \qquad (15\text{-}6)$$

As a result, we can simplify the circuit of Fig. 15-3 to that shown in Fig. 15-6. Furthermore the leakage reactances jx_1, jx_2 can be lumped together to create a single total leakage reactance jx. It is equal to the total leakage reactance of the motor referred to the stator side.

Figure 15-6
Equivalent circuit of a wound-rotor motor referred to the primary (stator) side.

The final equivalent circuit of the wound-rotor induction motor is shown in Fig. 15-7. Note that the value of R_2/s will vary from R_2 to infinity as the motor goes from start-up ($s = 1$) to synchronous speed ($s = 0$).

This equivalent circuit of a wound-rotor induction motor is so similar to that of a transformer that it is not surprising that the wound-rotor induction motor is sometimes called a rotary transformer.

The equivalent circuit of a squirrel-cage induction motor is the same, except that R_2 is the equivalent resistance of the rotor alone referred to the stator.

15.2 Power relationships

The equivalent circuit enables us to arrive at some basic electromechanical power relationships for the 3-phase induction motor. The following equations can be deduced by visual inspection of the equivalent circuit (Fig. 15-7):

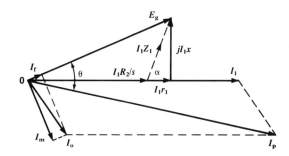

Figure 15-7
The primary and secondary leakage reactances x_1 and x_2 are combined to form an equivalent total leakage reactance x.

Figure 15-8
Phasor diagram of the voltages and currents in Fig. 15-7. The power factor of the motor is cos θ.

1. active power absorbed by the motor is:
$$P = E_g^2/R_m + I_1^2 r_1 + I_1^2 R_2/s$$

2. reactive power absorbed by the motor is:
$$Q = E_g^2/X_m + I_1^2 x$$

3. apparent power absorbed by the motor is:
$$S = \sqrt{P^2 + Q^2}$$

4. power factor of the motor is:
$$\cos θ = P/S$$

5. the line current is:
$$I_p = S/E_g$$

6. power supplied to the rotor is:
$$P_r = I_1^2 R_2/s$$

7. power dissipated as I^2R losses in the rotor is:
$$P_{jr} = I_1^2 R_2 = sP_r$$

8. mechanical power developed by the motor is:
$$P_m = P_r - P_{jr}$$
$$= P_r (1 - s)$$

9. torque developed by the motor is:
$$T = \frac{9.55 \, P_m}{n} = \frac{9.55 \, P_r(1 - s)}{n_s (1 - s)}$$
$$= \frac{9.55 \, P_r}{n_s}$$

10. efficiency of the motor is:
$$η = P_o/P_i = P_m/P$$

15.3 Phasor diagram of the induction motor

If we use current I_1 in Fig. 15-7 as the reference phasor, we obtain the complete phasor diagram shown in Figure 15-8. In this diagram (and also in future calculations) it is useful to define the impedance Z_1 and angle α as follows:

$$Z_1 = \sqrt{r_1^2 + x^2} \qquad (15\text{-}7)$$

$$α = \arctan x/r_1$$

where r_1 is the stator resistance and jx is the total leakage reactance of the motor referred to the stator. Because r_1 and jx are fixed, it follows that Z_1 and α are fixed, irrespective of the slip the motor may have.

In large motors above 1000 hp, jx is much larger than r_1 and so the angle α approaches 90°.

As far as mechanical power, torque and speed are concerned, we can neglect the magnetizing branch R_m and jX_m. Consequently, for these calculations the equivalent circuit and corresponding phasor diagram can be simplified to that shown in Fig. 15-9 and 15-10.

Figure 15-9
As far as mechanical power, torque and speed are concerned, we can neglect the magnetizing branch X_m and R_m. This yields a simpler circuit for the analysis of motor behavior.

Figure 15-10
Phasor diagram of the circuit of Fig. 15-9. Note that phasor I_1Z_1 is always α degrees ahead of phasor I_1.

15.4 Breakdown torque and speed

The breakdown torque is the maximum torque the motor can develop. To attain it, we must maximize $P_r = I_1^2 R_2/s$, the power to the rotor, because

$$T = 9.55\ P_r/n_s$$

Thus, we must maximize the power "dissipated" in R_2/s. According to a basic power transfer theorem, the power is maximum when

$$\frac{R_2}{s} = Z_1 \qquad (15\text{-}8)$$

that is when

$$I_1\ \frac{R_2}{s} = I_1Z_1$$

The phasor diagram corresponding to this special condition is shown in Fig. 15-11. It is a special case of the phasor diagram of Fig. 15-10. Simple geometry yields the following results:

1. phasors AB and BC have the same length and the angle between them is $(180 - \alpha)°$.

2. angle CAB = angle ACB = $\alpha/2$

Consequently,

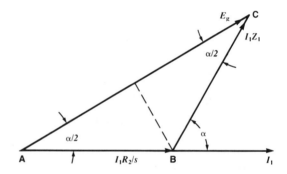

Figure 15-11
Phasor diagram when the motor develops its maximum torque. Under these conditions $R_2/s = Z_1$.

$$I_1\ \frac{R_2}{s}\ \cos\ \frac{\alpha}{2} = E_g/2$$

$$I_1Z_1\ \cos\ \frac{\alpha}{2} = E_g/2$$

The slip at breakdown torque is:

$$s_b = R_2/Z_1 \qquad (15\text{-}9)$$

The current at the breakdown torque is:

$$I_{1b} = E_g/(2\ Z_1\cos\ \alpha/2) \qquad (15\text{-}10)$$

The breakdown torque is:

$$T_b = \frac{9.55\ E_g^2}{n_s(4Z_1\cos^2\alpha/2)} \qquad (15\text{-}11)$$

We note that the magnitudes of both the breakdown torque T_b and the "breakdown" current I_{1b} are fixed, in the sense that they are independent of the rotor resistance R_2.

However, the slip at the breakdown torque depends upon R_2. Indeed, if $R_2 = Z_1$, the breakdown torque coincides with the starting torque because $s_b = 1$.

These conclusions are all borne out by Figure 13-18, Chapter 13.

In practice, the angle α lies between 80° and 89°. Furthermore, because r_1 is usually much smaller than x, Z_1 can be replaced by x. The larger angles correspond to medium- and high-power motors. In such machines the ratio R_2/Z_1 can be as low as 0.02. Consequently, the breakdown torque occurs at slips as small as 2%.

15.5 Equivalent circuit of two practical motors

The impedances and resulting equivalent circuits of two squirrel-cage motors, rated 5 hp and 5000 hp are given in Figures 15-12 and 15-13, together with the motor ratings. The motors are both connected in wye and the impedances are given per phase.

motor rating:
5 hp, 60 Hz, 1800 r/min, 440 V, 3-phase
full-load current: 7 A
locked-rotor current: 39 A

r_1 = stator resistance 1.5 Ω
r_2 = rotor resistance 1.2 Ω
jx = total leakage reactance 6 Ω
jX_m = magnetizing reactance 110 Ω
R_m = no-load losses resistance 900 Ω

(The no-load losses include the iron losses plus windage and friction losses)

motor rating:
5000 hp, 60 Hz, 600 r/min, 6900 V, 3-phase
full-load current: 358 A
locked-rotor current: 1616 A

r_1 = stator resistance 0.083 Ω
r_2 = rotor resistance 0.080 Ω
jx = total leakage reactance 2.6 Ω
jX_m = magnetizing reactance 46 Ω
R_m = no-load losses resistance 600 Ω

The no-load losses of 26.4 kW (per phase) consist of 15 kW for windage and friction and 11.4 kW for the iron losses.

Figure 15-12
Equivalent circuit of a 5 hp squirrel-cage induction motor. Because there is no external resistor in the rotor, $R_2 = r_2$.

Figure 15-13
Equivalent circuit of a 5000 hp squirrel-cage induction motor. Although this motor is 1000 times more powerful than the motor in Fig. 15-12, the circuit diagram remains the same.

15.6 Calculation of the breakdown torque

We will now calculate the breakdown torque T_b and the corresponding speed n_b and current I_{1b} for the 5 hp motor.

1. $Z_1 = \sqrt{r_1^2 + x^2} = \sqrt{1.5^2 + 6^2} = 6.18\ \Omega$

2. $\alpha = \arctan x/r_1 = \arctan 6/1.5 = 76°$

3. $\alpha/2 = 38°$; $\cos \alpha/2 = 0.788$

4. The slip at breakdown is:
   $$s_b = R_2/Z_1 = 1.2/6.18 = 0.194$$

5. The speed n_b at breakdown is:
   $$n_b = n_s(1 - s_b)$$
   $$= 1800(1 - 0.194)$$
   $$= 1450\ \text{r/min}$$

6. The current at breakdown is:
   $$I_{1b} = \frac{E_L}{2\ Z_1 \cos \alpha/2}$$

   $$= \frac{254}{2 \times 6.18 \times 0.788} = 26.1\ \text{A}$$

7. The power to the rotor is:
   $$P_r = I_1^2 Z_1$$
   $$= 26.1^2 \times 6.18 = 4210\ \text{W}$$

8. The breakdown torque T_b is:
   $$T_b = \frac{9.55\ P_r}{n_s}$$

   $$= \frac{9.55 \times 4210}{1800} = 22.3\ \text{N·m}$$

Note that this is the torque developed *per phase*. The total torque is therefore $3 \times 22.3 = 67$ N·m.

15.7 Torque-speed curve and other characteristics

We can determine the complete torque-speed curve of the 5 hp motor by selecting various values of slip and solving the circuit of Fig. 15-12. The results are listed in Table 15A, and the curve is given in Fig. 15-14.

The same calculations are made for the 5000hp motor. Table 15B lists the results and Fig. 15-15 shows the torque-speed curve. Note the relatively low starting torque for this

Figure 15-14
Torque-speed curve of a 5 hp motor.

TABLE 15A TORQUE-SPEED CHARACTERISTIC

5 hp, 440 V, 1800 r/min, 60 Hz
squirrel cage induction motor

s	I_1	P_r	T	n
-	A	W	N·m	r/min
0.0125	2.60	649	3.44	1777
0.025	5.09	1243	6.60	1755
0.026	5.29	1291	6.85	1753
0.05	9.7	2256	12.0	1710
0.1	17.2	3547	18.8	1620
0.2	26.4	4196	22.3	1440
0.4	33.9	3441	18.3	1080
0.6	36.6	2674	14.2	720
0.8	37.9	2150	11.4	360
1	38.6	1788	9.49	0

The rated power of 5 hp is developed at s = 0.026

Figure 15-15
Torque-speed curve of a 5000 hp motor.

TABLE 15B   TORQUE-SPEED AND
LOAD CHARACTERISTIC

5000 hp, 6900 V, 600 r/min, 60 Hz
squirrel-cage induction motor

s	torque	Total power	speed	cos θ	eff'cy	I_p
—	kN·m	hp	r/min	%	%	A
2	1.49	−377	−600	4.9	−29.8	1617
1	2.98	0	0	6.3	0	1616
0.6	4.95	500	240	8.2	23.4	1614
0.4	7.39	1120	360	10.6	40.8	1610
0.2	14.4	2921	480	17.7	64.7	1593
0.1	26.8	6095	540	30.8	80.4	1535
0.05	42.1	10114	570	51.7	89.5	1363
0.03077	47.0	11520	581.5	68.2	93.1	1133
0.02	43.1	10679	588	79.8	95.1	878
0.0067	19.9	50003	596	90.1	96.6	358
0.0033	10.2	2577	598	85.1	95.4	198

large motor, as well as the near-synchronous speed from no-load right up to the breakdown torque. These characteristics are typical for large squirrel-cage induction motors.

15.8 Properties of an asynchronous generator

We have already learned that a squirrel-cage induction motor can act as a generator if it is driven above synchronous speed. Now that we have the equivalent circuit for the 5 hp motor, we can calculate the power it can generate, together with its other properties as a generator.

Let us connect the motor to a 440 V, 3-phase line and drive it at a speed of 1845 r/min, which is 45 r/min above synchronous speed. The slip is

$$s = (n_s - n)/n_s$$
$$= (1800 - 1845)/1800$$
$$= -0.025$$

The value of R_2/s in the equivalent circuit is therefore:

$$R_2/s = 1.2/(-0.025)$$
$$= -48 \ \Omega$$

The negative resistance indicates that power is flowing from the rotor to the stator rather than from the stator to the rotor. Referring to Fig. 15-16 we make the following calculations:

Figure 15-16
Equivalent circuit of a 5 hp motor operating as an asynchronous generator. Note that a negative resistance is reflected into the primary circuit.

1. Net resistance of branch 1 - 2 - 3 - 4 is:
$$R_n = -48 + 1.5 = -46.5 \ \Omega$$

2. impedance of branch 1 - 2 - 3 - 4 is:
$$Z = \sqrt{R_n^2 + x^2}$$
$$= \sqrt{(-46.5)^2 + 6^2}$$
$$= 46.88 \ \Omega$$

3. current in branch 1 - 2 - 3 - 4 is:
$$I_1 = E/Z = 254/46.88$$
$$= 5.42 \ A$$

4. active power delivered to rotor is:
$$P_r = I_1^2 R_2/s = 5.42^2 \ (-48)$$
$$= -1410 \ W$$

this negative power means that 1410 W is flowing from the rotor to the stator.

5. the I^2R losses in the rotor are:
$$P_{jr} = I_1^2 r_2 = 5.42^2 \times 1.2$$
$$= 35.2 \ W$$

6. the mechanical power input to the shaft is equal to P_r plus the losses P_{jr} in the rotor:
$$P_m = P_r + P_{jr}$$
$$= 1410 + 35.2$$
$$= 1445 \ W$$

7. The I^2R losses in the stator are:
$$P_{js} = I_1^2 r_1 = 5.42^2 \times 1.5$$
$$= 44.1 \ W$$

8. the iron plus windage and friction losses are:
$$P_f + P_v = E^2/R_m = 254^2/900$$
$$= 71.7 \ W$$

9. the active power delivered to the line "feeding" the motor is:
$$P_e = \text{power delivered from rotor}$$
$$\text{to stator} - \text{losses}$$
$$= P_r - P_{js} - P_f - P_v$$
$$= 1410 - 44.1 - 71.7$$
$$= 1294 \ W$$

(P_e for the 3 phases = 3 × 1294 = 3882 W)

10. reactive power absorbed by the leakage reactance is:
$$Q_1 = I_1^2 x = 5.42^2 \times 6$$
$$= 176 \ var$$

11. reactive power absorbed by the magnetizing reactance is:
$$Q_2 = E^2/X_m = 254^2/110$$
$$= 586 \ var$$

12. total reactive power absorbed by the motor is:
$$Q = Q_1 + Q_2$$
$$= 176 + 586 = 762 \ vars$$

13. apparent power at the generator terminals A,B is:
$$S = \sqrt{P_e^2 + Q^2} = \sqrt{1294^2 + 762^2}$$
$$= 1502 \ VA$$

14. the line current I_p is:
$$I_p = S/E = 1502/254$$
$$= 5.91 \ A$$

15. the power factor at the generator terminals is:
$$\cos \theta = P_e/S = 1294/1502$$
$$= 0.861 = 86.1\%$$

16. the efficiency of the asynchronous generator is:
$$\eta = \frac{\text{useful electric power}}{\text{mechanical input}} = \frac{P_e}{P_m}$$
$$= \frac{1294}{1445} = 0.895 = 89.5\%$$

17. horsepower needed to drive the generator is:
$$P_m = 3P_m/746 = 3 \times 1445/746$$
$$= 5.81 \ hp$$

15.9 Tests to determine the equivalent circuit

The approximate values of r_1, r_2, X_m, R_m and x in the equivalent circuit can be found by means of the following tests.

No-load test

When an induction motor runs at no-load, the slip is exceedingly small. Referring to Fig. l5-6, this means that the value of R_2/s is very high and so current I_1 is negligible compared to I_0. Thus, at no-load the circuit consists essentially of the magnetizing branch X_m, R_m. Their values can be determined by measuring the voltage, current and power at no-load, as follows:

a) Measure the stator resistance R_{LL} between any two terminals. Assuming a wye connection, the value of r_1 is:
$$r_1 = R_{LL}/2$$

b) Run the motor at no-load using rated line-to-line voltage, E_{NL} (Fig. 15-17). Measure the no-load current I_{NL} and the total 3-phase active power P_{NL}.

Locked-rotor test

When the rotor of an induction motor is locked, the stator current I_p is almost 6 times its rated value. Furthermore, the slip s is equal to one. This means that r_2/s is equal to r_2, where r_2 is the resistance of the rotor reflected into the stator. Because I_p is much greater than the exciting current I_0, we can neglect the magnetizing branch. This leaves us with the circuit of Fig. 15-9, composed of the leakage reactance x, the stator resistance r_1 and the reflected rotor resistance r_2. Their values can be determined by measuring the voltage, current and power under locked-rotor conditions, as follows:

a) Apply reduced 3-phase voltage to the stator so that the stator current is about equal to its rated value.

b) Take readings of E_{LR} (line-to-line), I_{LR} and the total 3-phase power P_{LR} (Fig. 15-18).

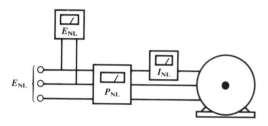

Figure 15-17
A no-load test permits the calculation of X_m and R_m of the magnetizing branch.

The following calculations of total apparent power S_{NL} and total reactive power Q_{NL} are then made:

$$S_{NL} = E_{NL}I_{NL}\sqrt{3}$$
$$Q_{NL} = \sqrt{S_{NL}^2 - P_{NL}^2}$$

$P_f + P_v$ = windage, friction and iron losses
$$= P_{NL} - 3I_{NL}^2 r_1$$

The resistance R_m representing $P_f + P_v$ is
$$R_m = 3E_{NL}^2/(P_f + P_v)$$

The magnetizing reactance is:
$$X_m = 3E_{NL}^2/Q_{NL}$$

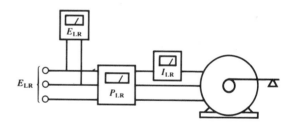

Figure 15-18
A locked-rotor test permits the calculation of the total leakage reactance x and the total resistance $(r_1 + r_2)$. From these results we can determine the equivalent circuit of the induction motor.

The following calculations are then made:
$$S_{LR} = E_{LR}I_{LR}\sqrt{3}$$
$$Q_{LR} = \sqrt{S_{LR}^2 - P_{LR}^2}$$
$$x = Q_{LR}/3I_{LR}^2$$
$$3I_{LR}^2(r_1 + r_2) = P_{LR}$$

hence

$$r_2 = P_{LR}/(3I_{LR}^2) - r_1$$

More elaborate tests are conducted on large machines, but the above-mentioned procedure gives results that are adequate in most cases.

Example 15-1:
A no-load test conducted on a 30 hp, 835 r/min, 440 V, 3-phase, 60 Hz squirrel-cage induction motor, yielded the following results:

No-load voltage (line-to-line)	440 V
No-load current	14 A
No-load power	1470 W
Resistance measured between two terminals:	0.5 Ω

The locked rotor test, conducted at reduced voltage, gave the following results:

Locked - rotor voltage (line-to-line)	163 V
Locked - rotor power	7200 W
Locked - rotor current	60 A

Determine the equivalent circuit of the motor.

Solution:
Assuming the stator windings are connected in wye, the resistance per phase is :

$$r_1 = 0.5 \ \Omega/2 = 0.25 \ \Omega$$

From the no- load test we find:

$$S_{NL} = E_{NL}I_{NL}\sqrt{3} = 440 \times 14\sqrt{3}$$
$$= 10669 \ VA$$

$$P_{NL} = 1470 \ W$$

$$Q_{NL} = \sqrt{S_{NL}^2 - P_{NL}^2} = \sqrt{10669^2 - 1470^2}$$
$$= 10568 \ var$$

$$X_m = E_{NL}^2/Q_{NL} = 440^2/10568$$
$$= 18.3 \ \Omega$$

$$R_m = E_{NL}^2/P_{NL} = 440^2/1470$$
$$= 131.7 \ \Omega$$

From the locked- rotor test we find:

$$S_{LR} = E_{LR}I_{LR}\sqrt{3} = 163 \times 60\sqrt{3}$$
$$= 16939 \ VA$$

$$P_{LR} = 7200 \ W$$

$$Q_{LR} = \sqrt{S_{LR}^2 - P_{LR}^2} = \sqrt{16939^2 - 7200^2}$$
$$= 15333 \ var$$

$$I_{LR} = 60 \ A$$

Total leakage reactance referred to stator is:

$$x = \frac{Q_{LR}}{3I_{LR}^2} = \frac{15333}{3 \times 60^2} = 1.42 \ \Omega$$

Total resistance referred to stator is:

$$r_1 + r_2 = P_{LR}/3I_{LR}^2 = 7200/(3 \times 60^2)$$
$$= 0.67 \ \Omega$$

$$r_1 = 0.25 \ \Omega$$

$$r_2 = 0.67 - 0.25 = 0.42 \ \Omega$$

(in a squirrel-cage motor, $R_2 = r_2$ because

$$R_x = 0 \ ; \ \text{see Eq. 15-1})$$

The equivalent circuit is shown in Fig. I5-19.

Figure 15-19
Determining the equivalent circuit of a squirrel cage induction motor (see Example 15-1)

QUESTIONS AND PROBLEMS

15-1 Without referring to the text, explain the meaning of the impedances, currents and voltages in Fig. 15-2.

15-2 A wye-connected squirrel-cage motor having a synchronous speed of 900 r/min has a stator resistance of

0.7 Ω and an equivalent rotor resistance of 0.5 Ω. If the total leakage reactance is 5 Ω, and the line-to-neutral voltage is 346 V, calculate:

a. the value of Z_1 and the angle α.

b. the speed when the breakdown torque is reached.

c. the current I_1 at the breakdown torque (see Fig. 15-9).

d. the value of the breakdown torque [N·m]

15-3 a. In problem 15-2, draw the equivalent circuit if the motor runs at 950 r/min in the same direction as the revolving flux. Does the machine operate as a generator? Calculate the torque of the machine.

b. Draw the equivalent circuit if the motor runs at 950 r/min opposite to the revolving flux. Does the machine operate as a generator? Calculate the torque.

15-4 A 550 V, 1780 r/min, 3-phase, 60 Hz squirrel-cage induction motor running at no-load draws a current of 12 A and a total power of 1500 W. Calculate the value of X_m and R_m per phase (see Fig. 15-2).

15-5 The motor in Problem 15-4 draws a current of 30 A and a power of 2.43 kW when connected to a 90 V, 3-phase line under locked-rotor conditions. The resistance between two stator terminals is 0.8 Ω. Calculate the values of r_1, r_2 and x and the locked-rotor torque [N·m] at rated voltage.

15-6 If the line voltage for the motor in Fig. 15-15 dropped to 6200 V, calculate the new breakdown torque and starting torque.

15-7 A 440 V, 3-phase, 1800 r/min squirrel-cage motor has the following characteristics:

$$r_1 = 1.5\ \Omega,$$
$$r_2 = 1.2\ \Omega,$$
$$x = 6\ \Omega.$$

If the magnetizing branch can be neglected, calculate the value of the starting torque and the breakdown torque if a 4.5 Ω resistor is connected in series with each line.

15-8 In Problem 15-7 calculate the starting torque and the breakdown torque if a 4.5 Ω reactor is connected in series with each line.

16

SYNCHRONOUS GENERATORS

Three-phase synchronous generators are the primary source of all the electrical energy we consume. These machines are the largest energy converters found in the world. They convert mechanical energy into electrical energy, in powers ranging up to 1500 MW. In this chapter, we shall study the construction and characteristics of these large, modern generators. They are based upon the elementary principles covered in Sec. 8.6, and the reader may wish to review this material before proceeding further.

16.1 Commercial synchronous generators

Commercial synchronous generators are built with either a stationary or a rotating dc magnetic field.

A *stationary-field* synchronous generator has the same outward appearance as a dc generator. The salient poles create the dc field which is cut by a revolving armature. The armature possesses a 3-phase winding whose terminals are connected to three slip rings mounted on the shaft. A set of brushes, sliding on the slip rings, enables us to connect the armature to an external 3-phase load. The armature is driven by a gasoline engine, or some other source of motive power. As it rotates, a 3-phase voltage is induced, whose value depends upon the speed of rotation and upon the dc exciting current in the stationary poles. The frequency of the voltage depends upon the speed and the number of poles on the field. Stationary-field generators are used when the power output is less than 5 kVA. However, for greater outputs, it is cheaper, safer and more practical to employ a revolving dc field.

A *revolving-field* synchronous generator has a stationary armature called a stator. The 3-phase stator winding is directly connected to the load, without going through large, unreliable slip rings and brushes. A stationary stator also makes it easier to insulate the windings because they are not subjected to centrifugal forces. Figure 16-1 is a schematic

Figure 16-1
Schematic diagram of a typical 500 MW synchronous generator and its 2400 kW dc exciter. The dc exciting current I_x (6000 A) flows through the commutator and two slip rings. The dc control current I_c from the pilot exciter permits variable field control of the main exciter, which, in turn, controls I_x.

diagram of such a generator, sometimes called an *alternator*. The field is excited by a dc generator, usually mounted on the same shaft. Note that the brushes on the commutator have to be connected to another set of brushes riding on slip rings to feed the dc current I_x into the revolving field.

16.2 Number of poles

The number of poles on a synchronous generator depends upon the speed of rotation and the frequency we wish to produce. Consider, for example, a stator conductor that is successively swept by the north and south poles of the rotor. If a *positive* voltage is induced when a north pole sweeps across the conductor, a similar *negative* voltage is induced when the south pole speeds by. Thus, every time a complete pair of poles crosses the conductor, the induced voltage goes through a *complete cycle*. The same is true for every other conductor on the stator; we can therefore deduce that the alternator frequency is given by:

$$f = \frac{pn}{120} \qquad (16\text{-}1)$$

where
$f = $ frequency of the induced voltage [Hz]
$p = $ number of poles on the rotor
$n = $ speed of the rotor [r/min]

Example 16-1:
A hydraulic turbine turning at 200 r/min is connected to a synchronous generator. If the induced voltage has a frequency of 60 Hz, how many poles does the rotor have?

Solution:
From Eq. 16-1, we have:
$p = $ 120 f/n
  $= $ 120 × 60/200
  $= $ 36 poles, or 18 pairs of poles

16.3 Main features of the stator

From an electrical standpoint, the stator of a synchronous generator is identical to that of a three-phase induction motor (Sec. 13.17). It is

Figure 16-2a
Stator of a 3-phase, 500 MVA, 0.95 power factor, 15 kV, 60 Hz, 200 r/min, generator. Internal diameter: 9250 mm; effective axial length of iron stacking: 2350 mm; 378 slots. (*Marine Industrie*)

composed of a cylindrical laminated core containing a set of slots that carry the 3-phase lap winding (Figs. 16-2, 16-3). The winding is always connected in wye and the neutral is connected to ground. A wye connection is preferred to a delta connection because:

1. The voltage per phase is only $1/\sqrt{3}$ or 58% of the voltage between the lines. This means that the highest voltage between a stator conductor and the grounded stator core is only 58% of the line voltage. We can therefore reduce the amount of insulation in the slots which, in turn, enables us to increase the cross section of the conductors. A larger conductor permits us to increase the current and hence, the power output of the machine.

2. When a synchronous generator is under load, the voltage induced in each phase becomes distorted, and the waveform is no longer sinusoidal. The distortion is mainly

Figure 16-2b
The copper bars connecting successive stator poles are designed to carry a current of 3200 A. The total output is 19 250 A per phase. (*Marine Industrie*)

Figure 16-2c
The stator is built up from toothed segments of high-quality silicon-iron steel laminations (0.5 mm thick), covered with an insulating varnish. The slots are 22.3 mm wide and 169 mm deep. The salient poles of the rotor are composed of much thicker (2 mm) iron laminations. These laminations are not insulated because the dc flux they carry does not vary. The width of the poles from tip-to-tip is 600 mm and the air gap length is 33 mm. The 8 round holes in the face of the salient pole carry the bars of a squirrel-cage winding.

Figure 16-3
Stator of a 3-phase, 722 MVA, 3600 r/min, 19 kV, 60 Hz steam-turbine generator during the construction phase. The windings are water-cooled. The stator will eventually be completely enclosed in a metal housing (see background). The housing contains hydrogen under pressure to further improve the cooling. *(courtesy ABB)*

Figure 16-4
This 36-pole rotor is being lowered into the stator shown in Fig. 16-2. The 2400 A dc exciting current is supplied by a 330 V, electronic rectifier. Other details are: mass: 600 t; moment of inertia: 4140 t·m²; air gap: 33 mm. (*Marine Industrie*)

Figure 16-5
This rotor winding for a 250 MVA salient-pole generator is made of 18 turns of bare copper bars having a width of 89 mm and a thickness of 9 mm.

Figure 16-6
Salient-pole of a 250 MVA generator showing 12 slots to carry the squirrel-cage winding.

Figure 16-7a
Rotor of a 3-phase steam-turbine generator rated 1530 MVA, 1500 r/min, 27 kV, 50 Hz. The 40 Hz slots are being milled out of the solid steel mass. They will carry the dc winding. Effective axial magnetic length: 7490 mm; diameter: 1800 mm.

Figure 16-7b
Rotor with its 4-pole dc winding. Total mass: 204 t; moment of inertia: 85 t·m²; air gap 120 mm. The dc exciting current of 11.2 kA is supplied by a 600 V dc brushless exciter, bolted to the end of the main shaft. (*Allis-Chalmers Power Systems Inc., West-Allis, Wisconsin*)

due to an undesired *third harmonic* voltage whose frequency is three times that of the fundamental frequency. With a wye connection, the distorting line-to-neutral harmonics do not appear between the lines because they effectively cancel each other. Consequently, the *line voltages* remain sinusoidal under all load conditions. Unfortunately, when a delta connection is used, the harmonic voltages do not cancel, but add up. Because the delta is closed on itself, they produce a third-harmonic circulating current, which increases the I^2R losses.

The nominal line voltage of a synchronous generator depends upon its kVA rating. In general, the greater the power rating, the higher the voltage. However, the nominal voltage seldom exceeds 25 kV because the increased slot insulation takes up valuable space at the expense of the copper conductors.

16.4 Main features of the rotor

Synchronous generators are built with two types of rotors: salient-pole rotors and smooth, cylindrical rotors. Salient-pole rotors are usually coupled to low-speed hydraulic turbines while cylindrical rotors are driven by high-speed steam turbines.

1. Salient pole rotors. Most hydraulic turbines have to turn at low speeds (between 50 and 300 r/min) in order to extract the maximum power from a water fall. Because the rotor is directly coupled to the water-wheel, and because a frequency of 60 Hz is required, a large number of poles are required on the rotor. Low-speed rotors always possess a large diameter to provide the necessary space for the poles. The salient poles are mounted on a large circular steel frame which is fixed to a revolving vertical shaft (Fig. 16-4). To ensure good cooling, the field coils are made of bare copper bars, with the turns insulated from each other by strips of mica (Fig. 16-5). The coils are connected in series so that adjacent poles have opposite polarities.

In addition to the dc field winding, we often add a squirrel-cage winding, embedded in the pole-faces (Fig. 16-6). Under normal conditions, this winding does not carry any current because the rotor turns at synchronous speed. However, when the load on the generator changes suddenly, the rotor speed begins to oscillate, producing momentary speed variations above and below synchronous speed. This induces a voltage in the squirrel-cage winding, causing a large current to flow. The current reacts with the magnetic field of the stator, producing forces which dampen the oscillation of the rotor. For this reason, the squirrel-cage winding is sometimes called a *damper winding*.

The damper winding also tends to maintain balanced three-phase voltages between the lines, even when the line currents are unequal due to unbalanced load conditions.

2. Cylindrical rotors. It is well known that high-speed steam turbines are smaller and more efficient than low-speed turbines are. The same is true of synchronous generators. However, to generate the required frequency, we cannot use less than 2 poles, and this fixes the highest possible speed. On a 60 Hz system, it is 3600 r/min. The next lower speed is 1800 r/min, corresponding to a 4-pole machine. Consequently, these so-called *steam-turbine generators* possess either 2 or 4 poles.

The rotor of a turbine-generator is a long, solid steel cylinder which contains a series of longitudinal slots milled out of the cylindrical mass (Fig. 16-7). Concentric field coils, firmly wedged into the slots and retained by high-strength end-rings,* serve to create the poles.

The high speed of rotation produces strong centrifugal forces which impose an upper limit on the diameter of the rotor. In the case of a rotor turning at 3600 r/min, the elastic limit of the steel requires the manufacturer to limit the diameter to a maximum of 1.2 m. On the other

---

* See Fig. 11-28

hand, to build the powerful 1000 MVA to 1500 MVA generators we have to use massive rotors. It follows that high-power, high-speed rotors have to be very long.

16.5 Field excitation and exciters

The field excitation of a large synchronous generator is an important part of its overall design. The reason is that the field must ensure not only a satisfactory ac voltage level, but must also respond to sudden load changes in order to maintain system stability. Quickness of response is one of the important features of the field excitation. In order to attain it, two dc generators are used: a main exciter and a so-called *pilot exciter*.

The main exciter feeds the exciting current to the field of the synchronous generator by way of brushes and slip rings. Under normal conditions, the exciter voltage lies between 125 V and 600 V. It is regulated manually or automatically, by control signals that vary the current I_c, produced by the pilot exciter (Fig. 16-1).

The power rating of the main exciter depends upon the capacity of the synchronous generator. Typically, a 25 kW exciter is needed to excite a 1000 kVA alternator (2.5% of its rating) whereas a 2500 kW exciter suffices for an alternator of 500 MW (only 0.5% of its rating).

Under normal conditions, the excitation is varied automatically. It responds to the load changes so as to maintain a constant ac line voltage or to control the reactive power delivered to the electric utility system. A serious disturbance on the system may produce a sudden voltage drop across the terminals of the alternator. The exciter must then react very quickly to keep the ac voltage constant. For example, the exciter voltage may have to rise to twice its normal value in as little as 300 to 400 milliseconds. This represents a very quick response, considering that the power of the exciter may be several thousand kilowatts.

16.6 Brushless excitation

Owing to brush wear and carbon dust, we constantly have to clean, repair and replace brushes, slip rings and commutators on conventional dc excitation systems. To eliminate the problem, *brushless excitation systems* have been developed. Such a system consists of a 3-phase stationary-field generator whose ac output is rectified by a group of rectifiers. The dc output from the rectifiers is fed directly into the field of the synchronous generator (Fig. 16-8).

The armature of the ac exciter and the rectifiers are mounted on the main shaft and turn together with the synchronous generator. In comparing the excitation system of Fig. 16-8 with that of Fig. 16-1, we can see they are identical, except that the 3-phase rectifier replaces the commutator, slip rings and brushes. In other words, the commutator (which is really a mechanical rectifier) is replaced by an electronic rectifier. The result is that the brushes and slip rings are no longer needed.

The dc control current I_c from the pilot exciter regulates the main exciter output I_x, as in the case of a conventional dc exciter. The frequency of the main exciter is generally two to three times the synchronous generator frequency (60 Hz). The increase in frequency is obtained by using more poles on the exciter than on the synchronous generator. Figure 16-9 shows the rotating portion of a typical brushless exciter.

16.7 Factors affecting the size of synchronous generators

The prodigious amount of energy generated by electrical utility companies has made them very conscious about the efficiency of their generators. For example, if the efficiency of a 1000 MW generating station improves by only 1%, it represents extra revenues of several thousand dollars per day. In this regard, the *size* of the generator is particularly important because its efficiency automatically improves as the power increases. For example, if a small 1 kilowatt synchronous generator has an efficiency of 50%, a larger, but similar model having a capacity of 10 MW *inevitably* has an efficiency of about 90%. This improvement in efficiency with size is the reason why synchronous generators of 1000 MW and up possess efficiencies of the order of 99%.

Another advantage of large machines is that the power output per kilogram increases as the power increases. Returning to our example, if the 1 kW generator weighs 20 kg (thus yielding 1000 W/20 kg = 50 W/kg), a 10 MW generator of similar construction will weigh only 20 000 kg, thus yielding 500 W/kg. From a power standpoint, large machines weigh relatively less than small machines; consequently, they are cheaper.

Everything, therefore, favors the large machines. However, as they increase in size, we run into serious cooling problems. In effect,

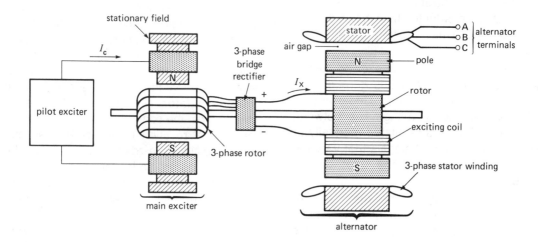

Figure 16-8
Typical brushless exciter system.

Figure 16-9
This brushless exciter provides the dc current for the rotor shown in Fig. 16-7. The exciter consists of a 7000 kVA generator and two sets of diodes. Each set, corresponding respectively to the positive and negative terminals, is housed in the circular rings mounted on the shaft, and visible in the center of the photograph. The ac exciter is seen to the right. The two round conductors protruding from the center of the shaft (foreground) lead the exciting current to the 1530 MVA generator. (*Allis-Chalmers Power Systems Inc., West Allis, Wisconsin*)

large machines inherently produce high power losses per unit surface area (W/m²); consequently, they tend to overheat. To prevent an unacceptable temperature rise, we must design efficient cooling systems that become ever more elaborate as the power increases. For example, a circulating cold-air system is adequate to cool synchronous generators whose rating is below 50 MW, but between 50 MW and 300 MW, we have to resort to hydrogen cooling. Very big generators in the 1000 MW range have to be equipped with hollow, water-cooled conductors. Ultimately, a point is reached where the increased cost of cooling exceeds the savings made elsewhere, and this fixes the upper limit to size.

To sum up, the evolution of big alternators has mainly been determined by the evolution of sophisticated cooling techniques (Figs. 16-10, 11). Other technological breakthroughs, such as better materials, and new windings have also played a major part in modifying the design of early machines (Fig. 16-12).

As regards speed, low-speed generators are always bigger than high-speed machines of equal power. Slow-speed bigness simplifies the cooling problem; a good air-cooling system, completed with a heat exchanger, usually suf-

Figure 16-10
Partial view of a 3-phase, salient-pole generator rated 87 MVA, 428 r/min, 50 Hz. Both the rotor and stator are water-cooled. The high resistivity of pure water and the use of insulating plastic tubing enables the water to be brought into direct contact with the live parts of the machine. *(courtesy ABB)*

turboreactor

alternator   hydraulic motor

Figure 16-11
The electrical energy needed on board the Concord aircraft is supplied by four 3-phase generators rated 60 kVA, 200/155 V, 12 000 r/min, 400 Hz. Each generator is driven by a hydraulic motor which absorbs a small portion of the enormous power developed by the turboreactor engines. The hydraulic fluid streaming from the hydraulic motor is used to cool the generator and then recycled. The generator itself weighs only 54.5 kg. (*Air France*)

fices. For example, the large, slow-speed 500 MVA, 200 r/min synchonous generators installed in a typical hydropower plant are air-cooled whereas the much smaller high-speed 500 MVA, 1800 r/min units installed in a steam plant have to be hydrogen-cooled.

16.8 No-load saturation curve

Figure 16-13a shows a two-pole synchronous generator operating at no-load. It is driven at constant speed by a turbine (not shown). The leads from the 3-phase, wye-connected stator are brought out to terminals A, B, C, N and a variable exciting current I_x produces the flux in the air gap.

Let us gradually increase the exciting current while observing the ac voltage E_o between terminal A, say, and the neutral N. For small values of I_x, the voltage increases in proportion to the exciting current. However, as the iron begins to saturate, the voltage rises much less for the same increase in I_x. If we plot the curve of E_o versus I_x, we obtain the *no-load saturation curve* of the synchronous generator. It is similar to that of a dc generator (Sec. 4-13).

Figure 16-13b shows the actual no-load saturation curve of a 36 MW, 3-phase generator having a nominal voltage of 12 kV (line to neutral). Up to about 9 kV, the voltage increases in proportion to the current, but then the iron begins to saturate. Thus, an exciting current of 100 A produces an output of 12 kV, but if the current is doubled, the voltage rises only to 15 kV.

Figure 16-13c is a schematic diagram of the generator showing the revolving rotor and the 3 phases on the stator.

Figure 16-12
This rotating-field generator was first installed in North America in 1888. It was used in a 1000-lamp street lighting system. The alternator was driven by an 1100 r/min steam engine and had a rated output of 2000 V, 30 A at a frequency of 110 Hz. It weighed 2320 kg, which represents 26 W/kg. A modern generator of equal speed and power produces about 140 W/kg and occupies only one-third the floor space.

(a)

(b)

nominal voltage (line-to-neutral)

Figure 16-13c
Electric circuit representing the generator of Fig. 16-13a.

Figure 16-13
a. Generator operating at no-load.
b. No-load saturation curve of a 36 MW, 21 kV, 3-phase generator.

16.9 Synchronous reactance-equivalent circuit of an ac generator

Consider a 3-phase synchronous generator having terminals A, B, C, feeding a balanced 3-phase load (Fig. 16-14). The generator is driven by a turbine (not shown), and is excited by a dc current I_x. The machine and its load are connected in wye, yielding the circuit of Fig. 16-15. Although neutrals N_1 and N_2 are not connected, they are at the same potential because the load is balanced. Consequently, we *could* connect them together (dotted line) without affecting the behavior of the voltages or currents in the circuit.

The revolving field carries an exciting current which produces a flux Φ. The flux induces, in the stator, three equal voltages E_o that are 120° out of phase (Fig. 16-16).

Each phase of the stator winding possesses a resistance R and a certain inductance L. Because this is an alternating-current machine, the inductance manifests itself as a reactance X_s, given by:

$$X_s = 2\pi f L$$

where

X_s = synchronous reactance, per phase [Ω]

f = generator frequency [Hz]

L = apparent inductance of the stator winding, per phase [H]

The so-called *synchronous reactance* of a generator is an internal impedance, just like its internal resistance R. The impedance is there,

Figure 16-14
Generator connected to a load.

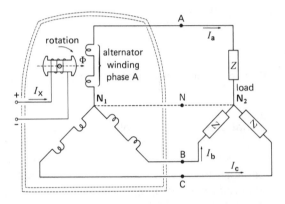

Figure 16-15
Electric circuit representing the installation of Fig. 16-14.

Figure 16-16
Voltages and impedances in a 3-phase generator and its connected load.

but it can neither be seen nor touched. The value of X_s is typically 10 to 100 times greater than R; consequently, we can always neglect the resistance, unless we are interested in efficiency or heating effects.

We can simplify the schematic diagram of Fig. 16-16 by showing only one phase of the stator. In effect, the two other phases are identical, except that their respective voltages (and currents) are out of phase by 120°. Furthermore, if we neglect the resistance of the windings, we obtain the very simple circuit of Fig. 16-17. A synchronous generator can therefore be represented by an equivalent circuit composed of an induced voltage E_o in series with a reactance X_s.

In this circuit, the exciting current I_x produces the flux Φ which induces the internal voltage E_o. Voltage E at the terminals of the generator depends upon E_o and the load Z. Note that E_o and E are line-to-neutral voltages and I is the line current.

Figure 16-17
Equivalent circuit of a 3-phase generator, showing only one phase.

16.10 Determining the value of X_s

We can determine the so-called unsaturated value of X_s by the following open-circuit and short-circuit test.

During the open-circuit test, the generator is driven at rated speed and the exciting current is raised until the rated line-to-neutral voltage E_n is attained. The corresponding exciting current I_{xn} is recorded.

The excitation is then reduced to zero and

the three terminals are short-circuited together. An ammeter is used to measure the resulting short-circuit current. With the generator again running at rated speed, the exciting current is raised to its original value I_{xn}.

The resulting short-circuit current I_{sc} is measured and X_s is calculated by using the expression

$$X_s = E_n/I_{sc} \quad * \qquad (16\text{-}2)$$

where

X_s = synchronous reactance, per phase [Ω]

E_n = rated open-circuit line-to-neutral voltage [V]

I_{sc} = short-circuit current, per phase, using the same exciting current I_x that was required to produce E_n [A]

Example 16-2:
A 3-phase synchronous generator produces an open-circuit line voltage of 6920 V when the dc exciting current is 50 A. The ac terminals are then short-circuited, and the three line currents are found to be 800 A.

a. Calculate the synchronous reactance per phase;

b. Calculate the terminal voltage if three 12 Ω resistors are connected in wye across the terminals.

Solution:
a. The line-to-neutral induced voltage is:
$$E_o = E_L/\sqrt{3} \qquad \text{Eq. 8-4}$$
$$= 6920/1.73$$
$$= 4000 \text{ V}$$

When the terminals are short-circuited, the only impedance limiting the current flow is

that due to the synchronous reactance. Consequently:
$$X_s = E_o/I = 4000/800$$
$$= 5 \ \Omega$$

The synchronous reactance per phase is therefore 5Ω.

b. The equivalent circuit per phase is shown in Fig. 16-18a.

Figure 16-18
a. See Example 16-2.
b. Actual line voltages and currents.

The impedance of the circuit is:
$$Z = \sqrt{R^2 + X_s^2} \qquad \text{Eq. 2-12}$$
$$= \sqrt{12^2 + 5^2}$$
$$= 13 \ \Omega$$

The current is:
$$I = E_o/Z = 4000/13 = 308 \text{ A}$$

The voltage across the load resistor is:
$$E = IR = 308 \times 12 = 3696 \text{ V}$$

The line voltage under load is:
$$E_L = \sqrt{3}E$$
$$= 1.73 \times 3696$$
$$= 6394 \text{ V}$$

The schematic diagram of Fig. 16-18b helps us visualize what is happening in the actual circuit.

---

* This value of X_S corresponds to the direct-axis synchronous reactance. It is widely used to describe synchronous machine behavior.

16.11 Base impedance, per unit X_s

We recall that when using the per-unit system, we select a base voltage and a base power. In the case of a synchronous generator we use the rated line-to-neutral voltage as the base voltage E_B, and the power *per phase* as the base power*. It follows that the base impedance Z_B is given by:

$$Z_B = \frac{E_B^2}{S_B} \tag{16-3}$$

where

Z_B = base impedance (line-to-neutral) of the generator $[\Omega]$

E_B = base voltage (line-to-neutral) [V]

S_B = base power per phase [VA]

The base impedance is used as a basis of comparison for other impedances that the generator possesses. Thus, the synchronous reactance may be expressed as a per unit value of Z_B. In general, $X_s(pu)$ lies between 0.8 and 2, depending upon the design of the machine.

Example 16-3:
A 30 MVA, 15 kV, 60 Hz ac generator has a synchronous reactance of 1.2 p.u. and a resistance of 0.02 p.u. Calculate:
a.   the base impedance of the generator;
b.   the value of the synchronous reactance;
c.   the winding resistance, per phase;
d.   the total full-load copper losses.

Solution:
a.   The base voltage is:

$$E_B = E_L/\sqrt{3} = 15000/\sqrt{3}$$
$$= 8660 \text{ V}$$

The base power is:
$$S_B = 30 \text{ MVA}/3 = 10 \text{ MVA}$$
$$= 10^7 \text{ VA}$$

* In many power studies the base power is selected to be equal to the rated power of the generator and the base voltage is the line-to-line voltage. This yields the same value Z_B for the base impedance.

The base impedance is:
$$Z_B = E_B^2/S_B \qquad \text{Eq. 16-3}$$
$$= 8660^2/10^7$$
$$= 7.5 \ \Omega$$

b.   The synchronous reactance is:
$$X_s = X_s(pu) \times Z_B$$
$$= 1.2 \ Z_B = 1.2 \times 7.5$$
$$= 9 \ \Omega$$

c.   The resistance per phase is:
$$R = R(pu) \times Z_B$$
$$= 0.02 \ Z_B = 0.02 \times 7.5$$
$$= 0.15 \ \Omega$$

Note that all impedance values are from line to neutral.

d.   The per-unit copper losses at full load are:
$$P(pu) = I^2(pu) \ R(pu)$$
$$= 1^2 \times 0.02 = 0.02$$

Note that at full-load the per-unit value of I is equal to 1.

The copper losses for all 3 phases are:
$$P = 0.02 S_B = 0.02 \times 30 = 0.6 \text{ MW}$$
$$= 600 \text{ kW}$$

The synchronous reactance does not remain constant, but varies with the degree of saturation. When the iron is heavily saturated, the value of X_s may be only half its unsaturated value. Despite this broad range, we usually take the unsaturated value for X_s because it yields sufficient accuracy in most cases of interest.

16.12 Short-circuit ratio

Instead of expressing the synchronous reactance as a per-unit value of Z_B, the so-called *short-circuit ratio* is sometimes used. It is the ratio of the field current I_{x1} needed to generate rated open-circuit armature voltage E_B to the field current I_{x2} needed to produce rated current I_B, on a sustained short-circuit. The short-circuit ratio (I_{x1}/I_{x2}) is exactly equal to the reciprocal of the per-unit value of X_s as defined in

Eq. 16-2. Thus, if the per-unit value of X_s is 1.2, the short-circuit ratio is 1/1.2 or 0.833.

16.13 Synchronous generator under load

The behavior of a synchronous generator depends upon the type of load it has to supply. There are many types of loads, but they can all be reduced to two basic categories:

1. isolated loads, supplied by a single generator;

2. the infinite bus.

We begin our study with isolated loads, leaving the discussion of the infinite bus to Sec. 16.16.

Consider a 3-phase generator that supplies power to a load having a lagging power factor. Figure 16-19 represents the equivalent circuit for one phase. In order to construct the phasor

Figure 16-19
Equivalent circuit of a generator under load.

diagram for this circuit, we list the following facts:

1. current I lags behind terminal voltage E by an angle θ.

2. Cosine θ = power factor of the load.

3. Voltage E_x across the synchronous reactance leads current I by 90°. It is given by the expression $E_x = jIX_s$.

4. Voltage E_o generated by the flux Φ is equal to the phasor sum of E plus E_x.

5. Both E_o and E_x are voltages that exist inside the synchronous generator windings and cannot be measured.

6. Flux Φ is that produced by the dc exciting current I_x.

The resulting phasor diagram is given in Fig. 16-20. Note that E_o leads E by δ degrees.

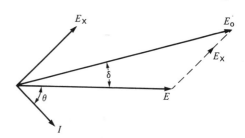

Figure 16-20
Phasor diagram for a lagging power factor load.

Furthermore the internally-generated voltage E_o is greater than the terminal voltage, as we would expect.

In some cases, the load is somewhat capacitive, so that current I leads the terminal voltage by an angle θ. What effect does this have on the phasor diagram? The answer is given in Fig. 16-21. The voltage E_x across the synchro-

Figure 16-21
Phasor diagram for a leading power factor load.

nous reactance is still 90° ahead of the current. Furthermore, E_o is again equal to the phasor sum of E and E_x. However, the terminal voltage is now *greater* than the induced voltage E_o, which is a very surprising result. In effect, the inductive reactance X_s enters into partial resonance with the capacitive reactance of the load. Although it may appear we are getting something for nothing, the higher terminal voltage does not yield any more power.

If the load is entirely capacitive, a very high terminal voltage can be produced with a small exciting current. However, in later chapters, we shall see that such under-excitation is undesirable.

Example 16-4:
A 36 MVA, 21 kV, 3-phase alternator has a synchronous reactance of 9 Ω, and a nominal current of 1 kA. The no-load saturation curve giving the relationship between E_o and I_x is given in Fig. 16-13. If the excitation is adjusted so that the terminal voltage remains fixed at 21 kV, calculate the exciting current required and draw the phasor diagram for the following conditions:
   a. no-load;
   b. resistive load of 36 MW;
   c. capacitive load of 12 Mvar.

Solution:
We shall immediately simplify the circuit to show only one phase. The line-to-neutral terminal voltage for all cases is fixed at:
$$E = 21/\sqrt{3} = 12 \text{ kV}$$

a. At no-load, there is no voltage drop in the synchronous reactance; consequently,
$$E_o = E = 12 \text{ kV}$$

The exciting current is:
$$I_x = 100 \quad \text{(see Fig. 16-13b)}$$

The phasor diagram is given in Fig. 16-22a.

With a resistive load of 36 MW:

b. The power per phase is:
$$P = 36/3 = 12 \text{ MW}$$

The full-load line current is:
$$\begin{aligned} I &= P/E = 12 \times 10^6/12\ 000 \\ &= 1000 \text{ A} \end{aligned}$$

The current is in phase with the terminal voltage.

The voltage across X_s is:
$$E_x = jIX_s = j1000 \times 9 = 9 \text{ kV} \angle 90°$$

This voltage is 90° ahead of I.

The voltage E_o generated by I_x is equal to the phasor sum of E and E_x. Referring to the phasor diagram, its value is given by:
$$\begin{aligned} E_o &= \sqrt{E^2 + E_x^2} = \sqrt{12^2 + 9^2} \\ &= 15 \text{ kV} \end{aligned}$$

The corresponding exciting current is:
$$I_x = 200 \text{ A} \quad \text{(see Fig. 16-13b)}$$

The phasor diagram is given in Fig. 16-22b.

Figure 16-22a
Phasor diagram at no-load.

Figure 16-22b
Phasor diagram with a unity power factor load.

With a capacitive load of 12 Mvar:

c. The reactive power per phase is:
$$Q = 12/3 = 4 \text{ Mvar}$$

The line current is:
$$\begin{aligned} I &= Q/E = 4 \times 10^6/12\ 000 \\ &= 333 \text{ A} \end{aligned}$$

The voltage across X_s is:
$$E_x = jIX_s = j333 \times 9 = 3 \text{ kV} \angle 90°$$

As before, E_x leads I by 90° (Fig. 16-22c).

The voltage E_o generated by I_x is equal to the phasor sum of E and E_x.

$$E_o = E + E_x = 12 + (-3)$$
$$= 9 \text{ kV}$$

The corresponding exciting current is:

$$I_x = 70 \text{ A} \quad \text{(see Fig. 16-13b)}$$

Note that E_o is again less than the terminal voltage E.

The phasor diagram for this capacitive load is given in Fig. 16-22c.

Figure 16-22c
Phasor diagram with a capacitive load.

16.14 Regulation curves

When a single synchronous generator feeds a variable load we are interested in knowing how the terminal voltage E changes as a function of the load current I. The relationship between E and I is called the regulation curve. Regulation curves are plotted with the field excitation fixed and for a given load power factor.

Figure 16-23 shows the regulation curves for the 36 MVA, 21 kV, 3-phase generator discussed in Example 16-4. They are given for loads having unity power factor, 0.9 power factor lagging and 0.9 power factor leading, respectively. These curves were derived using the method of Example 16-4, except that E_o was kept fixed instead of E. The starting point for all the curves is rated terminal voltage at rated power (kVA) output.

The change in voltage between no-load and full-load is expressed as a percent of the rated terminal voltage. The so-called percent regula-

Figure 16-23
Regulation curves of a synchronous generator at three different load power factors.

tion is given by the equation:

$$\text{percent regulation} = \frac{E_{NL} - E_B}{E_B} \times 100$$

where

$$E_{NL} = \text{no-load voltage [V]}$$
$$E_B = \text{rated voltage [V]}$$

Example 16-5

Calculate the percent regulation corresponding to the unity power factor curve in Fig. 16-23.

Solution:
The rated voltage at full-load is:
$$E_B = 12 \text{ kV}$$

The no-load terminal voltage is:
$$E_{NL} = 15 \text{ kV}$$

The percent regulation is:

$$\% \text{ regulation} = \frac{E_{NL} - E_B}{E_B} \times 100$$

$$= \frac{(15 - 12)}{12} \times 100 = 25 \%$$

We note that the percent regulation of a synchronous generator is much greater than that of a dc generator. The reason is the high impedance of the synchronous reactance.

Figure 16-24
Synchroscope. *(Lab-Volt)*

16.15 Synchronization of a generator

We often have to connect two or more generators in parallel to supply a common load. For example, as the power requirements of a large utility system build up during the day, generators are successively connected to the system to provide the extra power. Later, when the power demand falls, selected generators are temporarily disconnected from the system until power again builds up the following day. Synchronous generators are therefore regularly being connected and disconnected from a large power grid, in response to customer demand. Such a grid is said to be an *infinite bus* because it contains so many generators essentially connected in parallel that neither its voltage nor its frequency can be altered.

Before connecting a generator to an infinite bus (or in parallel with another generator), it must be *synchronized*. A generator is said to be

synchronized when it meets all the following conditions:

1. the generator frequency is equal to the system frequency;

2. the generator voltage is equal to the system voltage;

3. the generator voltage is in phase with the system voltage;

4. the phase sequence of the generator is the same as that of the system.

To synchronize an alternator, we proceed as follows:

1. adjust the speed regulator of the turbine so that the generator frequency is close to the system frequency;

2. adjust the excitation so that the generator voltage E_o is equal to the system voltage E;

3. observe the phase angle between E_o and E by means of a *synchroscope* (Fig. 16-24). This instrument has a pointer that continually indicates the phase angle between the two voltages, covering the entire range from zero to 360 degrees. Although the degrees are not shown, the dial has a zero marker to indicate when the voltages are in phase. In practice, when we synchronize an alternator, the pointer rotates slowly as it follows the phase angle between the alternator and system voltages. If the generator frequency is slightly higher than the system frequency, the pointer rotates clockwise, indicating that the generator has a tendency to lead the system frequency. Conversely, if the generator frequency is slightly low, the pointer rotates counterclockwise. The turbine speed-regulator is fine-tuned accordingly, so that the pointer barely creeps across the dial. A final check is made to

see that the alternator voltage is still equal to the system voltage. Then, at the moment the pointer crosses the zero marker...

4. the line circuit breaker is closed, connecting the generator to the system.

In modern generating stations, synchronization is usually done automatically.

16.16 Synchronous generator on an infinite bus

We seldom have to connect only two generators in parallel except in isolated locations (Fig. 16-25). As mentioned previously, it is much more common to connect a generator to a large power system (infinite bus) that already has dozens of alternators connected to it.

An infinite bus is a system so powerful that it imposes its own voltage and frequency upon any apparatus connected to its terminals. Once connected to a large system (infinite bus), a synchronous generator becomes part of a network comprising hundreds of other generators that deliver power to thousands of loads. It is impossible, therefore, to specify the nature of

the load (large or small, resistive or capacitive) connected to the terminals of this particular generator. What, then, determines the power the machine delivers? To answer this question, we must remember that both the value and the frequency of the terminal voltage across the generator is fixed. Consequently, we can vary only two machine parameters:

1. the exciting current I_x;

2. the mechanical torque of the turbine.

Let us see how a change in these parameters affects the performance of the machine.

16.17 Infinite bus - effect of varying the exciting current

Immediately after we synchronize a generator and connect it to an infinite bus, the induced voltage E_o is equal to, and in phase with, the terminal voltage E of the system (Fig. 16-26a). There is no difference of potential across the synchronous reactance and, consequently, the load current I is zero. Although the generator is connected to the system, it delivers no power; it is said to "float" on the line.

If we now increase the exciting current, the

Figure 16-25
This floating oil derrick provides its own energy needs. Four diesel-driven generators rated 1200 kVA, 440 V, 900 r/min, 60 Hz supply all the electrical energy. Although ac power is generated and distributed, all the motors on board are thyristor-controlled dc motors. (*Siemens*)

voltage E_o will increase and the synchronous reactance X_s will experience a difference of potential E_x given by:

$$E_x = E_o - E.$$

A current I will therefore circulate in the circuit given by:

$$I = (E_o - E)/X_s$$

Because the synchronous reactance is inductive, the current lags 90° behind E_x (Fig. 16-26b). The current is therefore 90° behind E, which means that the generator "sees" the system as if it were an inductive reactance. Consequently, when we over-excite a synchronous generator, it supplies reactive power to the infinite bus. The reactive power increases as we raise the dc exciting current. Contrary to what we might expect, it is impossible to make a generator deliver active power by raising its excitation.

Let us now decrease the exciting current so that E_o becomes smaller than E. As a result, phasor $E_x = E_o - E$ becomes negative and therefore points to the left (Fig. 16-26c). As always, current $I = E_x/X_s$ lags 90° behind E_x. However, this puts I 90° ahead of E, which means that the alternator "sees" the system as if it were a capacitor. Consequently, when we under-excite an alternator, it draws reactive power from the system. This reactive power produces part of the magnetic field required by the machine; the remainder is supplied by exciting current I_x.

16.18 Infinite bus - effect of varying the mechanical torque

Let us return to the situation with the synchronous generator floating on the line, E_o and E being equal and in phase. If we open the steam valve of the turbine driving the generator, the immediate result is an increase in mechanical torque (Fig. 16-27a). The rotor will accelerate and, consequently, E_o will attain its maximum

value a little sooner than before. Phasor E_o will slip ahead of phasor E, leading it by a phase angle δ. Although both voltages have the same value, the phase angle produces a difference of potential $E_x = E_o - E$ across the synchronous reactance (Fig. 16-27b).

A current I will flow (again lagging 90° behind E_x), but this time it is almost in phase with E. It follows that the generator feeds active power into the system. Under the driving force of the turbine, the rotor will continue to accelerate, the angle δ will continue to diverge, and the electrical power delivered to the system will gradually build up. However, as soon as the electrical power delivered to the system is equal to the mechanical power supplied by the turbine, the rotor will cease to accelerate. The generator will again run at synchronous speed, and the so-called *torque angle* δ between E_o and E will remain constant.

It is important to understand that a difference of potential is created when two equal voltages are out of phase. Thus, in Fig. 16-27, a potential difference of 4 kV exists between E_o and E, although both voltages have a value of 12 kV.

16.19 Physical interpretation of alternator behavior

The phasor diagram of Fig. 16-27b shows that when the phase angle between E_o and E increases, the value of E_x increases, and hence the value of I increases. But a larger current means that the active power delivered by the generator also increases. To understand the physical meaning of the diagram, let us examine the currents, fluxes, and position of the poles inside the machine.

Whenever 3-phase currents flow in the stator of a generator, they produce a rotating magnetic field, identical to that in an induction motor. In a synchronous generator, this field rotates at the same speed and in the same direction as the rotor. Furthermore, it has the same number of poles. The respective fields

$$E_o = E = 12 \text{ kV}$$

Figure 16-26a
Generator "floating" on an infinite bus.

Figure 16-26b
Over-excited generator on an infinite bus.

Figure 16-26c
Under-excited generator on an infinite bus.

(a)

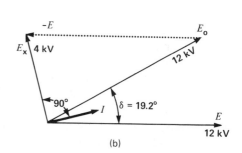

(b)

Figure 16-27
a. Turbine driving the generator.
b. Phasor diagram showing the torque angle δ.

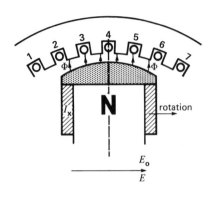

Figure 16-28a
The N poles of the rotor are lined up with the S poles of the stator.

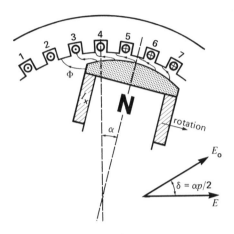

Figure 16-28b
The N poles of the rotor are ahead of the S poles of the stator.

produced by the rotor and stator are, therefore, stationary with respect to each other. Depending on the relative position of the stator poles on the one hand and the rotor poles on the other hand, powerful forces of attraction and repulsion may be set up between them. When the generator floats on the line, the stator current I is zero and so no forces are developed. The only flux is that created by the rotor, and it induces the voltage E_o (Fig. 16-28a). Note that the rotor and stator poles are lined up.

If a mechanical torque is applied to the generator (by admitting more steam to the turbine), the rotor accelerates and gradually advances by a mechanical angle α, compared to its original position (Fig. 16-28b). Stator currents immediately begin to flow, owing to the electrical phase angle δ between induced voltage E_o and terminal voltage E. The stator currents create a revolving field and a corresponding set of N and S poles. Forces of attraction and repulsion are developed between the stator poles and rotor poles, and these magnetic forces produce a torque that opposes the applied

mechanical torque. When the electromagnetic torque is equal to the mechanical torque, the mechanical angle will no longer increase, but will remain at a constant value α.

There is a direct relationship between the mechanical angle α and the phase angle δ, given by:

$$\delta = p\alpha/2 \qquad (16\text{-}4)$$

where

δ = torque angle between the terminal voltage E and the excitation voltage E_o [electrical degrees]

p = number of poles on the generator

α = mechanical angle between the centers of the stator and rotor poles [mechanical degrees]

Example 16-6:
The rotor poles of an 8-pole synchronous generator shift by 10 mechanical degrees, from no-load to full-load.

a. Calculate the torque angle between E_o and the terminal voltage E at full load;

b. Which voltage is leading?

Solution:
a. The torque angle is:
$$\delta = p\alpha/2 = 8 \times 10/2$$
$$= 40°$$

b. When a generator delivers active power, E_o *always* leads E.

16.20 Active power delivered by the generator

We can prove (see Sec. 16.23) that the active power delivered by a synchronous generator is given by the equation:

$$P = \frac{E_o E}{X_s} \sin \delta \qquad (16\text{-}5)$$

where
P = active power, per phase [W]
E_o = induced voltage, per phase [V]
E = terminal voltage, per phase [V]
X_s = synchronous reactance per phase [Ω]
δ = torque angle between E_o and E [°]

This equation can be used under all load conditions, including the case when the generator is connected to an infinite bus.

To understand its meaning, suppose a generator is connected to an infinite bus having a voltage E. Furthermore, suppose that the dc excitation of the generator is kept constant, so that E_o is constant. The term $E_o E/X_s$ is then fixed, and the active power which the alternator delivers to the bus will vary directly with sin δ, the sine of the torque angle. Thus, as we admit more steam, δ will increase and so, too, will the power. The relationship between the two is shown graphically in Fig. 16-29. Note that between zero degrees and 30° the power increases almost linearly with the torque angle. Rated power is typically attained at an angle of 30°.

However, there is an upper limit to the active power the generator can deliver. This limit is

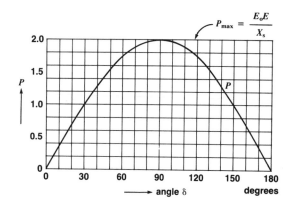

Figure 16-29
Graph showing the relationship between the active power delivered by a synchronous generator and the torque angle.

reached when δ is 90°. The peak power is then $P_{max} = E_o E/X_s$. If we try to exceed this limit (such as by admitting more steam to the turbine), the rotor will accelerate and lose synchronism with the infinite bus. The rotor will turn faster than the rotating field of the stator, and large, pulsating currents will flow in the stator. In practice, this condition is never reached because the circuit-breakers trip as soon as synchronism is lost. We then have to resynchronize the generator before it can again pick up the load.

Example 16-7:
A 36 MVA, 21 kV, 1800 r/min 3-phase generator connected to a power grid has a synchronous reactance of 9 Ω per phase.
If the exciting voltage is 12 kV (line-to-neutral), and the system voltage is 17.3 kV (line-to-line), calculate:
a. the active power which the machine delivers when the torque angle δ is 30° (electrical);
b. the peak power which the machine can deliver before it falls out of step.

Solution:
a.   We have:

E_o = 12 kV;
E = 17.3 kV/$\sqrt{3}$ = 10 kV;
δ = 30°

The active power delivered to the system is:

P = (E_oE/X_s) sin δ
   = (10 × 12/9) × 0.5
   = 6.67 MW

The total power delivered by all three phases is

(3 × 6.67) = 20 MW.

b.   The maximum power, per phase, is attained when δ = 90°:

P = (E_oE/X_s) sin 90
   = (10 × 12/9) × 1
   = 13.3 MW

The peak power of the alternator is therefore

(3 × 13.3) = 40 MW

16.21 Control of active power

When a synchronous generator is connected to a system, its speed is kept constant by an extremely sensitive governor. This device can detect speed changes as small as 0.01%. An automatic control system, sensitive to such small speed changes, immediately modifies the valve (or gate) opening of the turbine so as to maintain a constant speed and constant power output.

On a big utility network, the power delivered by each generator depends upon a program established in advance between the various generating stations. The station operators communicate with each other to modify the power delivered by each station so that the generation and transmission of energy is done as efficiently as possible. In more elaborate systems, the entire network is under the control of a computer.

In addition, individual overspeed detectors are always ready to respond to a large speed change, particularly if a generator, for one reason or another, should suddenly become disconnected from the system. Because the steam valves (or water gates) are still wide open, the generator will rapidly accelerate and may attain a speed 50% above normal in 4 to 5 seconds. The centrifugal forces at synchronous speed are already close to the limit the materials can withstand, so any excess speed can quickly create a very dangerous situation. Consequently, steam valves and water gates must immediately be closed off during such emergencies.

16.22 Transient reactance

We have seen that a generator connected to a system is subject to unpredictable load changes that sometimes occur very quickly. In such cases, the simple equivalent circuit shown in Fig. 16-17 does not reflect the behavior of the machine. This circuit is only valid under steady-state conditions, or when the load changes gradually.

For *sudden* load current changes, the synchronous reactance X_s must be replaced by another reactance X' whose value varies as a function of time. Figure 16-30 shows how X' varies when a generator is suddenly short-circuited. Prior to the short-circuit, the synchronous reactance is simply X_s. However, at the instant of short-circuit, the reactance immediately falls to a much lower value X'_d. It then increases gradually until it is again equal to X_s after a time interval T. The duration of the interval depends upon the size of the generator. For machines below 100 kVA it only lasts a fraction of a second, but for machines in the 1000 MVA range, it may last as long as 10 seconds.

The minimum reactance X'_d is called the *transient reactance* of the alternator. It may be as low as 15% of the synchronous reactance. Consequently, the initial short-circuit current is much higher than that corresponding to the

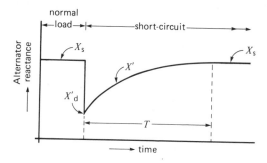

Figure 16-30
Variation of generator reactance following a short-circuit.

synchronous reactance X_s. This has a direct bearing on the capacity of the circuit breakers at the generator output. In effect, because they must interrupt a short-circuit in three to six *cycles*, it follows they have to interrupt a very high current.

On the other hand, the low transient reactance simplifies the voltage regulation problem when the generator load increases rapidly. First, the internal voltage drop due to X'_d is smaller than it would be if the synchronous reactance X_s were acting. Second, X' stays at a value far below X_s for a sufficiently long time to enable us to quickly raise the exciting current I_x. Raising the excitation increases E_o, which helps to stabilize the terminal voltage.

Example 16-8:
A 250 MVA, 25 kV, 3-phase steam-turbine generator has a synchronous reactance of 1.6 p.u. and a transient reactance X'_d of 0.23 p.u. It delivers its rated output at a power factor of 100%. A short-circuit suddenly occurs on the line, close to the generating station. Calculate:
a. the induced voltage E_o prior to the short-circuit;
b. the initial value of the short-circuit current;
c. the final value of the short-circuit current if the circuit breakers fail to open.

Solution:
a. The base impedance of the generator is:
$$Z_B = E_B^2/S_B$$
$$= 25\ 000^2/(250 \times 10^6)$$
$$= 2.5\ \Omega$$

The synchronous reactance is:
$$X_s = X_s(\text{pu})Z_B$$
$$= 1.6 \times 2.5$$
$$= 4\ \Omega$$

The rated terminal voltage per phase is:
$$E = 25/\sqrt{3} = 14.4\ \text{kV}$$

The rated load current per phase is:
$$I = S/\sqrt{3}E$$
$$= 250 \times 10^6/(1.73 \times 25\ 000)$$
$$= 5780\ \text{A}$$

The internal voltage drop E_x is:
$$E_x = IX_s = 5780 \times 4$$
$$= 23.1\ \text{kV}$$

The current is in phase with E because the power factor of the load is unity. Thus, referring to the phasor diagram (Fig. 16-31), E_o is:
$$E_o = \sqrt{E^2 + E_x^2}$$
$$= \sqrt{14.4^2 + 23.1^2}$$
$$= 27.2\ \text{kV}$$

b. The transient reactance is:
$$X'_d = X'_d(\text{pu})Z_B$$
$$= 0.23 \times 2.5$$
$$= 0.575\ \Omega$$

The initial short-circuit current is:
$$I_{sc} = E_o/X'_d$$
$$= 27.2/0.575$$
$$= 47.3\ \text{kA}$$

which is 8.2 times rated current.

Figure 16-31
See Example 16-7.

c.   If the short-circuit is sustained and the excitation is unchanged, the current will eventually level off at a steady-state value of:

$$I = E_o/X_s = 27.2/4$$
$$= 6.8 \text{ kA}$$

which is only 1.2 times rated current.

Figure 16-32 shows the generator current
prior to and during the short-circuit. We assume
a time interval T of 5 seconds. Note that in practice the circuit breakers would certainly trip
within 0.1 s after the short-circuit occurs. Consequently, they have to interrupt a current of
about 47 kA.

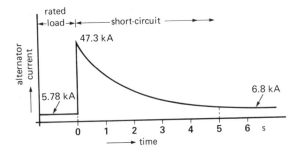

Figure 16-32
Change in current when a short-circuit occurs across
the terminals of a generator. See Example 16-7.

16.23 Power transfer between two sources

The circuit of Fig. 16-33 is used to represent
several important power devices and systems.
For example, it is encountered in the study of
generators, synchronous motors and transmission lines. In such circuits, we are often interested in the active power transmitted from a
source E_1 to a source E_2 or vice versa. Applying

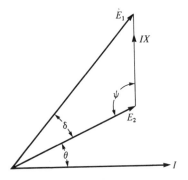

Figure 16-33
Power flow between two voltage sources.

Kirchhoff's voltage law to this circuit, we obtain
the equation:

$$E_1 = E_2 + jIX$$

If we assume that I lags behind E_2 by an arbitrary angle θ and E_1 leads E_2 by an angle δ, we
obtain the phasor diagram shown. Phasor IX
leads I by 90°. The active power absorbed by E_2
is:

$$P = E_2I \cos \theta \qquad (16\text{-}6)$$

From the sine law for triangles, we have:

$IX/\sin \delta = E_1/\sin \psi$
$\qquad\quad = E_1/\sin (90 + \theta)$
$\qquad\quad = E_1/\cos \theta$

Consequently, $I \cos \theta = E_1 \sin \delta/X$ (16-7)

Substituting in Eq. 16-6, we find:

$$P = \frac{E_1 E_2}{X} \sin \delta \qquad (16\text{-}8)$$

where

P = active power transmitted [W]
E_1 = voltage of source 1 [V]
E_2 = voltage of source 2 [V]
δ = phase angle between E_1 and E_2 [°]
X = reactance connecting the sources [Ω]

The active power received by E_2 is equal to that delivered by E_1, because the reactance consumes no active power. Its magnitude is determined by the phase angle between E_1 and E_2: the angle θ between E_2 and I does not have to be specified.

QUESTIONS AND PROBLEMS

Practical level

16-1 What are the advantages of using a stationary armature in large synchronous generators? Why is the stator always connected in wye?

16-2 State the main differences between steam-turbine generators and salient-pole generators. For a given power output, which of these machines is the larger?

16-3 In analyzing a hydropower site, it is found that the turbines should turn at close to 350 r/min. If the directly-coupled generators must generate a frequency of 60 Hz, calculate:
a. the number of poles on the rotor;
b. the exact turbine speed.

16-4 An isolated 3-phase generator produces a no-load line voltage of 13.2 kV. If a load having a lagging power factor of 0.8 is connected to the machine, must the excitation be increased or decreased in order to maintain the same line voltage?

16-5 What conditions must be met before a generator can be connected to a 3-phase system?

16-6 Calculate the number of poles on the generator in Fig. 16-12 using the information given.

16-7 Calculate the number of poles on the aircraft generator shown in Fig. 16-11.

16-8 A 3-phase generator turning at 1200 r/min generates a no-load voltage of 9 kV, 60 Hz. How will the terminal voltage be affected if the following loads are connected to its terminals?
a. resistive load;
b. inductive load;
c. capacitive load.

16-9 In Problem 16-8, if the field current is kept constant, calculate the no-load voltage and frequency if the speed is:
a. 1000 r/min   b. 5 r/min.

Intermediate level

16-10 What is meant by the synchronous reactance of a 3-phase generator? Draw the equivalent circuit of a generator, and explain the meaning of all the parameters.

16-11 State the advantages of brushless excitation systems over conventional systems. Using a schematic circuit diagram, show how the rotor in Fig. 16-7 is excited.

16-12 Referring to Fig. 16-13, calculate the exciting current needed to generate a no-load *line* voltage of: a. 24.2 kV;   b. 12.1 kV.

16-13 A 3-phase generator possesses a synchronous reactance of 6 Ω and the excitation voltage E_o is 3 kV per phase (ref. Fig. 16-19). Calculate the line-to-neutral voltage E for a resistive load of 8 Ω and draw the phasor diagram.

16-14 a. In Problem 16-13, draw the curve of E versus I for the following resistive loads: infinity, 24, 12, 6, 3, 0 ohms.
   b. Calculate the active power P per phase, in each case.

   c. Draw the curve of E versus P. For what value of load resistance is the power output a maximum?

16-15 Referring to Fig. 16-2, calculate the length of one pole pitch measured along the internal circumference of the stator.

16-16 The 3-phase generator shown in Fig. 16-16 has the following characteristics:
   $$E_o = 2400 \text{ V};$$
   $$X_s = 144 \ \Omega;$$
   $$R = 17 \ \Omega;$$
   load impedance $Z = 175 \ \Omega$ (resistive). Calculate:
   a. the synchronous impedance Z_s per phase;
   b. the total resistance of the circuit, per phase;
   c. the total reactance of the circuit, per phase;
   d. the line current;
   e. the line-to-neutral voltage across the load;
   f. the line voltage across the load;
   g. the power of the turbine driving the alternator;
   h. the phase angle between E_o and the voltage across the load.

16-17 A 3-phase generator rated 3000 KVA,

20 kV, 900 r/min, 60 Hz delivers power to a 2400 KVA, 16 kV load having a lagging power factor of 0.8. If the synchronous reactance is 100 Ω, calculate the value of E_o, per phase;

16-18 The generator in Fig. 16-2 has a synchronous reactance of 0.4 Ω, per phase. It is connected to an infinite bus having a line voltage of 14 kV, and the excitation voltage is adjusted to 1.14 p.u. Calculate:
   a. the torque angle δ when the generator delivers 420 MW;
   b. the mechanical displacement angle α
   c. the linear pole shift (measured along the inside stator circumference) corresponding to this displacement angle [in].

16-19 A test taken on the 500 MVA alternator of Fig. 16-2 yields the following results:
   1. open-circuit line voltage is 15 kV for a dc exciting current of 1400 A;
   2. using the same dc current, with the armature short-circuited, the resulting ac line current is 21 000 A.

   Calculate:
   a. the base impedance of the generator, per phase;
   b. the value of the synchronous reactance;
   c. the per unit value of X_s;
   d. the short-circuit ratio.

Advanced level

16-20 The synchronous generator in Fig. 16-2 has an efficiency of 98.4% when it delivers an output of 500 MW. Knowing that the dc exciting current is 2400 A at a dc voltage of 300 V, calculate:
   a. the total losses in the machine;
   b. the copper losses in the rotor;
   c. the torque developed by the turbine;
   d. the average difference in tempera-

ture between the cool incoming air and warm outgoing air, if the air flow is 280 m³/s.

16-21 Referring to Fig. 16-4, each coil on the rotor has 21.5 turns, and carries a dc current of 500 A. Knowing that the air gap length is 1.3 inches, calculate the flux density in the air gap at no-load. Neglect the mmf required for the iron portion of the magnetic circuit. (See Sec. 2.17).

16-22 Referring to Fig. 16-17, the following information is given about a generator:

E_o = 12 kV;
E = 14 kV;
X_s = 2 Ω;
E_o leads E by 30°.

a. Calculate the total active power output of the generator;
b. draw the phasor diagram for one phase;
c. calculate the power factor of the load.

16-23 The steam-turbine generator shown in Fig. 16-3 has a synchronous reactance of 1.3 p.u. The excitation voltage E_o is adjusted to 1.2 p.u. and the machine is connected to an infinite bus of 19 kV. If the torque angle δ is 20°, calculate:
a. the active power output;
b. the line current;
c. draw the phasor diagram, for one phase.

16-24 In Problem 16-23, calculate the active power output of the generator if the

steam valves are closed. Does the alternator receive or deliver reactive power and how much?

16-25 The generator in Problem 16-20 is driven by a hydraulic turbine whose moment of inertia is 54 × 10⁶ lb·ft². The rotor has a J of 4.14 × 10⁶ kg·m².
a. If the line circuit breakers suddenly trip, calculate the speed of the generating unit (turbine and alternator) 1 second later, assuming that the wicket gates remain wide open.
b. By how many mechanical degrees do the poles advance (with respect to their normal position) during the 1-second interval? By how many electrical degrees?

16-26 A 400 Hz alternator has a 2-hour rating of 75kVA, 1200 r/min, 3-phase, 450 V, 80 per cent power factor (Fig. 16-34). The stator possesses 180 slots and has an internal diameter of 22 inches and an axial length of 9.5 in. The rotor is designed for a field current of 31A at115 V. Calculate:
a. the number of poles on the rotor,
b. the number of coils on the stator,
c. the number of coils per phase group on the stator,
d. the length of one pole pitch , mea - sured along the circumference of the stator,
e. the resistance of the dc winding on the rotor and the power needed to excite it.

Figure 16-34
Rotor and stator of a 75 kVA, 1200 r/min, 3-phase, 450 V, 400Hz alternator for shipboard use. The alternator is driven by a 1200 r/min synchronous motor.

Figure 16-35
Stator and rotor of the 100 hp, 1200 r/min, 60 Hz synchronous motor. The stator is mounted on a bedplate that also serves as a base for the alternator. The rotor is equipped with a squirrel cage winding to permit starting as an induction motor. (*Courtesy Electro-Mécanik*)

17

SYNCHRONOUS MOTORS

The synchronous generators described in the previous chapter can operate either as generators or as motors. When operating as motors (by connecting them to a 3-phase source), they are called *synchronous motors*. As the name implies, synchronous motors run in synchronism with the revolving field. The speed of rotation is therefore tied to the frequency of the source. Because the frequency is fixed, the motor speed stays constant, irrespective of the load or voltage of the 3-phase line. However, synchronous motors are used not so much because they run at constant speed, but because they possess other unique electrical properties. We shall study these properties in this chapter.

Most synchronous motors are rated between 150 kW (200 hp) and 15 MW (20 000 hp) and turn at speeds ranging from 150 to 1800 r/min. Consequently, these machines are mainly used in heavy industry (Fig. 17-1). At the other end of the power spectrum, we find tiny single-phase synchronous motors used in control devices and electric clocks. They are discussed in Chapter 18.

17.1 Construction

Synchronous motors are identical in construction to salient-pole ac generators. The *stator* is composed of a slotted magnetic core which carries a 3-phase lap winding. Consequently, the winding is also identical to that of a 3-phase induction motor.

The *rotor* has a set of salient poles that are excited by a dc current (Fig. 17-2). The exciting coils are connected in series to two slip rings, and the dc current is fed into the winding from an external exciter. Slots are also punched out along the circumference of the salient poles. They carry a squirrel-cage winding similar to that in a 3-phase induction motor. This so-called *damper winding* serves to start the motor.

Modern synchronous motors often employ brushless excitation, similar to that used in synchronous generators. Referring to Fig. 17-3, a

Figure 17-1
Three-phase, unity power factor synchronous motor rated 3000 hp (2200 kW), 327 r/min, 4000 V, 60 Hz driving a compressor used in a pumping station on the Trans-Canada pipeline. Brushless excitation is provided by a 21 kW, 250 V alternator/rectifier which is mounted on the shaft between the bearing pedestal and the main rotor. (*General Electric*)

3-phase exciter and a rectifier are mounted at the end of the motor shaft. The dc output from the rectifier is fed directly into the salient-pole windings, without going through brushes and slip rings. The output of the exciter can be varied by controlling the small exciting current I_c that flows in the stationary field winding. Figure 17-4 shows how the exciter, rectifier and salient poles are mounted in a 3000 kW synchronous motor.

The rotor and stator always have the same number of poles. As in the case of a induction motor, the number of poles determines the synchronous speed of the motor:

$$n_s = 120 \, \frac{f}{p} \qquad (17\text{-}1)$$

where

n_s = motor speed [r/min]
f = frequency of the source [Hz]
p = number of poles

Figure 17-2
Rotor of a 50 Hz to 16 2/3 Hz frequency converter used to power an electric railway. The 4-pole rotor at the left is associated with a single-phase alternator rated 7000 kVA, 16 2/3 Hz, PF 85%. The rotor on the right is for a 6900 kVA, 50 Hz, 90% PF synchronous motor which drives the single-phase alternator. Both rotors are equipped with squirrel-cage windings. (ABB) *Boveri*)

1 - dc control source

2 - stationary exciter poles

3 - alternator (3-phase exciter)

4 - 3-phase connection

5 - bridge rectifier

6 - dc line

7 - rotor of synchronous motor

8 - stator of synchronous motor

9 - 3-phase input to stator

Figure 17-3
Diagram showing the main components of a brushless exciter for a synchronous motor. It is similar to that of a synchronous generator.

Figure 17-4a
Synchronous motor rated 4000 hp (3000 kW), 200 r/min, 6.9 kV, 60 Hz, 80% power factor designed to drive an ore crusher. The brushless exciter (alternator/rectifier) is mounted on the overhung shaft and is rated 50 kW, 250 V. (*General Electric*)

Figure 17-4b
Close-up of the 50 kW exciter, showing the armature winding and 5 of the 6 diodes used to rectify the ac current. (*General Electric*)

Example 17-1:
Calculate the number of salient poles on the rotor of the synchronous motor shown in Fig. 17-4a.

Solution:
The motor operates at 60 Hz and runs at 200 r/min; consequently:

$$n_s = 120 \, f/p$$
$$200 = (120 \times 60)/p$$
$$p = 36 \text{ poles}$$

The rotor possesses 18 north poles and 18 south poles.

17.2 Starting a synchronous motor

A synchronous motor cannot start by itself; consequently, the rotor is usually equipped with a squirrel-cage winding so that it can start up as an induction motor. When the stator is connected to the 3-phase line, the motor accelerates until it reaches a speed slightly below synchronous speed. The dc excitation is suppressed during this period.

While the rotor accelerates, the rotating flux created by the stator sweeps across the slower moving salient poles. Because the rotor coils possess a relatively large number of turns, a high voltage is induced in the rotor winding when it turns at low speeds. This voltage appears between the slip rings and it decreases as the rotor accelerates, eventually becoming negligible when the rotor approaches synchronous speed. To limit the voltage, and to improve the starting torque, we either short-circuit the dc winding or connect it to an auxiliary resistor during the starting period.

If the capacity of the supply line is limited, we sometimes have to apply reduced voltage to the stator. As in the case of induction motors, we use either autotransformers or series reactors to limit the starting current (see Chapter 20). Very large synchronous motors (20 MW and more) are sometimes brought up to speed by an auxiliary motor, called a *pony motor*. Finally, in some big installations, the motor may be brought up to speed by a variable-frequency electronic source.

17.3 Pull-in torque

As soon as the motor approaches synchronous speed, we excite the rotor with dc current. This produces alternate N and S poles around the circumference of the rotor (Fig. 17-5). If these poles face poles of *opposite* polarity on the stator at the moment the rotor is excited, a strong magnetic attraction is set up between them. The mutual attraction locks the rotor and stator together, and so the rotor is literally yanked into step with the revolving field. The torque developed at this moment is appropriately called the *pull-in torque* .

The pull-in torque of a synchronous motor is powerful, but the dc current must be applied at the right moment to produce the best effect. For example, if excitation is applied when the N poles of the rotor are opposite the N poles of the stator, the resulting magnetic repulsion produces a violent mechanical shock. The motor will immediately slow down and the circuit breakers will trip. In practice, starters for synchronous motors are designed to detect the precise moment when excitation should be applied. The motor then pulls automatically and smoothly into step with the revolving field.

Once the motor turns at synchronous speed, no voltage is induced in the squirrel-cage winding and so it carries no current. Consequently, the behavior of a synchronous motor is entirely different from that of an induction motor. Basically, a synchronous motor rotates because of the magnetic attraction between the poles of the rotor and the opposite poles of the stator.

To reverse the direction of rotation, we simply interchange any two lines connected to the stator.

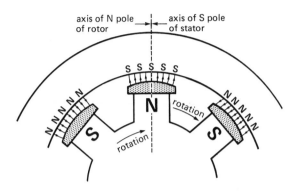

Figure 17-5
The poles of the rotor are attracted to the opposite poles on the stator. At no-load, the axes of the poles coincide.

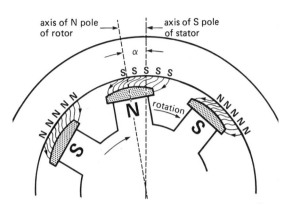

Figure 17-6
The rotor poles are displaced with respect to the stator poles when the motor delivers mechanical power.

17.4 Motor under load – general description

When a synchronous motor runs at no-load, the rotor poles are directly opposite the stator poles and their axes coincide (Fig. 17-5). However, if we apply a mechanical load, the rotor poles fall slightly behind the stator poles, but the rotor continues to turn at synchronous speed. The mechanical angle α between the poles increases progressively as we increase the load (Fig. 17-6). Nevertheless, the magnetic attraction keeps the rotor locked to the revolving field, and the motor develops an ever more powerful torque as the angle increases.

But there is a limit. If the mechanical load exceeds the so-called *pull-out torque* of the motor, the rotor poles suddenly pull away from the stator poles and the motor comes to a halt. A motor that pulls out of step creates a major disturbance on the line, and the circuit breakers immediately trip. This protects the motor because both the squirrel-cage and stator windings overheat rapidly when the machine ceases to run at synchronous speed.

The pull-out torque depends upon the magnetomotive force developed by the rotor and the stator poles. The mmf of the rotor poles depends upon the dc excitation I_x, while that of the stator depends upon the ac current flowing in the windings. The pull-out torque is usually 1.5 to 2.5 times the nominal full-load torque.

The mechanical angle α between the rotor and stator poles has a direct bearing on the stator current. As the angle increases, the current increases. This is to be expected because a larger angle corresponds to a bigger mechanical load, and the increased power can only come from the 3-phase ac source.

17.5 Motor under load – simple calculations

We can get a better understanding of the operation of a synchronous motor by referring to the equivalent circuit shown in Fig. 17-7a. It represents one phase of a wye-connected motor. It is identical to the equivalent circuit of an ac generator, because both machines are built the same way. Thus, the flux Φ created by the rotor induces a voltage E_o in the stator. This flux depends on the dc exciting current I_x. Consequently, E_o varies with the excitation.

As already mentioned, the rotor and stator poles are lined up at no-load. Under these conditions, induced voltage E_o is in phase with line voltage E (Fig. 17-7b). If, in addition, we adjust

Figure 17-7a
Equivalent circuit of a synchronous motor, showing one phase.

Figure 17-7b
Motor at no-load, with E_o adjusted to equal E.

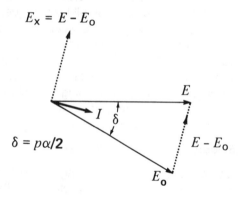

Figure 17-7c
Motor under load, E_o has the same value as in Fig. 17-7b, but it lags behind E.

the excitation so that $E_o = E$, the motor "floats" on the line, and the line current I is practically zero. In effect, the only current needed is to supply the small windage and friction losses in the motor, and so it is negligible.

What happens if we apply a mechanical load to the shaft? The motor will begin to slow down, causing the rotor poles to fall behind the stator

poles by an angle α. Owing to this mechanical shift, E_o reaches its maximum value a little later than before. Thus, referring to Fig. 17-7c, E_o is now δ electrical degrees behind E. The mechanical displacement α produces an *electrical* phase shift δ between E_o and E.

The phase shift produces a difference of potential across the synchronous reactance X_s given by:

$$E_x = E - E_o$$

Consequently, a current I must flow in the circuit, given by:

$$jIX_s = E_x$$

from which

$$
\begin{aligned}
I &= -jE_x/X_s \\
&= -j(E - E_o)/X_s
\end{aligned}
$$

The current lags 90° behind E_x because X_s is inductive. The phasor diagram under load is shown in Fig. 17-7c. Because I is nearly in phase with E, the motor absorbs active power. This power is entirely transformed into mechanical power, except for the relatively small copper and iron losses in the stator.

Example 17-2:
A synchronous motor connected to a 3980 V, 3-phase line generates an excitation voltage E_o of 1790 V (line-to-neutral) when the dc exciting current is 25 A. The synchronous reactance is 22 Ω and the torque angle between E_o and E is 30°. Calculate:
a.   the value of E_x;
b.   the ac line current;
c.   the power factor of the motor.

Solution:
This problem can best be solved by using complex notation.

a.   The voltage E (line-to-neutral) applied to the motor has a value
$$
\begin{aligned}
E &= E_L/\sqrt{3} = 3980/\sqrt{3} \\
&= 2300 \text{ V}
\end{aligned}
$$

Let us select E as the reference phasor, whose angle with respect to the horizontal axis is assumed to be zero. Thus,

$$E = 2300\angle0°$$

It follows that E_o is given by the phasor

$$E_o = 1790\angle-30°$$

The equivalent circuit per phase is given in Fig. 17-8a

Figure 17-8a
Equivalent circuit of a synchronous motor connected to a source E.

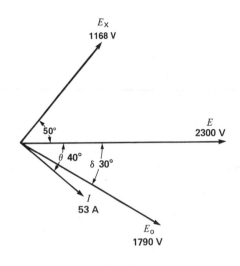

Figure 17-8b
See Example 17-2.

Moving clockwise around the circuit and applying Kirchhoff's voltage law we can write:

$$-E + E_x + E_o = 0$$

$$\begin{aligned}
E_x &= E - E_o \\
&= 2300\angle0° - 1790\angle-30° \\
&= 2300\,(\cos 0° + j \sin 0°) - \\
&\quad 1790\,(\cos - 30° + j \sin -30°) \\
&= 2300 - 1550 + j\,895 \\
&= 750 + j\,895 \\
&= 1168\angle50°
\end{aligned}$$

Thus, phasor E_x has a value of 1168 V and it leads phasor E by 50°.

b. The line current I is given by:

$$j22I = E_x$$

$$\begin{aligned}
I &= \frac{1168\angle50°}{22\angle90°} \\
&= 53\angle-40°
\end{aligned}$$

Thus, phasor I has a value of 53 A and it lags 40° behind phasor E.

c. The power factor of the motor is given by the cosine of the angle between the line-to-neutral voltage E across the motor terminals and the current I. Hence

$$\begin{aligned}
\text{power factor} &= \cos\theta = \cos 40° \\
&= 0.766 \text{ or } 76.6\ \%
\end{aligned}$$

The power factor is lagging because the current lags behind the voltage.

The complete phasor diagram is shown in Fig. 17-8b.

17.6 Power and torque

When a synchronous motor operates under load, it draws active power from the line. The power is given by the same equation we previously used for the synchronous generator in

Chapter 16:

$$P = (E_oE/X_s)\sin \delta \qquad 16\text{-}5$$

As in the case of a generator, the active power absorbed by the motor depends upon the line voltage E, the excitation voltage E_o and the phase angle δ between them. If we neglect the relatively small I^2R losses in the stator, all the power is transmitted across the air gap to the rotor. This is analogous to the power P_r transmitted across the air gap of an induction motor (Sec 13.13). However, in a synchronous motor, the rotor I^2R losses are entirely supplied by the dc source. Consequently, all the power transmitted across the air gap is available in the form of mechanical power. The mechanical power developed by a synchronous motor is therefore expressed by the equation:

$$P = \frac{E_oE}{X_s} \sin \delta \qquad (17\text{-}2)$$

where

P = mechanical power of the motor, per phase [W]

E_o = line-to-neutral voltage induced by I_x [V]

E = line-to-neutral voltage of the source [V]

X_s = synchronous reactance per phase [Ω]

δ = torque angle between E_o and E [electrical degrees]

This equation shows that the mechanical power increases with the torque angle, and its maximum value is reached when δ is 90°. The poles of the rotor are then midway between the N and S poles of the stator. The peak power P_{max} (per-phase) is given by

$$P_{max} = \frac{E_oE}{X_s} \qquad (17\text{-}3)$$

As far as torque is concerned, it is directly proportional to the mechanical power because the

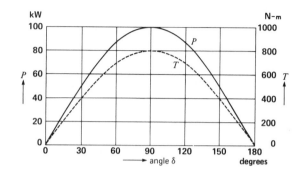

Figure 17-9
Power and torque per phase as a function of the torque angle δ. Synchronous motor rated 150 kW (200 hp), 1200 r/min, 3-phase, 60 Hz. See Example 17-3.

rotor speed is fixed. The torque is derived from Eq. 3-5:

$$T = \frac{9.55 \, P}{n_s} \qquad (17\text{-}4)$$

where

T = torque, per phase [N·m]

P = mechanical power, per phase [W]

n_s = synchronous speed [r/min]

9.55 = a constant [exact value = $60/2\pi$]

The maximum torque the motor can develop is called the pull-out torque, mentioned previously. It occurs when $\delta = 90°$ (Fig. 17-9).*

Example 17-3:
A 150 kW, 1200 r/min, 460 V, 3-phase synchronous motor has a synchronous reactance of 0.8 Ω, per phase. If the excitation voltage E_o is fixed at 300 V, per phase, determine:
a.   the power versus δ curve;
b.   the torque versus δ curve;
c.   the pull out torque of the motor.

---

* The remarks in this section apply to motors having smooth rotors. Most synchronous motors have salient poles; in this case the pull-out torque occurs at an angle of about 70°. Furthermore, the so-called reluctance torque created by the salient poles increases the pull-out torque (calculated by Eq. 17-4) by about 8%.

Solution:
a. The line- to- neutral voltage is:
$$E = E_L/\sqrt{3} = 460/1.73$$
$$= 266 \text{ V}$$

The mechanical power per phase is:
$$P = (E_oE/X_s) \sin \delta \qquad \text{Eq. 17-2}$$
$$= (266 \times 300/0.8) \sin \delta$$
$$= 99\,750 \sin \delta \text{ [W]}$$
$$= 100 \sin \delta \text{ [kW]}$$

By selecting different values for δ, we can calculate the corresponding values of P and T, per phase.

δ	P	T
[°]	[kW]	[N·m]
0	0	0
30	50	400
60	86.6	693
90	100	800
120	86.6	693
150	50	400
180	0	0

These values are plotted in Fig. 17-9.

b. The torque curve can be found by applying Eq. 17-4:
$$T = 9.55P/n_s$$
$$= 9.55P/1200$$
$$= P/125$$

c. The pull-out torque T_{max} coincides with the maximum power output:
$$T_{max} = 800 \text{ N·m}$$

The actual pull-out torque is 3 times as great (2400 N·m) because this is a 3-phase machine. Similarly, the power and torque values given in Fig. 17-9 must also be multiplied by 3. Consequently, this 150 kW motor can develop a maximum output of 300 kW, or about 400 hp.

17.7 Mechanical and electrical angles

As in the case of synchronous generators, there is a precise relationship between the mechanical angle α, the torque angle δ and the number of poles p. It is given by:
$$\delta = p\alpha/2 \qquad (17-5)$$

Example 17-4:
A 3-phase, 6000 kW, 4 kV, 180 r/min, 60 Hz motor has a synchronous reactance of 1.2 Ω. At full load, the rotor poles are displaced by a *mechanical* angle of 1° from their no-load position. If the line-to-neutral excitation E_o = 2.4 kV, calculate the mechanical power developed.

Solution:
The electrical torque angle is:
$$\delta = p\alpha/2 = (40 \times 1)/2 = 20°$$

Assuming a wye connection, the voltage E applied to the motor is:
$$E = E_L/\sqrt{3} = 4 \text{ kV}/1.73$$
$$= 2.3 \text{ kV}$$
$$= 2300 \text{ V}$$

and the excitation voltage is:
$$E_o = 2400 \text{ V}$$

The mechanical power developed per phase is:
$$P = (E_oE/X_s) \sin \delta \qquad \text{Eq. 17-2}$$
$$= (2400 \times 2300/1.2) \sin 20°$$
$$= 1\,573\,200$$
$$= 1573 \text{ kW}$$

Total power = 3×1573
= 4719 kW (~6300 hp)

17.8 Reluctance torque

If we gradually reduce the excitation of a synchronous motor when it is running at no-load, we find that the motor continues to run at synchronous speed even when the exciting current is zero. The reason is that the flux produced by the stator prefers to cross the short gap between the salient poles and the stator rather than the much longer air gap between the poles. In other words, because the reluctance of the magnetic circuit is less in the axis of the salient poles, the flux is concentrated as shown in Fig. 17-10a. On account of this phenomenon, the motor develops a so-called "reluctance torque".

If a mechanical load is applied to the shaft, the rotor poles will fall behind the stator poles, and the stator flux will have the shape shown in

oped between 45° and 90°, this is an unstable region of operation. The reason is that the angle increases while the power decreases.

As in the case of a conventional synchronous motor, the mechanical power curve has exactly the same shape as the torque curve. Thus, in the absence of dc excitation, the

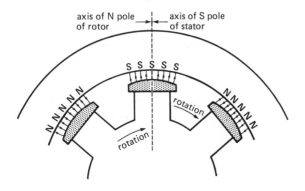

Figure 17-10a
The flux produced by the stator flows across the air gap through the salient poles.

Figure 17-10b
The salient poles are attracted to the stator poles, thus producing a reluctance torque.

Fig. 17-10b. Thus, a considerable reluctance torque can be developed without any dc excitation at all.

The reluctance torque becomes zero when the rotor poles are midway between the stator poles. The reason is that the N and S poles on the stator attract the salient poles in opposite directions (Fig. 17-10c). Consequently, the reluctance torque is zero precisely at that angle where the "regular" torque T attains its maximum value, namely at $\delta = 90°$.

Figure 17-11 shows the reluctance torque as a function of the angle δ. The torque reaches a maximum positive value at $\delta = 45°$. For larger angles, it attains a maximum *negative* value at $\delta = 135°$. Obviously, to run as a reluctance-torque motor, the angle must lie between zero and 45°. Although a positive torque is still devel-

Figure 17-10c
The reluctance torque is zero when the salient poles are midway between the stator poles.

Figure 17-11
Reluctance torque versus the torque angle.

mechanical power reaches a peak at $\delta = 45°$.

Does the saliency of the poles modify the power and torque curves shown in Fig. 17-9? The answer is yes. In effect, the curves shown in Fig. 17-9 are those of a smooth-rotor synchronous motor. The torque of a salient-pole motor is equal to the sum of the smooth-rotor component and the reluctance-torque component of Fig. 17-11. Thus, the true torque curve of a synchronous motor has the shape given in Figure 17-12.

The peak reluctance torque is about 25 percent of the peak "smooth-rotor" torque. As a

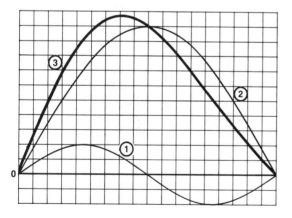

Figure 17-12
In a synchronous motor, the reluctance torque (1) plus the smooth-rotor torque (2) produce the resultant torque (3). Torque (2) is due to the dc excitation of the rotor.

result, the peak torque of a salient-pole motor is slightly greater than that of a smooth-rotor motor, as can be seen in Fig. 17-12. However, the difference is not very great, and for this reason we shall continue to use Eqs. 17-2 to 17-5 to describe synchronous motor behavior.

17.9 Losses and efficiency of a synchronous motor

In order to give the reader a sense of the order of magnitude of the pull-out torque, resistance, reactance and losses of a synchronous motor, we have drawn up Table 17A. It shows the characteristics of a 2000 hp and a 200 hp synchronous motor, respectively labelled Motor A and Motor B.

The following points should be noted:

1. The torque angle at full-load ranges between 27° and 37°. It corresponds to the electrical angle δ mentioned previously.

2. The power needed to excite the 2000 hp motor (4.2 kW) is only about twice that needed for the 200 hp motor (2.1 kW). In general, the larger the synchronous motor the smaller is the per-unit power needed to excite it.

3. The total losses of Motor A (38 kW) are only four times those of Motor B (9.5 kW) despite the fact that Motor A is ten times as powerful. This is another property of large motors: the more horsepower they develop, the smaller the relative losses are. As a result, the efficiencies improve with increase in power. Compare the efficiencies of the two motors: 97.5% versus 94.0%.

4. The synchronous reactance X_s per phase is much larger than the resistance of the stator winding. Note that for the 2000 hp motor X_s is 122 times larger than R_s. As a result, we can always neglect the effect of R_s as far as motor performance is concerned.

TABLE 17A   CHARACTERISTICS OF TWO SYNCHRONOUS MOTORS

NAMEPLATE RATING	MOTOR A	MOTOR B
power [hp]	2000 hp	200 hp
power [kW]	1492 kW	149 kW
line voltage	4000 V	440 V
line current	220 A	208 A
speed	1800 r/min	900 r/min
frequency	60 Hz	60 Hz
phases	3	3
LOAD CHARACTERISTICS		
power factor	1.0	1.0
pull-out torque (pu)	1.4	2.2
torque angle at full load	36.7°	27°
connection	wye	wye
dc exciter power	4.2 kW	2.1 kW
dc exciter voltage	125 V	125 V
air gap	10 mm	6 mm
LOSSES		
windage and friction	8.5 kW	1 kW
stator core loss	11 kW	2 kW
stray losses	4 kW	1 kW
stator I^2R	10.3 kW	3.5 kW
rotor I^2R	4.2 kW	2 kW
total losses	38 kW	9.5 kW
efficiency	97.5%	94.0%
IMPEDANCES AND VOLTAGES (line-to-neutral values)		
stator X_s	7.77 Ω	0.62 Ω
stator resistance R_s	0.0638 Ω	0.0262 Ω
ratio X_s/R_s	122	23
phase voltage E	2309 V	254 V
phase voltage E_o	2873 V	285 V

17.10 Excitation and reactive power

Consider a wye-connected synchronous motor connected to a 3-phase source whose line voltage E_L is fixed (Fig. 17-13). It follows that the line-to-neutral voltage E is also fixed. The line currents I produce a magnetomotive force U_a in the stator. On the other hand, the rotor produces a dc magnetomotive force U_r. The total flux Φ is therefore due to the combined action of U_a and U_r.

Figure 17-13
The total flux Φ is due to the mmf produced by the rotor (U_r) plus the mmf produced by the stator (U_a). For a given E_L, the flux Φ is essentially fixed.

Pursuing our reasoning, flux Φ induces a line-to-neutral voltage E_a in the stator. If we neglect the very small IR drop in the stator, we have $E_a = E$. Because E is fixed, it follows that Φ is also fixed, as in the case of a transformer (see Sec. 9.2).

The mmf needed to create the constant flux Φ may be produced either by the stator or the rotor, or by both. If the rotor exciting current I_x is zero, all the flux has to be produced by the stator. The stator must then absorb considerable reactive power from the 3-phase line (see Sec. 7.9). But if we excite the rotor with a dc current I_x, the rotor mmf helps produce part of the flux Φ. Consequently, less reactive power is drawn from the ac line. If we gradually raise the excitation, the rotor will eventually produce all the required flux by itself. The stator then draws no more reactive power, with the result that the power factor of the motor becomes unity (1.0).

What happens if we excite the motor above this critical level? The stator, instead of absorbing reactive power, actually *delivers* reactive power to the 3-phase line. The motor then behaves like a source of reactive power, just as if it were a capacitor. Thus, by varying the dc excitation, we can cause the motor to either absorb or deliver reactive power. Because of this important property, synchronous motors are sometimes used to correct the power factor of a plant at the same time as they furnish mechanical power.

17.11 Power factor rating

Most synchronous motors are designed to operate at unity power factor. However, if they also have to deliver reactive power, they are usually designed to operate at a full-load power factor of 0.8 (leading). A motor designed for a power factor of 0.8 can deliver reactive power equal to 75 percent of its rated mechanical load. Thus, the 3000 kW motor shown in Fig. 17-4 can supply 75% x 3000 = 2250 kvar to the line at the same time as it develops its rated mechanical output of 3000 kW. Motors designed to operate at leading power factors are bigger and more costly than unity power factor motors are. The reason is that both the dc exciting current and the stator current are higher.

Figure 17-14 is the schematic diagram of a unity power factor motor operating at full load. The line-to-neutral voltage is E_{ab} and the line current is I_p. The active power absorbed per phase is therefore:

$$P = E_{ab}I_p \qquad (17\text{-}6)$$

Figure 17-14
Unity power factor synchronous motor and phasor diagram at full-load.

The active power absorbed is very nearly equal to the mechanical power of the motor.

Figure 17-15 shows an 80% power factor motor also operating at full load. It develops the same mechanical power as the motor in Fig. 17-14. The line current I_s leads E_{ab} by arcos $0.8 = 36.87°$. This current can be broken

Figure 17-15
80 percent power factor synchronous motor and phasor diagram at full-load.

up into two components I_p and I_q and it is clear that

$$I_p = 0.8\,I_s \qquad (17\text{-}7)$$

$$I_q = 0.6\,I_s \qquad (17\text{-}8)$$

The active power P is given by:

$$P = E_{ab}I_p = 0.8\,E_{ab}I_s \qquad (17\text{-}9)$$

The reactive power delivered by the motor is:

$$Q = E_{ab}I_q = 0.6\,E_{ab}I_s \qquad (17\text{-}10)$$

It follows from Eqs. 17-9 and 17-10 that

$$Q = 0.75\,P$$
$$= 75\% \text{ of rated mechanical output}$$

as was stated previously.

If we compare I_p with I_s, we find that $I_s = 1.25\,I_p$. Thus, for the same mechanical power output, a motor designed for a leading power factor of 80% has to carry a line current that is 25% greater than one that operates at unity power factor.

17.12 V-curves

Suppose a synchronous motor is operating at its rated mechanical load. We wish to examine its behavior as the excitation is varied. Because a change in excitation does not affect the speed, the mechanical power remains fixed. Let us begin by adjusting the excitation so that the power factor is unity, thus yielding the phasor diagram shown in Fig. 17-16. We assume $I_x = 100$ A and $P = 800$ kW.

If we reduce the excitation to 70 A the motor will draw reactive power from the line in addition to the active power. We assume that S increases to $S = 1000$ kVA. As a result, the line current will increase from I_p to I_{s1} (Fig. 17-17). Note that the component of I_{s1} in phase with E_{ab} is the same as before because the motor is still developing the same mechanical power.

Current I_{s1} lags behind E_{ab}, and so the power factor of the motor is lagging. The field current I_x in the rotor is smaller than before, but the apparent power S absorbed by the stator is greater.

If we increase the excitation to $I_x = 200$ A, the motor *delivers* reactive power to the line to which it is connected (Fig. 17-18). The apparent power is again greater than in the unity power factor case. We assume $S = 1000$ kVA. The line current becomes I_{s2} and it leads E_{ab}. However, the in-phase component of I_{s2} is still equal to I_p because the mechanical power is the same.

By varying the excitation this way, we can plot the apparent power of the synchronous motor as a function of the dc exciting current. This yields a V-shaped curve (Fig. 17-19). The V-curve is always drawn for a fixed

(a)   (b)

Figure 17-16
a. Synchronous motor operating at unity power factor with a mechanical load of 800 kW. Field excitation is 100 A.
b. Phasor diagram shows current in phase with the voltage.

(a)   (b)

Figure 17-17
a. Field excitation reduced to 70 A but with same mechanical load. Motor absorbs reactive power from the line.
b. Phasor diagram shows current lagging behind the voltage.

(a)   (b)

Figure 17-18
a. Field excitation raised to 200 A but with same mechanical load. Motor delivers reactive power to the line.
b. Phasor diagram shows current leading the voltage.

mechanical load. In our case, the V-curve corresponds to full-load. The no-load V-curve is also shown, to illustrate the large reactive power that can be absorbed or delivered by simply changing the excitation.

Figure 17-19
No-load and full-load V-curves of a 1000 hp synchronous motor.

Example 17-5:
A 4000 hp (3000 kW), 6600 V, 60 Hz, 200 r/min synchronous motor operates at full-load at a leading power factor of 0.8. If the synchronous reactance is 11 Ω, calculate:
a. the apparent power of the motor, per phase;
b. the ac line current;
c. the value and phase of E_o;
d. draw the phasor diagram;
e. determine the torque angle δ.

Solution:
We shall immediately change the given values to correspond to one phase of a wye-connected motor.

a. The active power per phase is:
$$P = 3000/3 = 1000 \text{ kW}$$

The apparent power per phase is:
$$S = P/\cos\theta = 1000/0.8 \quad \text{Eq. 8-11}$$
$$= 1250 \text{ kVA}$$

b. The line-to-neutral voltage is:
$$E = E_L/\sqrt{3} = 6600/1.73$$
$$= 3815 \text{ V}$$

The line current is:
$$I = S/E = 1250 \times 1000/3815$$
$$= 328 \text{ A}$$

I leads E by an angle of arcos 0.8 = 36.9°

c. To determine the value and phase of the excitation voltage E_o we draw the equivalent circuit of one phase (Fig. 17-20). This will enable us to write the circuit equations. Furthermore, we select E as the reference phasor and so it is given by:

$$E = 3815\angle 0°$$

It follows that I is given by:

$$I = 328\angle 36.9°$$

Writing the equation for the circuit we find:

$$-E + jIX_s + E_o = 0$$

thus

$$
\begin{aligned}
E_o &= E - jIX_s \\
&= 3815\angle 0° - j\,(328\angle 36.9°)\,11 \\
&= 3815\angle 0° - 3608\angle(36.9° + 90°) \\
&= 3815\,(\cos 0° + j\sin 0°) - \\
&\quad 3608\,(\cos 126.9° + j\sin 126.9°) \\
&= 3815 + 2166 - j2885 \\
&= 5981 - j2885 \\
&= 6640\angle -26°
\end{aligned}
$$

d. Consequently, E_o lags 26° behind E, and the complete phasor diagram is shown in Fig. 17-21.

e. The torque angle δ is 26°.

17.13 Stopping synchronous motors

Owing to the inertia of the rotor and load, large synchronous motors may take several hours to stop after they are disconnected from the line. To reduce the stopping time, we use the following braking methods:

1. maintain full dc excitation with the armature in short-circuit;

2. maintain full dc excitation with the armature connected to 3 external resistors;

3. apply mechanical braking.

In methods (1) and (2), the motor slows

Figure 17-20
Circuit of a synchronous motor connected to a source E. Note the arbitrary (+) polarity marks and arbitrary direction of current flow. See Example 17-5.

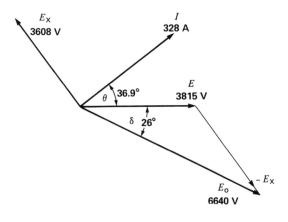

Figure 17-21
See Example 17-5.

down because it functions as a generator, dissipating its energy in the resistive elements of the circuit. Mechanical braking is usually applied after the motor has reached half speed or less, to prevent undue wear of the brake shoes.

Example 17-6:
A 1500 kW, 4600 V, 600 r/min, 60 Hz synchronous motor possesses a synchronous reactance of 16 Ω and a stator resistance of 0.2 Ω, per phase. The excitation voltage E_o is 2400 V

and the moment of inertia of the motor and its load is 275 kg·m². We wish to stop the motor by short-circuiting the armature while keeping the dc rotor current fixed. Calculate:

a. the power dissipated in the armature at 600 r/min;

b. the power dissipated in the armature at 150 r/min;

c. the kinetic energy at 600 r/min;

d. the kinetic energy at 150 r/min;

e. the time required for the speed to fall from 600 r/min to 150 r/min.

Solution:

a. Referring to Fig. 17-22a, the motor has just been disconnected from the line and is now operating as a generator in short-circuit. The speed is still 600 r/min and so the frequency

induced voltage E_o is proportional to the speed. Consequently, when the speed has dropped to 150 r/min:

$$E_o = 2400 \times (150/600) = 600 \text{ V}$$

The frequency is also proportional to the speed, and so:

$$f = 60 \times (150/600) = 15 \text{ Hz}$$

The synchronous reactance is proportional to the frequency; consequently,

$$X_s = 16 \times (150/600) = 4 \ \Omega$$

Referring to Fig. 17-22b, the new impedance per phase at 150 r/min is:

$$Z = \sqrt{0.2^2 + 4^2} = 4 \ \Omega$$

The current per phase is:

$$I = E_o/Z = 600/4 = 150 \text{ A}$$

16 Ω
150 A
2400 V
600 r/min
E_o
0.4 Ω

Figure 17-22a
Motor turning at 600 r/min (Example 17-6).

4 Ω
150 A
600 V
150 r/min
E_o
0.4 Ω

Figure 17-22b
Motor turning at 150 r/min (Example 17-6).

is 60 Hz. Consequently, the impedance per phase is:

$$Z = \sqrt{R^2 + X_L^2} \qquad \text{Eq. 2-12}$$
$$= \sqrt{0.2^2 + 16^2}$$
$$= 16 \ \Omega$$

The current per phase is:

$$I = E_o/Z = 2400/16$$
$$= 150 \text{ A}$$

The power dissipated in the 3 phases at 600 r/min is:

$$P = 3I^2R = 3 \times 150^2 \times 0.2$$
$$= 13.5 \text{ kW}$$

b. Because the exciting current is fixed, the

Thus, the short-circuit current remains unchanged as the motor decelerates from 600 r/min to 150 r/min. The power dissipated in the 3 phases is therefore the same as before:

$$P = 13.5 \text{ kW}$$

c. The kinetic energy at 600 r/min is:

$$E_{k1} = 5.48 \times 10^{-3}Jn^2 \qquad \text{Eq. 3-8}$$
$$= 5.48 \times 10^{-3} \times 275 \times 600^2$$
$$= 542.5 \text{ kJ}$$

d. The kinetic energy at 150 r/min is:

$$E_{k2} = 5.48 \times 10^{-3} \times 275 \times 150^2$$
$$= 33.9 \text{ kJ}$$

e. The loss in kinetic energy in decelerating from 600 r/min to 150 r/min is:

$$W = E_{k1} - E_{k2}$$
$$= 542.5 - 33.9$$
$$= 508.6 \text{ kJ}$$

This energy is lost as heat in the armature resistance. The time for the speed to drop from 600 r/min to 150 r/min is given by:

$$P = W/t \qquad \text{Eq. 3-4}$$
$$13.5 = 508.6/t$$

whence $t = 37.7 \text{ s}$

Note that the motor would stop much quicker if external resistors were connected across its terminals.

17.14 The synchronous motor versus the induction motor

We have already seen that induction motors have excellent properties for speeds above 600 r/min. But at lower speeds, they become heavy, costly, and have relatively low power factor and low efficiency.

Synchronous motors are particularly attractive for low-speed drives because the power factor can always be adjusted to 1.0 and the efficiency is high. Although more complex to build, their weight and cost are often less than those of induction motors of equal power and speed. This is particularly true for speeds below 300 r/min.

A synchronous motor can improve the power factor of a plant while carrying its rated load. Furthermore, its starting torque can be made considerably greater than that of an induction motor. The reason is that we can raise the resistance of the squirrel-cage winding without affecting the speed or efficiency at synchronous speed. Figure 17-23 compares the properties of a squirrel-cage induction motor and a synchronous motor having the same nominal rating. The biggest difference is in the starting torque.

High-power electronic converters gener-ating very low frequencies enable us to run synchronous motors at ultra-low speeds. Thus, huge motors in the 10 MW range drive crushers, rotary kilns, and variable-speed ball mills.

(a)

(b)

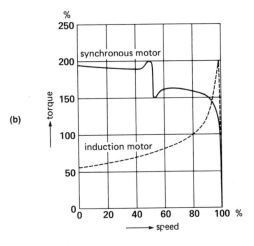

Figure 17-23
Comparison between the efficiency and starting torque of a squirrel-cage induction motor (a) and a synchronous motor (b), both rated at 4000 hp, 1800 r/min, 6.9 kV, 60 Hz.

17.15 Synchronous capacitor

A synchronous capacitor is essentially a synchronous motor running at no-load. Its only purpose is to absorb or deliver reactive power on a 3-phase system, in order to stabilize the voltage (see Chap. 25). The machine acts as an enormous 3-phase capacitor (or inductor) whose reactive power can be varied by changing the dc excitation.

Most synchronous capacitors have ratings that range from 20 Mvar to 200 Mvar and many are hydrogen-cooled (Fig. 17-24). They are started up like synchronous motors. However, if the system cannot furnish the required starting power, a pony motor is used to bring them up to synchronous speed. For example, in one installation, a 160 Mvar synchronous capacitor is started up by means of a 1270 kW wound-rotor motor.

Example 17-7:
A synchronous capacitor is rated at 160 Mvar, 16 kV, 1200 r/min, 60 Hz. It has a synchronous reactance of 0.8 p.u. and is connected to a 16 kV line. Calculate the value of E_o so that the machine:
    a. absorbs 160 Mvar;
    b. delivers 120 Mvar.

Solution:
a.   The nominal impedance of the machine is:

$$Z_n = E_n{}^2/S_n \qquad \text{Eq. 16-3}$$
$$= 16\ 000^2/(160 \times 10^6)$$
$$= 1.6\ \Omega$$

The synchronous reactance per phase is:

$$X_s = X_S(\text{pu})\ Z_n = 0.8 \times 1.6$$
$$= 1.28\ \Omega$$

The line current for a reactive load of 160 Mvar is:

$$I_n = S_n/(\sqrt{3}\ E_n)$$
$$= 160 \times 10^6/(1.73 \times 16\ 000)$$
$$= 5780\ A$$

Figure 17-24a
Three-phase, 16 kV, 900 r/min synchronous capacitor rated −200 Mvar (supplying reactive power) to + 300 Mvar (absorbing reactive power). It is used to regulate the voltage of a 735 kV transmission line. Other characteristics: mass of rotor: 143 t; rotor diameter: 2670 mm; axial length of stator iron: 3200 mm; air gap length: 39.7 mm.

Figure 17-24b
Synchronous capacitor enclosed in its steel housing containing hydrogen under pressure (300 kPa, or about 44 lbf/in²). (*Hydro-Québec*)

The drop across the synchronous reactance is:

$$E_x = IX_s = 5780 \times 1.28$$
$$= 7400 \text{ V}$$

The line-to-neutral voltage is:

$$E = E_L/\sqrt{3} = 16\,000/1.73$$
$$= 9250 \text{ V}$$

Selecting E as the reference phasor, we have:

$$E = 9250\angle 0°$$

The current I lags 90° behind E because the machine is absorbing reactive power; consequently:

$$I = 5780\angle -90°$$

From Fig. 17-25a we can write:

$$-E + jIX_s + E_o = 0$$

hence

$$E_o = E - jIX_s$$
$$= 9250\angle 0° - 5780 \times 1.28$$
$$\angle(90° - 90°)$$
$$= 1850\angle 0°$$

Note that the excitation voltage(1850 V) is much less than the line voltage(9250 V).

Figure 17-25a
Under-excited synchronous capacitor absorbs reactive power (Example 17-7)

b. The load current when the machine is delivering 120 Mvar is:

$$I = Q/(\sqrt{3}E_n)$$
$$= 120 \times 10^6/(1.73 \times 16\,000)$$
$$= 4335 \text{ A}$$

This time, I leads E by 90° and so:

$$I = 4335\angle 90°$$

From Fig. 17-25b we can write:

$$E_o = E - jIX_s$$
$$= 9250\angle 0° - 4335 \times 1.28\angle 180°$$
$$= (9250 + 5550)\angle 0°$$
$$= 14\,800\angle 0°$$

The excitation voltage (14 800 V) is now considerably greater than the line voltage (9250 V).

Figure 17-25b
Over-excited synchronous capacitor delivers reactive power (Example 17-7).

QUESTIONS AND PROBLEMS

Practical level

17-1 Compare the construction of a synchronous generator, a synchronous motor and a squirrel-cage induction motor.

17-2 Explain how a synchronous motor starts up. When should the dc excitation be applied?

17-3 Why does the speed of a synchronous motor remain constant even under variable load?

17-4 Name some of the advantages of a synchronous motor compared to a squirrel-cage induction motor.

17-5 What is meant by a synchronous capacitor and what is it used for?

17-6 a. What is meant by an "under-excited" synchronous motor?
b. If we over-excite a synchronous motor, does its mechanical power output increase?

17-7 A synchronous motor draws 2000 kVA at a power factor of 90% leading. Calculate the approximate power developed by the motor [hp].

17-8 A synchronous motor driving a pump operates at a power factor of 100%. What happens if the dc excitation is increased?

17-9 A 3-phase, 225 r/min synchronous motor connected to a 4 kV, 60 Hz line draws a current of 320 A and absorbs 2000 kW. Calculate:
a. the apparent power supplied to the motor;
b. the power factor;
c. the reactive power absorbed;
d. the number of poles on the rotor.

17-10 A synchronous motor draws 150 A from a 3-phase line. If the exciting current is raised, the current drops to 140 A. Was the motor over- or under-excited before the excitation was changed?

Intermediate level

17-11 a. Calculate the approximate full-load

current of the 3000 hp motor in Fig. 17-1, if it has an efficiency of 97%.
b. What is the value of the field resistance?

17-12 Referring to Fig. 17-2, at what speed must the rotor turn to generate the indicated frequencies?

17-13 A 3-phase synchronous motor rated 800 hp, 2.4 kV, 60 Hz operates at unity power factor. The line voltage suddenly drops to 1.8 kV, but the exciting current remains unchanged. Explain how the following quantities are affected:
a. motor speed and mechanical power output;
b. torque angle δ;
c. position of the rotor poles;
d. power factor;
e. stator current.

17-14 A synchronous motor has the following parameters, per phase (ref. Fig. 17-7a):

$E = 2.4$ kV; $E_o = 3$ kV; $X_s = 2\ \Omega$; $I = 900$ A.

Draw the phasor diagram and determine:

a. torque angle δ;
b. active power, per phase;
c. power factor of the motor;
d. reactive power absorbed (or delivered), per phase.

17-15 a. In Problem 17-14, calculate the line current and the new torque angle δ if the mechanical load is suddenly removed.
b. Calculate the new reactive power absorbed (or delivered) by the motor, per phase.

17-16 A 500 hp synchronous motor drives a compressor and its excitation is adjusted so that the power factor is unity. If the excitation is increased without making any other change, what is the effect upon:

a. the active power absorbed by the motor?

b. the line current?

c. the reactive power absorbed (or delivered) by the motor?

d. the torque angle?

Advanced level

17-17 The 4000 hp, 6.9 kV motor shown in Fig. 17-4 possesses a synchronous reactance of 10 Ω, per phase. The stator is connected in wye, and the motor operates at full load (4000 hp) with a leading power factor of 0.89. If the efficiency is 97%, calculate:

a. the apparent power;

b. the line current;

c. the value of E_o, per phase;

d. the mechanical displacement of the poles from their no-load position;

e. the total reactive power supplied to the electrical system;

f. the approximate maximum power the motor can develop, without pulling out of step [hp].

17-18 In Problem 17-17, we wish to adjust the power factor to unity. Calculate:

a. the exciting voltage E_o required, per phase;

b. the new torque angle.

17-19 A 3-phase, unity power factor synchronous motor rated 400 hp, 2300 V, 450 r/min, 80 A, 60 Hz, drives a compressor. The stator has a synchronous reactance of 0.88 p.u., and the excitation E_o is adjusted to 1.2 p.u. Calculate:

a. the value of X_s and of E_o, per phase;

b. the pull-out torque [ft·lbf];

c. the line current when the motor is about to pull out of synchronism.

17-20 The synchronous capacitor in Fig. 17-24 possesses a synchronous reactance of 0.6 Ω, per phase. The resistance per phase is 0.007 Ω. If the machine coasts to a stop, it will run for about 3 h. In order to shorten the stopping time, the stator is connected to three large 0.6 Ω braking resistors connected in wye. The dc excitation is fixed at 250 A so that the initial line voltage across the resistors is one tenth of its rated value, or 1600 V, at 900 r/min. Calculate:

a. the total braking power and braking torque at 900 r/min;

b. the braking power and braking torque at 450 r/min;

c. the average braking torque between 900 r/min and 450 r/min;

d. the time for the speed to fall from 900 r/min to 450 r/min, knowing that the moment of inertia of the rotor is 1.7×10^6 lb·ft^2.

18

SINGLE-PHASE MOTORS

Single-phase motors are the most familiar of all electric motors because they are used in home appliances and portable machine tools. In general, they are employed when 3-phase power is not available.

There are many kinds of single-phase motors on the market, each designed to meet a specific application. However, we shall limit our study to a few basic types, with particular emphasis on the widely used split-phase induction motor.

18.1 Construction of a single-phase induction motor

Single-phase induction motors are very similar to 3-phase induction motors. They are composed of a squirrel-cage rotor (identical to that in a 3-phase motor) and a stator (Fig. 18-1). The stator carries a so-called *main winding* which creates a set of N-S poles. It also carries a smaller *auxiliary winding* that only operates during the brief period when the motor starts up. The auxiliary winding has the same number of poles as the main winding has.

Figure 18-2 shows the progressive steps in winding a 4-pole, 36-slot stator. Starting with the bare stator, slot liners are first inserted and the main winding is then laid in the slots (Figs. 18-2a, 18-2b). The auxiliary winding is then imbedded so that it straddles the main winding (Fig. 18-2c).

Each pole of the main winding consists of a group of four concentric coils, connected in series (Fig. 18-3a). Adjacent poles are connected to produce alternate N, S polarities. The empty slot in the center of each pole (shown as a vertical dash), and the partially filled slots on either side of it, are used to lodge the auxiliary winding. The latter has 2 concentric coils per pole (Fig. 18-3b).

Figure 18-4 shows a 2-pole stator; the large main winding and the smaller auxiliary winding

Figure 18-1
Cutaway view of a 5 hp, 1725 r/min, 60 Hz single-phase capacitor-start motor. (*Gould*)

are displaced at right angles to each other. The reason for this arrangement will be explained shortly.

18.2 Synchronous speed

As in the case of 3-phase motors, the synchronous speed of all single-phase induction motors is given by the equation:

$$n_s = \frac{120f}{p} \qquad \text{Eq. 17-1}$$

where

n_s = synchronous speed [r/min]
f = frequency of the source [Hz]
p = number of poles

The rotor turns at slightly less than synchronous speed, and the full-load slip is typically 3% to 5% for fractional horsepower motors.

Example 18-1:
Calculate the speed of the 4-pole single-phase motor shown in Fig. 18-1 if the slip is 3.4%. The line frequency is 60 Hz.

Figure 18-2a
Bare, laminated stator of a 1/4 hp (187 W), single-phase motor. The 36 slots are insulated with a paper liner. The squirrel-cage rotor is identical to that of a 3-phase motor.

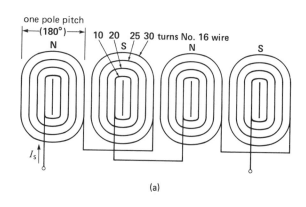

(a)

Figure 18-2b
Four poles of the main winding are inserted in the slots.

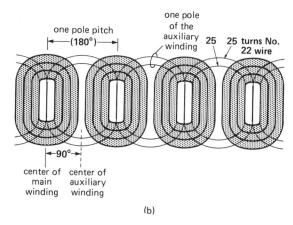

(b)

Figure 18-2c
Four poles of the auxiliary winding straddle the main winding. (*Lab-Volt*)

Figure 18-3
a. Main winding of a 4-pole motor, showing the number of turns per coil.
b. Position of the auxiliary winding with respect to the main winding.

Solution:

a. The motor has 4 poles, consequently,

$$n_s = 120f/p = (120 \times 60)/4$$
$$= 1800 \text{ r/min}$$

The speed is given by:

$$s = (n_s - n)/n_s \qquad \text{Eq. 17-2}$$
$$0.034 = (1800 - n)/1800$$
$$n = 1739 \text{ r/min}$$

However, if we spin the rotor in one direction or the other, it will continue to rotate in the direction of spin. As a matter of fact, the rotor quickly accelerates until it reaches a speed slightly below synchronous speed. The acceleration indicates that the motor develops a positive torque as soon as it begins to turn. Figure 18-6 shows the typical torque-speed curve when the main winding is excited. Although the starting torque is zero, the motor develops a powerful torque as it approaches synchronous speed.

Figure 18-4
Main and auxiliary windings in a 2-pole single-phase motor. The stationary contact in series with the auxiliary winding opens when the centrifugal switch, mounted on the shaft, reaches 75 percent of synchronous speed.

Figure 18-5
Currents in the rotor bars when the rotor is locked. The resulting forces cancel each other and no torque is produced.

18.3 Torque-speed characteristic

Figure 18-5 is a schematic diagram of the rotor and main winding of a 2-pole single-phase induction motor. Suppose the rotor is locked. If an ac voltage is applied to the stator, the resulting current I_s produces an ac flux Φ_s. The flux alternates back and forth but, unlike the flux in a 3-phase stator, no revolving field is produced. The flux induces an ac voltage in the stationary rotor which, in turn, creates large ac rotor currents. In effect, the rotor behaves like the short-circuited secondary of a transformer; consequently, the motor has no tendency to start by itself.

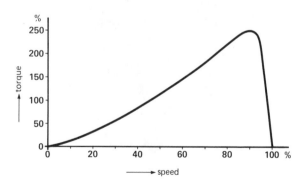

Figure 18-6
Typical torque-speed curve of a single-phase motor.

18.4 Principle of operation

The principle of operation of a single-phase induction motor is quite complex, and may be explained by the so-called *cross-field theory*.*

As soon as the rotor begins to turn, a "speed emf" E is induced in the rotor conductors, as they cut the stator flux Φ_s (Fig. 18-7). This voltage increases as the rotor speed increases. It causes currents I_r to flow in the rotor bars facing the stator poles. These cur-

Figure 18-7
Currents induced in the rotor bars due to rotation. They produce a flux that acts at right angles to the stator flux.

rents produce an ac flux Φ_r which acts at right angles to the stator flux Φ_s. Equally important is the fact that Φ_r does not reach its maximum value at the same time as Φ_s does. In effect, Φ_r lags almost 90° behind Φ_s, owing to the inductance of the rotor.

The combined action of Φ_s and Φ_r produces a revolving magnetic field, similar to that in a 3-phase motor. The value of Φ_r increases with increasing speed, becoming almost equal to Φ_s at synchronous speed. This explains, in part, why the torque increases as the motor speeds up.

We can understand how the revolving field is produced by referring to Fig. 18-8. It gives a snapshot of the currents and fluxes created respectively by the rotor and stator, at succes-

sive intervals of time. We assume that the motor is running far below synchronous speed, and so Φ_r is much smaller than Φ_s. By observing the flux in the successive pictures of Fig. 18-8, it is obvious that the combination of Φ_s and Φ_r produces a revolving field. Furthermore, the flux is strong horizontally and relatively weak vertically. Thus, the field strength at low speed follows the elliptic pattern shown in Fig. 18-8f. The flux rotates counterclockwise in the same direction as the rotor. Furthermore, it rotates at synchronous speed, irrespective of the actual speed of the rotor. As the motor approaches synchronous speed, Φ_r becomes almost equal to Φ_s and a nearly perfect revolving field is produced.

18.5 Locked-rotor torque

To produce a starting torque in a single-phase motor, we must somehow create a revolving field. This is done by adding an auxiliary winding, as shown in Fig. 18-9. When the main and auxiliary windings are connected to an ac source, the main winding produces a flux Φ_s, while the auxiliary winding produces a flux Φ_a. If the fluxes are out of phase, so that Φ_a either lags or leads Φ_s, a rotating field is set up.

The reader will immediately see that the auxiliary winding produces a strong flux Φ_a during the acceleration period when the rotor flux Φ_r (mentioned previously) is weak. As a result, Φ_a "strengthens" Φ_r, thereby producing a powerful torque both at standstill and at low speeds. The locked-rotor torque is given by:

$$T = kI_aI_s\sin\alpha \qquad (18-1)$$

where

T = locked-rotor torque [N·m]
I_a = locked-rotor current in the auxiliary winding [A]
I_s = locked-rotor current in the main winding [A]
α = phase angle between I_s and I_a [°]
k = a constant, depending on the design of the motor

The double revolving field theory (not discussed here) is also used to explain the behavior of the single-phase motor.

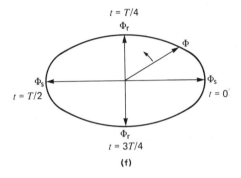

Figure 18-8
Instantaneous currents and flux in a single-phase motor with the main winding excited. The duration of 1 cycle is T seconds, and conditions are shown at successive quarter-cycle intervals.

a. Stator current I_s is maximum, rotor current I_r is zero.
b. Stator current is zero, rotor current is maximum; however, Φ_r is smaller than Φ_s.
c. Stator current is maximum, but negative.
d. Rotor current is maximum, but negative.
e. After one complete cycle ($t = T$), the conditions repeat.
f. Resulting flux Φ in the air gap rotates ccw at synchronous speed. Its amplitude varies from a maximum of Φ_s to a minimum Φ_r.

Figure 18-9
Currents and fluxes at standstill when the main and auxiliary windings are energized. An elliptical revolving field is produced.

(a)   (b)

Figure 18-10
a. Centrifugal switch in the closed, or stopped position. The stationary contact is closed.
b. Centrifugal switch in the open, or running position. Owing to centrifugal force, the rectangular weights have swung out against the restraining tension of the springs. This has caused the plastic collar to move to the left along the shaft, thus opening the stationary contact.

To obtain the desired phase shift between I_s and I_a (and hence between Φ_s and Φ_a), we add an impedance Z in series with the auxiliary winding. The impedance may be resistive, inductive, or capacitive, depending upon the desired starting torque. The choice of impedance gives rise to various types of so-called *split-phase* motors. In many cases, the desired impedance is incorporated in the auxiliary winding itself.

A special switch is also connected in series with the auxiliary winding. It disconnects the winding when the motor reaches about 75% of synchronous speed. A speed-sensitive centrifugal switch, mounted on the shaft, is often used for this purpose (Fig. 18-10).

18.6 Resistance split-phase motor

The main winding of a single-phase motor is always made of relatively large wire, to reduce the I^2R losses (Fig. 18-11a). The winding also has a relatively large number of turns. Consequently, under locked-rotor conditions, the inductive reactance is high and the resistance is low. As a result, the locked-rotor current I_s lags considerably behind the applied voltage E (Fig. 18-11b).

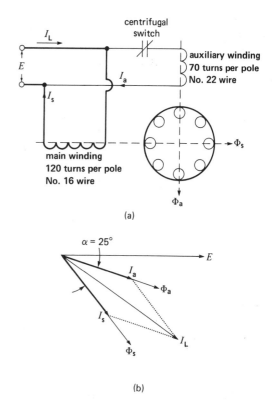

(a)

(b)

Figure 18-11
a. Resistance split-phase motor (1/4 hp, 115 V, 1725 r/min, 60 Hz) at standstill.
b. Corresponding phasor diagram. The current in the auxiliary winding leads the current in the main winding by 25°.

In a *resistance split-phase motor* (often simply called split-phase motor), the auxiliary winding has a relatively small number of turns of fine wire. Its resistance is higher and its reactance lower than that of the main winding, with the result that the locked rotor current I_a is more nearly in phase with E. The resulting phase angle α between I_a and I_s produces the starting torque.

The line current I_L is equal to the phasor sum of I_s and I_a. It is usually 6 to 7 times the nominal current of the motor.

Owing to the small wire used on the auxiliary winding, the current density is high and the winding heats up very quickly. If the starting period lasts for more than 5 seconds, the winding begins to smoke and may burn out, unless the motor is protected by a built-in thermal relay. This type of split-phase motor is well suited for infrequent starting of low-inertia loads.

Example 18-2:
A resistance split-phase motor is rated at 1/4 hp (187 W), 1725 r/min, 115 V, 60 Hz. When the rotor is locked, a test at reduced voltage on the main and auxiliary windings yields the following results.

	main winding	auxiliary winding
applied voltage	$E = 23$ V	$E = 23$ V
current	$I_s = 4$ A	$I_a = 1.5$ A
active power	$P_s = 60$ W	$P_a = 30$ W

Calculate:
a. the phase angle between I_a and I_s;
b. the locked-rotor current drawn from the line at 115 V.

Solution:
We first calculate the phase angle between I_s and E of the main winding.
a. The apparent power is:
$$S_s = EI_s = 23 \times 4 = 92 \text{ VA}$$

The power factor is:

$$\cos \phi_s = P_s/S_s = 60/92 = 0.65$$
$$\text{thus, } \phi_s = 49.2°$$

I_s lags 49.2° behind the voltage E.

We now calculate the phase angle between I_a and E of the auxiliary winding.

The apparent power is:
$$S_a = EI_a = 23 \times 1.5 = 34.5 \text{ VA}$$

The power factor is:
$$\cos \phi_a = P_a/S_a = 30/34.5 = 0.87$$
$$\text{thus, } \phi_a = 29.6°$$
I_a lags 29.6° behind the voltage.

The phase angle between I_s and I_a is:
$$\alpha = \phi_s - \phi_a = 49.2° - 29.6°$$
$$= 19.6°$$

b. To determine the total line current, we first calculate the total value of P and Q drawn by both windings, and then deduce the total apparent power S.

The total active power absorbed is:
$$P = P_s + P_a$$
$$= 60 + 30 = 90 \text{ W}$$

The reactive powers Q_s and Q_a of the main and auxiliary windings are:
$$Q_s = \sqrt{S_s^2 - P_s^2}$$
$$= \sqrt{92^2 - 60^2} = 69.7 \text{ var}$$
$$Q_a = \sqrt{S_a^2 - P_a^2}$$
$$= \sqrt{34.5^2 - 30^2} = 17.0 \text{ var}$$

The total reactive power absorbed by the motor is:
$$Q = Q_s + Q_a$$
$$= 69.7 + 17.0 = 86.7 \text{ var}$$

The total apparent power absorbed is:
$$S = \sqrt{P^2 + Q^2}$$
$$= \sqrt{90^2 + 86.7^2} = 125 \text{ VA}$$

The locked-rotor current at 23 V is:
$$I_1 = S/E = 125/23 = 5.44 \text{ A}$$

The current drawn at 115 V is:

$$I_L = 5.44 \times (115/23) = 27.2 \text{ A}$$

Owing to their low cost, resistance split-phase induction motors are the most popular single-phase motors. They are used where a moderate starting torque is required and where the starting periods are infrequent. They drive fans, pumps, washing machines, oil burners, small machine tools, and other devices too numerous to mention. The power rating usually lies between 60 W and 250 W (1/12 hp to 1/3 hp).

18.7 Capacitor-start motor

The capacitor-start motor is identical to a split-phase motor, except that the auxiliary winding has about as many turns as the main winding has. Furthermore, a capacitor and a centrifugal switch are connected in series with the auxiliary winding (Fig. 18-12a).

The capacitor is chosen so that I_a leads I_s by about 80°, which is considerably more than the 25° found in a split-phase motor. Consequently, for equal starting torques, the current in the auxiliary winding is only about half that in a split-phase motor. It follows that the auxiliary winding of a capacitor motor heats up less quickly. Furthermore, the locked-rotor line current I_L is smaller, being typically 4 to 5 times the rated full-load current.

Owing to the high starting torque and the relatively low value of I_a the capacitor-start motor is well suited to applications involving either frequent or prolonged starting periods. Although the *starting* characteristics of this motor are better than those of a split-phase motor, both machines possess the same characteristics under load. The reason is that the main windings are identical and the auxiliary winding is no longer in the circuit when the motor has come up to speed.

The wide use of capacitor-start motors is a direct result of the availability of small, reliable, low cost electrolytic capacitors. Prior to the development of these capacitors, repulsion-

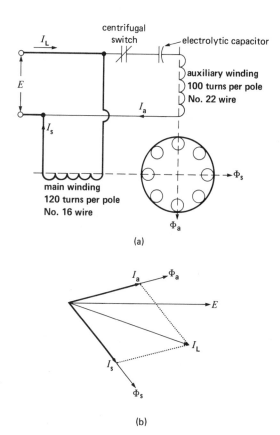

Figure 18-12
a. Capacitor-start motor.
b. Corresponding phasor diagram.

induction motors had to be installed whenever a high starting torque was required. Repulsion-induction motors possess a special commutator and brushes that require considerable maintenance. Most motor manufacturers have stopped making them.

Capacitor-start motors are used when a high starting torque is required. They are built in sizes ranging from 120 W to 7.5 kW (~1/6 hp to 10 hp). Typical loads are compressors, large fans, pumps, and high-inertia loads.

Table 18A gives the properties of a capacitor-start motor having a rating of 250 W(1/3 hp), 1760 r/min, 115 V, 60 Hz. Figure 18-13 shows the torque-speed curve for

Figure 18-13
Torque-speed curves of a capacitor-start motor, rated 1/3 hp (250 W), 1760 r/min, 115 V, 60 Hz, class A insulation.

TABLE 18A

CHARACTERISTICS OF A CAPACITOR-START MOTOR
RATING: 250 W, 1760 r/min, 115 V, 60 Hz, INSULATION CLASS 105°C

full load		no load	
voltage	— 115 V	voltage	— 115 V
power	— 250 W	current	— 4.0 A
current	— 5.3 A	losses	— 105 W
P.F.	— 64 %	**locked rotor**	
efficiency	— 63.9 %	voltage	— 115 V
speed	— 1760 r/min	current I_s	— 23 A
torque	— 1.35 N·m	current I_a	— 19 A
break down		current I_L	— 29 A
torque	— 3.4 N·m	torque	— 6 N·m
speed	— 1600 r/min	capacitor	— 320 μF
current	— 13 A		

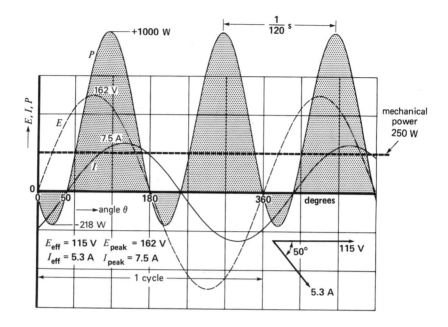

Figure 18-14
The instantaneous power absorbed by a single-phase motor varies between + 1000 W and − 218 W. The power output is constant at 250 W; consequently, vibrations are produced.

the same machine. Note that during the acceleration phase (0 to 1370 r/min), the main and auxiliary windings together produce a very high starting torque. When the rotor reaches 1370 r/min, the centrifugal switch snaps open, causing the motor to operate along the torque-speed curve of the main winding. The torque suddenly drops from 9.5 N·m to 2.8 N·m, but the motor continues to accelerate until it reaches 1760 r/min, the rated full-load speed.

18.8 Efficiency and power factor of single-phase induction motors

The efficiency and power factor of fractional horsepower single-phase motors are usually low. Thus, at full load, a 186 W motor (1/4 hp) has an efficiency and power factor of about 60 %. The low power factor is mainly due to the large magnetizing current, which ranges between 70% and 90% of full-load current. Consequently, even at no-load, these motors

reach temperatures close to full-load temperature.

The relatively low efficiency and power factor of these motors is a consequence of their fractional horsepower ratings. Integral horsepower single-phase motors can have efficiencies and power factors above 80 percent.

18.9 Vibration of single-phase motors

If we touch the stator of a single-phase motor, we note that it vibrates rapidly, whether it operates at full-load or no-load. These vibrations do not exist in 2-phase or 3-phase motors; consequently, single-phase motors are more noisy.

What causes this vibration? It is due to the fact that a single-phase motor always receives *pulsating* electric power whereas it delivers *constant* mechanical power. Consider the 250 W motor having the properties given in Table 18A. The full-load current is 5.3 A and it lags 50° behind the line voltage. If we draw the waveshapes of voltage and current, we can plot the instantaneous power P supplied to the motor (Fig. 18-14). We find that P oscillates between +1000 W and −218 W. When the power is positive, the motor receives energy from the line. Conversely, when it is negative, the motor returns energy to the line. However, whether the instantaneous electric power is positive, negative, or zero, the mechanical power delivered is a steady 250 W.

The motor will slow down during the brief periods when the electric power it receives is less than 250 W. On the other hand, it will accelerate whenever the electric power exceeds the mechanical output plus the losses. The acceleration intervals coincide with the positive peaks of the power curve. Similarly, the deceleration intervals coincide with the negative peaks. Consequently, the acceleration/deceleration intervals occur twice per cycle, or 120 times per second on a 60 Hz system. As a result, both the stator and rotor vibrate at twice the line frequency.

The stator vibrations are transmitted to the mounting base which, in turn, generates additional vibration and noise. To eliminate the

problem, the motor is often cradled in a resilient mounting (Fig. 18-15). It consists of two soft rubber rings placed between each end bell and a supporting metal bracket. Because the rotor also vibrates, a tubular rubber isolator is sometimes placed between the shaft and the mechanical load, particularly when the load is a fan.

Two-phase and 3-phase motors do not vibrate because the total instantaneous power they receive from *all* the phases is constant (see Sec. 8.7).

18.10 Capacitor-run motor

The capacitor-run motor is essentially a 2-phase motor that receives its power from a single-phase source. It has two windings, one of which is directly connected to the source. The other winding is also connected to the source, but in series with a paper capacitor (Fig. 18-16). The capacitor-fed winding has a large number of turns of relatively small wire, compared to the directly connected winding.

This particularly quiet motor is used to drive fixed loads in hospitals, studios, and other places where silence is important. It has a high power factor on account of the capacitor and no centrifugal switch is required. However, the starting torque is low.

The motor acts as a true 2-phase motor only when it operates at full load (Fig. 18-16b). Under these conditions, fluxes Φ_a and Φ_s created by the two windings are equal and out of phase by 90°. The motor is then vibration-free. Capacitor-run motors are usually rated below 500 W.

18.11 Reversing the direction of rotation

In order to reverse the direction of rotation of the motors we have discussed so far, we have to interchange the leads of either the auxiliary winding or the main winding.

However, if a single-phase motor is equipped with a centrifugal switch, its rotation cannot be reversed while the motor is running. If the main winding leads are interchanged the

Figure 18-15
Capacitor-start motor supported in a resilient mount (*Brook Crompton Parkinson*)

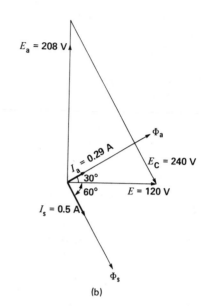

Figure 18-16
a. Capacitor-run motor having a NEMA rating of 30 millihorsepower.
b. Corresponding phasor diagram at full load.

motor will continue to turn in the same direction.

In the case of a capacitor-run motor (Fig. 18-16) the direction of rotation *can* be changed while the motor is running because both windings are in the circuit at all times. In the case of very small motors the rotation can be reversed by using a double-throw switch as shown in Fig. 18-17. In such a motor, the main and auxiliary windings are identical. When the switch is in position 1, winding A is directly across the line, while winding B is in series with the capacitor. With this connection, the motor turns clockwise. When the switch is thrown to position 2, the role of the windings is reversed and the motor will come to a halt and then run up to speed in the opposite direction.

18.12 Shaded-pole motor

The shaded-pole motor is very popular for ratings below 0.05 hp (~40 W) because of its extremely simple construction (Fig. 18-18). It is basically a small squirrel-cage motor in which the auxiliary winding is composed of a copper

Figure 18-17
Reversible single-phase motor using a two-pole switch and capacitor.

ring surrounding a portion of each pole.

The main winding is a simple coil connected to the ac source. The coil produces a total flux Φ that may be considered to be made up of three components Φ_1, Φ_2 and Φ_3, all in phase. Flux Φ_1 links the short-circuited ring on the left-hand pole, inducing a rather large current I_a. This current produces a flux Φ_a that lags behind Φ_1. Consequently, Φ_a also lags behind Φ_2 and Φ_3. The combined action of $(\Phi_2 + \Phi_3)$ and Φ_a produces a weak revolving field, which starts the motor. The direction of rotation is from the "unshaded" side to the "shaded" (ring side) of the pole. A similar torque is set up by the pole on the right. Flux Φ_2 induces a current I_b in the ring, and the resulting flux Φ_b lags behind Φ_2. As before, the combined action of $(\Phi_1 + \Phi_3)$ and Φ_b produces a weak revolving field that drives the rotor clockwise.

Although the starting torque, efficiency, and power factor are very low, the simple construction and absence of a centrifugal switch give this motor a marked advantage in low-power applications. The direction of rotation cannot be changed, because it is fixed by the position of the copper rings. Table 18B gives the typical properties of a shaded-pole motor having a rated output of 6 W.

Example 18-3:
Calculate the full-load efficiency and slip of the shaded-pole motor whose properties are listed in Table 18B.

Solution:
The efficiency is:
a. $\eta = (P_o/P_i) \times 1000$ Eq. 3-6
    $= (6/21) \times 100$
    $= 28.6\%$
  $s = (n_s - n)/n_s$
    $= (3600 - 2900)/3600$
    $= 0.194 = 19.4\%$

18.13 Universal motor

The single-phase universal motor is very simi-

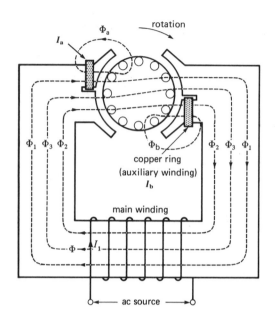

Figure 18-18a
Fluxes in a shaded-pole motor.

Figure 18-18b
Shaded-pole motor rated at 5 millihorsepower, 115 V, 60 Hz, 2900 r/min. (*Gould*)

TABLE 18B

Properties of a shaded-pole motor, rated 6 W, 115 V, 60 Hz.

No load

current	0.26 A
input power	15 W
speed	3550 r/min

Locked rotor

current	0.35 A
input power	24 W
torque	10 mN·m

Full load

current	0.33 A
input power	21 W
speed	2900 r/min
torque	19 mN·m
mechanical power	6 W
breakdown speed	2600 r/min
breakdown torque	21 mN·m

Figure 18-19
Alternating current series motor, also called universal motor.

lar to a dc series motor (Sec. 5.8). The basic construction of a small universal motor is shown in Fig. 18-19. The entire magnetic circuit is laminated to reduce eddy current losses. Such a motor can operate on either ac or dc, and the resulting torque-speed is about the same, in each case. That is why it is called a *universal* motor.

When the motor is connected to an ac source, the ac current flows through the armature and the series field. The field produces an ac flux Φ that reacts with the current flowing in the armature to produce a torque. Because the armature current and the flux reverse simultaneously, the torque always acts in the same direction. No revolving field is produced in this type of machine; the principle of operation is the same as that of a dc series motor and it possesses the same basic characteristics.

The main advantage of fractional horsepower universal motors is their high speed and high starting torque. They can therefore be used to drive high-speed centrifugal blowers in vacuum cleaners. The high speed and corresponding small size for a given power output is

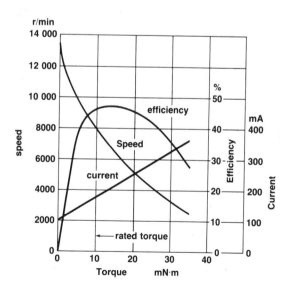

Figure 18-20
Characteristics of a small 115 V, 60 Hz universal motor having a full-load rating of 1/100 hp at 8000 r/min.

also an advantage in driving portable tools, such as electric saws and drills. No-load speeds as high as 5000 to 15 000 r/min are possible, but, as in any series motor, the speed drops rapidly with increasing load.

Series motors are built in many different sizes, starting from small toy motors to very large traction motors used in some electric locomotives.

Figure 18-20 gives the ac performance curves of a 115 V, 8000 r/min, universal motor rated at 1/100 hp. The full-load current is 175 mA.

18.14 Hysteresis motor

To understand the operating principle of a hysteresis motor, let us first consider Fig. 18-21. It shows a stationary rotor surrounded by a pair of N, S poles that can be rotated mechanically in a clockwise direction. The rotor is composed of a ceramic material of high coercive force. Thus, it is a permanent magnet material whose resistivity approaches that of an insulator. Consequently, it is impossible to set up eddy currents in such a rotor.

As the N, S field rotates, it magnetizes the rotor; consequently, poles of opposite polarity are continuously produced under the moving N, S poles. In effect, the revolving field is continuously reorienting the magnetic domains in the

rotor. Clearly, the domains go through a complete cycle (or hysteresis loop) each time the field makes one complete revolution. Hysteresis losses are therefore produced in the rotor, proportional to the area of the hysteresis loop (Sec. 2.26). These losses are dissipated as heat in the rotor.

Let us assume that the hysteresis loss per revolution is E_h joules and that the field rotates at n r/min. The energy dissipated in the rotor per minute is:

$$W = nE_h$$

The corresponding power (dissipated as heat) is:

$$\begin{aligned} P_h &= W/t && \text{Eq. 3-4} \\ &= nE_h/60 \ [W] \end{aligned}$$

However, the power dissipated in the rotor can only come from the mechanical power used to drive the N,S poles. This power is given by:

$$P = nT/9.55 \qquad \text{Eq. 3-5}$$

Because $P = P_h$, we have:

$$nT/9.55 = nE_h/60$$

whence

$$T = E_h/6.28 \qquad (18\text{-}2)$$

where

$T =$ torque exerted on the rotor [N·m]
$E_h =$ hysteresis energy dissipated in the rotor, per turn [J/r]
$6.28 =$ constant (exact value $= 2\pi$]

Equation 18-2 brings out the remarkable feature that the torque needed to drive the magnets (Fig. 18-21) is constant, irrespective of the speed of rotation. In other words, whether the poles just barely creep around the rotor or whether they move at high speed, the torque exerted on the rotor is always the same. It is this basic property that distinguishes hysteresis motors from all other motors.

In practice, the revolving field is produced by a 3-phase stator, or by a single-phase stator

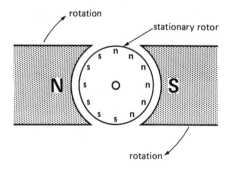

Figure 18-21
Permanent magnet rotor and a mechanically-driven revolving field.

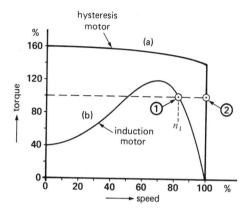

Figure 18-22
Typical torque-speed curves of two capacitor-run motors:
a. hysteresis motor; b. induction motor

Figure 18-23
Single-phase hysteresis clock motor having 32 poles and a ferrite rotor.

having an auxiliary winding. When a hysteresis rotor is placed inside such a stator, it immediately accelerates up to synchronous speed. The accelerating torque is essentially constant as shown by Fig. 18-22. This is entirely different from a squirrel-cage induction motor, whose torque falls toward zero as it approaches synchronous speed.

Thanks to the fixed frequency of large distribution systems, the hysteresis motor is employed in electric clocks, and other precise timing devices (Fig. 18-23).

It is also used to drive tape-decks, turntables and other precision audio equipment. In such devices, the constant speed is, of course, the feature we are looking for. However, the hysteresis motor is particularly well suited to drive such devices because of their high inertia. Inertia prevents many synchronous motors (such as reluctance motors)from coming up to speed because to reach synchronism, they have to *suddenly* lock with the revolving field. No such abrupt transition occurs in the hysteresis motor because it develops a constant torque right up to synchronous speed.

In turntable audio equipment, these salient features are further enhanced by designing

the motor to function as a vibration-free capacitor-run motor.

Example 18-4:
A small 60 Hz hysteresis clock motor possesses 32 poles. In making one complete turn with respect to the revolving field, the hysteresis loss in the rotor amounts to 0.8 J. Calculate:

a. the pull-in and pull-out torques,
b. the maximum power output before the motor stalls,
c. the rotor losses when the motor is stalled,
d. the rotor losses when the motor runs at synchronous speed.

Solution:
a. The pull-in and pull-out torques are about equal in a hysteresis motor:

$$T = E_h/6.28 = 0.8/6.28 \quad \text{Eq. 18-2}$$
$$= 0.127 \ \text{N·m}$$

b. The synchronous speed is:

$$n_s = 120f/p = 120 \times 60/32$$
$$= 225 \ \text{r/min}$$

The maximum power is:

$$P = nT/9.55 = (225 \times 0.127)/9.55$$
$$= 3W \ (\text{or } 3/746 = 1/250 \ \text{hp})$$

c. When the motor stalls, the rotating field moves at 225 r/min with respect to the rotor. The energy loss per minute is therefore:

$$W = 225 \times 0.8 = 180 \text{ J}$$

The power dissipated in the rotor is:

$$P = W/t = 180/60 = 3 \text{ W}$$

d. There is no energy loss in the rotor when the motor runs at synchronous speed because the magnetic domains no longer reverse.

18.15 Synchronous reluctance motor

We can build a synchronous motor by milling out a standard squirrel-cage rotor so as to create a number of salient poles. The number of poles must be equal to the number of poles on the stator. Figure 18-24 shows a rotor milled out to create four salient poles.

Figure 18-24
Rotor of a synchronous reluctance motor.

Such a *reluctance motor* starts up as a standard squirrel cage motor but, when it approaches synchronous speed, the salient poles lock with the revolving field, and so the motor runs at synchronous speed. Both the pull-in and pull-out torques are weak, compared to those of a hysteresis motor of equal size. Furthermore, reluctance motors cannot accelerate high-inertia loads to synchronous speed. The reason can be seen by referring to Fig. 18-22. Suppose the motor has reached a speed n_1 corresponding to full-load torque (operating point 1).

The stator poles are slipping past the rotor poles at a rate that corresponds to the slip. If the rotor is to lock with the revolving field, it must do so in the time it takes for one stator pole to sweep past a rotor pole. If pull-in is not achieved during this interval (Δt), it will never be achieved. The problem is that in going from speed n_1 to synchronous speed, the kinetic energy of the revolving parts must increase by an amount given by Eq. 3-8:

$$\Delta E_k = 5.48 \times 10^{-3} J(n_s^2 - n_1^2) \quad (18\text{-}3)$$

where J is the moment of inertia.

Furthermore, the interval is given by:

$$\Delta t = 60/(n_s - n_1)p \quad (18\text{-}4)$$

Consequently, to reach synchronous speed, the motor must develop an accelerating power P_a of at least:

$$
\begin{aligned}
P_a &= \Delta E_k/\Delta t \quad (18\text{-}5)\\
&= 0.9 \times 10^{-4} n_s (n_s - n_1)^2 J p \text{ (approx)}
\end{aligned}
$$

Furthermore, the motor must continue to supply the power P_L demanded by the load. If the sum of $P_a + P_L$ exceeds the power capacity of the motor, it will never pull into step. In essence, a reluctance motor can only synchronize when the slip ($n_s - n_1$) is small and the moment of inertia J is low.

Despite this drawback, the reluctance motor is cheaper than any other type of synchronous motor. It is particularly well adapted to variable-frequency electronic speed control. Inertia is then no problem because the speed of the revolving field always tracks with the speed of the rotor. Three-phase reluctance motors of several hundred horsepower have been built, using this approach.

18.16 Synchro drive

In some remote-control systems, we may have to move the position of a small rheostat that is one or two meters away. This problem is easily solved by using a flexible shaft. But if the rheostat is 100 m away, the flexible-shaft solution becomes impractical. We then employ an electrical "shaft" to tie the knob and rheostat together. How does such a shaft work?

Consider two conventional wound-rotor induction motors whose 3-phase stators are connected in parallel (Fig. 18-25). Two phases of the respective rotors are also connected in parallel, and energized from a *single-phase* source. The remarkable feature about this arrangement is that the rotor on one machine will automatically track with the rotor on the other. Thus, if we turn rotor A clockwise through 17 degrees, rotor B will move clockwise through 17°. Obviously, such a system enables us to control a rheostat from a remote location.

Two miniature wound-rotor motors are required. One (the transmitter) is coupled to the control knob, and the other (the receiver) is coupled to the rheostat. The 5-conductor cable (conductors a-b-c-1-2) linking the

transmitter and receiver constitutes the flexible electrical shaft.

The behavior of this so-called *selsyn* or *synchro* control system is explained as follows. Assume the transmitter and receiver are identical, and that the rotors are in identical positions. When the rotors are excited, they behave like the primaries of two transformers, inducing voltages in the respective stator windings. The voltages induced in the three stator windings of the transmitter are always unequal because the windings are displaced by 120°. The same is true for the voltages induced in the stator of the receiver.

Nevertheless, no matter what the respective stator voltages of the transmitter and receiver may be, they are identical in both synchros (phase by phase) when the rotors occupy the same position. The stator voltages balance each other and, consequently, no current flows in the lines connecting the stators. The rotors, however, carry a small exciting current I_o.

Now if we turn the rotor of the transmitter, its three stator voltages will change. They will no longer balance the stator voltages of the receiver; consequently, currents I_a, I_b, I_c will

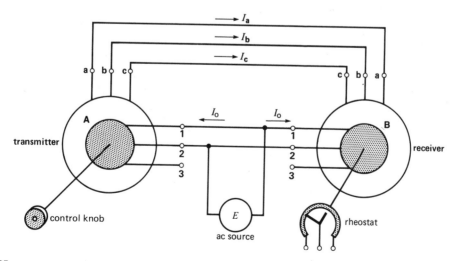

Figure 18-25
Components and connections of a synchro system.

flow in the lines connecting the two devices. These currents produce a torque on both rotors tending to line them up. Since the rotor of the receiver is free to move, it will line up with the transmitter. As soon as the rotors are aligned, the stator voltages are again in balance (phase by phase), and the torque-producing currents disappear.

Synchros are often employed to indicate the position of an antenna, a valve, a gun turret, and so on, with the result that the torque requirements are small. Such transmitters and receivers are built with watch-like precision to ensure they will track with as little error as possible.

QUESTIONS AND PROBLEMS

Practical level

18-1 A 6-pole single-phase motor is connected to a 60 Hz source. What is its synchronous speed?

18-2 What is the purpose of the auxiliary winding in a single-phase induction motor? How can we change the rotation of such a motor?

18-3 State the main difference between a split-phase motor and a capacitor-start motor. What are their relative advantages?

18-4 Explain briefly how a shaded-pole motor operates.

18-5 List some of the properties and advantages of a universal motor.

18-6 Why are some single-phase motors equipped with a resilient mounting? Is such a mounting necessary on 3-phase motors?

18-7 What is the main advantage of a capacitor-run motor?

18-8 Which of the motors discussed in this chapter is best suited to drive the following loads:
   a. a small portable drill;
   b. a 3/4 hp air compressor;
   c. a vacuum cleaner;
   d. a 1/100 hp blower;
   e. a 1/3 hp centrifugal pump;
   f. a 1/4 hp fan for use in a hospital ward;
   g. an electric timer;
   h. a hi-fi turn-table.

Intermediate level

18-9 Referring to Fig. 18-11, the effective impedance of the main and auxiliary windings under locked-rotor conditions are given as follows:

	Effective resistance	Effective reactance
Main winding	4 Ω	7.5 Ω
Auxiliary winding	7.5 Ω	4 Ω

If the line voltage is 119 V, calculate:
   a. the magnitude of I_a and I_s;
   b. the phase angle between I_a and I_s;
   c. the line current I_L;
   d. the power factor under locked-rotor conditions.

18-10 The palm of the human hand can just barely tolerate a temperature of 130°F. If the no-load temperature of the frame of a 1/4 hp motor is 64°C in an ambient temperature of 76°F;
   a. can a person keep his hand on the frame?
   b. is the motor running too hot?

18-11 Referring to Fig. 18-13, if the motor is connected to a load whose torque is constant at 4 N·m, explain the resulting behavior of the motor when it is switched on the line.

18-12 a. A single-phase motor vibrates at a frequency of 100 Hz. What is the frequency of the power line?

b. A capacitor-run motor does not have to be set in a resilient mounting. Why?

c. A 4-pole, 60 Hz single-phase hysteresis motor develops a torque of 6 in·lb when running at 1600 r/min. Calculate the hysteresis loss per revolution [J].

18-13 Referring to the 6 W shaded-pole motor in Table 18B, calculate:
a. the rated power output in milli-horsepower;
b. the full-load power factor;
c. the slip at the breakdown torque;
d. the per unit no-load current and locked-rotor current.

18-14 Referring again to Fig. 18-13, calculate:
a. the locked-rotor torque [ft·lbf];
b. the per-unit value of the LR torque;
c. the starting torque when only the main winding is excited;
d. the per-unit breakdown torque;
e. how are the torque-speed curves affected if the line voltage falls from 115 V to 100 V?

Advanced level

18-15 In Table 18A, calculate:
a. the voltage across the capacitor under locked-rotor conditions;
b. the corresponding phase angle between I_s and I_a.

18-16 Referring to Fig. 18-16, if the capacitor-run motor operates at full load, calculate:
a. the line current I_L;
b. the power factor of the motor;
c. the active power absorbed by each winding;
d. the efficiency of the motor.

18-17 The motor described in Table 18A, has a LR power factor of 0.9 lagging. It is installed in a workshop situated 600 ft from a home, where the main service entrance is located. The line is composed of a 2-conductor cable made of No. 12 gauge copper. The ambient temperature is 25°C and the service entrance voltage is 122 V. Using Table AX3 in the Appendix, calculate:
a. the resistance of the transmission line;
b. the starting current and the voltage at the motor terminals;
c. the starting torque [N·m].

19

STEPPER
MOTORS

Stepper motors are special motors that are used when motion and position have to be precisely controlled. As their name implies, stepper motors rotate in discrete steps, each step corresponding to a pulse that is supplied to one of its stator windings. Depending on its design, a stepper motor can advance by 90°, 45°, 18° or by as little as a fraction of a degree per pulse. By varying the pulse rate, the motor can be made to advance very slowly, one step at a time, or to rotate step-wise at speeds as high as 4000 r/min.

Stepper motors can turn clockwise or counterclockwise, depending upon the sequence of the pulses that are applied to the windings.

The behavior of a stepper motor depends greatly upon the power supply that drives it. The power supply generates the pulses which, in turn, are usually initiated by a microprocessor or computer. The pulses are counted and stored by the computer, clockwise (cw) pulses being (+) while counterclockwise (ccw) pulses are (−). As a

result, the net number of steps is known *exactly* at all times. It follows that the number of revolutions is always precisely known, to an accuracy of one step. This permits the motor to be used as a precise positioning device in machine-tools, X-Y plotters, typewriters, tape-decks, valves and printers.

In this chapter we will cover the operating principle of the more common stepper motors, together with their properties and limitations. We shall also discuss the types of drives used to actuate these machines.

19.1 Elementary stepper motor

A very simple stepper motor is shown in Figure 19-1. It consists of a stator having three salient poles and a 2-pole rotor made of soft iron. The windings can be successively connected to a dc power supply by means of three switches A, B, C.

When the switches are open, the rotor can take up any position. However, if switch A is closed, the resulting magnetic field created by

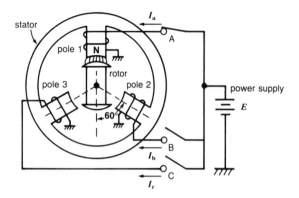

Figure 19-1
Simple stepper motor in which each step moves the rotor by 60°.

pole 1 will attract the rotor and so it will line up as shown. If we now open switch A and simultaneously close switch B, the rotor will line up with pole 2. In so doing, it will rotate ccw by 60°. Next, if we open switch B and simultaneously close switch C, the rotor will turn ccw by an additional 60°, this time lining up with pole 3.

Clearly, we can make the rotor advance ccw in 60° steps by closing and opening the switches in the sequence A, B, C, A, B, C, ... Furthermore, we can reverse the rotation by operating the switches in the reverse sequence A, C, B, A, C, B.... In order to fix the final position of the rotor, the last switch that was closed in a switching sequence must remain closed. This holds the rotor in its last position and prevents it from moving under the influence of external forces. In this stationary state, the motor will remain "locked" provided the external torque does not exceed the so-called holding torque of the motor.

In moving from one position to the next, the motion of the rotor will be influenced by the inertia and the frictional forces that come into play. We now examine the nature of these forces.

19.2 Effect of inertia

Suppose the motor operates at no-load and that the rotor has a low inertia and a small amount of bearing friction. It is initially facing

pole 1. Let this correspond to the zero degree (0°) angular position. At the moment switch A opens and switch B closes, the rotor will start accelerating ccw towards pole 2. It rapidly picks up speed and soon reaches the center line of pole 2 where it should come to rest. However, the rotor is now moving with considerable speed and so it will overshoot the center line. As it does so, the magnetic field of pole 2 will pull it in the opposite direction, thereby braking the rotor. The rotor will come to a halt and start moving in the opposite (cw) direction. Picking up speed, it will again overshoot the center line of pole 2, whereupon the magnetic field will exert a pull in the ccw direction.

The rotor will therefore oscillate like a pendulum around the center line of pole 2. The oscillations will gradually die out, because of bearing friction. Figure 19-2 shows the angular position of the rotor as a function of time. The rotor starts at 0° and reaches 60° (center line of pole 2) after 2 ms. It overshoots the center line by 30° before coming to a halt (at 3 ms). The rotor now moves in reverse, and again crosses the center line at t = 4 ms.

The oscillations continue this way, gradually diminishing in amplitude until the rotor comes to rest at t > 10 ms.

The reader will note that we have also drawn the instantaneous speed of the rotor as a function of time. The speed can be given in revolutions per second, but for stepper motors it is more meaningful to speak of degrees per second. The speed is momentarily zero at t = 3 ms, 5 ms, 7 ms and permanently zero at t > 10 ms. The speed is greatest whenever the rotor crosses the center line of pole 2. Clearly, the oscillations last a relatively long time before the rotor settles down.

Without making any other changes, suppose we increase the inertia of the rotor by mounting a flywheel on the shaft. We discover that both the period and the amplitude of the oscillations increase when the inertia increases. In Figure 19-3, for example, the time to reach the

Figure 19-2
In moving from pole 1 to pole 2, the rotor oscillates around its 60-degree position before coming to rest. The speed is zero whenever the rotor reaches the limit of its overshoot.

Figure 19-3
Same conditions as in Figure 19-2 except that the inertia is greater. The overshoot is greater and the rotor takes longer to settle down.

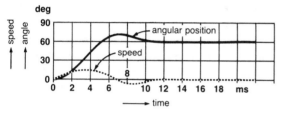

Figure 19-4.
Same conditions as in Figure 19-3 except that viscous damping has been added.

Figure 19-5
Same conditions as in Figure 19-2 except that the rotor is coupled to a mechanical load.

60° position has increased from 2 ms to 4 ms. Furthermore, the amplitude of the oscillations has increased. The rotor also takes a longer time to settle down (18 ms instead of 10 ms).

The oscillations can be damped by increasing the friction. For example, if the bearing friction is raised sufficiently, the oscillations shown in Figure 19-3 can be suppressed so as to give only a single overshoot, shown in Figure 19-4. In practice, the damping is accomplished by using an eddy-current brake or a viscous damper. A viscous damper uses a fluid such as oil or air to brake the rotor whenever it is moving. Viscous damping means that the braking effect is proportional to speed; it is therefore zero when the rotor is at rest.

19.3 Effect of a mechanical load

Let us return to the condition shown in Figure 19-2 where the rotor has low inertia and only a small amount of viscous damping due to bearing friction. If the rotor is coupled to a mechanical load while it is moving, the effect is shown in Figure 19-5. As we would expect, it takes longer for the motor to attain the 60° position (compare 2 ms in Figure 19-2 with 4 ms in Figure 19-5). Furthermore, the overshoot is smaller and the oscillations are damped more quickly.

In summary, both the mechanical load and the inertia increase the stepping time. The oscillations also prolong the time before the rotor settles down. Therefore, in order to obtain fast stepping response, the inertia of the rotor (and its load) should be as small as possible and the oscillations should be suppressed by using a viscous damper.

The time to move from one position to the next can also be reduced by increasing the current in the winding. However, thermal limitations due to I^2R losses dictate the maximum current that can be used.

Returning to Fig. 19-1, let us excite the windings in succession so that the motor rotates. Figure 19-6 shows the current pulses I_a, I_b, I_c and the instantaneous position of the rotor (as well as its speed) when the motor makes one half a revolution. We assume the stepper motor has some inertia and that it is driving a mechanical load. Note that the speed of the rotor is zero at the beginning and at the end of each pulse. In this figure, the pulses have a duration of 8 ms. Consequently, the stepping rate is 1000/8 = 125 steps per second. One revolution requires 6 steps and so it takes 6 ÷ 125 = 0.048 s to complete one turn. The *average* speed is therefore 60/0.048 = 1250 revolutions per minute. However, the stepper motor rotates in start-stop jumps and not smoothly as an ordinary motor would.

19.4 Torque versus current

As mentioned previously, the torque developed by a stepper motor depends upon the current. Figure 19-7 shows the relationship between the two for a typical stepping motor. When the current is 8 A, the motor develops a torque of 3 N·m. This is the torque that the motor can exert in moving from one position to the next and so it is called the pull-over torque.

When the motor is at rest, a holding current must continue to flow in the last winding that was excited, so that the rotor remains locked in place.

19.5 Start-stop stepping rate

When the stepping motor inches along in the start-stop fashion shown in Figure 19-6, there is an upper limit to the permissible stepping rate. If the pulse rate of the current in the windings is too fast, the rotor is unable to accurately follow the pulses, and steps will be lost. This defeats the whole purpose of the motor which is to correlate its instantaneous position (steps) with the number of net (+ and −) pulses. In order to maintain synchronism, the rotor must settle down before advancing to the next position. Referring to Figure 19-6, this means that the interval between successive

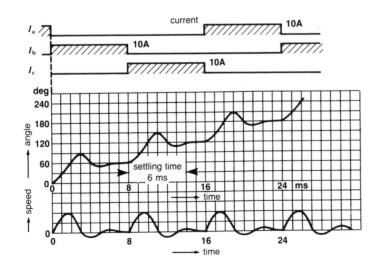

Figure 19-6
Graph of current pulses, angular position and instantaneous speed of rotor during the first 4 steps. Three steps (24 ms) produce one half revolution.

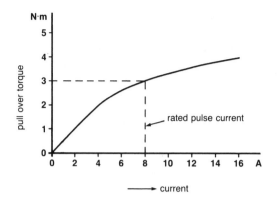

Figure 19-7
Graph of pull over torque versus current of a stepper motor; diameter: 3.4 inches; length: 3.7 in; weight: 5.2 lbm.

steps must be at least 6 ms, which corresponds to a maximum of 1000/6 = 167 steps per second (sps).

Bearing in mind what was said in Sec. 19.2, it is clear that the maximum number of steps per second depends upon the load torque and the inertia of the system. The higher the load and the greater the inertia, the lower will be the allowable number of steps per second.

The start-stop stepping mode is sometimes referred to as the "start-without-error" mode. A start-without-error characteristic is shown by dotted curve 1 in Figure 19-8. It shows that if the load torque is, say, 1.4 N·m, the maximum possible stepping rate, without losing count, is 500 steps per second.

But if the same load also has a high inertia, the permissible start-stop rate drops to about 400 steps per second for the same load torque (curve 2).

19.6 Slew speed

A stepper motor can be made to run at uniform speed without starting and stopping at every step. When the motor runs this way it is said to be *slewing*. Because the motor runs essentially at uniform speed, the inertia effect is absent. Consequently, for a given stepping rate, the

motor can carry a greater load torque when it is slewing. The solid curve 3 in Figure 19-8 shows the relationship between the load torque and the steps per second when the motor is slewing. For example, the motor can develop a torque of 2.2 N.m when it runs at 500 steps per second. However, if the load torque should exceed 2.2 N.m when the stepping rate is 500 sps, the motor will fall out of step and the position (steps) of the rotor will no longer correspond to the net number of pulses provided to its windings.

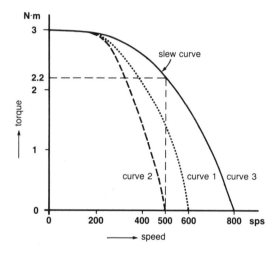

Figure 19-8
Start-stop and slewing characteristic of a typical stepper motor. Each step corresponds to an advance of 1.8 degrees.
curve 1: Start-stop curve with only stepper motor inertia
curve 2: same conditions as curve 1 but with an additional load inertia of 2 kg·cm² curve 3: slewing curve

Figure 19-9 shows the difference between the start-stop mode and slewing. Suppose the motor is turning at an average speed of 250 steps per second in both cases. The motor will cover the same number of steps per second, namely 1 step every 4 ms. However, the angle (position) increases smoothly with time when the motor is slewing and this is shown by the uniform slope of line 0A (Fig. 19-9a). The corresponding slew speed is constant (Fig. 19-9b).

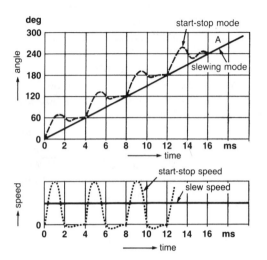

Figure 19-9

a. Angular position versus time curve when the stepper motor operates in the start-stop mode and the slewing mode. Stepping rate is the same in both cases.

b. Instantaneous speed versus time curve when the stepper motor operates in the start-stop and the slewing mode.

On the other hand, the angle increases stepwise in the start-stop mode. Consequently, the speed continually varies between a maximum and zero and its *average* value is equal to the slew speed (Fig. 19-9b).

19.7 Ramping

When a motor is carrying a load, it cannot suddenly go from zero to a stepping rate of say, 5000 sps. In the same way, a motor that is slewing at 5000 sps cannot be brought to a dead stop in one step. Thus, to bring a motor up to speed it must be accelerated gradually. Similarly, to stop a motor that is running at high speed, it must be decelerated gradually — always subject to the condition that the instantaneous position of the rotor must correspond to the number of pulses. The process whereby a motor is accelerated and decelerated is called *ramping*. During the acceleration phase, ramping consists of a progressive increase in the number of driving pulses per second.

The ramping phase is usually completed in a fraction of a second. The ramp is generated by the power supply that drives the stepper motor. Furthermore, it is programmed to retain precise position control over the motor and its load.

19.8 Types of stepper motors

There are 3 main types of stepper motors:

- variable reluctance stepper motors
- permanent magnet stepper motors
- hybrid stepper motors

Variable reluctance stepper motors are based upon the principle illustrated in Figure 19-1. However, the stator often has four main poles (instead of 3) and the pole-faces are slotted so as to produce a number of teeth. The rotor is also toothed, each tooth corresponding to a salient pole. The number of teeth (poles) on the rotor and stator determines the angular motion per step. Steps of 18°, 15°, 7.5°, 5° and 1.8° are common.

Permanent magnet stepper motors are similar to variable reluctance motors except that the rotor has N and S poles. Figure 19-10 shows a permanent-magnet motor having 4 stator poles and 6 rotor poles, the latter being permanent magnets. Owing to the permanent magnets, the rotor remains lined up with the last pair of stator poles that were excited by the driver. In effect, the motor develops a so-called *detent torque* which keeps the rotor in place even when no current flows in the stator windings.

Coils A1, A2 are connected in series, as are coils B1, B2. Starting from the position shown, if coils B are excited, the rotor will move through an angle of 30°. However, the direction of rotation depends upon the direction of current flow. Thus, if the current in coils B produces N and S poles as shown in Figure 19-10, the rotor will turn ccw. Stepper motors that have to develop considerable power are usually equipped with permanent magnets.

Figure 19-10
Permanent-magnet stepper motor that advances 30° per step.

Hybrid stepper motors have two identical soft-iron armatures mounted on the same shaft. The armatures are indexed so that the salient poles interlap. Figure 19-11a shows two 5-pole armatures that are driven by a 4-pole stator. This arrangement makes the motor look like a variable reluctance motor. However, a permanent magnet PM is sandwiched between the armatures (Fig. 19-11b). It produces an axial magnetic field, with the result that all the poles on armature 1 are N poles while those on armature 2 are S poles.

Stator coils A1, A2 are connected in series, and so are stator coils B1, B2. The motor develops a small detent torque because of the permanent magnet and so the rotor will remain in the position shown in Figure 19-11a. If we now excite coils B, the rotor will rotate by 18° and line up with stator poles B. The direction of rotation will again depend upon the direction of current flow in coils B.

Figure 19-12 shows an exploded view of a

(a)

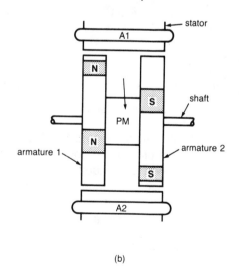

(b)

Figure 19-11a
Hybrid motor having a 4-pole stator and two 5-pole armatures mounted on the same shaft. The salient poles on the first armature are all N poles, while these on the second armature are all S poles. Each step produces an advance of 18°.

Figure 19-11b
Side view of the rotor, showing the permanent magnet PM sandwiched between the two armatures. The 4-pole stator is common to both armatures.

Figure 19-12
Exploded view of a standard hybrid stepping motor. The rotor is composed of two soft-iron armatures having 50 salient poles each. A short permanent magnet is sandwiched between the armatures. The stator has 8 poles, each of which has 5 salient poles in the pole face. Outside diameter of motor: 2.2 in; axial length 1.5 in; weight: 0.8 lbm. (*Photo courtesy of Pacific Scientific, Motor and Control Division, Rockford, IL*)

hybrid stepping motor. Figure 19-13 shows the special construction of a stator in which permanent magnets are embedded in the stator slots, in addition to the permanent magnet on the rotor.

Figure 19-14a shows another type of hybrid motor and Fig. 19-14b is a cross section view of its construction. Figures 19-14c and 19-14d respectively show the specifications and torque-speed characteristics of this motor. Note that the pull out characteristic corresponds to the slewing curve while the pull in characteristic corresponds to the start-without-error curve.

It should be noted that the number of poles on the stator of a stepper motor is never equal to the number of poles on the rotor. This feature is totally different from that in any other type of motor we have studied so far. Indeed, it is the difference in the number of poles that enables the motors to step as they do.

19.9 Motor windings and associated drives

Stepper motors use either a *bipolar* or a *unipolar* winding on the stator.

Bipolar winding. In a 4-pole stator, the bipolar winding consists of the two coil sets A1, A2 and B1, B2 shown in Figure 19-15. The current I_a in

Figure 19-13
Stator lamination details and construction of an enhanced
motor stator lamination stack assembly. Rare earth permanent magnets are fitted into the stator slots in addition to the
permanent magnet of the hybrid rotor. (*Courtesy of Pacific
Scientific, Motor and Control Division*)

Figure 19-14a
External view of a hybrid stepper motor. It is equipped with
bipolar windings rated to operate at 5 V. External diameter
of motor: 1.65 in; axial length: 0.86 in; weight 5.1 oz. (*Courtesy of AIRPAX© Corporate*)

Figure 19-14b
Cross section view of the hybrid stepper motor shown in
Fig. 19-14a. (*Courtesy of AIRPAX© Corporate*)

Specifications

Ordering Part No. (Add Suffix)	L82401		L82402	
	Unipolar		Bipolar	
Suffix Designation	– P1	– P2	– P1	– P2
DC Operating Voltage	5	12	5	12
Res. per Winding Ω	9.1	52.4	9.1	52.4
Ind. per Winding mH	7.5	46.8	14.3	77.9
Holding Torque mNm/oz-in*	73.4/10.4		87.5/12.4	
Rotor Moment of Inertia g·m²	12.5×10^{-4}			
Detent Torque mNm/oz-in	9.2/1.3			
Step Angle	7.5°			
Step Angle Tolerance*	.5°			
Steps per Rev.	48			
Max Operating Temp	100°C			
Ambient Temp Range Operating Storage	–20°C to 70°C –40°C to 85°C			
Bearing Type	Bronze sleeve			
Insulation Res. at 500Vdc	100 megohms max			
Dielectric Withstanding Voltage	650 ± 50 VRMS 60 Hz for 1 to 2 seconds			
Weight g/oz	144/5.1			
Lead Wires	26 AWG			

*Measured with 2 phases energized.

Figure 19-14c
Specifications of the hybrid stepper motor shown in Fig. 19-14a. The motor can be built for either unipolar or bipolar operation at a rated driving voltage of either 5 V or 12 V. (*Courtesy of AIRPAX© Corporate*)

UNIPOLAR

BIPOLAR

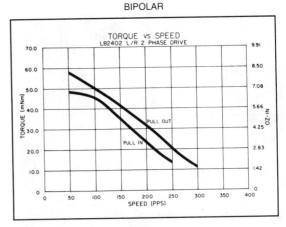

NOTE: The above curves are typical.

Figure 19-14d
Typical torque-speed characteristics of the hybrid stepper motor shown in Fig. 19-14a. The pull out curve corresponds to the slewing characteristic; the pull in curve corresponds to the start-without-error characteristic. (*Courtesy AIRPAX© Corporate*)

Figure 19-15
Schematic diagram showing how the stator coils A1, A2 and B1, B2 are connected to the common dc source by means of switches Q1 to Q8. The dc source is shown twice to simplify the connection diagram.

coil set A reverses periodically and the same is true for current I_b in coil set B. The coils are excited by a common dc source and because the current pulses I_a, I_b must alternate, a switching means is required. The switches are represented by the contacts Q1 to Q8. In practice, transistors are used as switches because they can turn the current on and off at precise instants of time.

The coils can be excited sequentially in three different ways: (1) wave drive, (2) normal drive and (3) half-step drive.

In the *wave drive* only one set of coils is excited at a time. The switching sequence for cw rotation is given in Table 19A and the resulting current pulses I_a, I_b are shown in Figure 19-16. Note that the flux produced by I_a and I_b rotates by 90° per step.

TABLE 19A Wave Switching Sequence for cw Rotation

STEP		1	2	3	4	1
Q1	Q2	on	—	—	—	on
Q3	Q4	—	—	on	—	—
Q5	Q6	—	on	—	—	—
Q7	Q8	—	—	—	on	—

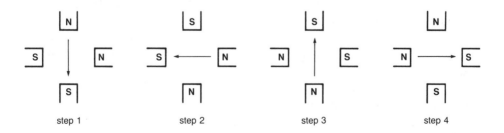

Figure 19-16
Current pulses in a wave drive and the resulting flux positions at each step. See Table 19A for switching sequence.

STEP		1	2	3	4	1
Q1	Q2	on	—	—	on	on
Q3	Q4	—	on	on	—	—
Q5	Q6	on	on	—	—	on
Q7	Q8	—	—	on	on	—

TABLE 19B Normal Switching Sequence for cw Rotation

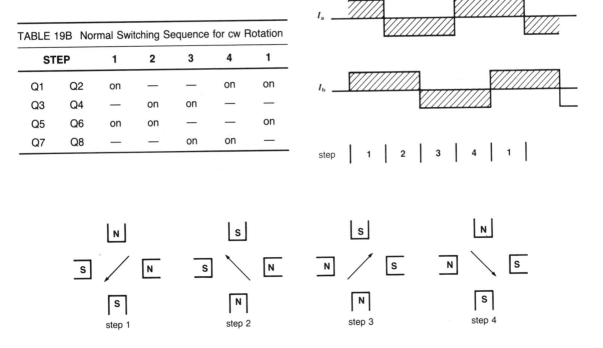

Figure 19-17
Current pulses in a normal drive and resulting flux positions at each step. See Table 19B for switching sequence.

In the *normal drive*, both sets of coils are excited at a time. The switching sequence for cw rotation is given in Table 19B and the resulting current pulses I_a, I_b are shown in Figure 19-17. Note that the flux is oriented midway between the poles at each step. However, it still rotates by 90° per step. The normal drive develops a slightly greater torque then the wave drive.

The *half-step drive* is obtained by combining the wave drive and the normal drive. The switching sequence for cw rotation is given in Table 19C and the resulting current pulses I_a, I_b are shown in Figure 19-18. The flux now rotates only 45 degrees per step. The main advantage of the half-step drive is that it improves the resolution of position and it tends to reduce the problem of resonance.

Unipolar Winding. The unipolar winding consists of two coils per pole instead of only one (Fig. 19-19a). The coil set A1, A2 produces flux in the opposite direction to coil set 1A, 2A. Consequently, when they are operated in sequence, an alternating flux is produced. The advantage of the unipolar winding is that the number of switching transistors drops from 8 to 4 and the transient response is slightly faster. Figure 19-19b shows the schematic diagram of the windings and the switching sequence for a wave drive. The flux rotates in exactly the same way as shown in Figure 19-16.

19.10 High-speed operation

So far, we have assumed that the current pulse in a winding rises immediately to its rated value *I* at the beginning of the pulse and drops immediately to zero at the end (Fig. 19-20a). In prac-

TABLE 19C   Half-Step Switching Sequence for cw Rotation

STEP		1	2	3	4	5	6	7	8	1
Q1	Q2	on	on	—	—	—	—	—	on	on
Q3	Q4	—	—	—	on	on	on	—	—	—
Q5	Q6	—	on	on	on	—	—	—	—	—
Q7	Q8	—	—	—	—	—	on	on	on	—

Figure 19-18
Current pulses in a half-step drive and resulting flux positions at each step. See Table 19C for switching sequence

(a)

(b)

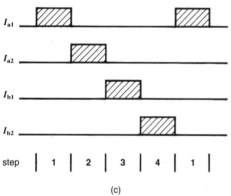

(c)

TABLE 19D	Wave Switching Sequence for cw Rotation				
STEP	**1**	**2**	**3**	**4**	**1**
Q1	on	—	—	—	on
Q2	—	—	on	—	—
Q3	—	on	—	—	—
Q4	—	—	—	on	—

Figure 19-19
a. Coil arrangement in a 4-pole unipolar winding.
b. Schematic diagram of coils, switches and power supply in a unipolar drive.
c. Current pulses in a wave drive using a unipolar winding. The flux rotates in the same way as in a bipolar winding. See Table 19D for switching sequence.

tice, this does not happen because of the inductance of the windings. If a winding has an inductance of L henrys and a resistance of R ohms, its time constant T_o is equal to L/R seconds.

Let the coil be connected to a dc source of E volts by means of a transistor (Fig. 19-20b). A diode (D) is connected across the windings to prevent the high induced voltage from destroying the switching transistor at the moment it interrupts the current flow. The resulting current has the shape given in Figure 19-20d.

How can we explain this pulse shape? When

Figure 19-20a
Ideal current pulse in a winding.

Figure 19-20d
Real current pulse.

Figure 19-20b
Typical circuit of a switching transistor and coil connected to a dc source. The diode protects the transistor against overvoltage.

Figure 19-20e
Shortest possible current pulse that still attains the rated current I.

the transistor is switched on, the transient current i_1 only reaches its rated value $I = E/R$ after about $3T_o$ seconds. Then, when the transistor switches the line current off, the transient current i_2 continues to flow in the coil for about $3T_o$ seconds (Fig. 19-20c). If this current pulse is compared with the ideal current pulse shown in Figure 19-20a, we observe two important facts:

1. because the current does not immediately rise to its final value when the transistor is turned on, the torque developed by the stepping motor is smaller than normal. As a result, the rotor does not move as quickly as we would expect.

2. when the transistor is turned off, current i_2 continues to circulate in the coil/diode loop. As a result, the effective length of the pulse is $T_p + 3T_o$ instead of T_p. The pulse being thus prolonged by the component $3T_o$ means that we cannot switch from one coil to the next as quickly as we would have thought.

Figure 19-20c
Transient current in coil and diode when transistor is switched off.

The shortest possible pulse has a length therefore of $3T_o$ (current rises to its rated value I) plus $3T_o$ (current drops from I to zero) giving a total of $6T_o$ seconds (Fig. 19-20e). It so happens that the windings of stepper motors have time constants T_o ranging from about 1 ms to 8 ms. Thus, the duration of one step can be no shorter than about 6×1 ms = 6 ms. This corresponds to a maximum stepping rate of about 1000/6 = 166 steps per second. Such stepping rates are considered to be slow and so various means are used to speed them up.

19.11 Modifying the time constant

One way to quicken the stepping rate is to reduce the time constant T_o. This can be done by adding an external resistance to the motor windings and raising the dc voltage so that the same rated current I will flow. Such an arrangement is shown in Figure 19-21. The external resistor has a value 4 times that of the coil resistance R, and the dc voltage is raised from E to $5E$ volts. As a result, the time constant drops by a factor of five (L/R to $L/5R$). This means that the maximum stepping rate can be increased by the same factor. Thus, stepping rates of the order of 1000 per second become feasible.

The only draw-back to this solution is

1. the power supply is more expensive because it has to deliver 5 times as much power (the voltage is $5E$ instead of E).

2. a lot of power is wasted in the external resistor which means that the efficiency of the system is very low. Low efficiency is not too important in small stepping motors that develop only a few watts of mechanical power. But fast-acting stepper motors in the 100 W range must be driven by other means.

19.12 Bilevel drive

Bilevel drives enable us to obtain fast rise and fall times of current without using external resistors. The principle of a bilevel drive can be understood by referring to Figure 19-22a. Switches Q1 and Q2 represent transistors that open and close the circuit in the manner explained below. Numerical values will be used

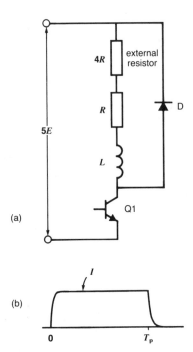

(a)

(b)

Figure 19-21
a. Circuit to increase the rate of growth and decay of current in the coil.
b. Resulting current pulse. Compare with Figure 19-20d.

to explain how the circuit behaves. Thus, the winding is assumed to have a resistance of 0.3 Ω, an inductance of 2.4 mH and a rated current of 10 A. The power supply is 60 V with a tap at 3 V. Thus, if the voltage were applied permanently, the resulting current in the winding would be 60 V/0.3 Ω = 200 A. This is much greater than the rated current of 10 A.

Switch Q1 is initially closed. The current pulse is initiated by closing Q2. Current then starts flowing as shown in Figure 19-22b.

The time constant of this circuit is T_o = 2.4 mH/0.3 Ω = 8 ms. The initial rate of rise of current corresponds therefore to a straight line OP that reaches 200 A in 8 ms. Thus, the current in the coil rises at a rate of 200 A/8 ms = 25 000 A/s. The time to reach 10 A is therefore 10/25 000 = 0.4 ms (Fig. 19-22c).

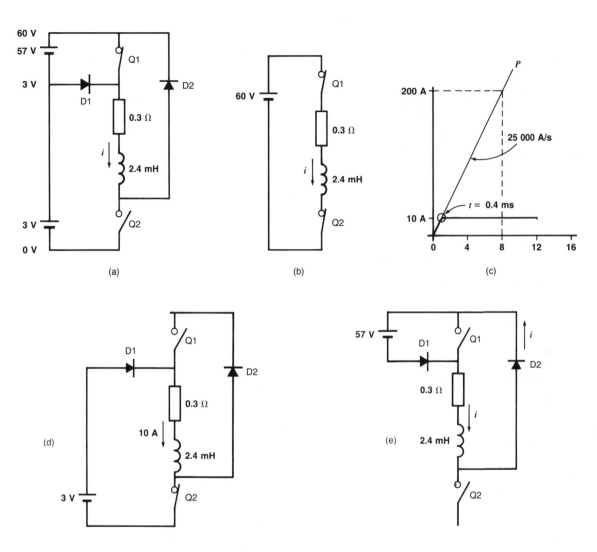

Figure 19-22
a. Circuit of a bilevel drive when current in coil is zero.
b. Equivalent circuit when current in coil is increasing.
c. Rate of increase of current and time to reach 10 A.

d. Equivalent circuit when current in coil is constant.
e. Equivalent circuit when current in coil is decreasing.

As soon as the current reaches this rated value, switch Q1 opens, which forces the current to follow the new path shown in Figure 19-22d. The current is now fed by the 3 V source and so it remains fixed at 3 V/0.3 Ω = 10 A.

The current will stay at this value until we want to end the pulse, say after 5 ms. To terminate the pulse we open Q2 which forces the current to follow the path shown in Figure 19-22e. The 57 V source now tries to drive a current through the coil that is opposite to i. Consequently, i will decrease. The time constant of the circuit is again 8 ms, and so the current will decrease at a rate of 57/60 × 25 000 = 23 750 A/s. It will therefore become zero after an interval of 10/23 750 = 0.42 ms. The moment the current reaches zero, Q1 closes. This forces the current to remain zero until the next pulse is initiated. The resulting pulse shape is shown in Figure 19-22f, together with the Q1, Q2 switching sequence that produces it.

Figure 19-22f
Pulse waveshape using a bilevel drive. Note the switching sequence of Q1 and Q2 that creates it.(x= closed, o= open).

So-called chopper drives are also used. Their principle of operation is similar to the bilevel method except that the current is kept constant during the flat portion of the pulse by repeated on-off switching of the high voltage (60 V) rather than by using a low fixed dc voltage (3V). Choppers are described in Chapter 21.

Electronic drives for stepper motors have become very sophisticated. Some of these circuit-board drives are shown in Figure 19-23, together with the motors they control.

19.13 Instability and resonance

When a stepper motor is operating at certain slewing speeds, it tends to be unstable. The rotor may turn erratically or simply chatter without rotating any more. This instability, often called resonance, is due to the natural vibration of the stepper motor which manifests itself at one or more range of speeds. For example, the range of instability may lie between 2000 sps and 8000 sps. It is possible to ramp through this range without losing step and thereby attain stable slewing speeds between 8000 and 15 000 sps.

19.14 Stepper motors and linear drives

Most stepper motors are coupled to a lead screw of some kind which permits the rotary motion to be converted to a linear displacement. Suppose, for example, that a stepper motor having 200 steps per revolution is coupled to a lead screw having a pitch of 5 threads per inch. The motor has to make 200 × 5 = 1000 steps to produce a linear motion of 1 inch. Consequently, each step produces a displacement of 0.001 in. By counting the pulses precisely, we can position a machine tool, X-Y arm, etc. to a precision of one thousandth of an inch over the full length of the desired movement.

This great precision without benefit of feedback is the reason why stepper motors are so useful in control systems.

QUESTIONS AND PROBLEMS

Practical Level

19-1   What is the main use of stepper motors?
19-2   What is the difference between a reluctance and a permanent magnet stepper motor?
19-3   Describe the construction of a hybrid stepper motor
19-4   A stepper motor advances 2.5° per step. How many pulses are needed to complete 8 revolutions?
19-5   Explain what is meant by normal drive, wave drive and half-step drive.

Intermediate Level

19-6   The 2-pole rotor in Figure 19-1 is

Figure 19-23
Typical electronic drives and the stepper motors they control. (*Courtesy Pacific Scientific, Motor and Control Division*)

replaced by a 4-pole rotor. Calculate the new angular motion per pulse.

19-7   Why is viscous damping employed in stepper motors?

19-8   When a stepper motor is ramping or slewing properly, every pulse corresponds to a precise angle of rotation. True or false?

19-9   The stepper motor in Figure 19-10 is driven by a series of pulses having a duration of 20 ms. How long will it take for the rotor to make one complete revolution?

19-10 A stepper motor rotates 1.8° per step. It drives a lead screw having a pitch of 20 threads per inch. The lead screw, in turn, produces a linear motion of a cutting tool. If the motor is pulsed 7 times, by how much does the cutting tool move?

19-11 A stepper motor advances 7.5° per pulse. If its torque-speed characteristic is given by Figure 19-8, calculate the power [watts] it develops when it is slewing
a. at 500 steps per second.
b. at 200 steps per second.

19-12 The stepper motor shown in Figure 19-14 has a unipolar winding. If it operates in the start-stop mode at a pulse rate of 150 per second,
  a. What is the maximum torque it can develop?
  b. How much mechanical power (millihorsepower) does it develop?
  c. How much mechanical energy [J] does it produce in 3 seconds?

19-13 For a given load torque, the stepping rate can be increased by increasing the rate of rise and rate of fall of the current in the windings. Name two ways whereby this can be accomplished.

19-14 Referring to Figure 19-14d what is the maximum slew speed of the unipolar motor, expressed in revolutions per minute?

Advanced Level

19-15 a. Referring to the stepper motor properties listed in Figure 19-14c, calculate the time constant of a bipolar winding rated at 12 V.
  b. If the 12 V are applied to the winding, approximately how long will it take for the current to reach its final value?
  c. What is the final value of the current in the winding?

19-16 The two armatures on a hybrid stepper motor each have 50 salient poles (teeth). Calculate
  a. the angle between two successive teeth on an armature.

  b. the angle between one tooth on one armature and the next tooth on the other armature.
  c. the angle of advance per pulse.

19-17 Why can a stepper motor develop a larger torque when it is slewing than when it is operating in the stop-start mode?

19-18 A powerful permanent magnet stepper motor used for positioning a valve has the following specifications:
  winding: bipolar
  current: 13 A
  winding resistance: 60 mΩ
  winding inductance: 0.77 mH
  detent torque: 0.16 N·m
  holding torque: 9.5 N·m
  torque at 50 sps: 8 N·m
  steps per revolution: 200
  rotor inertia: 0.7×10^{-3} kg·m^2

  Motor diameter: 4.2 in
  Motor axial length: 7.0 in
  Motor weight: 9 kg

  The motor is chopper-driven at 65 V and it develops a torque of 2.2 N·m at 10 000 sps. Calculate
  a. the speed [r/min] and power [hp] of the motor when it is running at 10 000 sps.
  b. the time constant of the windings [ms]
  c. the time to reach 13 A when 65 V is applied to the winding [μs].

PART III

**ELECTRICAL
AND ELECTRONIC
DRIVES**

20

BASICS OF INDUSTRIAL MOTOR CONTROL

Industrial control, in its broadest sense, encompasses all the methods used to control the performance of an electrical system. When applied to machinery, it involves the starting, acceleration, reversal, deceleration and stopping of a motor and its load. In this chapter, we shall study the electrical (but not electronic) control of 3-phase alternating current motors. Our study is limited to elementary circuits because industrial circuits are usually too intricate to explain in a few words. However, the basic principles covered here apply to any system of control, no matter how complex it may appear to be.

20.1 Control devices

Every control circuit is composed of a number of basic components connected together to achieve the desired performance. The size of the components varies with the power of the motor, but the principle of operation remains the same. Using only a dozen basic compo-

nents, we can design control systems that are very complex. The basic components are:

1. Disconnecting switches
2. Manual circuit breakers
3. Cam switches
4. Pushbuttons
5. Relays
6. Magnetic contactors
7. Thermal relays and fuses
8. Pilot lights
9. Limit switches and other special switches
10. Resistors, reactors, transformers and capacitors.

Table 20A illustrates these devices, and states their main purpose and application. The symbols for these and other devices are given in Table 20B.

In order to understand the sections that follow, the legends in Table 20A must be read before proceeding further.

Disconnecting switches

A *disconnecting switch* isolates the motor from the power source. It consists of 3 knife-switches and 3 line fuses enclosed in a metallic box. The knife-switches open and close simultaneously be means of an external handle. An interlocking mechanism prevents the hinged cover from opening when the switch is closed. Disconnecting switches (and their fuses) are selected to carry the nominal full-load current of the motor, and to withstand short-circuit currents for brief intervals.

Figure 20-1
Three-phase, fused disconnecting switch rated 600 V, 30 A. (*Square D*)

Manual circuit breakers

A *manual circuit breaker* opens and closes a circuit, like a toggle switch. It trips (opens) automatically when the current exceeds a predetermined limit. After tripping, it can be reset manually. Manual circuit breakers are often used instead of disconnecting switches because no fuses have to be replaced.

Figure 20-2
Three-phase circuit breaker, 600 V, 100 A. (*Square D*)

Cam switches

A *cam switch* has a group of fixed contacts and an equal number of moveable contacts. The contacts can be made to open and close in a preset sequence by rotating a handle or knob. Cam switches are used to control the motion and position of hoists, callenders, machine tools, etc.

Figure 20-3
Three-phase surface-mounted cam switch, 230 V, 2 kW. (*Klockner-Moeller*)

TABLE 20A	BASIC COMPONENTS FOR CONTROL CIRCUITS

Pushbuttons

A *pushbutton* is a switch activated by finger pressure. Two or more contacts open or close when the button is depressed. Pushbuttons are usually spring loaded so as to return to their normal position when pressure is removed.

Figure 20-4
Mechanical-interlocked pushbuttons with NO and NC contacts; rated to interrupt an ac current of 6 A one million times. (*Siemens*)

Control relays

A *control relay* is an electromagnetic switch that opens and closes a set of contacts when a so-called *relay coil* is energized. The relay coil produces a strong magnetic field which attracts a moveable armature bearing the contacts. Control relays are mainly used in low-power circuits. They include time-delay relays whose contacts open or close after a definite time interval. Thus, a time-delay *closing* relay actuates its contacts after the relay coil has been energized. On the other hand, a time-delay *opening* relay actuates its contacts some time after the relay coil has been de-energized.

Figure 20-5
Single-phase relays: 25 A, 115/230 V and 5 A, 115 V. (*Potter and Brumfield*)

Thermal relays

A *thermal relay* (or overload relay) is a temperature-sensitive device whose contacts open or close when the motor current exceeds a preset limit. The current flows through a small, calibrated heating element which raises the temperature of the relay. Thermal relays are inherent time-delay devices because the temperature cannot follow the instantaneous changes in current.

Figure 20-6
Three-phase thermal relay with variable current setting, 6 A to 10 A. (*Klockner-Moeller*)

TABLE 20A	BASIC COMPONENTS FOR CONTROL CIRCUITS

Magnetic contactors

A *magnetic contactor* is basically a large control relay designed to open and close a power circuit. It possesses a relay coil which activates a set of contacts. Magnetic contactors are used to control motors ranging from 0.5 hp to several hundred horsepower. The size, dimensions and performance of contactors are standardized.

Figure 20-7a
Three-phase magnetic contactor rated 50 hp, 575 V, 60 Hz. Width: 158 mm; height: 155 mm; depth: 107 mm; weight: 3.5 kg. (*Siemens*)

Figure 20-7b
Three-phase magnetic contactor with mercury contacts, 480 V, 60 A. The contactor is particularly quiet on opening and closing. (*Davis Controls Ltd.*)

Pilot lights

A *pilot light* indicates the on/off state of a remote component in a control system.

Figure 20-8
Pilot light, 120 V, 3 W mounted in a start-stop pushbutton station. (*Siemens*)

TABLE 20A   BASIC COMPONENTS FOR CONTROL CIRCUITS

Limit switches and special switches

A *limit switch* is a low-power snap-action device which opens or closes a contact, depending upon the position of a mechanical part. Other limit switches are sensitive to pressure, temperature, liquid level, direction of rotation, and so on.

Figure 20-9a
Limit switch with one NC contact; rated for ten million operations; position accuracy: 0.5 mm. (*Square D*)

Figure 20-9b
Liquid level switch. (*Square D*)

(a)   (b)

20.2 Normally open and normally closed contacts

Control circuit diagrams always show components in a state of rest, that is, when they are not energized (electrically) or activated (mechanically). In this state, some electrical contacts are open while others are closed. They are respectively called *normally open contacts* (NO) *and normally closed contacts* (NC) and are designated by the following symbols:

normally open contact (NO)   ─┤├─

normally closed contact (NC) ─╫├─

20.3 Relay coil exciting current

When a magnetic contactor is in its de-energized or open position, the magnetic circuit has a very long air gap, compared to when the contactor is closed. Consequently, in the case of an ac contactor the inductive reactance of the relay coil is much lower when the contactor is open than when it is closed. Because the coil is excited by a fixed ac voltage, the magnetizing current is much higher in the open than in the closed contactor position. In other words, a considerable inrush current is drawn by the relay coil at the moment it is excited. This places a heavier than expected duty on auxiliary contacts that energize the coil.

Example 20-1:

A 3-phase NEMA size 5 magnetic contactor rated at 270 A, 460 V possesses a 120 V, 60 Hz relay coil. The coil absorbs an apparent power of 2970 VA and 212 VA, respectively, in the open and closed contactor position. Calculate:
a. the inrush exciting current;
b. the normal, sealed exciting current;
c. the control power needed to actuate the relay coil compared to the power handled by the contactor.

Solution:
a. The inrush current in the relay coil is:
$$I = S/E = 2970/120 = 24.75 \text{ A}$$

b. The normal relay coil current when the contactor is sealed (closed) is:
$$I = S/E = 212/120 = 1.77 \text{ A}$$

c. The steady-state control power needed to actuate the relay coil is 212 VA. The power which the contactor can handle is:
$$P = EI \sqrt{3} = 460 \times 270 \sqrt{3}$$
$$= 215\ 120 \text{ VA}$$

Thus, the small control power (212 VA) can control a load whose power is $215\ 120/212 = 1015$ times greater.

20.4 Control diagrams

A control system can be represented by four types of circuit diagrams. They are listed as follows, in order of increasing detail and completeness:

- block diagram
- one-line diagram*
- wiring diagram
- schematic diagram

A *block diagram* is composed of a set of rectangles, each representing a control device, together with a brief description of its function. The rectangles are connected by arrows which indicate the direction of power flow (Fig. 20-10).

A *one-line diagram* is similar to a block diagram except that the components are shown by their symbols, rather than by rectangles. The symbols give us an idea of the nature of the components; consequently, one-line diagrams yield more information. The lines connecting the various components represent two or more conductors (Fig. 20-11).

A *wiring diagram* shows the connections between the components, taking into account the physical location of the terminals and even

* Also called single-line diagram

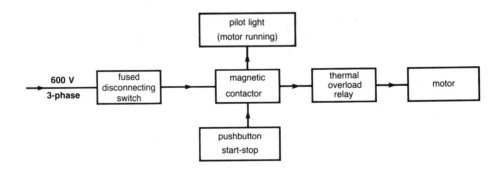

Figure 20-10
Block diagram of a combination starter.

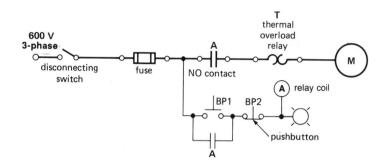

Figure 20-11
One-line diagram of a combination starter.

Figure 20-12
Wiring diagram of a combination starter.

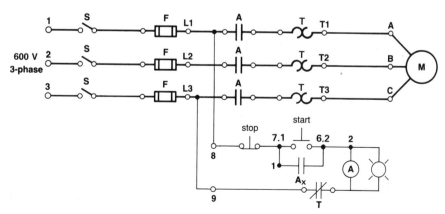

Figure 20-13
Schematic diagram of a combination starter.

TABLE 20B   GRAPHIC SYMBOLS FOR ELECTRICAL DIAGRAMS

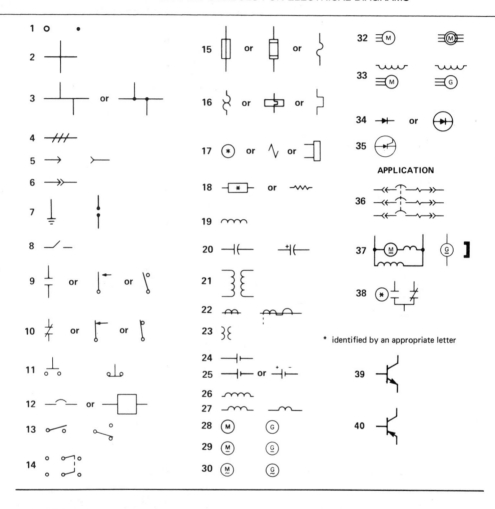

* identified by an appropriate letter

1. terminal; connection   2. conductors crossing   3. conductors connected   4. three conductors   5. plug; receptacle
6. separable connector   7. ground connection; arrester   8. disconnecting switch   9. normally open contact (NO)
10. normally closed contact (NC)   11. pushbutton NO; NC   12. circuit-breaker   13. single pole switch; three-way switch
14. double pole double throw switch   15. fuse   16. thermal overload element   17. relay coil   18. resistor
19. winding, inductor or reactor   20. capacitor; electrolytic capacitor   21. transformer   22. current transformer; bushing type
23. potential transformer   24 dc source (general)   25. cell   26. shunt winding   27. series winding; commutating pole or compensating winding
28. motor; generator (general symbols)   29. dc motor; dc generator (general symbols)   30. ac motor; ac generator (general symbols)
32. 3-phase squirrel-cage induction motor; 3-phase wound-rotor motor   33. synchronous motor; 3-phase alternator   34. diode
35. thyristor or SCR   36. 3-pole circuit breaker with magnetic overload device, drawout type
37. dc shunt motor with commutating winding; permanent magnet dc generator   38. magnetic relay with one NO and one NC contact.
39. NPN transistor   40. PNP transistor

For a complete list of graphic symbols and references see *"IEEE Standard and American National Standard Graphic Symbols for Electrical and Electronics Diagrams"* (ANSI Y32.2/IEEE No. 315) published by the Institute of Electrical and Electronic Engineers, Inc., New York, N.Y. 10017. Essentially the same symbols are used in Canada and several other countries.

the color of wire. These diagrams are employed when installing equipment or when troubleshooting a circuit (Fig. 20-12).

A *schematic diagram* shows all the electrical connections between components, without regard to their physical location or terminal arrangement. This type of diagram is indispensable when troubleshooting a circuit or analyzing its mode of operation (Fig. 20-13). In the sections that follow, this is the kind of diagram we shall be using.

The reader should note that the four diagrams in Fig. 20-10 to 20-13 all relate to the same control circuit. The symbols used to designate the various components are given in Table 20B.

20.5 Starting methods

Three-phase squirrel-cage motors are started either by connecting them directly across the line or by applying reduced voltage to the stator. The starting method depends upon the power capacity of the supply line and the type of load.

Across-the-line starting is simple and inexpensive. The main disadvantage is the high starting current, which is 5 to 6 times the rated full-load current. It can produce a significant line voltage drop, which may affect other customers connected to the same line. Voltage-sensitive devices such as incandescent lamps, television sets, and high-precision machine tools respond badly to such voltage dips.

Mechanical shock is another problem that should not be overlooked. Equipment can be seriously damaged if full-voltage starting produces a hammer-blow impulse. Conveyor belts are another example where sudden starting may not be acceptable.

In large industrial installations, we can sometimes tolerate across-the-line starting even for motors rated up to 10 000 hp. Obviously, the fuses and circuit-breakers must be designed to carry the starting current during the acceleration period.

A motor control circuit contains two basic components: a disconnecting switch and a starter. The disconnecting switch is always placed between the supply line and the starter. The switch and starter are sometimes mounted in the same enclosure to make what is called a *combination starter*. The fuses in the disconnecting switch are rated at about 3.5 times full-load current; consequently, they do not protect the motor against sustained overloads. Their primary function is to protect the motor and supply line against catastrophic currents resulting from a short-circuit in the motor or starter, or a failure to start up. The fuse rating, in amperes, must comply with the requirements of the National Electric Code.

In some cases the disconnecting switch and its fuses are replaced by a manual circuit breaker.

20.6 Manual across-the-line starters

Manual 3-phase starters are composed of a circuit breaker and either two or three thermal relays, all mounted in an appropriate enclosure. They are used for small motors (10 hp or less) at voltages ranging from 120 V to 600 V. The thermal relays trip the circuit-breaker whenever the current in one of the phases exceeds the rated value for a significant length of time.

Single-phase manual starters (Fig. 20-14) are built along the same principles but they contain only one thermal relay. The thermal relays are selected for the particular motor that is connected to the starter.

20.7 Magnetic across-the-line starters

Magnetic across-the-line starters are employed whenever we have to control a motor from a remote location. They are also used whenever the power rating exceeds 10 kW.

Figure 20-15 shows a typical magnetic starter and its associated connection diagram. The starter has three main components: a magnetic contactor, a thermal relay, and a

Figure 20-14
Manual starters for single-phase motors rated 1 hp (0.75 kW); left: surface mounted; center: flush mounted; right: waterproof enclosure. (*Siemens*)

control station.We now describe these components.

1. The *magnetic contactor* A possesses three heavy contacts A and one small auxiliary contact A_x. As can be seen, these contacts are normally open. Contacts A must be big enough to carry the starting current and the nominal full-load current, without overheating. Contact A_x is much smaller because it only carries the current of relay coil A.

   The relay coil is represented by the same symbol (A) as the contacts it controls. Contacts A and A_x remain closed as long as the coil is energized.

2. The *thermal relay* T protects the motor against sustained overloads.* The relay comprises three individual heating elements, respectively connected in series with the three phases. A small, normally

closed contact T forms part of the relay assembly. It opens when the thermal relay gets too hot, and stays open until the relay is manually reset.

Figure 20-15a
Three-phase across the line magnetic starter, 30 hp, 600 V, 60 Hz. (*Klockner-Moeller*)

---

*   The thermal relay is often designated by the letters OL (overload).

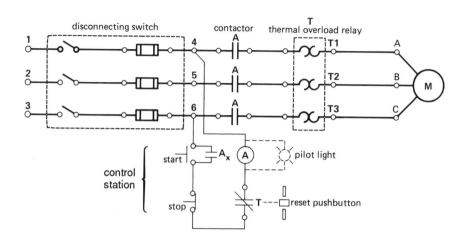

Figure 20-15b
Schematic diagram of a 3-phase across-the-line magnetic starter.

The current rating of the thermal relay is chosen to protect the motor against sustained overloads. Contact T opens after a period of time which depends upon the magnitude of the overload current. Thus, Figure 20-16 shows the tripping time as a multiple of the rated relay current. At rated current (multiple 1), the relay never trips, but at twice rated current, it trips after an interval of 40 s. The thermal relay is equipped with a reset button enabling us to reclose contact T following an overload. It is preferable to wait a few minutes before pushing the button, to allow the relay to cool down.

3. The *control station*, composed of start-stop pushbuttons, may be located either close to, or far away from the starter. The pilot light is optional.

Referring to Figure 20-15b, to start the motor, we first close the disconnecting switch and then depress the *start* button. Coil A is immediately energized causing contacts A and A_x to close. The full line voltage appears across the motor and the pilot light is on. When the pushbutton is released, it returns to its normal position, but the relay coil remains excited because auxiliary con-

Figure 20-16
Typical curve of a thermal overload relay, showing tripping time versus line current. The tripping time is measured from cold-start conditions. If the motor has been operating at full load for one hour or more, the tripping time is reduced by about 30 percent.

tact A_x is now closed. Contact A_x is said to be a *self-sealing* contact.

To stop the motor, we simply push the *stop* button, which opens the circuit to the coil. In case of a sustained overload, the opening of contact T produces the same effect.

It sometimes happens that a thermal relay will trip for no apparent reason. This condition can occur when the ambient temperature around the starter is too high. We can remedy the situation by changing the location of the starter, or by replacing the relay by another one having a higher current rating. Care must be taken before making such a change because if the ambient temperature around the motor is also too high, the occasional tripping may actually serve as a warning.

Figure 20-17 shows a typical combination starter. Figure 20-18 shows another combination starter equipped with a small step-down transformer to excite the control circuit. Such transformers are always used on high-voltage starters (above 600 V) because they permit the use of standard control components, such as pushbuttons and pilot lights and reduce the shock hazard to operating personnel.

20.8 Inching and jogging

In some mechanical systems, we have to adjust the position of a motorized part very precisely. To accomplish this, we energize the motor in short spurts so that it barely starts before it again comes to a halt. A double-contact pushbutton J is added to the usual start/stop circuit, as shown in Fig. 20-19. This arrangement permits conventional start-stop control as well as so-called *jogging*, or *inching*. The following description shows how the control circuit operates.

If the jog button J is in its normal position (not depressed) relay coil A is excited as soon as the start button is depressed. Sealing contact A_x in the main contactor closes and so the

Figure 20-17
Three-phase across-the-line combination starter, 150 hp, 575 V, 60 Hz. The protruding knob controls the disconnecting switch; the pushbutton station is set in the transparent polycarbonate cover. (*Klockner-Moeller*)

Figure 20-18
Three-phase across-the-line combination starter rated 100 hp, 575 V, 60 Hz. The isolating circuit breaker is controlled by an external handle. The magnetic contactor is mounted in the bottom left-hand corner of the water proof enclosure. The small 600 V/120 V transformer in the lower right-hand corner supplies low-voltage power for the control circuit. (*Square D*)

motor will continue to turn after the start button is released. Thus, the control circuit operates in the same way as in Fig. 20-15b.

Suppose now that the motor is stopped, and we depress the jog button. This closes contacts 3, 4 and relay coil A is excited. Contact A_x closes, but contacts 1, 2 are now open and so the closure of A_x has no effect. The motor will pick up speed so long as the jog button is depressed. However, if it is released, coil A will become de-energized, and contactor A will drop out, causing A_x to open. Thus, when contacts 1, 2 are again bridged, the motor will nevertheless come to a halt. Thus, by momentarily

Figure 20-19
Control circuit and pushbutton station for start-stop jog operation. Terminal 8, L3 correspond to terminals 8, L3 in Fig. 20-13.

depressing the jog button we can briefly excite the motor.

Jogging imposes severe duty on the main power contacts A because they continually make *and break* currents that are 6 times greater than normal. It is estimated that each impulse corresponds to 30 normal start/stop operations. Thus, a contactor that can normally start and stop a motor 3 million times, can only jog the motor 100 000 times before the contacts have to be replaced.* Furthermore, jogging should not be repeated too quickly because the intense heat of the breaking arc may cause the main contacts to weld together.

---

* A magnetic contactor has an estimated *mechanical* life of about 20 million open/close cycles, but the electrical contacts should be replaced after 3 million normal cycles.

When jogging is required, the contactor is usually selected to be one NEMA size bigger than for normal duty.

20.9 Reversing the direction of rotation

We can reverse the direction of rotation of a 3-phase motor by interchanging any two lines. This can be done by using two magnetic contactors A and B and a manual 3-position cam switch (Fig. 20-20). When contactor A is closed, lines L1, L2, L3 are connected to terminals A, B, C of the motor. But when contactor B is closed, the same lines are connected to motor terminals C, B, A.

In the forward direction, the cam switch engages contact 1, which energizes relay coil A, causing contactor A to close.*

To reverse the rotation, we move the cam switch to position 2. However, in doing so, we have to move past the off position (0). Consequently, it is impossible to energize coils A and B simultaneously. Occasionally, however, a mechanical defect may prevent a contactor from dropping out, even after its relay coil is de-energized. This is a serious situation, because when the *other* contactor closes, a short-circuit results across the line. The short-circuit current could easily be 50 to 500 times greater than normal, and both contactors could be severely damaged. To eliminate this danger, the contactors are mounted side by side and mechanically interlocked, so as to make it physically impossible for both to be closed at the same time. The interlock is a simple steel bar, pivoted at the center, whose extremities are tied to the moveable armature of each contactor.

During an emergency, push-button U, equipped with a large red bull's-eye, can be used to stop the motor (Fig. 20-20c). In prac-

---

* The contacts and relay coils may be designated by any appropriate letters. Thus, the letters F and R are often used to designate "forward" and "reverse" operating components. In this book, we have adopted the letters A and B mainly for reasons of continuity from one circuit to the next.

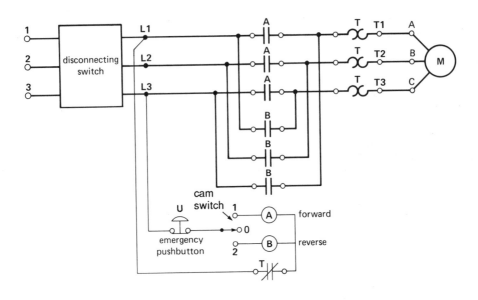

Figure 20-20a
Simplified schematic diagram of a reversible magnetic starter.

Figure 20-20b
Three-position cam switch in Fig. 20-20a. (*Siemens*)

Figure 20-20c
Emergency stop push-button in Fig. 20-20a. (*Square D*)

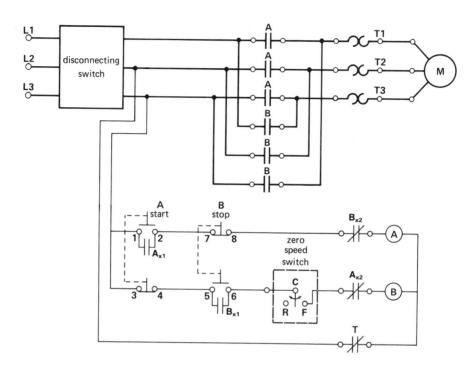

Figure 20-21a
Simplified schematic diagram of a starter with plugging control.

tice, operators find it easier to hit a large button than to turn a cam switch to the off position.

20.10 Plugging

We have already seen that an induction motor can be brought to a rapid stop by reversing two of the lines (Sec. 14.8). However, to prevent the motor from turning in reverse, a zero-speed switch must open the line as soon as the machine has come to rest. The circuit of Fig. 20-21a shows the basic elements of such a plugging circuit. The circuit operates as follows:

1. Contactor A is used to start the motor. In addition to its 3 main contacts A, it has 2 small auxiliary contacts A_{x1} and A_{x2}.

2. The start pushbutton has one NO contact 1,2 and one NC contact 3,4 which operate together. Thus, contact 3, 4 opens before

contact 1,2 closes.

3. Contactor B is used to stop the motor. It is identical to contactor A, having 2 auxiliary contacts B_{x1} and B_{x2} in addition to the 3 main contacts B.

4. The stop pushbutton is identical to the start pushbutton. Thus, when it is depressed contact 7,8 opens before contact 5,6 closes.

5. Contact F-C of the zero-speed switch is normally open, but it closes as soon as the motor turns in the forward direction. This "prepares" the plugging circuit for the eventual operation of coil B.

6. Contacts A_{x1} and B_{x1} are sealing contacts so that pushbuttons A and B have only to be pressed momentarily to start or stop the motor.

7. Contacts A_{x2} and B_{x2} are electrical interlocks to prevent the relay coils A and B from being excited at the same time. Thus, when the motor is running, contact A_{x2} is open. Consequently, relay coil B cannot become excited by depressing pushbutton B until contactor A has dropped out, causing contact A_{x2} to reclose.

Several types of zero-speed switches are on the market and Fig. 20-21b shows one that operates on the principle of an induction motor. It consists of a small permanent magnet rotor N, S and a bronze ring, supported on bearings, that is free to pivot between stationary contacts F and R. The permanent magnet is coupled to the shaft of the main motor. As soon as the motor turns clockwise, the permanent magnet drags the ring along in the same direction, closing contacts F-C. When the motor stops turning, the brass ring returns to the off position. Because of its function and shape, the ring is often called a "drag-cup".

Figure 20-21c shows another zero-speed switch that operates on the principle of centrifugal force.

20.11 Reduced-voltage starting

Some industrial loads have to be started very gradually. Examples are coil winders, printing presses, and other machines that process fragile products. In other industrial applications, we cannot connect a motor directly to the line because the starting current is too high. In all these cases, we have to reduce the voltage applied to the motor either by connecting resistors (or reactors) in series with the line or by employing an autotransformer. In reducing the voltage, we recall that:

1. The locked-rotor current is proportional to the voltage: reducing the voltage by half reduces the current by half;
2. The locked-rotor torque is proportional to the *square* of the voltage: reducing the voltage by half reduces the torque by a factor of four.

Figure 20-21b
Typical zero-speed switch for use in Fig. 20-21a.

Figure 20-21c
Zero-speed switch, centrifugal type (*Hubbel*).

Figure 20-22a
Simplified schematic diagram of the power section of a reduced-voltage primary resistor stator.

20.12 Primary resistance starting

Primary resistance starting consists of placing three resistors in series with the motor during the start-up period (Fig. 20-22a). Contactor A closes first and when the motor has nearly reached synchronous speed, a second contactor B short-circuits the resistors. This method gives a very smooth start with complete absence of mechanical shock. The voltage drop across the resistors is high at first, but gradually diminishes as the motor picks up speed and the current falls. Consequently, the voltage across the motor increases with speed, and so the electrical and mechanical shock is negligible when full voltage is finally applied (closure of contactor B). The resistors are short-circuited after a delay that depends upon the setting of a time-delay relay.

The schematic control diagram (Fig. 20-22b) reveals the following circuit elements:

A, B - magnetic contactor relay coils and the associated auxiliary contact A_x

RT - time-delay relay that closes the circuit of coil B after a preset interval of time

As soon as the start pushbutton is depressed, relay coils A and RT are excited. This causes the contacts A and A_x to close immediately. However, the contact RT only closes after a certain time delay and so the relay coil of contactor B is only excited a few seconds later.

If the magnetic contactors A, B are particularly large, the inrush exciting currents could damage the start pushbutton contacts if they are connected as shown in Figure 20-22b. In such cases, it is better to add an auxiliary relay having more robust contacts. Thus, in Fig. 20-22c, the purpose of auxiliary relay RA is to carry the exciting currents of relay coils A and B. Note that the start pushbutton contacts carry only the exciting current of relay coils RA and RT. Other circuit components are straightfor-

Figure 20-22b
Control circuit of Fig. 20-22a.

Figure 20-22c
Control circuit of Fig. 20-22a using an auxiliary relay RA.

ward, and the reader should have no difficulty in analyzing the operation of the circuit.

How are the starting characteristics affected when resistors are inserted in series with the stator? Figure 20-23a shows the torque-speed curve (1) when full voltage is applied to a typical 3-phase, 1800 r/min induction motor. Corresponding curve (2) shows what happens when resistors are inserted in series with the line. The resistors are chosen so that the locked-rotor voltage across the stator is 0.65 p.u. The locked-rotor torque is therefore $(0.65)^2 = 0.42$ p.u. or only 42 percent of full-load torque. This

means that the motor must be started at light load.

Figure 20-23b shows the current versus speed curve (1) when full voltage is applied to the stator. Curve (2) shows the current when the resistors are in the circuit. When the speed reaches about 1700 r/min, the resistors are short-circuited. The current jumps from about 1.8 p.u. to 2.5 p.u. - a very moderate jump.

Example 20-2:
A 150 kW (200 hp), 460 V, 3-phase, 3520 r/min, 60 Hz induction motor has a locked-rotor torque of 600 N·m and a locked-rotor current of 1400 A. Three resistors are connected in series with the line so as to reduce the voltage across the motor to 0.65 p.u. Calculate:

a. the apparent power absorbed by the motor under full-voltage, locked-rotor conditions;

b. the apparent power absorbed by the motor when the resistors are in the circuit;

c. the apparent power drawn from the line, with the resistors in the circuit;

d. the locked-rotor torque developed by the motor.

Figure 20-23a
Typical torque-speed curves of a 3-phase squirrel-cage induction motor: (1) full voltage starting; (2) primary resistance starting with voltage reduced to 0.65 p.u.

Figure 20-23b
Typical current-speed curves of a 3-phase squirrel-cage induction motor: (1) full voltage starting; (2) primary resistance starting with voltage reduced to 0.65 p.u.

Solution:

a. At full voltage, the locked-rotor apparent power is:

$$S = \sqrt{3}EI \qquad \text{Eq. 8-9}$$
$$= 1.73 \times 460 \times 1400$$
$$= 1114 \text{ kVA}$$

b. The voltage across the motor at 0.65 p.u. is:

$$E = 0.65 \times 460 = 299 \text{ V}$$

The current drawn by the motor decreases in proportion to the voltage:

$$I = 0.65 \times 1400 = 910 \text{ A}$$

The apparent power drawn by the motor is:

$$S_m = \sqrt{3}EI$$
$$= 1.73 \times 299 \times 910$$
$$= 471 \text{ kVA}$$

c. The apparent power drawn from the line is:

$$S_L = \sqrt{3}EI$$
$$= 1.73 \times 460 \times 910$$
$$= 724 \text{ kVA}$$

Thus, the apparent power is only 724 kVA/ 1114 kVA = 65% of the apparent power under full-voltage conditions.

d. The torque varies as the square of the voltage:

$$T = 0.65^2 \times 600$$
$$= 0.42 \times 600$$
$$= 252 \text{ N·m} (\sim 186 \text{ ft·lbf})$$

The results of these calculations are summarized in Fig. 20-24.

Figure 20-24
See Example 20-2.

Example 20-3:
In Example 20-2, if the locked-rotor power factor of the motor alone is 0.35, calculate the required resistance of the series resistors and the power they dissipate.

Solution:
We will solve this problem by considering active and reactive powers and using the power triangle method.

a. The apparent power drawn by the motor at reduced voltage is:

$$S_m = 471 \text{ kVA (from Example 20-2)}$$

The corresponding apparent power drawn by the line is:

$$S_L = 724 \text{ kVA (from Example 20-2)}$$

The active power drawn by the motor is:

$$P_m = S_m \cos\theta = 471 \times 0.35$$
$$= 165 \text{ kW}$$

The reactive power absorbed by the motor is:

$$Q_m = \sqrt{S_m^2 - P_m^2} = \sqrt{471^2 - 165^2}$$
$$= 441 \text{ kvar}$$

The resistors can only absorb active power in the circuit. Consequently, the reactive power supplied by the line *must* be equal to that absorbed by the motor:

$$Q_L = 441 \text{ kvar}$$

The active power supplied by the line is:

$$P_L = \sqrt{S_L^2 - Q_L^2} = \sqrt{724^2 - 441^2}$$
$$= 574 \text{ kW}$$

The active power absorbed by the three resistors is:

$$P_R = P_L - P_m$$
$$= 574 - 165$$
$$= 409 \text{ kW}$$

The active power per resistor is:

$$P = P_R/3 = 409/3 = 136 \text{ kW}$$

The current in each resistor is

$$I = 910 \text{ A (from Example 20-2)}$$

The value of each resistor is:

$$P = I^2R$$
$$136\ 000 = 910^2R$$
$$R = 0.164\ \Omega$$

The three resistors must therefore have a resistance of 0.164Ω and a short-term rating of 136 kW.

This is an interesting example of the usefulness of the power triangle method in solving a relatively difficult problem. The results are summarized in Fig. 20-25.

S_L = 724 kVA S_m = 471 kVA
P_L = 574 kW P_m = 165 kW
Q_L = 441 kvar Q_m = 441 kvar   locked rotor
series resistors
460 V R 299 V   M
910 A   910 A

Figure 20-25
See Example 20-3.

20.13 Autotransformer starting

The advantage of an autotransformer starter is that for a given torque it draws a much lower line current than does resistance starting. The disadvantage is that autotransformers cost more, and the transition from reduced-voltage to full-voltage is not quite as smooth.

Autotransformers usually have taps to give output voltages of 0.8, 0.65, and 0.5 p.u. The corresponding starting torques are respectively 0.64, 0.42, and 0.25 of the full-voltage starting torque. Furthermore, the starting currents on the *line* side are also reduced to 0.64, 0.42, and 0.25 of the full-voltage locked-rotor current.

Figure 20-26 shows a starter using two autotransformers connected in open delta. A simplified circuit diagram of such a starter is given in Fig. 20-27. Contactor A has five NO contacts A and one small NO contact A_x. This contactor is in operation only during the brief period that the motor is starting up.

Contactor B has 3 NO contacts B. It is in service while the motor is running.

The autotransformers are set on the 65% tap. The time-delay relay RT possesses three contacts RT1, RT2, RT3. The contact RT1 in parallel with the start button closes as soon as coil RT is energized. The other two contacts RT2, RT3 operate after a delay that depends upon the RT relay setting. Contactors A and B are mechanically interlocked to prevent them from closing simultaneously.

Contactor A closes as soon as the start button is depressed. This excites the autotransformer and reduced voltage appears across the motor terminals. A few seconds later, contact RT2 in series with coil A opens, causing contactor A to open. At the same time, contact RT3 causes contactor B to close. Thus, contactor A drops out, followed almost immediately by the closure of contactor B. This action applies full voltage to the motor, and simultane-

Figure 20-26
Reduced voltage autotransformer starter, 100 hp, 575 V, 60 Hz. (*Square D*)

Figure 20-27
Simplified schematic diagram of an autotransformer starter.

ously disconnects the autotransformer from the line.

In transferring from contactor A to contactor B, the motor is disconnected from the line for a fraction of a second. This creates a problem because when contactor B closes, a large transient current is drawn from the line. This transient surge is hard on the contacts and also produces a mechanical shock. For this reason, we sometimes employ more elaborate circuits in which the motor is never completely disconnected from the line.

Figure 20-28a and 20-28b compare the graphs of torque and line current when autotransformer starting (3) and resistance starting (2) is used. The locked-rotor voltage in each case is 0.65 p.u. The reader will note that the locked-rotor torques are identical, but the locked-rotor line current is much lower using an autotransformer (2.7 versus 4.2 p.u.).

However, once the motor reaches 90% of synchronous speed, resistance starting produces a higher torque because the terminal voltage is slightly higher than the 65% value that existed at the moment of start-up. On the other hand, the line current at all speeds is smaller when using an autotransformer.

Because the autotransformers and resistors operate for very short periods, they can be wound with much smaller wire than continuously-rated devices. This enables us to drastically reduce the size, weight and cost of these components.

Example 20-4:
A 200 hp (150 kW), 460 V, 3-phase, 3520 r/min, 60 Hz induction motor has a locked rotor torque of 600 N·m and a locked-rotor current of 1400 A (Same motor as in Example 20-2). Two autotransformers, connected in open delta, and having a 65% tap, are

Figure 20-28a
Typical reduced voltage (0.65 p.u.) torque-speed curves of a 3-phase squirrel-cage induction motor: (2) primary resistance starting; (3) autotransformer starting.

Figure 20-28b
Typical reduced voltage (0.65 p.u.) current-speed curves of a 3-phase squirrel-cage induction motor: (2) primary resistance starting; (3) autotransformer starting.

employed to provide reduced-voltage starting.

Calculate:
a. the apparent power absorbed by the motor;
b. the apparent power supplied by the 460 V line;
c. the current supplied by the 460 V line;
d. the locked-rotor torque.

Solution:
a. The voltage across the motor is:
$$E = 0.65 \times 460 = 299 \text{ V}$$

The current drawn by the motor is:
$$I = 0.65 \times 1400 = 910 \text{ A}$$

The apparent power drawn by the motor is:
$$S_m = \sqrt{3}EI$$
$$= 1.73 \times 299 \times 910$$
$$= 471 \text{ kVA}$$

Figure 20-29
See Example 20-4.

b. The apparent power supplied by the line is equal to that absorbed by the motor because the active and reactive power consumed by the autotransformers is negligible (Sec. 12.1). Consequently,
$$S_L = S_m = 471 \text{ kVA}$$

c. The current drawn from the line is:
$$I = S_L/(\sqrt{3}E) \qquad \text{Eq. 8-9}$$
$$= 471\,000/(1.73 \times 460)$$
$$= 592 \text{ A}$$

Note that this current is considerably smaller than the line current (910 A) with resistance starting.

d. The locked-rotor torque varies as the square of the motor voltage:
$$T = 0.65^2 \times 600$$
$$= 252 \text{ N·m}$$

The results of these calculations are summarized in Fig. 20-29. It is worthwhile comparing them with the results in Fig. 20-24.

20.14 Other starting methods

Several other methods are employed to limit the current and torque when starting induction motors. Some only require a change in the stator winding connections. The so-called "part-winding" starting method can be used when the induction motor has two identical 3-phase windings that operate in parallel when the motor is running. During the starting phase, only one of these 3-phase windings is used. As a result, the impedance is higher than if the two windings were connected in parallel. After the motor has picked up speed, the second 3-phase winding is connected so that the two windings operate in parallel. Figure 20-30 shows how two 3-pole contactors A and B can be arranged for part-winding starting. Contactor A closes first, thus energizing windings 1, 2, 3. Shortly after, contactor B closes, bringing winding 7, 8, 9 in parallel with winding 1, 2, 3.

There are many different types of part winding connections and some larger motors have specially designed windings so that the starting performance is optimized.

In *wye-delta* starting, all six stator leads are brought out to the terminal box. The windings are connected in wye during start-up, and in delta during normal running conditions. This starting method gives the same results as an autotransformer starter having a 58% tap. The reason is that the voltage across each wye-connected winding is only $1/\sqrt{3}$ (= 0.58) of its rated value.

Finally, to start wound-rotor motors, we progressively short-circuit the external *rotor* resistors in one, two, or more steps. The number of steps depends upon the size of the machine and the nature of the load.

20.15 Cam switches

Some industrial operations have to be under the continuous control of an operator. In hoists, for example, an operator has to vary the lifting

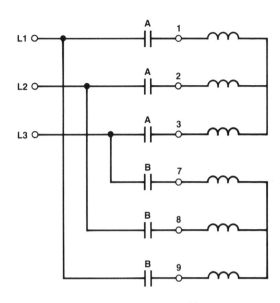

Figure 20-30
Part-winding starting of an induction motor.

and lowering rate, and the load has to be carefully set down at the proper place. Such a supervized control sequence can be done with cam switches.

Figure 20-31 shows a 3-position cam switch designed for the *forward, reverse* and *stop* operation of a 3-phase induction motor. For each position of the knob, some contacts are closed while others are open. This information is given in a table, usually glued to the side of the switch. A cross **X** designates a closed contact, while a blank space is an open contact. In the "forward" position, for example, contacts 2, 4, and 5 are closed and contacts 1 and 3 are open. When the knob is turned to the stop position, all contacts are open. Figure 20-31b shows the shape of the cam that controls the opening and closing of contact 1.

The schematic diagram (Fig. 20-32) shows how to connect the cam switch to a 3-phase motor. The state of the contacts (open or closed) is shown directly on the diagram for

contact	forward	stop	reverse
1			X
2	X		
3			X
4	X		
5	X		X

cam shown in off position

Figure 20-31
a. Cam switch external appearance;
b. detail of the cam controlling contact 1, in the "stop" position;
c. table listing the on-off state of the 5 contacts.

each position of the knob. The 3-phase line and motor are connected to the appropriate cam-switch terminals. Note that jumpers J1,J2,J3,J4 are also required to complete the connections. The reader should analyze the circuit connections and resulting current flow for each position of the switch. For example, when the switch is in the forward position, L1 is connected to T1, L2 to T2 and L3 to T3.

Some cam switches are designed to carry several hundred amperes, but we often prefer to use magnetic contactors to handle large currents. In such cases, a small cam switch is employed to control the relay coils of the contactors. Very elaborate control schemes can be designed with multicontact cam switches.

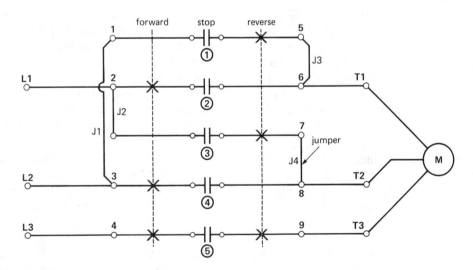

Figure 20-32
Schematic diagram of a cam switch permitting forward-reverse and stop operation of a 3-phase motor.

ELECTRIC DRIVES

20.16 Fundamentals of electric drives*

In this chapter, we have seen the basic control equipement that is used to start and stop induction motors.However,some industrial drives require a motor to function at various torques and speeds, both in forward and reverse. In addition to operating as a motor, the machine often has to function for brief periods as a generator or brake. In electric locomotives, for example, the motor may run clockwise or counterclockwise and the torque may act either with, or against, the direction of rotation. In other words, the speed and torque may be positive or negative.

In describing industrial drives, the various operating modes can best be shown in graphical form. The positive and negative speeds are plotted on a horizontal axis, and the positive and negative torques on a vertical axis (Fig. 20-33). This gives rise to four operating quadrants, labelled respectively quadrants 1, 2, 3, and 4.

If a machine operates in quadrant 1, both the torque and speed are positive, meaning that they act in the same direction (clockwise, say). Consequently, a machine operating in this quadrant functions as a motor. As such, it delivers mechanical power to the load. The machine also operates as a motor in quadrant 3, except that both the torque and speed are reversed.

A machine that operates in quadrant 2 develops a positive torque but its speed is negative. In other words, the torque acts clockwise while the machine turns counterclockwise. In this quadrant, the machine *absorbs* mechanical power from the load; consequently, it functions basically as a generator. The mechanical power is converted into electric power and the

latter is usually fed back into the line. However, the electric power may be dissipated in an external resistor, such as in dynamic braking.

Depending on the way it is connected, a machine may also function as a *brake* when operating in quadrant 2. The mechanical power absorbed is again converted to electric power, but the latter is immediately and unavoidably converted into heat. In effect, when a machine functions as a brake, it absorbs electric power from the supply line at the same time as it absorbs mechanical power from the shaft. Both power inputs are dissipated as heat – often

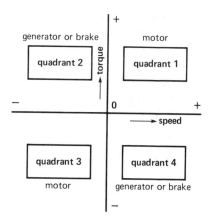

Figure 20-33
Electric drives can operate in four distinct quadrants.

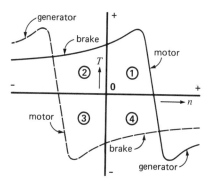

Figure 20-34
Typical torque-speed curve of a squirrel-cage induction motor.

---

* An electric drive is a system consisting of one or several electric motors and of the entire electric control equipment designed to govern the performance of these motors.(*IEEE Standard Dictionary of Electrical and Electronics Terms*)

inside the machine itself. For example, whenever a machine is plugged, it operates as a brake. In large power drives, we seldom favor the brake mode of operation because it is very inefficient. Consequently, the circuit is usually arranged so that the machine functions as a generator when operating in quadrant 2.

Quadrant 4 is identical to quadrant 2, except that the torque and speed are reversed; consequently, the same remarks apply.

20.17 Typical torque-speed curves

The torque-speed curve of a 3-phase induction motor is an excellent example of the motor-generator-brake behavior of an electrical machine. Referring to the solid curve in Fig. 20-34, the machine acts as a motor in quadrant 1, as a brake in quadrant 2 and as a generator in quadrant 4 (Sec. 14.16). If the stator leads are reversed, another torque-speed curve is obtained. This dotted curve shows that the machine now operates as a motor in quadrant 3, as a generator in quadrant 2, and as a brake in quadrant 4. Note that the machine can function either as a generator or brake in quadrants 2 and 4. On the other hand, it always runs as a motor in quadrants 1 and 3.

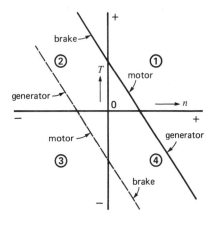

Figure 20-35
Typical torque-speed curve of a dc motor.

To give another example, Figure 20-35 shows the complete torque-speed curve of a dc shunt motor. The motor-generator-brake modes are again apparent. If the armature leads are reversed, we obtain the dotted curve.

In designing variable-speed electric drives, we try to vary the speed and torque in a smooth, continuous way to satisfy the load requirements. This is usually done by shifting the entire torque-speed characteristic back and forth along the horizontal axis. For example, the torque-speed curve of the dc motor (Fig. 20-35) may be shifted back and forth by varying the armature voltage. Similarly, we can shift the curve of an induction motor by varying the voltage and frequency applied to the stator.

To better understand the basic principles of variable speed control, we will first show how variable frequency affects the behavior of a squirrel-cage induction motor.

20.18 Shape of the torque-speed curve

The torque-speed curve of a 3-phase squirrel-cage induction motor depends upon the voltage and frequency applied to the stator. We already know that if the frequency is fixed, the torque varies as the square of the applied voltage. We also know that the synchronous speed depends on the frequency. The question now arises, how is the torque-speed curve affected when both the voltage and frequency are varied? In practice, they are varied in the same proportion so as to maintain a constant flux in the air gap. Thus, when the frequency is doubled, the stator voltage is doubled. Under these conditions, the shape of the torque-speed curve remains the same, but its position along the speed axis shifts with the frequency.

Figure 20-36 shows the torque-speed curve of a 15 hp, (11 kW) 3-phase, 460 V, 60 Hz squirrel-cage induction motor. The full-load speed and torque are respectively 1725 r/min and 60 N·m; the breakdown torque is 160 N·m and the locked rotor torque is 80 N·m.

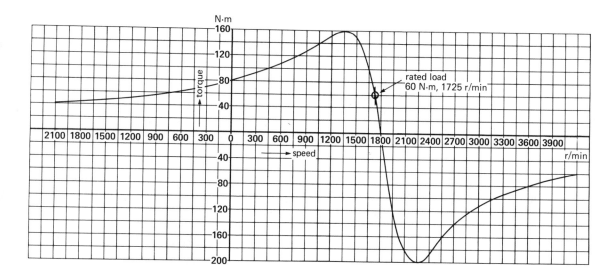

Figure 20-36
Torque-speed curve of a 15 hp, 460 V, 60 Hz, 3-phase squirrel-cage induction motor.

Figure 20-37
Torque-speed curve at three different frequencies.

If we reduce both the voltage and frequency to one fourth their original value (115 V and 15 Hz), the new torque-speed curve is shifted towards the left. The curve crosses the axis at a synchronous speed of 1800/4 = 450 r/min

(Fig. 20-37). Similarly, if we raise the voltage and frequency by 50 percent (690 V, 90 Hz), the curve is shifted to the right and the new synchronous speed is 2700 r/min.

Even if we bring the frequency down to

zero (dc), the torque-speed curve retains essentially the same shape. Current can be circulated in any two lines of the stator while leaving the third line open. The motor develops a symmetrical braking torque that increases with increasing speed, reaching a maximum in both directions, as shown in Figure 20-38. In this figure, the dc current in the windings is adjusted to produce the rated breakdown torque.

Because the shape of the torque-speed curve is the same at all frequencies, it follows that the torque developed by an induction motor is the same whenever the slip speed (in r/min) is the same.

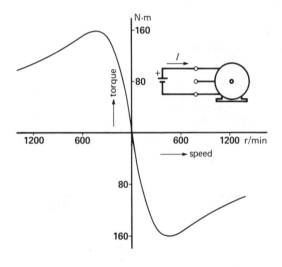

Figure 20-38
Stator excited by dc current.

Example 20-5:
A standard 3-phase, 10 hp, 575 V, 1750 r/min, 60 Hz squirrel-cage induction motor develops a torque of 110 N·m at a speed of 1440 r/min. If the motor is excited at a frequency of 25 Hz, calculate:
a. the required stator voltage to maintain the same flux in the machine;
b. the new speed at a torque of 110 N·m.

Solution:
a. The voltage must be reduced in proportion to the frequency.
$$E = (25/60) \times 575 = 240 \text{ V}$$

b. The synchronous speed of the 4-pole, 60 Hz motor is obviously 1800 r/min. Consequently, the slip speed at a torque of 110 N·m is:
$$n_1 = n_s - n$$
$$= 1800 - 1440 = 360 \text{ r/min}$$

The slip speed is the same for the same torque, irrespective of the frequency. The synchronous speed at 25 Hz is:
$$n_s = (25/60) \times 1800 = 750 \text{ r/min}$$

The new speed at 110 N·m is:
$$n = 750 - 360 = 390 \text{ r/min}$$

20.19 Current-speed curves

The current-speed characteristic of an induction motor is a V-shaped curve having a minimum value at synchronous speed. The minimum is equal to the magnetizing current needed to create the flux in the machine. Because the flux is kept constant, the magnetizing current is the same at all synchronous speeds.

Figure 20-39 shows the current-speed curve of the 15 hp, 460 V, 60 Hz squirrel-cage induction motor mentioned previously. We have plotted the effective values of current; consequently, the current is always positive. The locked rotor current is 120 A and the corresponding torque is 80 N.m.

If the stator voltage and frequency are varied in the same proportion, the current-speed curve retains the same shape, but shifts along the horizontal axis. In effect, the torque-speed and current-speed curves move in unison as the frequency is varied. This produces a very important result.

Suppose, for example, that the voltage and frequency are reduced by 75 percent to 115 V, 15 Hz. The locked rotor current decreases to

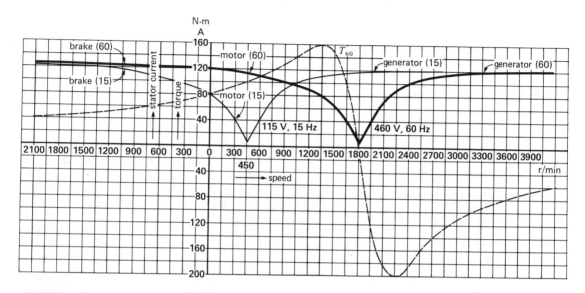

Figure 20-39
Current-speed curve at 60 Hz and 15 Hz. Also T-n curve at 460 V, 60 Hz.

80 A, but the corresponding torque *increases* to 160 N.m, equal to the full breakdown torque. Thus, by reducing the frequency, we obtain a larger torque with a smaller current (Fig. 20-40). This is one of the big advantages of frequency control. In effect, we can gradually accelerate a motor and its load by progressively increasing the voltage and frequency. During the start-up period, the voltage and frequency can be varied automatically so that the motor develops close to its breakdown torque all the way from zero to rated speed. This ensures a rapid acceleration at practically constant current.

In conclusion, an induction motor has excellent characteristics under variable frequency conditions. For a given frequency, the speed changes very little with increasing load. In many ways, the torque-speed characteristic resembles that of a dc shunt motor with variable armature-voltage control.

Example 20-6:
Using the information revealed by the 60 Hz torque-speed and current-speed curves of

Figure 20-40
The starting torque increases and the current decreases with decreasing frequency.

Fig. 20-39, calculate the voltage and frequency required so that the machine will run at 3200 r/min while developing a torque of 100 N.m. What is the corresponding stator current?

Solution:
a. We first have to find the slip speed corresponding to a torque of 100 N.m.

According to Figure 20-39, when the motor operates at 60 Hz and a torque of 100 N.m, the speed is 1650 r/min. Consequently, the slip speed is:

$$n_1 = n_s - n$$
$$= 1800 - 1650 = 150 \text{ r/min}$$

The slip speed is the same when the motor develops 100 N·m at 3200 r/min. Consequently, the synchronous speed must be:

$$n_s = 3200 + 150 = 3350 \text{ r/min}$$

The corresponding frequency is therefore:

$$f = (3350/1800) \times 60 = 111.7 \text{ Hz}$$

The corresponding stator voltage is:

$$E = (111.7/60) \times 460 = 856 \text{ V}$$

The given current-speed curve (Fig. 20-39) shows that the stator current is 40 A when the torque is 100 N.m. Because the current-speed curve shifts along with the torque-speed curve, the current is again 40 A at 3200 r/min and 100 N.m.

20.20 Regenerative braking

A further advantage of frequency control is that it permits regenerative braking. Referring to Fig. 20-41, suppose the motor is connected to a 460 V, 60 Hz line. It is running at 1650 r/min driving a load of constant torque $T_L = 100$ N·m (operating point 1). If we suddenly reduce the frequency and voltage by 50 percent, the motor will immediately operate along the 30 Hz, 230 V torque-speed curve. Because the speed cannot change instantaneously (owing to inertia), we suddenly find ourselves at operating point 2 on the new torque-speed curve. The motor torque is negative; consequently, the speed will drop very quickly, following the 50 percent curve until we reach torque T_L (operating point 4). The interesting feature is that in moving along the curve from point 2 to point 3, energy is returned to the ac line, because the motor acts as an asynchronous generator during this interval.

The ability to develop a high torque from zero to full speed, together with the economy of regenerative braking, is the main reason why frequency-controlled induction motor drives are becoming so popular. These electronically-controlled drives are covered in Chapter 23.

Figure 20-41
Effect of suddenly changing the stator frequency.

QUESTIONS AND PROBLEMS

Practical level

20-1   Name four types of circuit diagrams and describe the purpose of each.

20-2   Without referring to the text, describe the operation of the starter shown in Fig. 20-15b, and state the use of each component.

20-3   Give the symbols for a NO and a NC contact, and for a thermal relay.

20-4   Identify all the components shown in Fig. 20-20a using the equipment list given in Table 20A. Where are contact T and coil A situated physically?

20-5 If the start and stop pushbuttons in Fig. 20-21a are pushed simultaneously, what will happen?

20-6 Referring to Fig. 20-13, if contact A_x in parallel with the start pushbutton were removed, what effect would it have on the operation of the starter?

20-7 If a short-circuit occurs in motor M of Fig. 20-13, which device will open the circuit?

20-8 A partial short-circuit between the turns of the stator winding in Fig. 20-13 produces a 50% increase in the line current of one phase. Which device will shut down the motor?

20-9 Under what circumstances is reduced-voltage starting required?

20-10 Referring to Fig. 20-36, in which quadrants do the following torque-speed operating points occur?
a. + 1650 r/min, + 100 N·m;
b. + 3150 r/min, − 100 N·m.

20-11 Referring to Fig. 20-36, calculate the mechanical power [hp] of the motor when it runs at 450 r/min.

20-12 A standard 3-phase, 4-pole squirrel-cage induction motor is rated at 208 V, 60 Hz. We want the motor to turn at a no-load speed of about 225 r/min while maintaining the same flux in the air gap. Calculate the required voltage and frequency to be applied to the stator.

20-13 Referring to Fig. 20-39, what is the current in the stator under the following conditions, knowing that the stator is energized at 460 V, 60 Hz?
a. machine running as a motor at 1650 r/min and developing a torque of 100 N·m;
b. machine running as a brake at 300 r/min;

c. machine driven as an asynchronous generator at a torque of 120 N·m.

20-14 State in which quadrants a machine operates:
a. as a brake;
b. as a motor;
c. as a generator.

20-15 A machine is turning clockwise in quadrant 3. Does it develop a clockwise or counterclockwise torque?

Intermediate level

20-16 A thermal relay having the tripping curve given in Fig. 20-16 has to protect a 40 hp, 575 V, 3-phase, 720 r/min induction motor having a nominal current rating of 40 A. If the relay is set to 40 A, how long will it take to trip if the motor current is (a) 60 A; (b) 240 A?

20-17 a. If the control circuit of Fig. 20-19 is used in place of that shown in Fig. 20-13, show that the motor will start and continue to run if we momentarily press the start button.
b. Show that if we press the jog button, the motor only runs for as long as the button is depressed.

20-18 A magnetic contactor can make 3 million "normal" circuit interruptions before its contacts need to be replaced. If an operator jogs the motor so that it starts and stops once per minute, after how many working days (approx) will the contacts have to be replaced assuming the operator works an 8-hour day?

20-19 a. Referring to Fig. 20-21a, and assuming the motor is initially at rest, explain the operation of the circuit when the "start" button is momentarily depressed.
b. If the motor is running normally, what happens if we momentarily press the "stop" button?

20-20 a. Explain the sequence of events that takes place when the start button in Fig. 20-22c is momentarily depressed, knowing that relay RT is adjusted for a delay of 10 s.

b. With the motor running, explain what happens when the stop button is depressed.

20-21 Referring to Fig. 20-27, describe the sequence of events that takes place when the start button is momentarily depressed, knowing that relay RT is set for a delay of 5 s. Draw the actual circuit connections, in sequence, until the motor reaches its final speed.

20-22 A 100 hp, 460 V, 3-phase induction motor possesses the characteristics given by curve 1, Figs. 20-23a and 20-23b. The full-load current is 120 A, and the thermal relays are set to this value. If the relay tripping curve is given by Fig. 20-16, calculate the approximate tripping time if the load current suddenly rises to 240 A. (Assume the motor had been running for several hours at full load).

20-23 Referring to Fig. 20-36, and neglecting windage and friction losses, calculate the power P_r supplied to the rotor when the machine runs:
a. as a motor at 1650 r/min;
b. as a brake at 750 r/min;
c. as a generator at 2550 r/min.

20-24 In Problem 20-23, calculate the value of the rotor I^2R losses in each case.

20-25 Referring to Fig. 20-36, calculate the voltage and frequency that must be applied to the machine so that it runs as a high-efficiency motor:
a. At a speed of 1200 r/min developing a torque of 100 N·m;

b. At a speed of 2400 r/min developing a torque of 60 N·m.

20-26 Referring to Fig. 20-39, calculate the voltage and frequency to be applied to the stator so that the locked rotor torque is 100 N·m at a current of 40 A.

20-27 a. It is impossible for a machine to instantaneously change from a point in quadrant 1 to point in quadrant 2. Why?

b. Can it move instantaneously from quadrant 1 to quadrant 4?

20-28 A 4-pole, shunt-wound dc motor has an armature circuit resistance of 4 Ω. It is connected to a 240 V dc source and the no-load speed is 1800 r/min; the corresponding armature current is negligible. Assuming constant field excitation and assuming that armature reaction effects can be neglected, calculate:
a. the armature current at 900 r/min;
b. the mechanical power output [hp] at 1200 r/min;
c. the torque [N·m] at 300 r/min;
d. the starting torque [ft·lbf].
e. Draw the torque-speed curve that passes through quadrants 1, 2 and 4 (see Fig. 20-35).

20-29 a. In Problem 20-28, draw the torque-speed curve if 60 V is applied to the armature, while maintaining the same field excitation;

b. What is the frequency of the current in the armature coils at a speed of 300 r/min?

Advanced level

20-30 a. The curves in Fig. 20-23 relate to a 100 hp, 460 V, 1765 r/min, 3-phase, 60 Hz induction motor, whose full-load current is 120 A. Calculate the breakdown torque for curves 1 and 2 [ft·lbf].

b. Calculate the torque developed when the resistors are in the circuit and the line current is 480 A [ft·lbf].

20-31 The motor having the T-n characteristic given in Fig. 20-36 is running at a no-load speed of 1800 r/min. The total moment of inertia of the rotor and its load is 90 lb·ft^2. The speed has to be reduced to a no-load value of 1200 r/min, by suddenly changing the voltage and frequency applied to the stator. Calculate:
a. the voltage and frequency required;
b. the initial kinetic energy stored in the moving parts;
c. the final kinetic energy in the moving parts;
d. Is all the "lost" kinetic energy returned to the 3-phase line? Explain.

20-32 A 15 hp, 460 V, 3-phase, 60 Hz induction motor has the torque-speed characteristic given in Fig. 20-36.
a. What is the new shape of the curve if we apply 230 V, 60 Hz to the stator?
b. Calculate the new breakdown torque [ft·lbf].

20-33 In Problem 20-32, calculate the stator voltage needed to reduce the breakdown torque to 60 N·m.

21

FUNDAMENTAL ELEMENTS OF POWER ELECTRONICS

Electronic systems and controls have gained wide acceptance in power technology; consequently, it has become almost indispensable to know something about power electronics. Naturally, we cannot cover all aspects of this broad subject in a single chapter. Nevertheless, we can explain in simple terms the behavior of a large number of electronic power circuits, including those most commonly used today.

As far as electronic devices are concerned, we shall limit the discussion to diodes and thyristors. They are found in all electronic systems that involve the conversion of ac power to dc power and vice versa. Other devices such as triacs, gate turn-off thyristors, and switching power transistors are, of course, important, but their action on a circuit is basically no different from that of a thyristor and its associated switching circuitry. All these devices are basically ultra-high-speed switches, so much so, that much of power electronics can be explained by the opening and closing of circuits at precise instants of time. However, we should

not conclude that circuits containing these components and devices are simple – they are not – but the circuits can be understood without having an extensive background in semiconductor theory.

21.1 Potential level

In Chapter 2, Secs. 2.4 and 2.5 we described two ways of representing voltages in a circuit. We now introduce a third method that is particularly useful in circuits dealing with power electronics. The method is based upon the concept of potential levels.

To understand the operation of electronic circuits, it is useful to imagine that individual terminals have a potential level with respect to a reference terminal. The reference terminal is any convenient point chosen in a circuit; it is assumed to have zero electric potential. The potential level of all others points is then measured with respect to this zero reference terminal. In graphs, the reference level is shown as a horizontal line having a potential of 0 V.

Figure 21-1
Potential level method of representing voltages.

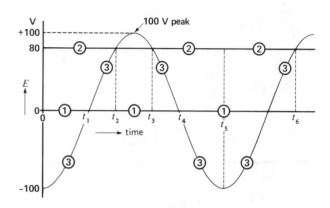

Figure 21-2
Potential levels of terminals 1, 2 and 3.

Consider, for example, the circuit of Fig. 21-1, composed of an 80 V battery connected in series with an ac source having a peak voltage of 100 V. Of the three possible terminals, let us choose terminal 1 as the reference point. The potential level of this terminal is therefore shown by a horizontal line 1 (Fig. 21-2).

Consider now the potential level of terminal 2. The difference of potential between terminals 1 and 2 is always 80 V and terminal 2 is positive with respect to terminal 1. The level of this terminal is therefore indicated by a second horizontal line **2** placed 80 V above line **1**.

Now consider terminal 3. Voltage E between terminals 1 and 3 is alternating and we assume that its initial value is 100 V, with terminal 3 negative with respect to terminal 1. Because E is alternating, the potential of terminal 3 is first negative, then positive, with respect to terminal 1. The changing level is shown by curve **3**. Thus, during the interval from 0 to t_1, the level of point **3** is below the level of point **1** which indicates that terminal 3 is negative with respect to terminal 1. During the interval t_1 to t_4, the polarity reverses and so the level of **3** is now above line **1**. Terminal 1 is therefore negative with respect to terminal 3, because line **1** is below curve **3**.

This potential-level method now enables us to determine the instantaneous voltages between any two terminals in a circuit, as well as their relative polarities. For example, during the interval from t_2 to t_3; terminal 3 is positive with respect to terminal 2, because curve **3** is above line **2**. The voltage between these terminals reaches a maximum of 20 V during this inverval. Then from t_3 to t_6, terminal 3 is negative with respect to terminal 2 and the voltage between them reaches a maximum value of 180 V at instant t_5.

We could have chosen another terminal as a reference terminal. Thus, in Fig. 21-3, we chose terminal 3 and, as before, we represent the zero potential of this reference terminal by a horizontal line **3** (Fig. 21-4). Knowing that E is an alternating voltage and that terminal 1 is initially positive with respect to terminal 3 (Fig. 21-2), we can draw curve **1**.

To determine the level of terminal 2, we know that it is always 80 V positive with respect to terminal 1. Consequently, we draw curve **2**, so that it is always 80 V above curve **1**. By so doing, we automatically establish the level of terminal 2 with respect to terminal 3.

Figures 21-2 and 21-4 do not have the same appearance; however, at every instant, the rel-

Figure 21-3
Changing the reference terminal.

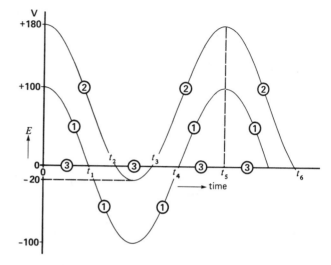

Figure 21-4
The relative potential levels are the same as in Fig. 21-2.

ative polarities and potential differences between terminals are identical. From an electrical point of view, the two figures are identical and we invite the reader to check the truth of this statement.

In electronic circuits, we usually try to select a reference terminal so that it is easy to follow the voltages we are interested in.

21.2 Voltage across some circuit elements

Let us first look at the voltage levels that appear across some active and passive circuit elements commonly found in electronic circuits. Specifically, we examine sources, switches, resistors, coils and capacitors.

1. **Sources**. By definition, ac and dc voltage sources have zero internal impedance. Consequently, they impose "rigid" potential levels; nothing that happens in a circuit can modify these levels. On the other hand, ac and dc *current* sources have infinite internal impedance. Consequently, they deliver a constant current, and the voltage levels must respond accordingly.

2. **Potential across a switch**. When a switch is open (Fig. 21-5), the voltage across its terminals depends exclusively upon the external elements that make up the circuit. On the other hand, when the switch is closed, the potential level of both terminals must be the same. Thus, if we happen to know the level of terminal 2, then the level of terminal 1 is also known. This simple rule also applies to thyristors and diodes, because they behave like switches.

Figure 21-5
Potential across a switch.

3. **Potential across a resistor**. If no current flows in a resistor, its terminals 3, 4 must be at the same potential, because the *IR* drop is zero (Fig. 21-6). Consequently, if we happen to know the potential level of one of the terminals, the level of the other is also known. On the other hand, if the resistor carries a

Figure 21-6
Potential across a resistor.

current I, the IR drop produces a corresponding potential difference between the terminals. For example, if current flows in the direction shown in Fig. 21-6, the potential of terminal 3 is above that of terminal 4, by an amount equal to IR.

4. **Potential across a coil or inductance**. The terminals of a coil are at the same potential only during those moments when the current is not changing. If the current varies, the potential difference is given by

$$E = L \, (\Delta I / \Delta t) \qquad \text{Eq. 2-27}$$

Thus, if the current in Fig. 21-7 is *increasing* while flowing in the direction shown, the potential of terminal 5 is *above* that of terminal 6, by an amount equal to $L \, \Delta I / \Delta t$.

Figure 21-7
Potential across an inductor.

5. **Potential across a capacitor**. The terminals of a capacitor are at the same potential only when the capacitor is completely discharged. Furthermore, the potential difference between the terminals remains constant whenever current I is zero (Fig. 21-8).

Figure 21-8
Potential across a capacitor.

6. **Initial potential level**. A final rule regarding potential levels is worth remembering. Unless we know otherwise, we assume the following initial conditions:

a. all currents in the circuit are zero;
b. all capacitors are discharged.

These assumed starting conditions enable us to analyze the behavior of any circuit from the moment power is applied.

21.3 The diode

A diode is an electronic device possessing two terminals, respectively called anode (A) and cathode (K) (Fig. 21-9). Although it has no moving parts, a diode acts like a high-speed switch whose contacts open and close according to the following rules:

Rule 1. When no voltage is applied across a diode, it acts like an open switch. The circuit is therefore open between terminals A and K (Fig. 21-9a).

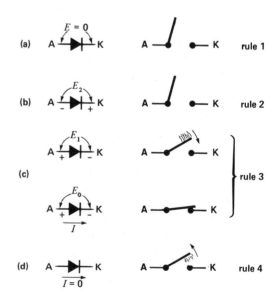

Figure 21-9
Basic rules governing diode behavior.

Rule 2. If we apply an *inverse voltage* E_2 across the diode, so that the anode is *negative* with respect to the cathode, the diode continues to act as an open switch (Fig. 21-9b).

Rule 3. If a momentary *forward voltage* E_1 is applied across the terminals, so that anode A is slightly *positive* with respect to the cathode, the terminals become short-circuited. The diode acts like a closed switch and a current immediately begins to flow from anode to cathode (Fig. 21-9c). Only a few tenths of a volt are needed to trigger the diode so that its "contacts" close.

While the diode conducts, a small voltage drop appears across its terminals. However, the voltage drop is less that 1.5 V, so we can neglect it in most electronic circuits. It is precisely because of this small drop that we can assume the diode is essentially a closed switch when it conducts.

Rule 4. As long as current flows, the diode acts like a closed switch. However, if it stops flowing for even as little as 10 μs, the diode immediately returns to its original open state (Fig. 21-9d). Conduction will only resume when the anode again becomes slightly positive with respect to the cathode (Rule 3).

In conclusion, a diode behaves like a normally open switch whose contacts close as soon as the anode *voltage* becomes slightly positive with respect to the cathode. Its contacts only reopen when the *current* (not the voltage) has fallen to zero. This simple rule is crucially important to an understanding of circuits involving diodes and thyristors.

Symbol for a diode. The symbol for a diode bears an arrow that indicates the direction of conventional current flow when the diode conducts.

(c)

(a)   (b)   (d)

Figure 21-10
a. Average current: 4 A; PIV: 400 V; body length: 10 mm; diameter: 5.6 mm.
b. Average current: 15 A; PIV: 500 V; stud type; length less thread: 25 mm; diameter: 17 mm.
c. Average current: 500 A; PIV: 2000 V; length less thread: 244 mm; diameter: 40 mm.
d. Average current: 2600 A; PIV: 2500 V; Hockey Puk; distance between pole faces: 35 mm; overall diameter: 98 mm.
   (*Photos courtesy of International Rectifier*)

21.4 Principal characteristics of a diode

Peak inverse voltage. A diode can withstand only so much inverse voltage before it breaks down. The *peak inverse voltage* (PIV), ranges from 50 V to 2000 V, depending on the construction. If we exceed the rated PIV, the diode begins to conduct in reverse and, in many cases, it is immediately destroyed.

Maximum average current. There is also a limit to the current a diode can carry. The maximum current may range from a few hundred milliamperes to over 2000 A, depending on the construction and size of the diode.

Maximum temperature. A diode must never operate beyond its rated temperature. Most silicon diodes can operate satisfactorily provided the internal temperature lies between $-50°C$ and $+200°C$. The temperature of a diode can change very quickly owing to its small size and small mass. To improve heat transfer, diodes

are usually mounted on thick metallic supports, called *heat sinks*. Furthermore, in large installations, the diodes may be cooled by fans, by oil, or by a continuous flow of deionized water. Table 21A gives the specifications of some typical diodes. Figure 21-10 shows a range of low power to very high power diodes. Diodes have many applications and to illustrate their properties, we now analyze a few simple circuits.

21.5 Battery charger with series resistor

The circuit of Fig. 21-11a represents a battery charger. Transformer T, connected to a 120 V ac supply, furnishes a sinusoidal secondary voltage having a peak of 100 V. A 60 V battery, a 1 Ω resistor, and a diode D are connected in series across the secondary.

To explain the operation of the circuit, let us choose point 1 as the reference terminal. The potential of this terminal is therefore a straight horizontal line. The potential of terminal 2 swings sinusoidally above and below point 1,

TABLE 21 A			PROPERTIES OF SOME TYPICAL DIODES					
relative	I_0	E_0	I_{cr}	E_2	I_2	T_J	d	l
power	A	V	A	V	mA	°C	mm	mm
low	1	0.8	30	1000	0.05	175	3.8	4.6
medium	12	0.6	240	1000	0.6	200	11	32
high	100	0.6	1 600	1000	4.5	200	25	54
very high	1000	1.1	10 000	2000	50	200	47	26

I_0 - average dc current
E_0 - voltage drop corresponding to I_0
I_{cr} - peak value of surge current for 1 cycle
E_2 - peak inverse voltage
I_2 - reverse leakage current corresponding to E_2
T_J - maximum junction temperature (inside the diode)

d - diameter
l - length

(a)

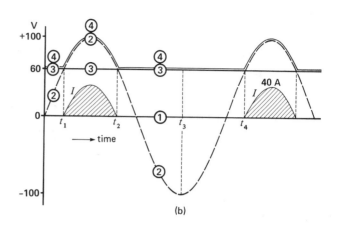

(b)

Figure 21-11
a. Simple battery charger circuit

b. Corresponding voltage and current waveforms.

according to whether 2 is positive or negative with respect to 1. The level of terminal 3 is always 60 V above terminal 1, because the battery voltage is constant. The potential levels are shown in Fig. 21-11b.

Circuit analysis

a. Prior to $t = 0$, we assume all currents are zero. The potential of point 4 is therefore at the same level as point 3.
b. During the interval from 0 to t_1, anode 2 is negative with respect to cathode 4; consequently, the diode cannot conduct (Rule 2);
c. At instant t_1, terminal 2 becomes positive with respect to 4, and the diode begins to conduct (Rule 3). From this moment on, the diode acts like a closed switch. Consequently, point 4 must follow point 2.
d. From t_1 to t_2 the current in the circuit is given by:

$$I = E_{43}/1 \ \Omega$$

The current reaches a peak of 40 A when $E_{43} = +40$ V. As long as current flows, the diode behaves like a closed switch;
e. At instant t_2, the current is zero, and the diode immediately "opens" the circuit (Rule 4). From this moment on, point 4 must follow point 3.

f. From t_2 to t_4, point 2 is negative with respect to point 3. Because point 4 follows point 3 (no IR drop in the resistor), the PIV across the diode reaches a maximum of 160 V, at instant t_3;
g. Finally, from t_4 on, the cycle (c to f above) repeats itself.

This circuit produces a pulsating current that always flows into the positive terminal of the battery. Consequently, the latter receives energy and progressively charges up.

21.6 Battery charger with series inductor

The current flow in the battery charger of Fig. 21-11 is limited by resistor R. Unfortunately, this produces I^2R losses and a corresponding poor efficiency. We can get around this problem by replacing the resistor by an inductor, as shown in Fig. 21-12a. Let us analyze the operation of this circuit.

a. As in the previous example, the diode begins to conduct at instant t_1 when anode 2 becomes positive with respect to cathode 4. From this moment on, point 4 follows point 2, and voltage E_{43} appears across the inductor (Fig. 21-12b). The latter begins to "accumu-

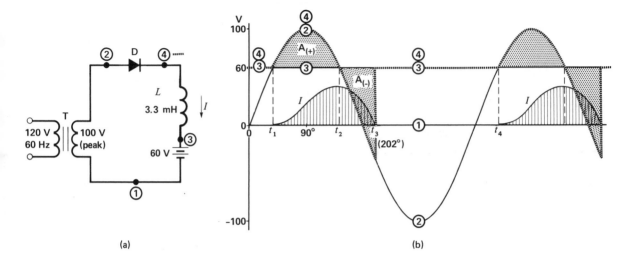

(a)

(b)

Figure 21-12
a. Battery charger using a series inductor.

b. Corresponding voltage and current waveforms.

late" volt seconds and the current increases gradually until it reaches a maximum given by

$$I_{max} = A_{(+)}/L \qquad \text{Eq. 2-28}$$

where

$A_{(+)}$ = dotted area between t_1 and t_2 [V·s]
L = inductance [H]

Note that the current reaches its peak at instant t_2 whereas it was zero at this moment, when a resistor was used. This is consistent with the fact that current through an inductor lags behind the applied voltage.

b. The current decreases between t_2 and t_3, becoming zero at t_3 when dotted area $A_{(-)}$ is equal to dotted area $A_{(+)}$.

c. As soon as the current is zero, the diode "opens" the circuit, whereupon point 4 must jump to the level of point 3. It stays at this level until instant t_4 whereupon the whole cycle repeats.

This is an interesting example of the use of an inductor to store and release electrical

energy. During the interval from t_1 to t_2, the inductor stores energy and, from t_2 to t_3, it returns it again to the circuit (see Sec. 2.13).

Example 21-1:
The coil in Fig. 21-12 has an inductance of 3.3 mH and the battery voltage is 60 V. Calculate the peak current if the supply frequency is 60 Hz.

Solution:
a. To calculate the peak current, we must find the value of area $A_{(+)}$. This can be done by integral calculus, but we shall employ a much simpler graphical method. Thus, referring to Fig. 21-12c, we have redrawn the voltage levels using graph paper. The 360-degree voltage cycle is divided into 24 equal parts, each representing an interval

$$\Delta t = (1/24) \times (1/60) = 1/1440 \text{ s}$$

Similarly, the ordinates are scaled off in 10 V intervals. Consequently, each small square represents an area of $(1/1440) \times 10 = 0.006944$ V·s = 6.944 mV·s.

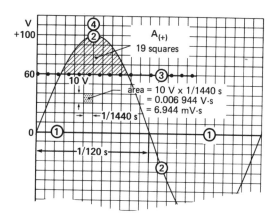

Figure 21-12c
See Example 21-1.

b. By counting squares, we find that $A_{(+)}$ contains 19 squares; consequently, its area corresponds to:

$$A_{(+)} = 19 \times 6.944 = 132 \text{ mV·s}$$
$$= 0.132 \text{ V·s}$$

c. The peak current is:

$$I_{max} = A_{(+)}/L = 0.132/0.0033 = 40 \text{ A}$$

Thus, the current reaches the same peak with an inductor of 3.3 mH as it did with a resistor of 1 Ω. However, the big advantage of the inductor is that it has essentially no losses. The conversion of ac power to dc power is therefore much more efficient.

21.7 Single-phase bridge rectifier

The circuit of Fig. 21-13a enables us to rectify both the positive and negative half cycles of an ac source, to supply a dc load R. The four diodes together make up what is called a *single-phase bridge rectifier*.

The circuit operates as follows. When source voltage E_{12} is positive, terminal 1 is positive with respect to terminal 2 and current i_a flows through R by way of diodes A1 and A2. Consequently, point 3 follows point 1 and point 4 follows point 2 during this conduction interval. Conduction ceases when i_a falls to zero at instant t_1 (Fig. 21-13b). The polarity then reverses and E_{21} becomes positive, meaning that terminal 2 is positive with respect to terminal 1. Current i_b now flows through R in the same direction as before, but this time by way of diodes B1 and B2. Consequently, point 3

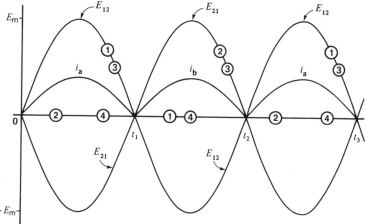

Figure 21-13
a. Single-phase bridge rectifier.

b. Voltage levels.

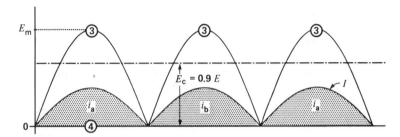

Figure 21-13c
Voltage and current waveforms
in load R.

now follows point 2 while point 4 follows point 1. Voltage E_{34} across the load is therefore composed of a series of half-cycle sine waves that are always positive. The voltage pulsates between zero and a maximum value E_m equal to the peak voltage of the source. The average value of this rectified voltage is given by:

$$E_d = 0.90E \qquad (21\text{-}1)$$

where

E_d = dc voltage of a single-phase bridge rectifier [V]

E = effective value of the ac line voltage [V]

0.90 = constant [exact value = $(2\sqrt{2})/\pi$]

Referring to Fig. 21-13b, in addition to drawing the curve E_{12} we have also drawn curve E_{21}. This enables us to use the potential levels of either terminals 1 or 2 as zero reference potentials. Thus, we can select terminal 2 as the reference during the first half cycle, terminal 1 during the second half cycle, and so on, on alternate half cycles. By using this technique, terminal 4 always remains at zero potential while terminal 3 follows the positive part of the sine waves. It then becomes evident that E_{34} across the load is a pulsating dc voltage. Figure 21-13c shows the rectified voltage and current of the load.

In addition to its dc value, load voltage E_{34} contains an ac component whose fundamental frequency is twice the line frequency. In effect, the voltage across the load pulsates between zero and $+E_m$, twice per cycle. Consequently, the so-called peak-to-peak *ripple* is equal to E_m.

In the case of a resistive load, the current has the same waveshape as the voltage; its average, or dc value, is given by:

$$I_d = E_d/R$$

Example 21-2:
The ac source in Fig. 21-13a has an effective voltage of 120 V, 60 Hz. The load draws a dc current of 20 A. Calculate:
a. the dc voltage across the load;
b. the average dc current in each diode;

Solution:
a. The dc voltage across the load is given by Eq. 21-1:
$$\begin{aligned} E_d &= 0.90E \\ &= 0.90 \times 120 \\ &= 108 \text{ V} \end{aligned}$$

b. The dc current in the load is 20 A, but the diodes only carry the current on alternate half cycles. Consequently, the average dc current in each diode is:
$$I = I_d/2 = 20/2 = 10 \text{ A}$$

21.8 Filters

The rectifier circuits we have studied so far produce pulsating voltages and currents. In some types of loads, we cannot tolerate such pulsations, and *filters* must be used to smooth out the valleys and peaks. The basic purpose of a dc filter is to produce a smooth power flow into a load. Consequently, a filter must absorb energy whenever the dc voltage or current tends to rise, and release energy whenever the voltage or current tends to fall. In this way, the filter tends to maintain a constant voltage and current in the load.

(b)

Figure 21-14
a.   Rectifier with inductive filter.
b.   Rectifier with capacitive filter.

The most common filters are inductors and capacitors. Inductors store energy in their magnetic field. They tend to maintain a constant current; consequently, they are placed *in series* with the load (Fig. 21-14a). Capacitors store energy in their electric field. They tend to maintain a constant voltage; consequently, they are placed *in parallel* with the load (Fig. 21-14b).

The filtering action improves as we increase the amount of energy stored in the filter. In the case of a bridge rectifier using an inductor, we obtain good current smoothing (peak-to-peak ripple less than 5 % of the dc current) provided that:

$$W_L > P/f \qquad (21\text{-}2)$$

where

W_L = dc energy stored in the smoothing inductor [J]

P = dc power absorbed by the load [W]

f = frequency of the source [Hz]

The current in Fig. 21-14a is much more constant than in Fig. 21-13a. The voltage between terminals 3 and 4 pulsates between zero and E_m as before, but it is very smooth

across the load (Fig. 21-15). The dc voltage across the load is again given by Eq. 21-1. This is to be expected because the dc *IR* drop across the inductor is negligibly small.

Bridge rectifiers are used to provide dc current for relays, electromagnets, motors, and many other magnetic devices. In most cases, the self-inductance of the coil is sufficient to provide good filtering. Thus, although the voltage across the coil may pulsate very strongly, the dc current through it is smooth. Consequently, the magnetic field pulsates very little.

Example 21-3:
We wish to build a 135 V, 20 A dc power supply using a single-phase bridge rectifier and an inductive filter. The peak-to-peak current ripple should be less than 5%. If a 60 Hz ac source is available, calculate:
a.   the effective value of the ac voltage;
b.   the energy stored in the inductor;
c.   the inductance of the inductor;
d.   the peak-to-peak current ripple.

Solution:
a.   The ac voltage is given by Eq. 21-1:
$$E_d = 0.9E$$
$$135 = 0.9\,E$$
$$E = 150 \text{ V}$$

b.   The dc power output of the rectifier is:
$$P = E_d I_d$$
$$= 135 \times 20$$
$$= 2700 \text{ W}$$

Figure 21-15
Current and voltage waveforms with inductive filter.

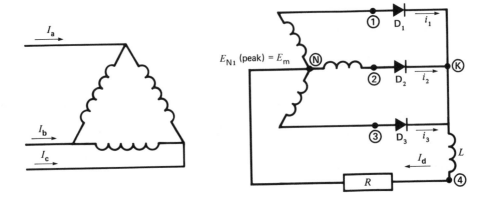

Figure 21-16
Three-phase, 3-pulse rectifier with inductive filter.

The energy to be stored in the inductor (or "choke") is given by:

$$W_L > P/f$$ Eq. 21-2

$$W_L > 2700/60$$

$$> 45 \text{ J}$$

Consequently, to keep the current ripple within 5%, the inductor must store at least 45 J in its magnetic field.

c. The inductance of the choke can be calculated from:

$$W_L = \frac{1}{2} L I_d^2$$ Eq. 2-8

$$45 = \frac{1}{2} L(20)^2$$

$$L = 0.225 \text{ H}$$

d. The peak-to-peak ripple is about 5% of the dc current:

$$I_{peak-to-peak} = 0.05 \times 20 = 1 \text{ A}$$

The dc output current therefore pulsates between 19.5 A and 20.5 A.

21.9 Three-phase, 3-pulse rectifier

The simplest 3-phase rectifier is composed of three diodes connected in series with the sec-

ondaries of a 3-phase, delta-wye transformer (Fig. 21-16). The line-to-neutral voltage has a peak value E_m. A very large filter inductance L is connected in series with the load, so that current I_d remains ripple-free. Although the load is represented by a resistance R, in reality it is always a useful energy-consuming device, and not a heat-dissipating resistor. Thus, the load may be a dc motor, a large magnet or an electroplating bath. This simple rectifier has some serious drawbacks, but it provides a good introduction to 3-phase rectifiers in general. We now analyze its behavior.

1. **Voltage across the load.** Choosing the transformer neutral as a zero reference point, the secondary terminals follow the voltage levels 1, 2, 3 shown in Fig. 21-17. These potential levels are rigidly fixed by the ac source and they reach a peak value E_m.

   Before we energize the circuit, points K, 4, N are at the same level because I is zero. However, the moment we apply power ($t = 0$) the potential of point 1 suddenly becomes positive with respect to K. This immediately initiates conduction in diode D1 (Sec. 21-3, Rule 3). Current i_1 increases rapidly, attaining a final value I which depends upon load R. During this interval, K

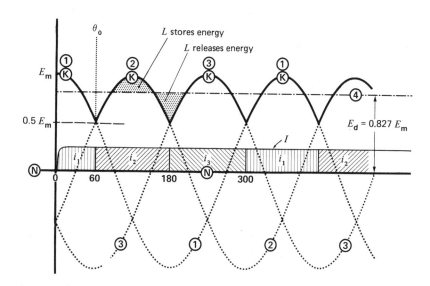

Figure 21-17
Voltage and current waveforms in a 3-phase, 3-pulse rectifier.

is at the same level as point 1 because the diode is conducting.

As points K and 1 move together in time, they eventually reach a critical moment, corresponding to an angle θ_0 of 60° (Fig. 21-17). The moment is critical because immediately later, terminal 2 becomes positive with respect to K and 1. According to Rule 3, this initiates conduction in diode D2, so that *it* begins to carry current I. At the same time that conduction starts in diode D2, it ceases in diode D1. Consequently, beyond 60°, point K follows the level of point 2.

The sudden switchover from one diode to another is called *commutation*. When the switchover takes place automatically (as it does in our example), it is called *natural* commutation. Commutation from one diode to another does not really take place instantaneously, as we have indicated. Owing to transformer leakage reactance, the current gradually increases in diode D2 while it decreases in diode D1. This gradual transition continues until all the load current is carried by diode D2. However, the commutation period is short (less than 2 ms on a 60 Hz system) and, for our purposes, we

shall assume it occurs instantaneously.

The next critical moment occurs at 180°, because terminal 3 then becomes positive with respect to point 2 (and point K). Commutation again takes place as the load current switches from diode D2 to diode D3. Point K therefore follows the positive peaks of waves 1, 2, and 3, and each diode carries the full load current for equal intervals of time (120°). The diode currents i_1, i_2, i_3 have a rectangular waveshape composed of positive current intervals of 120° followed by zero current intervals of 240°.

Voltage E_{kN} across the load and inductor pulsates between $0.5E_m$ and E_m. The ripple voltage is therefore smaller than that produced by a single-phase bridge rectifier (Fig. 21-15). Furthermore, the fundamental ripple frequency is three times the supply frequency, which makes it easier to achieve good filtering. The dc voltage across the load is given by:

$$E_d = 0.675E \qquad (21-3)$$

where

$E_d =$ average or dc voltage of a 3-pulse rectifier [V]

E = effective ac line voltage [V]

0.675 = a constant [exact value

= $3/(\pi \sqrt{2})$]

Note that if we reverse the diodes in Fig. 21-16, the rectifier operates the same way, except that the load currents reverse. Voltage E_{KN} becomes negative, and point K follows the negative peaks of waves 1, 2 and 3.

2. **Line currents**. Currents i_1, i_2, i_3 which flow in the diodes also flow in the secondary windings of the transformer. These currents have a chopped rectangular waveshape which is quite different from the sinusoidal currents we are familiar with. Furthermore, the currents flow for only 1/3 of the time in a given winding. Owing to this intermittent flow, the nominal rating of the transformer cannot be harnessed to deliver an equivalent amount of dc power. For example, if the transformer in Fig. 21-16 has a rating of 100 kVA it can only deliver 74 kW of dc power without overheating.

The chopped secondary currents are reflected into the primary windings, with the result that the transmission line currents feeding the transformer also change very abruptly. The sudden jumps in currents I_a, I_b and I_c produce rapid fluctuations in the magnetic field surrounding the transmission line.

These fluctuations can induce substantial voltages and noise in nearby telephone lines.

Owing to these drawbacks, we try to design rectifiers so that the line currents change more gradually and so that the transformer windings carry current for more than one-third of the time. This is achieved by using 3-phase, 6-pulse rectifiers.

21.10 Three-phase, 6-pulse rectifier*

Consider the circuit of Fig. 21-18 in which a transformer T (identical to the one shown in Fig. 21-16), supplies power to 6 diodes and their associated loads R_1 and R_2. The upper set of diodes together with inductor L_1 and load R_1 are identical to the 3-phase, 3-pulse rectifier we have just studied. Thus, load current I_{d1} flows in the neutral line, as shown. The lower set of diodes, together with R_2 and L_2, also constitute a 3-phase, 3-pulse rectifier. The corresponding load current I_{d2} flows in the neutral as shown. The two 3-phase rectifiers operate quite independently of each other, K following the positive peaks of points 1, 2, 3, while A follows the negative peaks. All diodes conduct during 120° intervals.

If we make $R_1 = R_2$, then $I_{d1} = I_{d2}$ and the current in the neutral becomes zero. Conse-

* Also called 3-phase bridge rectifier.

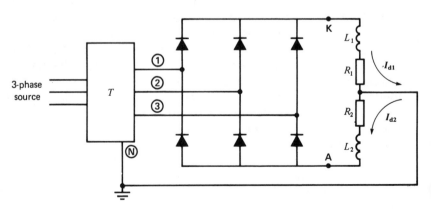

Figure 21-18
Dual 3-phase, 3-pulse rectifier.

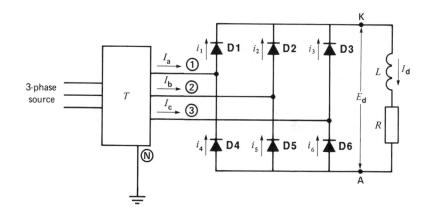

Figure 21-19
Three-phase, 6-pulse rectifier
with inductive filter.

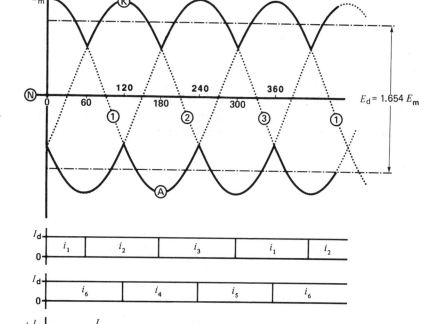

Figure 21-20
Voltage and current waveforms
in Fig. 21-19.

quently, we can remove the neutral conductor, yielding the circuit of Fig. 21-19. The two loads and the two inductors are simply combined into one, shown as R and L, respectively. The 6 diodes constitute what is called a 3-phase, 6-pulse rectifier. It is called 6-pulse because the currents flowing in the 6 diodes start at six different moments during each cycle of the line frequency. However, each diode still conducts for only 120°.

The line currents I_a, I_b, I_c supplied by the transformer are given by Kirchhoff's law:

$$I_a = i_1 - i_4$$
$$I_b = i_2 - i_5$$
$$I_c = i_3 - i_6$$

They consist of three identical rectangular waves that are out of phase by 120° (Fig. 21-20). The currents now flow for 2/3 of the time in the secondary windings. Consequently, a 100 kVA transformer can deliver 95 kW of dc power without overheating.

The average dc output voltage is twice that of a 3-phase, 3-pulse rectifier, and its value is given by:

$$E_d = 1.35E \qquad (21-4)$$

where

E_d = dc voltage of a 6-pulse rectifier [V]
E = effective line voltage [V]
1.35 = a constant [exact value = $3\sqrt{2}/\pi$]

It is much easier to visualize the shape of the output voltage E_{KA}, by using A as a reference point. Thus, in Fig. 21-21, we show the line voltages, E_{12}, E_{23}, E_{31} (and E_{21}, E_{32}, E_{13}) rather than the line-to-neutral voltages used in Fig. 21-20. This approach is identical to that used in the single-phase rectifier of Figure 21-13b. The level of K follows the tops of the successive sine waves while A remains at zero potential. The output voltage fluctuates between 1.414E and 1.225E, where E is the effective value of the line voltage.* The average value of E_{KA} is 1.35E, as given by Eq. 21-4.

The peak-to-peak ripple is only (1.414 − 1.225) E = 0.189E and the fundamental ripple frequency is six times the line frequency. Consequently, the ripple is much easier to filter. In

* $\sqrt{2}E = 1.414E$; $\sqrt{2}E \cos 30° = 1.225\ E$.

Figure 21-21
Another way of showing E_{KA} using line voltage potentials. Note also the position of E_{2N} with respect to the line voltages.

effect, we obtain good inductor filtering (peak-to-peak current ripple less than 5%) provided that:

$$W_{\mathrm{L}} > 0.03P/f \qquad (21\text{-}5)$$

where

W_{L} = dc energy stored in the inductor [J]
P = dc power absorbed by the load [W]
f = frequency of the source [Hz]

Figure 21-21 shows that the inductor stores energy whenever the rectifier voltage exceeds the average value E_{d}. This energy is then released during the brief interval when the rectifier voltage is less than E_{d}.

The PIV across each diode is equal to the peak value of the line voltage, or $\sqrt{2E}$.

The 3-phase, 6-pulse rectifier is a big improvement over the 3-phase, 3-pulse rectifier. It constitutes the basic building block of most larger rectifier installations.

Another way of looking at the 3-phase bridge rectifier is to consider the diodes to be in a box (Fig. 21-22). The box is fed by three ac lines and it has two output terminals K and A. The diodes act like automatic switches that successively connect these terminals to the ac lines. The connections can be made in six distinct ways, as shown in Fig. 21-22. It follows that the output voltage E_{KA} is composed of segments of the ac line voltages. That is why we draw line voltages in Fig. 21-21 instead of line-to-neutral voltages.

Each dotted connection in Fig. 21-22 represents a diode that is conducting. The successive 60-degree intervals correspond to those in Fig. 21-20. For example, in the interval from 300° to 360°, diodes D1 and D5 are conducting. It follows from Fig. 21-19 that K is effectively connected to line 1 while A is effectively connected to line 2.

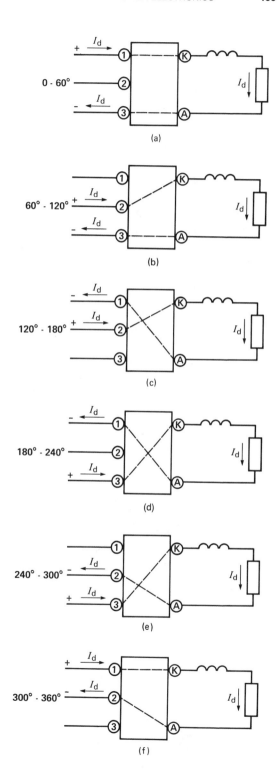

(a)

(b)

(c)

(d)

(e)

(f)

Figure 21-22
Successive diode connections between the 3-phase input and dc output terminals of a 3-phase, 6-pulse rectifier.

Because the diode voltage drop is small, we can assume that each dotted line represents a loss-free connection. The dc power absorbed by the load must therefore be equal to the active power drawn from the 3-phase source.

Example 21-4:
A 3-phase bridge rectifier has to supply power to a 360 kW, 240 V dc load. If a 600 V, 3-phase, 60 Hz feeder is available, calculate:
a. voltage rating of the 3-phase transformer;
b. dc current per diode;
c. PIV across each diode;
d. peak-to-peak ripple in the output voltage and its frequency.

Solution:
a. Secondary line voltage is:
$$E = E_d/1.35 = 240/1.35$$
$$= 177 \text{ V}$$

Thus, a 3-phase, transformer having a line voltage ratio of 600 V/177 V would be satisfactory. The primary and secondary windings may be connected either in wye or in delta.

b. dc load current I_d = 360 kW/240 V
$$= 1500 \text{ A}$$

dc current per diode = 1500/3 = 500 A

peak current in each diode = 1500 A

c. PIV across each diode
$$= \sqrt{2}E = 1.414 \times 177$$
$$= 250 \text{ V}$$

d. The output voltage E_{KA} fluctuates between 1.225E and 1.414E (Fig. 21-21). In other words, the voltage fluctuates between
$$E_{min} = 1.225 \times 177 = 217 \text{ V and}$$
$$E_{max} = 1.414 \times 1.77 = 250 \text{ V}$$

The peak-to-peak ripple is therefore:
$$E_{peak\text{-}to\text{-}peak} = 250 - 217 = 33 \text{ V}$$

Fundamental ripple frequency
$$= 6 \times 60 \text{ Hz} = 360 \text{ Hz}$$

Example 21-5:
a. Calculate the inductance of the smoothing choke required in Example 21-4, if the peak-to-peak ripple in the current is not to exceed 5%.
b. Does the presence of the choke modify the peak-to-peak ripple in the output voltage E_{KA}?

Solution:
a. Using Eq. 21-5, we have:
$$W_L > 0.03P/f$$
$$> 0.03 \times 360\ 000/60$$
$$> 180 \text{ J}$$

Consequently, the inductor must store 180 J in its magnetic field. The inductance is found from:

$$W_L = \frac{1}{2}LI^2_d \qquad \text{Eq. 2-8}$$

$$180 = \frac{1}{2}L\ (1500)^2$$

$$L = 1.6 \times 10^{-4}$$

$$= 0.16 \text{ mH}$$

b. No, the choke does not affect the voltage ripple between K and A.

21.11 Effective line current, fundamental line current and harmonics

We saw in Fig. 21-20 that the ac line currents consist of 120-degree rectangular waves having an amplitude I_d, where I_d is the dc current flowing in the load. Let us direct our attention to the current I_b flowing in line 2, and the corresponding line-to-neutral voltage E_{2N}. They are reproduced in Fig. 21-23 and it can be seen that the rectangular current wave is symmetrically located with respect to the sinusoidal voltage. In other words, the center of the positive current pulse coincides with the peak of the positive voltage "pulse". Thus, I_b is in phase with E_{2N}.

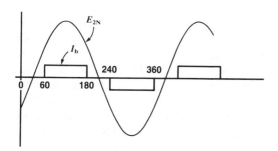

Figure 21-23
Line-to-neutral voltage and line current in phase 2 of Fig. 21-20.

The effective value I of the line current is given by the relationship

$$I^2 \times 180° = I_d^2 \times 120°$$

whence

$$I = (\sqrt{120/180})\, I_d$$
$$= 0.816\, I_d \qquad (21\text{-}6)$$

This effective current is composed of a fundamental rms component I_F plus all the harmonic components. As we have seen, I_F is in phase with the line to neutral voltage.

What is the value of I_F? To calculate it we reason as follows.

The dc power to the load is

$$P_d = E_d I_d$$

The active power supplied to the rectifier (and its load) is

$$P_{ac} = \sqrt{3}\, E I_F \qquad \text{Eq. 8-9}$$

Because no power is lost or stored in our ideal rectifier, it follows that $P_{ac} = P_d$. We can therefore write:

$$P_{ac} = P_d$$
$$\sqrt{3}\, E I_F = E_d I_d$$
$$= 1.35\, E I_d$$

and so

$$I_F = 0.779 I_d \qquad (21\text{-}7)$$

Combining Eqs 21-6 and 21-7 we find

$$I_F = 0.955 I \qquad (21\text{-}8)$$

Thus, owing to the presence of harmonics, the fundamental component I_F is slightly less than the effective current I.

21.12 Distortion power factor

We have just seen that the fundamental component I_F is in phase with the corresponding line to neutral voltage (Fig. 21-23). Consequently, we would be inclined to say that the power factor of the 3-phase, 6-pulse rectifier is 100 percent. However, by definition, power factor is given by the expression:

$$\text{power factor} = \frac{\text{active power}}{\text{apparent power}}$$
$$= \frac{P_{ac}}{E I \sqrt{3}}$$
$$= \frac{E I_F \sqrt{3}}{E I \sqrt{3}}$$
$$= \frac{I_F}{I}$$

Thus, in our example

$$\text{power factor} = 0.955 \qquad (21\text{-}9)$$

Thus, the true power factor is not 100% but only 95.5%. The reason is that the line current is rectangular and not sinusoidal. The power factor of 95.5% is due to distortion in the current and that is why this power factor is often called *distortion power factor*. The distortion power factor is always given by the ratio

$$\begin{array}{l}\text{distortion} \\ \text{power} \\ \text{factor}\end{array} = \frac{\left[\begin{array}{l}\text{rms value of the fundamental} \\ \text{component of the line current}\end{array}\right]}{[\text{rms value of the line current}]}$$

Although the total power factor of our rectifier is

less than 100%, the fundamental component of current is nevertheless in phase with the line to neutral voltage. Consequently, this ideal rectifier absorbs no reactive power from the line.

21.13 The thyristor*

A thyristor may be considered to be a diode whose moment of conduction can be controlled. Like a diode, a thyristor possesses an anode and a cathode, plus a third terminal called gate (Fig. 21-24). If the gate is con-

Figure 21-24
Symbol of a thyristor or SCR.

nected to the cathode, the thyristor will not conduct, even if the anode is positive†. The thyristor is said to be *blocked* (Fig. 21-25a). To initiate conduction, two conditions have to be met:

a. the anode must be positive;
b. a current I_g must flow into the gate for a few microseconds. In practice, the current is injected by applying a short, positive voltage pulse E_g to the gate (Fig. 21-25b).

---

* Thyristor is a generic term that applies to all four-layer controllable semiconductor devices. However, we use it in this book to refer specifically to the reverse-blocking triode thyristor, commonly called SCR (semiconductor controlled rectifier). This is in response to a seemingly general trend in the literature to use the terms SCR and thyristor interchangeably.

† To simplify the wording, we adopt the following terms:
  1. When the anode is positive with respect to the cathode, we simply say the anode is positive;
  2. When the gate is positive with respect to the cathode, we simply say the gate is positive.

As soon as conduction starts, the gate loses all further control. Conduction will only stop when anode current I falls to zero, after which the gate again exerts control.

Basically, a thyristor behaves the same way a diode does except that the gate enables us to initiate conduction precisely when we want to. This seemingly slight advantage is of profound importance. It enables us not only to convert ac power into dc power, but also to do the reverse: convert dc power into ac power. Thanks to the development of reliable thyristors (SCRs), we are witnessing a fundamental change in the control of large blocks of power. Table 21B lists some of the properties of typical thyristors. See also Figure 21-26.

21.14 Principles of gate firing

Consider Fig. 21-27a in which a thyristor and a resistor are connected in series across an ac source. A number of short positive pulses E_g is applied to the gate, of sufficient amplitude to initiate conduction. These pulses may be generated by a manual switch or an electronic control circuit.

Referring to Fig. 21-27b, the gate pulses occur at angle θ_1, θ_2, θ_3, θ_4 and θ_5. Table 21C explains how the circuit reacts to these pulses. The reader should follow these explanations carefully.

(a)   (b)

Figure 21-25
a. A thyristor does not conduct when the gate is connected to the cathode.
b. A thyristor conducts when the anode is positive and a current pulse is injected into the gate.

TABLE 21B			PROPERTIES	OF	SOME	TYPICAL	THYRISTORS		
relative	I_1	I_{cr}	E_2	E_P	I_G	E_G	T_J	d	l
power	A	A	V	V	mA	V	°C	mm	mm
medium	8	60	500	- 10	50	2.5	105	11	33
high	110	1 500	1200	- 5	50	1.25	125	27	62
very high	1200	10 000	1200	- 20	50	1.5	125	58	27

I_1 - maximum effective current during conduction

I_{cr} - peak value of surge current for 1 cycle

E_2 - peak inverse anode voltage

E_P - peak inverse gate voltage

E_G - positive gate voltage to initiate conduction

I_G - gate current corresponding to E_G

T_J - maximum junction temperature

d - diameter

l - length

(b)

(a)

(c)

Figure 21-26
Range of SCRs from medium to very high power capacity.
a.   Average current: 50 A; voltage: 400 V; length less thread: 31mm; diameter: 17 mm.
b.   Average current: 285 A; voltage 1200 V; length less thread: 244 mm; diameter: 37 mm.
c.   Average current: 1000 A; voltage: 1200 V; distance between pole faces: 27 mm; overall diameter: 73 mm. (*Photos courtesy of International Rectifier*)

(a)

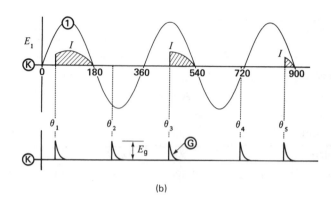

(b)

Figure 21-27
a. Thyristor and resistor connected to an ac source. (see description below).
b. Thyristor behavior depends on the timing of the gate pulses.

TABLE 21C	DESCRIPTION OF THYRISTOR BEHAVIOR (see Fig. 21-27)
angle or time interval	**Explanation of circuit operation**
zero to θ_1	Although the anode is positive, conduction is impossible because the gate voltage is zero. The thyristor behaves like an open switch.
angle θ_1	Conduction starts because both the anode and gate are positive.
θ_1 to 180°	Conduction continues even though the gate voltage has fallen to zero. Gate pulses have no further effect once the thyristor conducts. The anode to cathode voltage drop is less than 1.5 V; consequently, we can consider that the anode and cathode are shorted. The thyristor behaves like a closed switch.
angle 180°	The thyristor current is zero, conduction ceases and the gate regains control.
180° to 360°	Conduction is impossible because the anode is negative. Although the gate is triggered at angle θ_2, it produces no effect. The thyristor experiences an inverse voltage during this half cycle.
360° to 540°	Conduction starts at θ_3 and ceases again as soon as the current is zero. The gate pulse is delayed more than during the first positive half-cycle. Consequently, the anode current flows for a shorter time.
720° to 900°	Conduction starts at angle θ_5, but the resulting anode current is very small because of the long delay in firing the gate.

To summarize, we can control the current in an ac circuit by delaying the gate pulses with respect to the start of each positive half-cycle. If the pulses occur at the very beginning of each half-cycle, conduction lasts for 180°, and the thyristor behaves like an ordinary diode. On the other hand with a resistive load, if the pulses are delayed, say by 150°, current only flows during the remaining 30° of each half-cycle.

21.15 Power gain of a thyristor

When a voltage pulse is applied to the gate, a certain gate current flows. Because the pulses last only a few microseconds, the average power supplied to the gate is very small, compared to the average power supplied to the load. The ratio of the two powers, called power gain, may exceed one million. Thus, a gate input of only 1 W may control a load of 1000 kW.

An SCR does not, of course, have the magical property of turning one watt into a million watts. The large power actually comes from an appropriate power source, and the SCR only serves *to control* the power flow. Thus, in the same way that a small power input to the accelerator of an automobile produces a tremendous increase in motive power, so does a small input to the gate of an SCR produce a tremendous increase in electrical power.

21.16 Current interruption and forced commutation

A thyristor ceases to conduct and the gate regains control only after the anode current falls to zero. The current may cease flowing quite naturally (as it did at the end of each cycle in Fig. 21-27) or we can force it to zero artificially. Such *forced commutation* is required in some circuits where the anode current has to be interrupted at a specific instant.

Consider Fig. 21-28a in which a thyristor and a load resistor R are connected in series across a dc source E. If we apply a single positive pulse to the gate, the resulting dc load current I_1 will

flow indefinitely. However, we can stop conduction in the SCR in one of 3 ways:
1. momentarily reduce the supply voltage E to zero;
2. open the load circuit by means of a switch;
3. force the anode current to zero for a brief period.

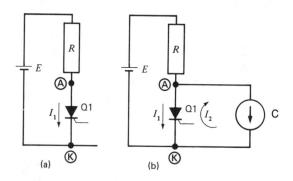

Figure 21-28
a.   Thyristor connected to a dc source.
b.   Forced commutation.

The first two solutions are trivial, so let us examine the third method. In Fig. 21-28b, a variable current source C delivering a current I_2 is connected in parallel with thyristor Q1. As we gradually increase I_2, the net current $(I_1 - I_2)$ flowing in the thyristor decreases. However, so long as the net current is not zero, the thyristor continues to conduct, with the result that the current flowing in the resistor is unchanged. But if we increase I_2 until it is equal to I_1, the thyristor stops conducting, and the gate regains control. In practice, I_2 is a brief current pulse, usually supplied by triggering a second thyristor. For example, in Fig. 21-29a, a load R can be switched on and off by alternately firing thyristors Q1 and Q2.

To understand how the circuit operates, suppose Q1 is conducting and Q2 is not. We assume that the circuit has been in operation long enough so that the voltages and currents have reached their steady-state values. It follows that capacitor C is charged to the supply voltage E, with polarities as shown in Fig. 21-29a. Neglecting the voltage drop across Q1, full voltage appears across load R. Thus, $I_1 = E/R$ and $I_2 = 0$

To stop conduction in R, we trigger thyristor Q2 (Fig 21-29b). This causes the capacitor to discharge in the circuit formed by C, Q1 and Q2. The discharge current I_C forces Q1 to stop conducting and produces the condition shown in Fig. 21-29c. The level of point 1 drops to E volts below the level of point K, with the result that I_1 reaches a momentary peak of $I_1 = 2E/R$. Current I_1 will quickly charge the capacitor in the opposite way and so point 1 will eventually reach the level of point 3. When the transients have subsided, the circuit appears as shown in Fig. 21-29d, but with $I_1 = 0$. Current I_2 can be made much smaller than load current I_1 by using a relatively high resistance R_0.

To restart conduction in the load, we fire Q1, producing the condition shown in Fig. 21-29e. Discharge current I_C now causes the extinction of Q2 and the capacitor charges up with the opposite polarity as shown in Fig. 21-29f. When conditions become stable, the circuit reverts to the one we started with, namely Fig. 21-29a.

The behavior of the voltages and currents is shown graphically in Fig. 21-29g.

This type of forced commutation, using a commutating capacitor, is employed in some converters* that generate their own frequency.

21.17 Basic thyristor power circuits

Thyristors are used in many different ways. However, in power electronics, seven basic circuits cover about 90 percent of all industrial applications. These circuits, and some of their

---

* A converter is any device that converts power of one frequency into power of another frequency. A converter may be a rectifier, an inverter, a cycloconverter or even a rotating machine.

applications, are listed in Table 21D. They are labelled circuit 1, circuit 2, circuit 3, - - - circuit 7.

To explain the principle of operation of these basic circuits, we shall use single-phase sources. In practice, 3-phase sources are mainly used, but single-phase examples are less complex, and they enable us to focus attention on the essential principles involved.

21.18 Controlled rectifier supplying a passive load (Circuit 1)

By definition, a passive load is one that contains no inherent source of energy. The simplest passive load is a resistor.

Figure 21-30a shows a resistive load and a thyristor connected in series across a single-phase source. The source produces a sinusoidal voltage having a peak value E_m. The gate pulses are synchronized with the line frequency and, in our example, they are delayed by an angle of 90°. Conduction is therefore initiated every time the ac voltage reaches its maximum positive value. Based upon explanations given in Sec. 21-13, it is obvious that current will flow for 90°.

In Fig. 21-30b, the current "lags" behind the voltage because it only flows during the final 90 degrees. This lag (also called displacement) produces the same effect as an inductive load. Consequently, the ac source has to supply reactive power Q in addition to the active power P (see Sec. 7-15). The power factor decreases as we delay the triggering pulse. On the other hand, if the SCR is triggered at zero degrees, no reactive power is absorbed by the rectifier.

21.19 Controlled rectifier supplying an active load (Circuit 2)

Figure 21-31 shows an ac source E_m and a dc load E_d connected by an SCR in series with an inductor. The load (represented by a battery), receives energy because when the thyristor conducts, current I enters the positive terminal. Smoothing inductor L limits the peak current to a value within the SCR rating. Gate pulses E_g

Figure 21-29
A discharging capacitor C and an auxiliary thyristor Q2 can force-commutate
the main thyristor Q1. Thus, the current in load R can be switched on and off by
triggering Q1 and Q2 in succession.

TABLE 21D   SOME BASIC THYRISTOR POWER CIRCUITS

Circuit No.	Thyristor circuit	Typical applications
1	Controlled rectifier supplying a passive load	Electroplating, dc arc welding, electrolysis
2	Controlled rectifier supplying an active load	Battery charger, dc motor control, dc transmission line
3	Naturally commutated inverter supplying an active ac load	DC motor control, wound-rotor motor speed control, dc transmission line
4	AC electronic contactor	Spot welding, lighting control, ac motor speed control
5	Cycloconverter	Low-speed synchronous motor control, electroslag refining of metals
6	Self-commutated inverter	Induction motor control, portable ac sources
7	DC chopper	Electric traction

(a)

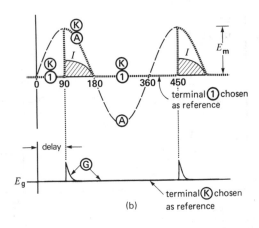

(b)

Figure 21-30
a. SCR supplying a passive load.

b. Voltage and current waveforms

initiate conduction at an angle θ_1 (Fig. 21-31b).

Using terminal 1 as a zero reference potential, it follows that the potential of terminal 2 lies E_d volts above it. Furthermore, the potential of terminal A oscillates sinusoidally around the level of terminal 1.

If the SCR were replaced by a diode, conduction would begin at angle θ_0 because this is the instant when the anode becomes positive. However, in our example, conduction only begins when the gate is fired, at θ_1 degrees. As soon as conduction starts, point K jumps from level 2 to the level of point A, and voltage E_{A2} appears across the inductor. The latter begins storing volt seconds, and current I increases accordingly. The volt seconds reach a maxi-

mum at θ_2, where area $A_{(+)}$ is maximum. The corresponding peak current is given by:

$$I_{max} = A_{(+)}/L \qquad \text{Eq. 2-28}$$

The current then gradually decreases and becomes zero at angle θ_3 where $A_{(-)}$ is equal to $A_{(+)}$. As soon as conduction stops, point K jumps from level A to the level of point 2 and stays there until the next gate pulse. The level of K is shown by the dotted line.

As in circuit 1 (Fig. 21-30), the load current lags (is displaced) behind voltage E_m; consequently, the source again has to supply reactive power Q as well as active power P.

If we reduce the so-called firing angle α, area $A_{(+)}$ increases, and so does current I. We can therefore vary the active power supplied to the load from zero ($\alpha = \alpha_1$) to a maximum ($\alpha = 0$).

From a practical point of view, the circuit could be used as a variable battery charger. Another application is to control the speed and torque of a dc motor. In this case, E_d represents the counter emf of the armature, and L the armature inductance.

21.20 Naturally-commutated inverter (Circuit 3)

An *inverter*, by definition, changes dc power into ac power. It performs the reverse operation of a rectifier, which converts ac power into dc power. There are two main types of inverters:
1. self-commutated inverters (or oscillators) in which the commutation elements are included within the power inverter;
2. externally commutated inverters (or naturally-commutated inverters) in which the means of commutation is not included within the power inverter.

In this section, we examine the operating principle of a naturally-commutated inverter. The circuit of such an inverter is identical to that of a controlled rectifier, except that the battery terminals are reversed (Fig. 21-32). Thus, the potential of terminal 2 lies below that of ter-

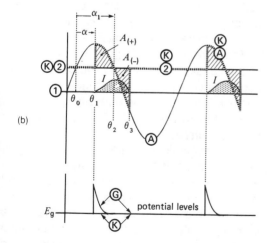

Figure 21-31
a.   SCR supplying an active load.
b.   Voltage and current waveforms

Figure 21-32
a. Naturally-commutated inverter.

b. Voltage and current waveforms.

minal 1 (Fig. 21-32b). Because current can only flow from anode to cathode, the dc source E_d *delivers* power whenever the thyristor conducts.

On the other hand, this power must be absorbed by the ac terminals because no losses occur in the inductor or the thyristor. Consequently, the circuit of Fig. 21-32 is potentially able to convert dc power into ac power.

To achieve this power conversion, the thyristor has to be triggered within a precisely defined range. First, to initiate conduction, A must be positive with respect to K. Since K is at the level of terminal 2, the triggering pulse must therefore be applied either prior to θ_0 or after θ_3 (Fig. 21-32b). For reasons that will soon become clear, the gate must be triggered *prior* to θ_0. Suppose the SCR is triggered at θ_1 degrees. Point K immediately jumps from level 2 to level A and the inductor accumulates positive volt seconds until angle θ_0 is reached. Thus, the resulting current reaches a peak at θ_0 where area $A+$ is maximum. It then gradually falls to zero and conduction stops at θ_2, when $A- = A+$.

To increase the current and hence the active power flow, we simply advance the firing angle

θ_1. However, we cannot carry this process too far. In order for conduction to cease, $A-$ must equal $A+$. However, the maximum area that $A-$ can have is that bounded by the trough of the sine wave and the level of point 2 (Fig. 21-32b). As we advance the firing angle, $A+$ becomes bigger and bigger; but if it should exceed the maximum available value $A-$, conduction will never stop. The dc current will then build up with each cycle, until the circuit breakers trip. For the same reason, conduction must never be initiated immediately after angle θ_3.

The current peaks lag behind the positive voltage peaks and so the ac source still has to supply reactive power Q to the inverter. Consequently, P and Q flow in opposite directions in an inverter. In our example, $P = E_d \times$ (average value of current I).

The current flowing into the ac terminals is far from sinusoidal and a stiff (low impedance) ac system is needed to maintain a sinusoidal voltage. In practice, we add appropriate filters to ensure that the current flowing into the ac line is reasonably sinusoidal. We should also remember that naturally-commutated inverters always involve 3-phase systems and not the "simple" single-phase circuit of Fig. 21-32.

21.21 AC electronic contactor (Circuit 4, Table 21D)

An ac electronic contactor is composed of two thyristors connected in anti-parallel (back-to-back), so that current can flow in both directions. Thus, in Fig. 21-33, a sinusoidal ac current flows through load resistor R when gates g1 and g2 are fired at 0° and 180° respectively. In contrast to a magnetic contactor, an electronic contactor is absolutely silent and its "contacts" never wear out.

A further advantage of this "contactor" is that the ac current can be precisely controlled. Thus, if gate pulses g1 and g2 are synchro-nized with the line frequency, a greater or lesser ac current will flow in the load, depending upon the firing angle α. However, such delayed firing will draw reactive power from the line, even if the load is purely resistive. The reason is that the current is displaced behind the voltage.

21.22 Cycloconverter (Circuit 5)

A cycloconverter produces low-frequency ac power directly from a higher-frequency ac source. A simple cycloconverter is shown in Fig. 21-34. It consists of three groups of thyristors, mounted back-to-back, and connected to a 3-phase source. They supply single-phase power to a resistive load R.

To understand the operation of the circuit, suppose all thyristors are initially blocked (nonconducting). Then, for an interval T, the gates of thyristors Q1, Q2 and Q3 are triggered by 4 successive pulses g1, g2, g3, g1, in such a way that the thyristors function as if they were ordinary diodes. As a result, the circuit behaves

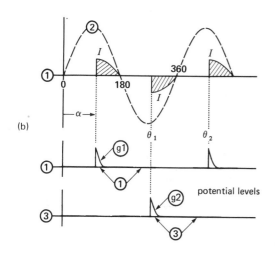

Figure 21-33
a. Electronic contactor.
b. Waveforms with a resistive load.

Figure 21-33c
Single-phase water-cooled contactor composed of two Hockey Puk thyristors. Continuous current rating: 1200 A (RMS) at 2000 V; cooling water requirements: 4.5 L/min at 35° max. For intermittent (10% duty) spot welding applications, this unit can handle 2140 A for 20 cycles. Width: 175 mm; length: 278 mm; depth: 114 mm.
(*Photo courtesy of International Rectifier*)

like a 3-pulse rectifier and terminal 4 is positive with respect to N (Fig. 21-35). The waveshape of E_{4N} is identical to that of Fig. 21-17.

During the next interval T, thyristors Q4, Q5, Q6, are fired by 4 similar pulses g4, g5, g6, g4. This makes terminal 4 *negative* with respect to N. The firing process is then repeated for the Q1, Q2, Q3 thyristors, and so on, with the result that a low-frequency ac voltage appears across the load. The duration of 1 cycle is $2\,T$ seconds. Compared to a sine wave, the low-frequency waveshape is rather poor. It is flat-topped and contains a large 180 Hz ripple when the 3-phase frequency is 60 Hz. However, this is of secondary importance because means are available to improve it.

Returning to Fig. 21-35 and assuming a 60 Hz source, we can show that each half-cycle corresponds to 540°, on a 60 Hz base. The duration of T is therefore $(540/360) \times (1/60) = 0.025$s, which corresponds to a frequency of $1/(2 \times 0.025) = 20$ Hz.

Obviously, by repeating the firing sequence g1, g2, g3, g1, . . ., we can keep terminal 4 positive for as long as we wish, followed by an equally long negative period, when g4, g5, g6, g4 . . . are fired. In this way, we can generate frequencies as low as we please. The high end of the frequency spectrum is limited to about 40 percent of the supply frequency. The reader should also note that this cycloconverter can supply a single-phase load from a 3-phase sys-

Figure 21-34
Elementary cycloconverter.

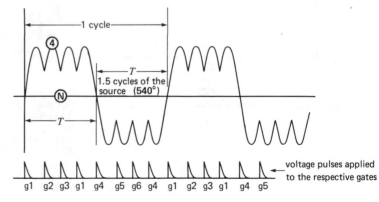

Figure 21-35
Typical voltage output of a cycloconverter.

tem, without unbalancing the 3-phase lines.

Later, we will encounter cycloconverters that can produce a sinusoidal, low-frequency 3-phase output from a 60 Hz, 3-phase input.

21.23 Self-commutated inverter (Circuit 6)

A self-commutated inverter changes dc power into ac power. There are many types of self-commutated inverters, but all are based on forced commutation. The output frequency can be as high as 20 kHz, depending on the switching capability of the thyristors. The load may be passive, such as resistor or inductor, or active, such as an ac motor.

Figure 21-36
Typical self-commutated inverter.

As an example of a simple self-commutated inverter, consider the circuit of Fig. 21-36. It consists of the following components:
1. dc source E;
2. thyristors Q1 and Q2;
3. smoothing inductor L to maintain a *constant* source current;
4. commutating capacitor C;
5. transformer having two primaries T_1 and T_2 and one secondary T_3. The transformer is assumed to be ideal;

6. load R;
7. external control system (not shown) that triggers gates g1 and g2, to obtain the desired output frequency.

To understand how the inverter operates, suppose that initially Q1 is conducting and Q2 is blocked. The constant current I_1 is flowing in primary winding T_1 and capacitor C is fully charged with polarities as shown in Fig. 21-37a. Constant current I_1, induces a constant current I_a in the secondary winding. This condition is stable until we apply a pulse to gate g2, initiating conduction in Q2.

As soon as Q2 conducts, the level of point 4 drops to point 1. Simultaneously, the capacitor discharges, producing a commutating current I_c that flows opposite to I_1 (Fig. 21-37b). Current I_c rises very fast and as soon as $I_c = I_1$, the current in Q1 becomes zero whereupon it acts like an open switch (Fig. 21-37c).

With Q1 blocked, the capacitor charges up rapidly via primary T_1; consequently, terminal 3 quickly becomes positive with respect to terminal 4. As charging current I_1 decreases in winding T_1, current I_2 builds up in winding T_2. When conditions stabilize, T_2 carries the same current as T_1 carried before. This is accompanied by a secondary current I_b.

The next pulse initiates conduction in Q1, causing the level of point 3 to drop to point 1. The capacitor again discharges, thus blocking Q2 and bringing us back to the starting point (Fig. 21-37a). In effect, the alternate gate pulses transfer conduction from one thyristor to the other, owing to the presence of commutating capacitor C. The action is similar to that in Fig. 21-29.

With regard to the transformer, when Q1 conducts, primary current I_1 induces a secondary current I_a. Based on the polarity marks, I_a must flow in the direction shown (Fig. 21-37a). Similarly, when Q2 conducts, primary current I_2 induces a secondary current I_b. The result is that secondary currents I_a and I_b are equal, but

(a)

(b)

(c)

Figure 21-37
Commutating sequence in a self-commutated inverter.

flow in opposite directions. Consequently, load R carries an alternating current.

To change the inverter frequency, we simply vary the frequency of the gate pulses. In this way, we can generate frequencies ranging from 4 Hz to 5000 Hz, depending on the design of the transformer and the properties of the SCRs.

It is important to note that self-commuted inverters can supply ac power to either resistive or inductive loads. Thus, they can deliver both active and reactive power to a load. This is in sharp contrast to naturally-commuted inverters which inherently absorb reactive power from the ac side.

21.24 DC chopper (Circuit 7, Table 21D)

In some electronic circuits, power has to be transferred from a high-voltage dc source E_s to a lower-voltage dc load E_O. One solution is to connect the source and load by a resistor, as shown in Fig. 21-38. However, the large I^2R loss renders this solution inefficient. Another solution is to connect an inductor and a switch between the source and load (Fig. 21-39). In order to follow the transfer of energy, assume the switch closes for a time T_1. During this interval, the voltage across the inductor is $E_s - E_0$, as shown in Fig. 21-40. The inductor accumulates volt seconds, and the resulting current i increases at a constant rate. At time T_1 (when the switch is about to open), the current is:

$$I_a = A_{(+)}/L$$

$$= (E_s - E_0)T_1/L \qquad (21\text{-}10)$$

The corresponding energy stored in the inductor is:

$$W = \frac{1}{2} LI_a^2 \qquad \text{Eq. 2-8}$$

When the switch opens, the current collapses and all the stored energy is dissipated in the arc across the switch. At the same time, a high voltage e_L is induced across the inductor

Figure 21-38
Inefficient power transfer.

Figure 21-39
Energy transfer using an inductor.

Figure 21-40
Energy is stored in the inductor.

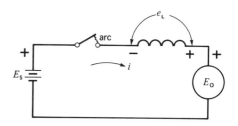

Figure 21-41
Energy is dissipated in the arc. Note the polarity of e_L.

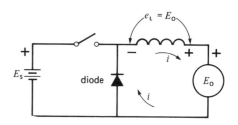

Figure 21-42
Energy transferred without loss.

Figure 21-43
E and I in the inductor of Fig. 21-42.

because the current is collapsing so quickly. The polarity of this voltage is *opposite* to what is was when the current was increasing (Fig. 21-41).

Although some energy is transferred from E_s to E_0 while the switch is closed, there is a great loss of energy every time the switch opens. The efficiency is therefore poor.

We can prevent this energy loss by adding a diode to the circuit as shown in Fig. 21-42. When the switch opens, current i again begins to fall, inducing a voltage e_L. However, e_L cannot jump to the high value it had before because

as soon as it exceeds E_0, the diode begins to conduct. Assuming the diode voltage drop is negligible, it follows that $e_L = E_0$. Because E_0 is constant, the voltage across the inductor is also constant. Current i therefore falls at a uniform rate given by

$$\frac{\Delta i}{\Delta t} = \frac{E_0}{L}$$

Eq. 2-27

The current eventually becomes zero after a time T_2 (Fig. 21-43).

We can calculate T_2 because the volt seconds accumulated during the "charging" period T_1 must equal the volt seconds released during the "discharge" interval T_2. Referring to Fig. 21-43, we have:

$$\frac{\text{V·s during}}{\text{charge period}} = \frac{\text{V·s during}}{\text{discharge period}}$$

$$A+ = A-$$

$$(E_s - E_0)\,T_1 = E_0 T_2$$

Consequently,

$$T_2 = (E_s - E_0)T_1/E_0 \qquad (21\text{-}11)$$

When the current is zero, the inductor will have supplied all its energy to load E_0. Simultaneously, the diode will cease to conduct. We can therefore reclose the switch for another interval T_1 and repeat the cycle indefinitely. Consequently, this circuit enables us to transfer energy from a high-voltage dc source to a lower voltage dc load without incurring any losses. In effect, the inductor absorbs energy at a relatively high voltage $(E_s - E_0)$ and delivers it at a lower voltage E_0.

The diode is sometimes called a freewheeling diode because it automatically starts conducting as soon as the switch opens. The switch is actually a thyristor whose on/off state is controlled by a means similar to that shown in Fig. 21-29.

Rapid chopping

Instead of letting the load current swing between zero and I_a, we can close the switch for a relatively long period T_1 until the current reaches the desired value I_a. We then open and close the switch rapidly so that the current increases and decreases by small increments. Referring to Figs. 21-44a and 21-44b, the switch is closed for an interval T_a and open during an interval T_b. When the switch is open, the load current falls from its peak value I_a to a final value I_b. During this interval, current flows in

Figure 21-44a
Currents in a chopper circuit.

Figure 21-44b
Current in the load.

Figure 21-44c
Current drawn from the source.

the inductor, the load and the freewheeling diode. The current decreases at the same rate as it did in Figure 21-43.

When the current has fallen to a value I_b, the switch recloses. The current in the diode immediately stops flowing, and the source now supplies current I_b. The current then builds up and when it reaches the value I_a, (after a time T_a) the switch reopens. The freewheeling diode again comes into play and the cycle repeats. The current supplied to the load oscillates therefore between I_a and the slightly lower value I_b (Fig. 21-44b). Its average or dc value is nearly equal to I_a but the exact value is given by:

$$I_0 = (I_a + I_b)/2 \qquad (21\text{-}12)$$

Whereas the current in the load is essentially constant, the current supplied by the source is composed of a series of sharp pulses, as shown in Fig. 21-44c. What is the average value of these pulses? It is easily found by noting that the average current during each *pulse* is $(I_a + I_b)/2 = I_0$. Consequently, the average current I_s during one *cycle* (time T) is:

$$I_s = I_0(T_a/T)$$

that is

$$I_s = I_0 f T_a \qquad (21\text{-}13)$$

where

I_s = dc current drawn from the source [A]
I_0 = dc current absorbed by the load [A]
T_a = "on" time of the switch [s]
T = period of one cycle [s]
f = switching frequency ($= 1/T$) [Hz]

The circuit of Fig. 21-44a clearly shows the current waveshapes in the source, the load, and the diode. Although the waveshapes are choppy and discontinuous, they still obey Kirchoff's current law, instant by instant.

Example 21-6:
The switch in Fig. 21-44a opens and closes at a frequency of 20 Hz and remains closed for 4 ms per cycle. A dc ammeter connected in series with the load E_0 indicates a current of 50 A.
a. If a dc ammeter is connected in series with the source, what current will it indicate?
b. What is the average current per pulse?

Solution:
a. Using Eq. 21-13, we have:

$$\begin{aligned} I_s &= I_0 f T_a \\ &= 50 \times 20 \times 0.004 \\ &= 4 \text{ A} \end{aligned}$$

b. The average current per pulse is 50 A.

Considering that the average current is only 4 A, the source may have to be specially designed to supply such a high current pulse. In most cases, a capacitor is connected across the terminals of the source. It can readily furnish the high current pulses as it discharges.

Turning our attention to the power aspects, the dc power drawn from the source must equal the dc power absorbed by the load because, ideally, there is no power loss in either the switch, the inductor or the freewheeling diode. We can therefore write:

$$E_s I_s = E_0 I_0$$

If we substitute Eq. 21-13 in the above equation, we obtain the useful relationship:

$$E_0 = E_s f T_a = E_s \frac{T_a}{T} \qquad (21\text{-}14)$$

where
E_0 = dc output voltage of the chopper [V]
Es = dc voltage of the source [V]
f = switching frequency [Hz]
T_a = "on" time of the switch [s]

Equation 21-14 indicates that we can control the dc output voltage E_0 simply by varying either the switching frequency f or the "on" time T_a. Thus, the chopper behaves like a highly efficient dc transformer in which the "turns ratio" is T_a/T. This ratio can be changed at will by varying the on/off time of the switch.

In practice, the mechanical switch is replaced by a thyristor that is turned on and off at a frequency f. The frequency may range from 30 Hz to 2000 Hz. Forced commutation is used to turn the thyristor off at the end of each cycle. The combination of the thyristor, inductor and diode is called a dc *chopper*. It is widely used in electric transit and trolley buses and wherever dc-to-dc energy conversion is needed.

(a)

(b)

(c)

(d)

Figure 21-45
a. Circuit of Example 21-7.
b. Current in the load.

Figure 21-45
c. Current drawn from the source.
d. Current in the freewheeling diode.

Example 21-7:
We wish to charge a 120 V battery from a 600 V dc source using a dc chopper. The average battery current should be 20 A, with a peak-to-peak ripple of 2 A. If the chopper "on" time is fixed at 1 ms, calculate:
a. the dc current drawn from the source;
b. the dc current in the diode;
c. the chopper frequency;
d. the value of the inductor.

Solution:
The circuit diagram is shown in Fig. 21-45a and the desired battery current is given in Fig. 21-45b. It fluctuates between 19 A and 21 A, thus yielding an average of 20 A with a peak-to-peak ripple of 2 A.

a. The power supplied to the battery is:
$$P = 120 \text{ V} \times 20 \text{ A} = 2400 \text{ W}$$

The power supplied by the source is therefore 2400 W.

The dc current from the source is:
$$I_s = P/E_s = 2400/600 = 4 \text{ A}$$

b. To calculate the average current in the diode, we refer to Fig. 21-45a. Current I_0 is 20 A and I_s was found to be 4 A. By applying Kirchoff's current law to the diode/inductor junction, the diode current I_D is
$$I_D = I_0 - I_s$$
$$= 20 - 4$$
$$= 16 \text{ A}$$

c. We have $E_0 = 120$ V,
$$E_s = 600 \text{ V, and}$$
$$T_a = 1 \text{ ms}$$
$$E_0 = E_s f T_a \qquad \text{Eq. 21-14}$$
$$120 = 600 f (0.001)$$
$$f = 200 \text{ Hz}$$

The chopper frequency is therefore 200 Hz.
$$T = 1/f = 1/200 = 5 \text{ ms}$$

Consequently, the "off" time T_b is:
$$T_b = T - T_a = 5 - 1 = 4 \text{ ms}$$

The waveshapes of I_s and I_D are shown in Figs. 21-45c and 21-45d, respectively. Note the sharp pulses delivered by the source.

d. During interval T_a, the average voltage across the inductor is $(600 - 120) = 480$ V.

The volt seconds accumulated by the inductor during this interval is 480×1 ms $= 480$ mV·s. The change in current during the interval is:

$$\Delta I = (21 - 19) = 2 \text{ A}.$$
$$\Delta I = A/L \qquad \text{Eq.2-28}$$
$$2 = 0.48/L$$
$$L = 0.24 \text{ H}$$

Consequently, the inductor has to have an inductance of 0.24 H. If a larger inductance were used, the current ripple would be smaller, but the dc voltages and currents would remain the same.

Impedance transformation

So far, we have assumed the chopper feeds power to an active load E_0. However, it may also be used to connect a high-voltage dc source E_s to a low-voltage load resistor R_0 (Fig. 21-46). Equations 21-13 and 21-14 still

Figure 21-46
A chopper can make a fixed resistor R_0 appear as a variable resistance between terminals 1-2.

apply, but we now have the additional relationship $E_0 = I_0 R_0$. Furthermore, the apparent dc resistance R_s across the terminals of the source is given by

$$R_s = E_s/I_s$$

We can therefore write

$$R_s = E_s/I_s$$

$$= \frac{E_0/fT_a}{I_0 fT_a}$$

$$= \frac{E_0}{I_0 f^2 T_a^2}$$

$$\therefore R_s = \frac{R_0}{f^2 T_a^2}$$

Consequently,

$$R_s = R_0/f^2 T_a^2 = R_0 \left[\frac{T}{T_a} \right]^2 \qquad (21\text{-}15)$$

where

R_s = apparent dc resistance across the source [Ω]
R_0 = actual load resistance [Ω]
f = chopper frequency [Hz]
T_a = "on" time [s]

The chopper, therefore, has the ability to change the apparent resistance of a fixed resistor by simply varying either the chopper frequency or the "on" time. Thus, the chopper again behaves like a dc transformer in which the "turns ratio" is T/T_a.

Example 21-8:
The chopper in Fig. 21-46 operates at a frequency of 30 Hz and the "on" time is 200 μs. Calculate the apparent resistance across the source, knowing that $R_0 = 36$ mΩ.

Solution:
Applying Equation 21-15, we have:

$$R_s = R_0/f^2 T_a^2$$
$$= 0.036/30^2 \times (200 \times 10^{-6})^2$$
$$= 1000 \ \Omega$$

This example shows that the actual value of a resistor can be increased many times by using a chopper.

This completes our review of the seven basic thyristor power circuits listed in Table 21D. We shall encounter them again in later sections and see how they are applied.

21.25 3-phase, 6-pulse converter

The 3-phase, 6-pulse thyristor converter is the most widely used rectifier/inverter unit in power electronics. Owing to its practical importance, we shall explain how it operates in some detail.

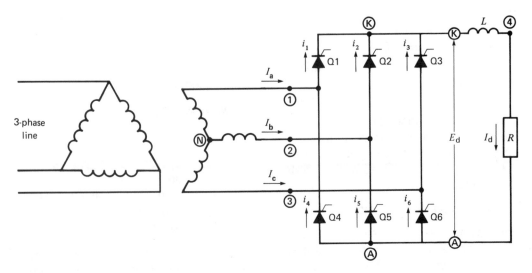

Figure 21-47
Three-phase, 6-pulse thyristor converter.

As in all 3-phase converters, the waveforms become rather complex, although not particularly difficult to understand. Even the simplest circuits yield chopped voltages and currents that pile on top of each other, and it taxes the mind to keep track of everything that is going on. Consequently, we shall keep the waveforms as simple as possible, so as to highlight the basic principle of operation.

Three-phase, 6-pulse converters have 6 thyristors connected to the secondaries of a 3-phase transformer (Fig. 21-47). The arrangement is identical to the rectifier circuit of Fig. 21-19, except that the diodes are replaced by thyristors. Because we can initiate conduction whenever we please, the thyristors enable us to vary the dc output voltage when the converter operates in the rectifier mode. The converter can also function as an inverter, provided that a dc source is used in place of the load resistor R.

21.26 Basic principle of operation

We can gain a basic understanding of how the converter works in the rectifier mode by referring to Fig. 21-48. In this figure, the SCRs

are assumed to be enclosed in a box, where they successively switch the output terminals K, A to the ac supply lines 1, 2, 3. The load is represented by a resistor in series with an inductor L. The inductor is assumed to have a very large inductance, so that the load current I_d remains constant. In Fig. 21-48a, the two thyristors Q1, Q5 located between terminals K-1 and A-2 are conducting. A moment later, the thyristors Q2, Q4 between K-2 and A-1 conduct (Fig. 21-48b). The other thyristors are similarly switched, in sequence. When these steps have been completed, the entire switching cycle repeats. The reader will note that the dc current I_d flows in the ac lines. However, Fig. 21-48 shows that the current in each line reverses periodically, and so it is a true ac current of amplitude I_d. It is also evident that the current in a particular line is zero for brief intervals. Thus, there is momentarily no current in line 3 of Fig. 21-48.

The switching sequence we have just described is similar to that of the diode bridge rectifier of Fig. 21-22. There is, however, an important difference. The thyristors can be made to conduct at precise moments on the ac

(a)

(a)

(b)

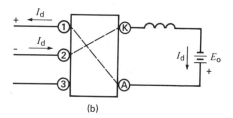

(b)

Figure 21-48
Rectifier mode (see Fig. 21-47)
a. Q1 and Q5 conducting.
b. Q2 and Q4 conducting.

Figure 21-49
Inverter mode (see Fig. 21-47)
a. Q1 and Q5 conducting.
b. Q2 and Q4 conducting.

voltage cycle. Thus, conduction can be initiated when the instantaneous voltage between the ac lines is either high or low. If the voltage is low, the dc output voltage will also be low. Conversely, if the thyristors conduct when the ac line voltage is momentarily near its peak, the dc output voltage will be high. In effect, the output voltage E_{KA} is composed of short 60-degree segments of the ac line voltage. The average value of E_{KA} is the dc output voltage E_d.

In examining Fig. 21-48, it can be seen that the line current always flows out of a line that is momentarily positive. This must be so because the line delivers active power to the load. For example, in Fig. 21-48a, e_{12} is positive when I_d flows in the direction shown.

Now that we know how the thyristor converter behaves as a rectifier, the question arises; how can it be made to operate as an inverter? Three basic conditions have to be met.

First, we must have a source of dc current I_d. Such a current source can be provided if a vol-

tage source E_0 is connected in series with a large inductance (Fig. 21-49a).

Second, the converter must be connected to a 3-phase line that can maintain an undistorted sinusoidal voltage, even when the line current is nonsinusoidal. The voltage may be taken from a power utility, or generated by a local alternator.

Third, the thyristors must be switched so that current I_d flows into an ac line that is momentarily positive. The gate firing must therefore be precisely synchronized with the line frequency.

The inverter operation can best be understood by referring to Fig. 21-49. The SCRs enclosed in the box are arranged the same way as in Fig. 21-48. In other words, the converters in the two figures are absolutely identical. Looking first at the dc side, the dc current I_d must flow in the same direction as before because SCRs cannot conduct in reverse. On the other hand, because we want the dc source E_0 to deliver power, I_d must flow out of the positive terminal, as shown. In other words, the positive side of E_0 must be connected to terminal A. On the ac side, the 3-phase line is simply connected to terminals 1, 2, 3.

We are now ready to fire the thyristors. However, the firing must be properly timed so that the ac line receives power. This is consistently done in Fig. 21-49 because current I_d always flows into an ac terminal that is momentarily positive. Note that the line voltage polarities in the inverter mode are consistently opposite to those in the rectifier mode.

The reader can see that the line current alternates as before, and it has a peak value equal to I_d. Indeed, the waveshape of the ac line currents is the same in Figs. 21-48 and 21-49; it is only the instantaneous line voltages that differ.

If the dc supply voltage E_0 is low, the thyristors must be fired when the instantaneous ac voltage is low. Conversely, if the dc voltage is high, the thyristors must be triggered when the ac line voltage is near its peak.

We wish to make one final important observation. The voltage that appears between terminals K and A is composed of 60-degree segments of the ac line voltages. Consequently, E_{KA} is a fluctuating voltage whose average value is E_d. This average voltage must be equal to E_0 because the dc voltage drop across the inductor is negligible. In addition to keeping current I_d constant, the inductor serves as a buffer between the fluctuating voltage E_{KA} and the constant voltage E_0.

21.27 Three-phase, 6-pulse rectifier feeding an active load

Consider the circuit of Fig. 21-50 in which a 3-phase, 6-pulse converter supplies power to a load. The load is composed of a dc voltage E_0 and a resistor R, in series with a smoothing inductor. The converter is fed from a 3-phase transformer. The gates of thyristors Q1 to Q6 are triggered in succession, at 60-degree intervals. We assume that the converter has been in operation for some time, so that conditions are

stable. Initially, suppose thyristors Q5 and Q6 are conducting, carrying load current I_d (Fig. 21-51a). Then, at the 0° point (θ_0), thyristor Q1 is triggered by gate pulse g1. Commutation occurs and Q1 starts conducting, taking over from Q5.

At 60°, thyristor Q2 is fired and the resulting commutation transfers the load current from Q6 to Q2. This switching process continues indefinitely and, as in Fig. 21-21, point K follows the peaks of the successive waves. The thyristors are labelled according to the order in which they are fired. Two SCR's conduct at a time; the conduction pairs are therefore Q1-Q2, Q2-Q3, Q3-Q4, and so on. Thus, by referring to Fig. 21-50, we can tell at a glance which thyristors are conducting at any given time.

The converter acts as a rectifier and the average or dc voltage between K and A is $E_d = 1.35E$. Because there is no appreciable dc voltage drop in an inductor, the dc voltage between points 4 and A is also $1.35E$. Consequently, the dc current I_d is given by:

$$I_d = (E_d - E_0)/R \qquad (21-16)$$

The triggering time has to be very precise to obtain the rectified voltage shown in Fig. 21-51a. Thus, if g1 fires slightly ahead of θ_0, conduction cannot start because anode 1 is then negative. On the other hand, if g1 fires after θ_0, Q5 (along with Q6) will continue to conduct until g1 is fired. In practice, the triggering pulses are made wide enough to ensure that commutation occurs at the desired instant.

21.28 Delayed triggering – rectifier mode

Let us now delay all triggering pulses by an angle α of 15° (Fig. 21-51b). Current I_d, instead of switching over to Q1 at θ_0, will continue to flow in Q5 until gate pulse g1 triggers Q1. Commutation occurs, and the potential of point K jumps from line 3 to line 1. A similar switching action takes place (but at later times) for the other thyristors. The resulting choppy

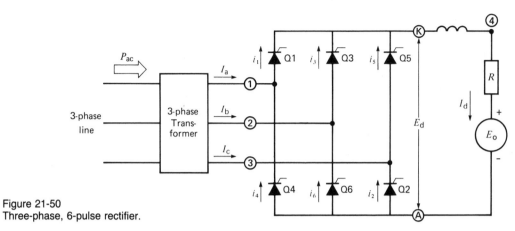

Figure 21-50
Three-phase, 6-pulse rectifier.

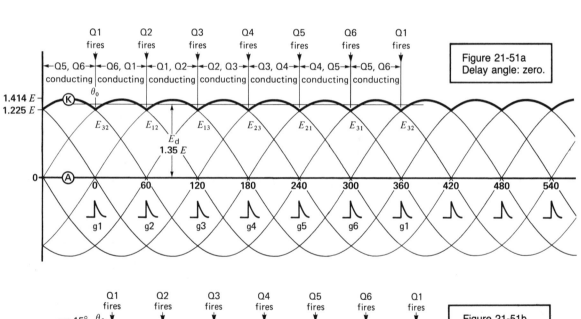

Figure 21-51a
Delay angle: zero.

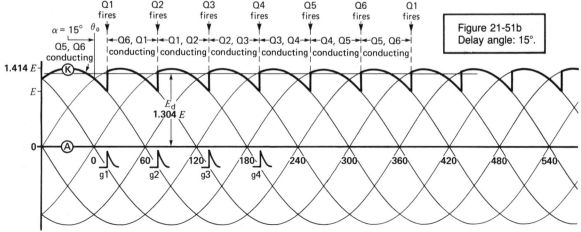

Figure 21-51b
Delay angle: 15°.

waveshape between terminals K and A is shown in Fig. 21-51b.

Note that the triggering delay does not shorten the conduction period; each thyristor still conducts for a full 120° and each voltage segment has a duration of 60 degrees. Furthermore, the current remains constant and ripple-free, owing to the presence of the big inductor. The level of point K follows the tops of the individual sine waves, but the average voltage E_d, between K and A, is obviously smaller than before. We can prove that it is given by:

$$E_d = 1.35E\cos \alpha \qquad (21\text{-}17)$$

where

E_d = dc voltage produced by the 3-phase, 6-pulse converter [V]

E = effective value of the ac line-to-line voltage [V]

α = firing angle [°]

According to Eq. 21-17, E_d becomes smaller and smaller as α increases. However, if E_d becomes equal to or less than E_0, the load current ceases to flow. Ordinarily, the current would reverse when E_d is smaller than E_0. However, this is impossible, because the SCRs can only conduct in the forward direction.

Figures 21-51c and 21-51d show the waveform between K and A for $\alpha = 45°$ and 75°, respectively. The ac component in E_{KA} is now very large, compared to the dc component.

Example 21-9.
The 3-phase converter of Fig. 21-50 is connected to a 3-phase 480 V, 60 Hz source. The load consists of a 500 V dc source having an internal resistance of 2 Ω. Calculate the power supplied to the load for triggering delays of a. 15°, b. 75°.

Solution:
a. The dc output voltage of the converter is:
$$E_d = 1.35E\cos\alpha$$
$$= 1.35 \times 480 \times \cos 15°$$
$$= 626 \text{ V}$$

Because the dc voltage drop across the inductor is negligible, the *IR* drop across the 2 Ω internal resistance is:
$$E = E_d - E_0$$
$$= 626 - 500 = 126 \text{ V}$$

The dc load current is therefore:
$$I_d = E/R = 126/2 = 63 \text{ A}$$

The power supplied to the load is:
$$P = E_d I_d$$
$$= 626 \times 63 = 39.4 \text{ kW}$$

b. With a phase angle delay of 75°, the converter voltage is:
$$E_d = 1.35E \cos \alpha$$
$$= 1.35 \times 480 \times \cos 75°$$
$$= 167.7 \text{ V}$$

Because E_d is less than E_0, the current tends to flow in reverse. This is impossible, and consequently the current is simply zero, and so, too, is the power.

21.29 Delayed triggering – inverter mode

If triggering is delayed by more than 90°, the voltage E_d developed by the converter becomes negative according to Eq. 21-17. This does not produce a negative current because, as we said, SCRs conduct in only one direction. Consequently, the load current is simply zero. However, we can *force* a current to flow by connecting a dc voltage of proper magnitude and polarity across the converter terminals. This external voltage E_0 must be slightly greater than E_d in order for current to flow (Fig. 21-52). The load current is given by:

$$I = (E_0 - E_d)/R$$

Because current flows out of the positive terminal of E_0, the "load" is actually a source, delivering a power output $P = E_0 I_d$. Part of this power is dissipated as heat in the circuit resistance R and the remainder is delivered to the secondaries of the 3-phase transformer. If we

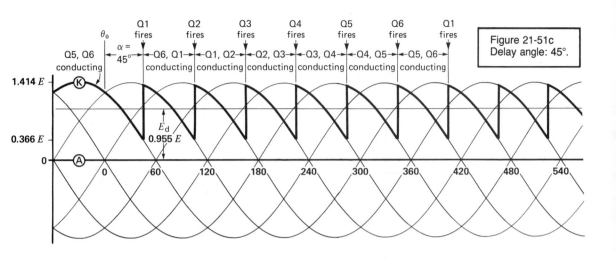

Figure 21-51c
Delay angle: 45°.

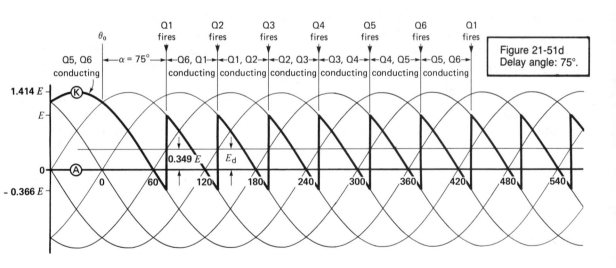

Figure 21-51d
Delay angle: 75°.

subtract the small transformer losses and the virtually negligible SCR losses, we are left with a net active power P_{ac} that is delivered to the 3-phase line.

The original rectifier has now become an *inverter*, converting dc power into ac power. The transition from rectifier to inverter is smooth, and requires no change in the converter connections. In the rectifier mode, the firing angle lies between 0° and 90°, and the load may be active or passive. In the inverter mode, the firing angle lies between 90° and 180°, and a dc source of proper polarity *must* be provided.

Figure 21-53 shows the waveshapes at fir-

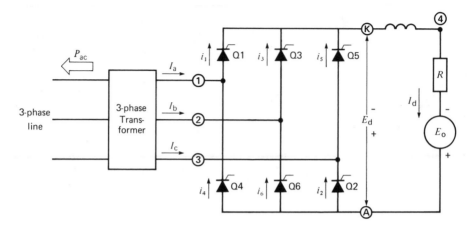

Figure 21-52
Three-phase, 6-pulse converter in the inverter mode.

ing angles of 105°, 135° and 165°. The dc voltage E_d generated by the inverter is still given by Eq. 21-17. It reaches a maximum value of $E_d = -1.35E$ at a firing angle of 180°.

21.30 Triggering range

The triggering angle of a given thyristor is usually kept between 15° and 165°. The thyristor acts as a rectifier between 15° and 90° and as an inverter between 90° and 165°. Under these conditions, the dc voltage developed reaches its maximum value at 15° and 165°; it is zero at 90°.

The triggering angle is seldom less than 8° in the rectifier mode. The reason is that sudden line voltage changes might cause a thyristor to

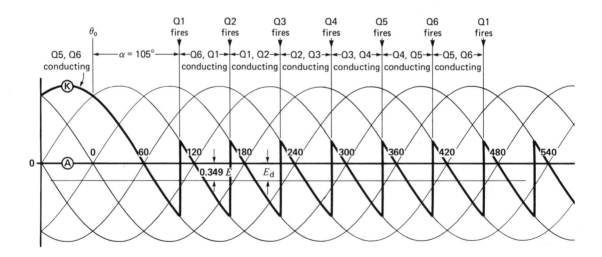

Figure 21-53a
Triggering sequence and waveforms with a delay angle of 105°

misfire, thus producing a discontinuity in the dc output current.

In the inverter mode, we never permit the firing angle to exceed 165°. If we go beyond this point, the inverter begins to lose its ability to switch reliably from one thyristor to the next. As a result the currents build up quickly until the circuit breakers trip. In some cases, the firing angle is not allowed to exceed 150°, to ensure an adequate safety margin.

Figure 21-54 shows the allowed and forbidden gate firing zones for a particular thyristor in

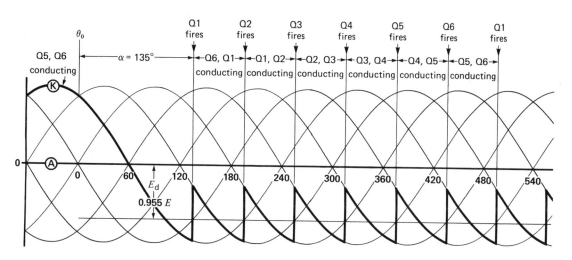

Figure 21-53b
Triggering sequence and waveforms with a delay angle of 135°

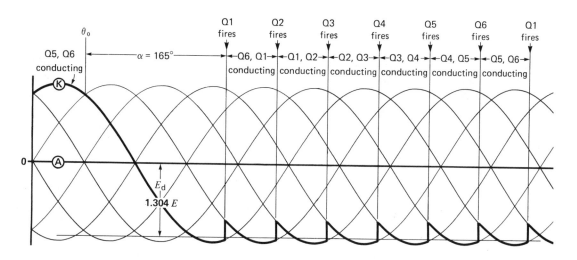

Figure 21-53c
Triggering sequence and waveforms with a delay angle of 165°.

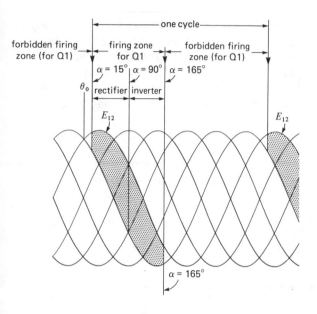

Figure 21-54
Permitted gate firing zones for thyristor Q1.

- e_c is the ac voltage generated by the converter (mainly 6th and 12th harmonics)
- D is a diode to remind us that current can flow in only one direction
- The dotted line between E_{ac} and E_d indicates that active power can flow between the ac and dc systems.
- Unlike a motor/generator set, the dc and ac systems are not electrically isolated from each other.

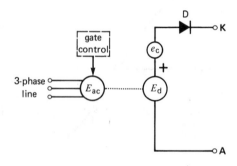

Figure 21-55
Equivalent circuit of a converter.

a 3-phase, 6-pulse converter. Specifically, it refers to Q1 in Fig. 21-50. The other thyristors have similar firing zones, but they occur at different times.

21.31 Equivalent circuit of a converter

We may think of a converter as being a static ac/dc motor-generator set whose dc output voltage E_d changes both in magnitude and polarity depending upon the gate pulse delay. However, the dc "generator" has some special properties:
1. it can carry current in only one direction;
2. it generates an increasingly large ac ripple voltage as the dc voltage decreases.

The analogy may be represented by the circuit of Fig. 21-55, in which:

- E_{ac} represents the 3-phase line voltage
- E_d is the dc voltage generated by the converter

When the converter is operating as a rectifier, the equivalent circuit is shown in Fig. 21-56. When operating as in inverter, the circuit is given by Fig. 21-57. The ac voltage generated by the converter appears across inductor L. Its inductance is sufficiently large to ensure an almost ripple-free dc current.

21.32 Currents in a 3-phase, 6-pulse converter

Figure 21-58 shows the voltage and current waveshapes when the converter functions as a rectifier at a firing angle of 45°. The current i_1, i_2, i_3, i_4, i_5, i_6 in each thyristor flows for 120°, and its peak value is equal to the dc current I_d. This holds true for any firing angle between zero and 180°. Consequently, the currents in a

Figure 21-56
Equivalent circuit of a 3-phase converter in the rectifier mode.

Figure 21-57
Equivalent circuit of a 3-phase converter in the inverter mode.

thyristor converter are identical to those in a plain 3-phase diode rectifier (Fig. 21-20). The only difference is that they flow later in the cycle.

The waveshapes of the corresponding ac line currents are easily found because they are equal to the difference between the respective thyristor currents. Thus, referring to Fig. 21-52, line current $I_a = i_1 - i_4$. These line currents also have a peak value I_d, but they flow in positive and negative pulses of 120°.

The heating effect of the ac line currents is important because they usually flow in the windings of a converter transformer. The I^2R loss depends upon the effective value I of the current. From Eq. 21-6 we know that

$$I = 0.816 I_d \qquad \text{Eq. 21-6}$$

The effective value of the ac line current is therefore directly related to the dc output current, and is unaffected by the firing angle. Clearly, the same is true when the converter operates as an inverter.

21.33 Power factor

We recall that in the 3-phase, 3-pulse rectifier (Figs. 21-16 and 21-17) the currents in lines 1, 2, and 3 are symmetrical with respect to the line-to-neutral voltages. Thus, rectangular current i_2 is exactly in the middle of the positive E_{2N} wave. In essence (and in actual fact), i_2 is in phase with E_{2N}. This is also true for the currents in the other two lines, as regards their respective voltages. This condition is reflected back into the primary of the transformer, and from there, to the 3-phase feeder. Because the currents are in phase with the voltages, the so-called *displacement power factor* is 100%. As a result, the rectifier draws no reactive power from the line. The same remarks apply to a 3-phase, 6-pulse rectifier (Fig. 21-20)

Referring now to Fig. 21-58, where triggering has been delayed by 45°, we note that the thyristor currents have all been shifted ("displaced") by 45°, to the right. Consequently, the line currents lag the respective voltages by 45°; the displacement power factor is no longer unity but only 0.707 (cos 45° = 0.707). This means that a converter absorbs reactive power from the ac system to which it is connected. This is true whether the converter operates as a rectifier or inverter. The reactive power is given by:

$$Q = P \tan \alpha \qquad (21\text{-}18)$$

where

Q = reactive power absorbed by the converter [var]

P = dc power of the converter (positive for a rectifier, negative for an inverter) [W]

α = triggering angle [°]

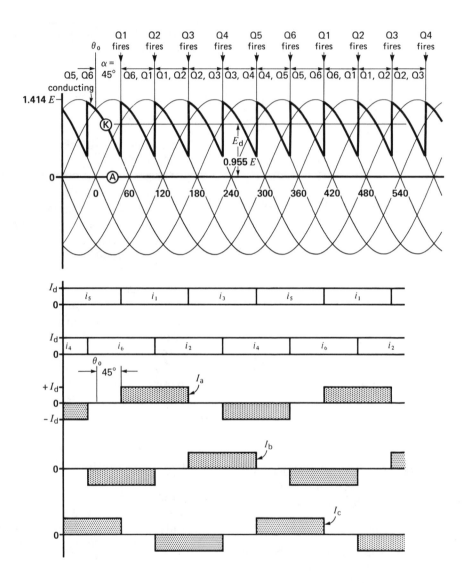

Figure 21-58
Voltage and current waveforms in the converter of Figure 21-50 with a delay angle of 45°.

The reader will note that the waveshape of the current is the same as in a conventional 6-pulse rectifier (Fig. 21-20).

Example 21-10:
In Example 21-9, and for a triggering angle of 15°, calculate
a. the displacement power factor
b. the reactive power absorbed by the converter.

Solution:
a. The displacement between the fundamental line current and the line-to-neutral voltage is $\alpha = 15°$.

The displacement power factor is
$$\cos \alpha = \cos 15° = 0.966 \text{ or } 96.6\%$$

b. The active power supplied to the converter is:
$$P = E_d I_d = 39.4 \text{ kW}$$

Hence
$$Q = P \tan \alpha$$
$$= 39.4 \tan 15$$
$$= 10.6 \text{ kvar}$$

Example 21-11:
A 16 kV dc source having an internal resistance of 1 Ω supplies 900 A to a 12 kV, 3-phase, 6-pulse, 60 Hz inverter (Fig. 21-59). Calculate:
a. the dc current carried by each SCR;
b. the dc voltage generated by the inverter;
c. the required firing angle α;
d. the effective value of the ac line currents;
e. the reactive power absorbed by the inverter;

Solution:
a. Each SCR carries the load current for one third of the time. The dc current is therefore:
$$I = I_d/3 = 900/3$$
$$= 300 \text{ A}$$

b. the voltage E_d generated by the inverter is equal to E_0 less the IR drop. Thus,

$$E_d = E_0 - I_d R.$$
$$= 16\ 000 - 900 \times 1$$
$$= 15\ 100 \text{ V}$$

c. Knowing that the effective ac line voltage is 12 000 V, the firing angle can be found from Eq. 21-17:
$$E_d = 1.35\ E \cos \alpha$$
$$15\ 100 = 1.35 \times 12\ 000 \cos \alpha$$
$$\cos \alpha = 0.932$$
$$\alpha = 21.2°$$

This is the firing angle that would be required if the converter operated as a rectifier. However, because it is in the inverter mode, the actual firing angle is:
$$\alpha = 180 - 21.2 = 158.8°$$

d. The effective value of the ac line current is:
$$I = 0.816\ I_d \qquad \text{Eq. 21-6}$$
$$= 0.816 \times 900$$
$$= 735 \text{ A}$$

e. The dc power absorbed by the inverter is:
$$P = E_d I_d$$
$$= 15\ 100 \times 900$$
$$= 13.6 \text{ MW}$$

P is actually negative because the inverter *absorbs* dc power; hence,
$$P = -13.6 \text{ MW}.$$

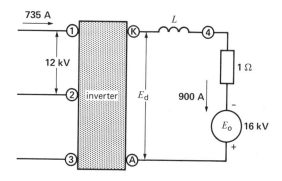

Figure 21-59
See Example 21-11.

The reactive power absorbed by the inverter is:

$$Q = P \tan \alpha \qquad \text{Eq. 23-16}$$
$$= -13.6 \tan 158.8$$
$$= 5.27 \text{ Mvar}$$

In practice, the actual reactive power is higher than the calculated value, owing to commutation overlap.

21.34 Commutation overlap

We mentioned in Section 21-9, that the current in a three-pulse rectifier cannot switch instantaneously from one diode to the next. The commutation process takes time and this is also true for thyristors. Thus, in a six-pulse converter, the commutation from Q1 to Q3 to Q5 is not instantaneous (as shown in Figure 21-58), but is more like that shown in Figure 21-60b.

The transfer of I_d from one thyristor to the next is effected during the so-called commutation overlap period, defined by angle u. The amount of overlap varies with the current I_d. At full load, u lies typically between 20° and 30°. At light load it can be as small at 5°.

On account of commutation overlap, the current in each thyristor flows for a period of $120 + u$ degrees instead of 120°, as we have assumed so far. The commutation overlap modifies the waveshape of E_{AK}, but we do not have to examine this aspect of converter behavior.

The commutation overlap delays the current buildup by angle u. It also delays the current cutoff by the same angle. Owing to these delays, the effective firing angle is somewhat greater than α. This reduces the power factor of the converter in both the rectifier and inverter mode. It also reduces the average dc voltage E_{dc}. This topic is covered in more detail in Chapter 28.

21.35 Extinction angle

We have seen that when a converter operates in the inverter mode, it is very important that conduction be initiated prior to $\alpha = 180°$. Because the current in an ideal inverter flows for 120°, the conduction must also cease before the angle of $(180 + 120) = 300°$ is reached. The interval between the end of commutation and 300° is called the *extinction angle* γ (Fig. 21-61). The extinction angle permits thyristor Q1 to recover its blocking ability before its anode (1) again becomes positive with respect to the cathode K. The value of γ lies typically between 15° and 20°.

In the case of an inverter, we often define the firing instant by the *angle of advance* β, rather than by the angle of delay α. It is easy to show that the following relationships exist between the commutation angle u, the delay angle α, the angle of advance β and the extinction angle γ:

$$\beta = 180 - \alpha \qquad (21\text{-}19)$$
$$\beta = u + \gamma \qquad (21\text{-}20)$$

Figure 21-60
a. Instantaneous commutation in a rectifier when $\alpha = 45°$ (see Fig. 21-58).
b. Same conditions with commutation overlap of 30°, showing current waveshapes in Q1, Q3, Q5.

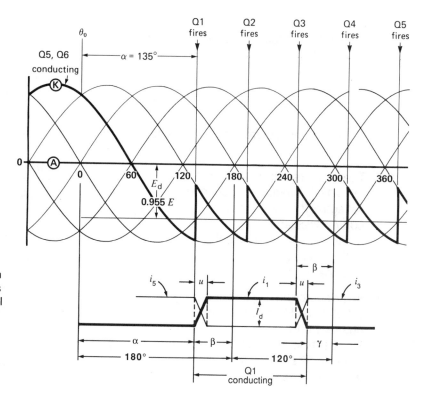

Figure 21-61
Waveshape of i_1 in thyristor Q1 for a delay angle α. The extinction angle γ permits Q1 to establish its blocking ability before the critical angle of 300° is reached. At 300° the anode of Q1 becomes positive with respect to its cathode. The figure also shows the relationship between angles α, β, γ and u.

QUESTIONS AND PROBLEMS

Practical level

21-1 State the basic properties of a diode.

21-2 State the basic properties of a thyristor.

21-3 What is the approximate voltage drop across a diode or SCR when it conducts?

21-4 What is the approximate maximum permissible operating temperature of a thyristor?

21-5 Explain the meaning of the following terms:

    anode
    cathode
    gate
    choke

    filter
    chopper
    peak inverse voltage
    rectifier
    self-commutated
    inverter
    harmonic
    commutation
    natural commutation
    converter
    inverter
    cycloconverter
    bridge rectifier
    forced commutation
    displacement power factor

21-6 The 3-phase transformer shown in Fig. 21-16 produces a secondary line voltage of 2.4 kV. The dc load current I_d is 600 A. Calculate:
   a. the dc voltage across the load;
   b. the average current carried by each diode;
   c. the maximum current carried by each diode.

21-7 The 3-phase transtormer shown in Fig. 21-19 produces a secondary line voltage of 2.4 kV. If the dc load current is 600 A, calculate:
   a. the dc voltage across the load;
   b. the average current carried by each diode.

21-8 An ac source having an effective voltage of 600 V, 60 Hz is connected to a single-phase bridge rectifier as shown in Fig. 21-14a. The load resistor has a value of 30 Ω. Calculate:
   a. the dc voltage E_{34};
   b. the dc voltage E_{54};
   c. the dc load current I;
   d. the average current carried by each diode;
   e. the active power supplied by the ac source.

21-9 The chopper shown in Fig. 21-46 is connected to a 3000 V dc source. The chopper frequency is 50 Hz and the on-time is 1 ms. Calculate:
   a. the voltage across resistor R_0;
   b. the value of I_s if $R_0 = 2$ Ω.

21-10 In Problem 21-9, if we double the on-time calculate the new power absorbed by the load.

Intermediate level

21-11 a. In Problem 21-7 calculate the power dissipated by the six diodes if the average voltage drop during the conduction period is 0.6 V.

b. What is the efficiency of the rectifier alone?

21-12 a. The current in Fig. 21-6 has a value of −6 A. What is the polarity of E_{34}?
   b. The current in Fig. 21-7 has a value of +6 A and E_{65} is negative. Is the current increasing or decreasing?

21-13 The single-phase bridge rectifier shown in Fig. 21-14a is connected to a 120 V, 60 Hz source. If the load resistance is 3 Ω, calculate:
   a. the dc load current;
   b. the PIV across the diodes;
   c. the energy that must be stored in the choke so that the peak-to-peak ripple is about 5% of the dc current;
   d. the inductance of the choke;
   e. the peak-to-peak ripple across the choke.

21-14 The line voltage is 240 V, 60 Hz on the secondary side of the converter transformer in Fig. 21-19. The dc load draws a current of 750 A. Calculate:
   a. the dc voltage produced by the rectifier;
   b. the active power supplied by the 3-phase source;
   c. the peak current in each diode;
   d. the duration of current flow in each diode [ms];
   e. the effective value of the secondary line current;
   f. the reactive power absorbed by the converter;
   g. the peak-to-peak ripple across the inductor.

21-15 The chopper shown in Fig. 21-46 is connected to a 2000 V dc source, and the load resistor R_0 has a value of 0.15 Ω. The on-time is fixed at 100 μs and the dc voltage across the resistor is 60 V. Calculate:
   a. the power supplied to the load;

b. the power drawn from the source;

c. the dc current drawn from the source;

d. the peak value of I_s;

e. the chopper frequency;

f. the apparent resistance across the dc source;

g. draw the waveshapes of I_s, I_0 and I_D.

21-16 The 3-phase, 6-pulse converter shown in Fig. 21-47 is directly connected to a 3-phase, 208 V line. Calculate:

a. the dc output voltage for a firing angle of 90°;

b. the firing angle needed to generate 60 V (rectifier mode);

c. the firing angle needed to generate 60 V (inverter mode).

21-17 The converter shown in Fig. 21-47 is connected to a transformer that produces a secondary line voltage of 40 kV, 60 Hz. The load draws a dc current of 450 A. If the delay angle is 75°, calculate:

a. the dc output voltage;

b. the active power drawn from the ac line;

c. the effective value of the secondary line current;

d. the reactive power absorbed by the converter.

21-18 In Problem 21-17, calculate:

a. the peak positive value of E_{KA};

b. the peak negative value of E_{KA};

c. the peak-to-peak ripple across the inductor.

21-19 The electronic contactor in Fig. 21-33 controls the power to a 15 Ω heater. The sinusoidal supply voltage has an effective value of 600 V at a frequency of 60 Hz. If the firing angle $\alpha = 0°$, calculate

a. The effective current

b. The power supplied to the heater

c. The effective value of the harmonic currents

d. the displacement power factor

e. The reactive power furnished by the ac line.

Advanced level

21-20 In Problem 21-19, the firing angle is increased to 120° and the effective current I is found to be 17.68 A. The rms value of the fundamental component is known to be $I_F = 12.34$. Calculate:

a. The rms value of all the harmonic currents.

b. The active power dissipated in heater [W]

c. The distortion power factor.

d. The apparent power supplied by the source

e. The total power factor.

f. The displacement power factor.

g. The reactive power supplied by the line [var].

21-21 The rectifier shown in Fig. 21-19 produces a dc output of 1000 A at 250 V. Inductor L reduces the current ripple, but an additional purpose is to limit the rapid build-up of dc current should load R become short-circuited. This enables the circuit breakers to trip before the dc current becomes too large. Assuming the initial current is 1000 A, calculate the minimum value of L so that the short-circuit current does not exceed 3000 A after 50 ms.

21-22 A diode having a PIV rating of 600 V is used in a battery charger similar to the one shown in Fig. 21-11a. The battery voltage is 120 V and $R = 10$ Ω.

a. Calculate the maximum effective secondary voltage of the transformer so that the diode will not break down.

b. For how many electrical degrees will the diode conduct if the RMS secondary voltage is 300 V?

c. What is the peak current in the diode?

21-23 The cycloconverter in Fig. 21-34 is connected to a 60 Hz source. Calculate the time interval between the firing of gates g1 and g4 if we wish to generate an output frequency of 12 Hz. Draw the waveshape of the voltage across the load resistance.

21-24 a. Referring to Fig. 21-15, and recognizing that area $A -$ is almost triangular, calculate the approximate value of $A +$ if the effective voltage produced by the source is 2000 V, 60 Hz [V·s].

b. If the peak-to-peak current ripple must not exceed 7 A, calculate the inductance of choke L.

21-25 The chopper shown in Fig. 21-42 transfers power from a 400 V source to a 100 V load. The inductor has an inductance of 5 H. If the chopper is "on" for 2 s and "off" for 10 s, calculate:

a. the maximum current in the inductor;
b. the energy transferred to the load, per cycle [J];
c. the average power delivered to the load [W];
d. draw the waveshape of i as a function of time and compare it with Figure 21-43.

21-26 In Problem 21-25, if the chopper is on for 2 s and off for 2 s, what is the value of the current:

a. after 2 s?
b. after 4 s?
c. after 6 s?
d. after 8 s?
e. Will anything prevent the current from building up indefinitely?

21-27 A 3-phase, 6-pulse converter shown in Fig. 21-52 is to be used as an inverter. The dc side is connected to a 120 V battery and $R = 10$ mΩ. The ac side is connected to a 3-phase, 120 V, 60 Hz line. If the battery delivers a current of 500 A, calculate:

a. the firing angle required;
b. the active power delivered to the ac line;
c. the reactive power absorbed by the converter.

22

ELECTRONIC CONTROL OF DIRECT CURRENT MOTORS

High-speed, reliable and inexpensive semiconductor devices have produced a dramatic change in the control of dc motors. In this chapter, we examine some of the basic principles of such electronic controls. The circuits involve rectifiers and inverters already covered in Chapter 21. The reader should therefore review this chapter before proceeding further. Sections 20.16, 20.17 on Electric Drives should also be consulted.

In describing the various methods of control, we shall only study the behavior of power circuits. Consequently, the many ingenious ways of shaping and controlling triggering pulses are not covered here. The reason is that they constitute, by themselves, a complex subject involving sophisticated electronics, logic circuits, integrated circuits and microprocessors. Nevertheless, the omission of this important subject does not detract from the thrust of this chapter, which is to explain the fundamentals of electronic dc drives.

22.1 First quadrant speed control

We begin our study with a variable speed drive for a dc shunt motor. We assume its operation is restricted to quadrant 1. The field excitation is fixed, and the speed is varied by changing the armature voltage. A 3-phase, 6-pulse converter is connected between the armature and a 3-phase source (Fig. 22-1). The field is separately excited by a single-phase bridge rectifier. External inductor L ensures a relatively smooth armature current. The armature inductance L_a is usually large enough, and so the external inductor can often be dispensed with. The armature is initially at rest and the disconnecting switch S is open.

A gate triggering processor receives external inputs such as actual speed, actual current, actual torque, etc. These inputs are picked off the power circuit by means of suitable transducers. In addition, the processor can be set for any desired motor speed and torque. The actual values are compared with the desired values, and the processor automatically gener-

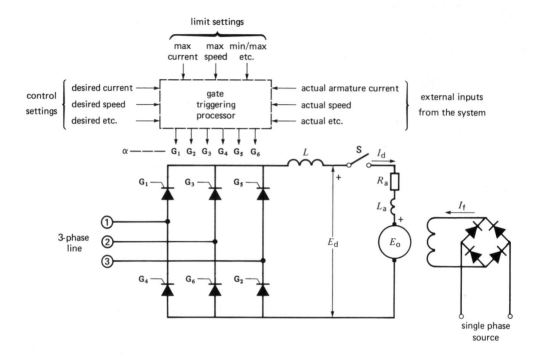

Figure 22-1
Armature torque and speed control of a dc motor using a thyristor converter.

ates gate pulses to bring them as close together as possible. Limit settings are also incorporated so that the motor never operates beyond acceptable values of current, voltage and speed.

Gate pulses are initially delayed by an angle $\alpha = 90°$ so that converter output voltage E_d is zero. Switch S is then closed and α is gradually reduced so that E_d begins to build up. Armature current I_d starts flowing and the motor gradually accelerates. During the starting period, the current is monitored automatically. Furthermore, the gate triggering processor is preset so that the pulses can never produce a current in excess of 1.6 p.u., say.

Three features deserve our attention as regards the start-up period:

1. no armature resistors are needed; consequently, there are no I^2R losses except those in the armature itself;

2. the power loss in the thyristors is negligible; consequently, all the active power drawn from the ac source is available to drive the load;

3. even if an inexperienced operator tried to start the motor too quickly, the current-limit setting would override the manual command. In effect, the armature current can never exceed the allowable preset value.

4. The converter absorbs a great deal of reactive power when the motor runs at low speed while developing its rated torque. Furthermore, the reactive power diminishes continually as the motor picks up speed. As a result, power factor correction is difficult to apply during the start-up phase.

When the motor reaches full speed, the firing angle is usually between 15° and 20°. Converter voltage E_d is slightly greater than induced voltage E_0 by an amount equal to the armature circuit IR_a drop. The converter voltage is given by the equation:

$$E_d = 1.35 \, E \cos \alpha \qquad \text{Eq. 21-17}$$

To reduce the speed, we increase α so that E_d becomes less than E_0. In a Ward-Leonard system, this would immediately cause the armature current to reverse (see Sec. 5.5). Unfortunately, the current cannot reverse in Fig. 22-1 because the SCRs conduct in only one direction. As a result, when we increase α, the current is simply cut off and the motor *coasts* to the lower speed. During this interval, E_0 gradually falls and, when it eventually becomes less than the new setting of E_d, the armature current again starts to flow. The torque quickly builds up and the motor will continue to run at the lower speed.

The efficiency at the lower speed is still high because the SCR losses are small. However, the ripple voltage generated by the converter is greater than under rated full-load conditions because α is greater (Sec. 21-28). Consequently, the armature current is not as smooth as before, which tends to increase the armature copper losses and iron losses. An equally serious problem is the large reactive power absorbed by the converter as the firing angle is increased. For example, when the firing angle is 45°, the converter absorbs as much reactive power from the 3-phase line as it does active power.

To stop the motor, we delay the pulses by 90° so that $E_d = 0$ V. The motor will coast to a stop at a rate that depends on the mechanical load and the inertia of the revolving parts.

Example 22-1:
A 750 hp, 250 V, 1200 r/min dc motor is connected to a 208 V, 3-phase, 60 Hz line using a 3-phase bridge converter (Fig. 22-2a). The full-load armature current is 2500 A and the armature resistance is 4 mΩ. Calculate:
a. the required firing angle α under rated full-load conditions;

Figure 22-2a
See Example 22-1.

b. the firing angle required so that the motor develops its rated torque at 400 r/min.

Solution:
a. At full load, the converter must develop a dc output of 250 V:
$$E_d = 1.35 \, E \cos \alpha \qquad \text{Eq. 21-14}$$
$$250 = 1.35 \times 208 \cos \alpha$$
$$\cos \alpha = 0.89$$
$$\alpha = 27°$$

Armature IR drop at rated current:
$$IR = 2500 \text{ A} \times 0.004 \ \Omega = 10 \text{ V}$$

Counter-emf (cemf) at 1200 r/min:
$$E_0 = 250 - 10 = 240 \text{ V}$$

b. To develop rated torque at 400 r/min, the armature current must still be 2500 A. cemf at 400 r/min:

$$E_0 = (400/1200) \times 240 = 80 \text{ V}$$

Armature IR drop = 10 V

Armature terminal voltage is:

$$E_d = 80 + 10 = 90 \text{ V}$$

The converter must therefore generate 90 V.

To determine the firing angle, we have:
$$E_d = 1.35 \, E \cos \alpha \qquad \text{Eq. 21-14}$$
$$90 = 1.35 \times 208 \cos \alpha$$
$$\alpha = 71° \text{ (see Fig. 22-2b)}$$

Figure 22-2b
Rated torque at 600r/min.

Example 22-2:
Referring to Example 22-1, calculate the reactive power absorbed by the converter when the motor develops full torque at 400 r/min.

Solution:
The load condition is given in Fig. 22-2b. The dc power absorbed by the motor is:

$$P = E_d I_d = 90 \times 2500 = 225 \text{ kW}$$

If we neglect the relatively small converter losses , the active power supplied by the ac source is also 225 kW.

The reactive power drawn from the ac source is given by:

$$Q = P \tan \alpha \qquad \text{Eq. 21-18}$$
$$Q = 225 \tan 71°$$
$$= 653 \text{ kvar}$$

(compare this with the active power of 225 kW)

This example shows that a large amount of reactive power is required as the firing angle is increased. It even exceeds the active power needed at full load. Capacitors should be installed on the ac side of the converter to reduce the burden on the three-phase feeder line. Alternatively, a variable-tap transformer could be placed between the 3-phase source and the converter. By reducing the ac voltage at the lower speeds, the reactive power can be reduced considerably. The reason is that the firing angle can then be kept between 15° and 20°. However, this may not be a feasible solution if the speed of the motor has to be continu-

ally varied. The tap-changing becomes too frequent. Later sections in this chapter (Secs. 22.8, 22.9) show other ways to reduce the reactive power demand.

22.2 Two-quadrant control – field reversal

We cannot always tolerate a situation where a motor simply coasts to a lower speed. To obtain a quicker response, we have to modify the circuit so that the motor acts temporarily as a generator. By controlling the generator output, we can make the speed fall as fast as we please. We often resort to dynamic braking using a resistor. However, the converter can also be made to operate as an inverter, feeding power back into the 3-phase line. Such *regenerative braking* is preferred because the kinetic energy is not lost. Furthermore, the generator output can be precisely controlled to obtain the desired rate of change in speed.

To make the converter act as an inverter, the polarity of E_d must be reversed as shown in Fig. 22-3. This means we must also reverse the polarity of E_0. Finally, E_d must be adjusted to be slightly less than E_0 to obtain the desired braking current I_d (Fig. 22-3).

These changes are not as simple as they may first appear. The polarity of E_d can be changed almost instantaneously by delaying the gate pulses by more than 90°. However, to change the polarity of E_0, we must reverse either the field or the armature, and this requires additional equipment. Reversing the

Figure 22-3
Motor control by field reversal.

armature or field also takes a significant length of time. Furthermore, after the generator (braking) phase is over, we must again reverse the armature or field so that the machine runs as a motor. Bearing these conditions in mind, we now list the steps to be taken when *field reversal* is employed.

Step 1: Delay the gate pulses by nearly 180° so that E_d becomes quite large and negative. This operation takes a few milliseconds, following which current I_d is zero.

Step 2: Reverse the field current as quickly as possible so as to reverse the polarity of E_0. The total reversing time may last from 1 to 2 seconds, owing to the high inductance of the shunt field. The armature current is still zero during this interval.

Step 3: Reduce α so that E_d becomes slightly less than E_0, enabling the desired armature current to flow. The motor now acts as a generator, feeding power back into the ac line. Its speed drops rapidly towards the lower setting.

What do we do when the lower speed is reached? We quickly rearrange the circuit so the dc machine again runs as a motor. This involves the following steps:

Step 4: Delay the gate pulse by nearly 180° so that E_d becomes quite large and negative. This operation takes a few milliseconds, after which current I_d is again zero.

Step 5: Reverse the field current as quickly as possible so as to make E_0 positive. The reversing time again lasts from 1 to 2 seconds. During this interval, the armature current is zero.

Step 6: Reduce α so that E_d becomes positive and slightly *greater* than E_0, enabling the desired armature current to flow. The machine now acts as a motor, and the converter is back to the rectifier mode.

22.3 Two-quadrant control – armature reversal

In some industrial drives, the long delay associated with field reversal is unacceptable. In such cases, we reverse the armature instead of the

Figure 22-4
Motor control by armature reversal.

field. This requires a high-speed reversing switch designed to carry the full armature current. The control system is arranged so that switching occurs only when the armature current is zero. Although this reduces contact wear and arcing, the switch still has to be fairly large to carry a current, say, of several thousand amperes.

Owing to its low inductance, the armature can be reversed in about 150 ms, which is about 10 times faster than reversing the field. Figure 22-4 is a simplified circuit showing a dc shunt motor connected to the converter by means of a reversing contactor. To reduce speed, the same steps are followed as in the case of field reversal, except that the armature is reversed instead of the field.

22.4 Two-quadrant control – two converters

When speed control has to be even faster, we use two identical converters connected in reverse parallel. Both are connected to the armature, but only one operates at a given time, acting either as a rectifier or inverter (Fig. 22-5). The other converter is on "standby", ready to take over whenever power to the armature has to be reversed. Consequently, there is no need to reverse the armature or field. The time to switch from one converter to the other is typically 10 ms. Reliability is considerably improved, and maintenance is reduced. Balanced against these advantages are higher cost and increased complexity of the triggering source.

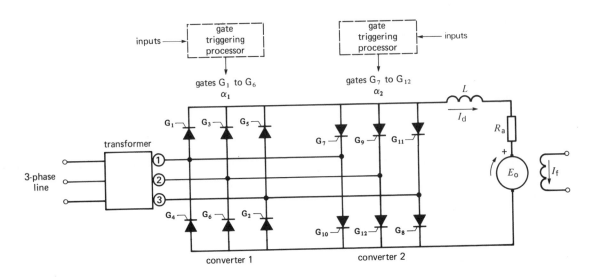

Figure 22-5
Two-quadrant control using two converters without circulating currents.

Because one converter is always ready to take over from the other, the respective converter voltages are close to the existing armature voltage, both in value and polarity. Thus, in Fig. 22-6a, converter 1 acts as a rectifier, supplying power to the motor at a voltage slightly higher than the cemf E_0. During this period, gate pulses are withheld from converter 2 so that it is inactive. Nevertheless, the control circuit continues to generate pulses having a delay α_2 so that E_{d2} would be equal to E_{d1} if the pulses were allowed to reach the gates (G7 to G12, Fig. 22-5).

To reduce the motor speed, gate pulses α_1 are delayed and, as soon as the armature current has fallen to zero, the control circuit withholds the pulses to converter 1 and simultaneously unblocks the pulses to converter 2. Converter 1 becomes inactive and the delay angle α_2 is then reduced so that E_{d2} becomes slightly less than E_0, thus permitting reverse current I_{d2} to flow (Fig. 22-6b).

This current reverses the torque, and the motor speed decreases rapidly. During the deceleration phase, α_2 is varied automatically so that E_{d2} follows the rapidly decreasing value of E_0. In some cases, α_2 is varied to maintain a constant braking current. During this period, the control circuit continues to generate gate pulses for converter 1 and the delay angle α_1 tracks α_2 so that E_{d1} would be equal to E_{d2} if the pulses were allowed to reach the gates (G1 to G6).

If the motor only operates in quadrants 1 and 4, it never reverses. Consequently, converters 1 and 2 always act respectively as rectifier and inverter.

22.5 Two-quadrant control – two converters with circulating current

Some industrial drives require precise speed and torque control right down to zero speed. This means that the converter voltage may at times be close to zero. Unfortunately, the converter current is discontinuous under these circumstances. In other words, the current in each thyristor no longer flows for 120°. Thus, at low

Figure 22-6a
Converter 1 in operation, converter 2 blocked.

Figure 22-6b
Converter 2 in operation, converter 1 blocked.

speeds, the torque and speed tend to be erratic, and precise control is difficult to achieve.

To get around this problem, we use two converters that function *simultaneously*. They are connected back-to-back across the armature (Fig. 22-8). When one functions as a rectifier, the other functions as an inverter, and vice versa. The armature current I is the difference between currents I_{d1} and I_{d2} flowing in the two converters. With this arrangement, the currents in both converters flow for 120°, even when $I =$ O. Obviously, with two converters continuously in operation, there is no delay at all in switching from one to the other. The armature current can be reversed almost instantaneously; consequently, this represents the most sophisticated control system available. It is also the most expensive. The reason is that when converters operate simultaneously, each must be provided with a large series inductor (L_1, L_2) to limit the ac circulating currents. Furthermore, the converters must be fed from separate sources, such as the isolated secondary windings of a 3-phase transformer. A typical circuit composed of a delta-connected primary and two wye-connected secondaries is shown in Fig. 22-8. Other transformer circuits are sometimes used to optimize performance, to reduce cost, to enhance reliability or to limit short-circuit currents.

Example 22-3:
The dc motor in Fig. 22-8 has an armature voltage of 450 V while drawing a current of 1500 A. Converter 1 delivers a current I_{d1} of 1800 A, and converter 2 absorbs 300 A. If the ac line voltage for each converter is 360 V, calculate:

a. the dc power associated with converters 1 and 2;

b. the active power drawn from the incoming 3-phase line;

c. the firing angles for converters 1 and 2;

d. the reactive power drawn from the incoming 3-phase line.

Figure 22-7
Precise 4-quadrant electronic speed control is provided in a modern steel mill. (*Siemens*)

Solution:

a. The dc power delivered by converter 1 is:

$$P_1 = E_{d1}I_{d1}$$
$$= 450 \times 1800$$
$$= 810 \text{ kW}$$

The power absorbed by converter 2 (operating as an inverter) is:

$$P_2 = E_{d2}I_{d2}$$
$$= 450 \times 300$$
$$= 135 \text{ kW}$$

Note that the converter voltages are essentially identical because the voltage drops in L_1 and L_2 are negligible. This means that the respective triggering of converters 1 and 2 cannot be voltage-controlled (by E_0,

for example). In effect, the triggering is current-controlled and is made to depend on the desired converter currents I_{d1} and I_{d2}.

b. The active power drawn from the incoming ac line is:

$$P = P_1 - P_2$$
$$= 810 - 135$$
$$= 675 \text{ kW}$$

Secondary winding 1, 2, 3 delivers 810 kW while secondary winding 7, 8, 9 receives 135 kW. It follows that the net active power drawn from the line (neglecting losses) is 675 kW.

c. The firing angle for converter 1 can be

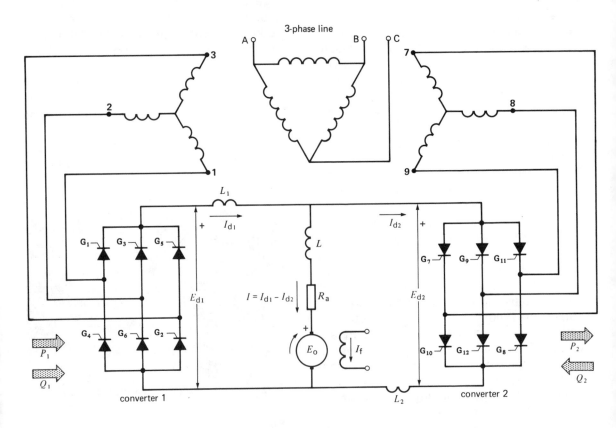

Figure 22-8
Two-quadrant control of a dc motor using two converters with circulating currents.

found from Eq. 21-17:

$$E_{d1} = 1.35 \, E \cos \alpha_1$$
$$450 = 1.35 \times 360 \cos \alpha_1$$
$$\cos \alpha_1 = 0.926$$
$$\alpha_1 = 22.2°$$

The firing angle for converter 2 is found to have the same value. However, because it operates as an inverter, the angle is:

$$\alpha_2 = 180 - \alpha_1$$
$$= 180 - 22.2$$
$$= 157.8°$$

Note that it would take a very small change in either α_1 or α_2 to make a very big change in I_{d1} and I_{d2}. This is why α_1 and α_2 are current-controlled.

d. The reactive power drawn by converter 1 is:

$$Q_1 = P_1 \tan \alpha_1$$
$$= 810 \tan 22.2$$
$$= 330 \text{ kvar}$$

The reactive power drawn by converter 2 is:

$$Q_2 = P_2 \tan \alpha_2$$
$$= - 135 \tan 157.8$$
$$= 55 \text{ kvar}$$

Consequently, the reactive power drawn from the incoming 3-phase line is:

$$Q = Q_1 + Q_2$$
$$= 330 + 55$$
$$= 385 \text{ kvar}$$

It is interesting to note that whereas the active powers substract, $(P = P_1 - P_2)$, the reactive powers add: $(Q = Q_1 + Q_2)$. The reason is that a naturally-commutated converter always absorbs reactive power, whether it functions as a rectifier or inverter.

22.6 Two-quadrant control with positive torque

So far, we have discussed various ways to obtain torque-speed control when the torque reverses. However, many industrial drives involve torques that always act in one direction, even when the speed reverses. Hoists and elevators fall into this category because gravity always acts downwards whether the load moves up or down. Operation is therefore in quadrants 1 and 2.

Consider a hoist driven by a shunt motor having constant field excitation. The armature is connected to the output of a 3-phase, 6-pulse converter. When the load is being raised, the motor absorbs power from the converter. Consequently, the converter acts as a rectifier (Fig. 22-9). The lifting speed depends directly upon converter voltage E_d. The armature current depends upon the weight of the load.

When the weight is being lowered, the motor reverses, which changes the polarity of E_0. However, the descending weight delivers power to the motor, and so it becomes a generator. We can feed the electric power into the ac line by making the converter act as an inverter. The gate pulses are simply delayed by more

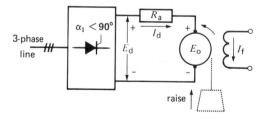

Figure 22-9
Hoist raising a load.

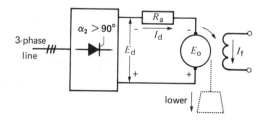

Figure 22-10
Hoist lowering a load.

than 90°, and E_d is adjusted to obtain the desired current flow (Fig. 22-10).

Hoisting and lowering can therefore be done in a stepless manner, and no field or armature reversal is required. However, the empty hook may not descend by itself. The downward motion must then be produced by the motor, which means that either the field or armature has to be reversed.

22.7 Four-quadrant control

We can readily achieve 4-quadrant control of a dc machine by using a *single* converter, combined with either field or armature reversal. However, a great deal of switching may be required. Four-quadrant control is possible without field or armature reversal by using two converters operating back-to-back. They may function either alternately or simultaneously, as previously described.

The following example illustrates 4-quadrant control of an industrial drive.

Example 22-4:
An industrial drive has to develop the torque-speed characteristic given in Fig. 22-11. A dc shunt motor is used, powered by two converters operating back-to-back. The converters function alternately (only one at a time). Determine the state of each converter over the 26-second operating period, and indicate the

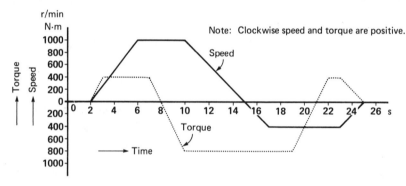

Figure 22-11
Torque-speed characteristic of an industrial drive.

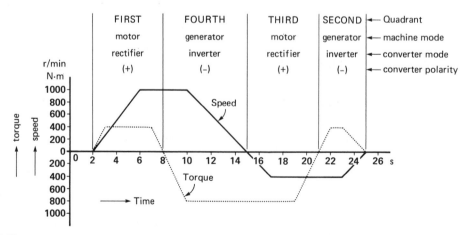

Figure 22-12
See Example 22-4.

polarity at the terminals of the dc machine. The speed and torque are considered positive when acting clockwise.

Solution:

The analysis of such a drive is simplified by subdividing the torque-speed curve into the respective 4 quadrants. In doing so, we look for those moments when either the torque or speed pass through zero. These moments always coincide with the transition from one quadrant to another. Referring to Fig. 22-11, the speed or torque passes through zero at 2, 8, 15, 21, and 25 s.

We draw vertical lines through these points (Fig. 22-12). We then examine whether the torque and speed are positive or negative during each interval. Knowing the respective signs, we can immediately state in which quadrant the motor is operating. For example, during the interval from 2 s to 8 s, both the torque and speed are positive. Consequently, the machine is operating in quadrant 1. On the other hand, in the interval from 21 s to 25 s, the speed is negative and the torque positive, indicating operation in quadrant 2.

Knowing the quadrant, we know whether the machine functions as a motor or generator. Finally, assuming that a positive (clockwise) speed corresponds to a "positive" armature voltage (Fig. 22-13a), we can deduce the required direction of current flow. This tells us which converter is in operation, and whether it acts as a rectifier or inverter.

Thus, taking the interval from 21 to 25 seconds, it is clear that the machine acts as a generator. Consequently, one of the two converters *must* function as an inverter. But which one? To answer the question, we first look at the polarity of the armature. Because the speed is negative, the armature polarity is negative, as shown in Fig. 22-13b. Current flows out of the positive terminal because the machine acts

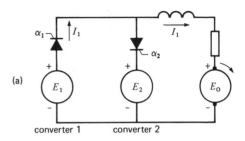

Figure 22-13a
Polarities when the speed is positive.

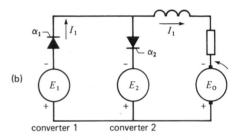

Figure 22-13b
Interval from 21 s to 25 s.

as a generator. Only converter 1 can carry this direction of current flow, and so it is the one in operation.

A similar line of reasoning enables us to determine the operating mode of each converter for the other intervals. The results are tabulated below; we encourage the reader to verify them.

Time interval	Operating mode	
	converter 1	converter 2
2 - 8 s	rectifier	off
8 - 15 s	off	inverter
15 - 21 s	off	rectifier
21 - 25 s	inverter	off

22.8 Six-pulse converter with freewheeling diode

When a dc motor is started up, we can significantly reduce the reactive power the converter absorbs by placing a diode across the converter output (Fig. 22-15). The usefulness of this so-called freewheeling diode is best illustrated by a numerical example.

Suppose a dc motor has the following characteristics:

  rated power: 100 hp
  rated armature voltage: 240 V
  rated armature current: 320 A
  armature resistance: 25 mΩ
  armature inductance: 1.7 mH

We begin our analysis using a conventional 6-pulse converter to drive the motor. The converter (rectifier) is fed by a 3-phase, 184 V, 60 Hz line. We will analyze the voltages and currents when the motor is at standstill with rated current flowing in the armature. As a result, the motor will develop rated torque. The circuit of the motor and converter is shown in Fig. 22-14a.

The motor is stalled and so the value of E_d is only that needed to supply the armature IR drop.

$$E_d = IR = 320 \text{ A} \times 25 \text{ m}\Omega$$
$$= 8 \text{ V}$$

To develop this dc voltage, the required firing angle is given by:

$$E_d = 1.35 \, E \cos \alpha$$
$$8 = 1.35 \times 184 \cos \alpha$$

whence

$$\cos \alpha = 0.0322$$

and so

$$\alpha = 88.15°$$

The active ac power P supplied to the converter is necessarily equal to that absorbed by the armature. Thus,

$$P = E_d I_d = 8 \times 320 = 2560 \text{ W}$$

The reactive power Q absorbed by the converter is therefore:

$$Q = P \tan \alpha = 2560 \tan 88.15°$$
$$= 79.26 \text{ kvar}$$

Note that the reactive power is 31 times

Figure 22-14a
Conventional rectifier supplying rated current to a stalled dc armature. Firing angle is 88.15°.

greater than the active power.

The line-to-neutral voltage E_{1N} and the corresponding line current I_a are shown in Fig. 22-14b. Note that Q1 is triggered 88.15° after θ_0. As a result, the center of the positive current pulse lags 88.15° behind the positive voltage peak E_{1N}. Thus, I_a lags almost 90° behind E_{1N} and that is why the reactive power is so large. The same remarks apply to I_b, I_c and their respective line-to-neutral voltages.

The effective value of the line current is:

$$I = 0.816\, I_d \qquad \text{Eq. 21-6}$$
$$= 0.816 \times 320$$
$$= 261 \text{ A}$$

Figure 22-14b also shows that voltage E_{KA} across the armature is a saw-tooth wave. The peak value oscillates between +137 V and −123 V, but the average value of the wave is only 8 V.

It would be useful if we could remove the negative voltages in E_{KA}. The reason is that for a given firing angle (such as 88.15°) the average voltage E_d will then increase. The negative voltages can be suppressed by placing a diode between terminals K and A (Fig. 22-15a). In effect, as soon as K becomes negative with respect to A, the diode starts conducting. During the conduction period all currents cease to flow in the converter, and so the line currents I_a, I_b, I_c are also zero. Note that I_a is now composed of a double positive pulse of current followed by a double negative pulse.

The presence of the diode produces the E_{KA} waveshape shown in Fig. 22-15a. The negative voltages are clipped off, and so the dc voltage across the armature becomes much larger than 8 V. As a result, the armature current also becomes much greater than 320 A. In order to make E_d = 8 V, we must increase the firing

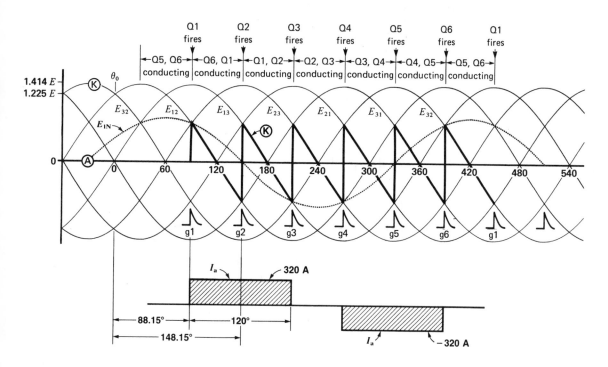

Figure 22-14b
Voltage and current waveshapes when armature is stalled while developing rated torque.

Figure 22-15a
Conventional rectifier and freewheeling diode supplying current to a stalled dc armature. Firing angle is 88.15°.

angle α. It can be proved that the dc voltage of a 3-phase 6-pulse converter equipped with such a freewheeling diode is given by:

$$E_d = 1.35\, E\, (1 - \cos(120 - \alpha)) \qquad (22\text{-}1)$$

where

E_d = dc voltage [V]

E = effective value of line-to-line voltage [V]

α = firing angle (between 60° and 120°) [°]

1.35 = a constant that applies only when α lies between 60° and 120° [exact value = $3\sqrt{2}/\pi$]

In our example, the dc voltage with a firing angle of 88.15° is:

$$E_d = 1.35\, E\, (1 - \cos(120 - \alpha))$$
$$= 1.35 \times 184\,(1 - \cos(120 - 88.15))$$

$$= 1.35 \times 184\,(1 - \cos 31.85°)$$
$$= 37.4\ \text{V}$$

The resulting armature current would therefore be:

$$I = 37.4/0.025 = 1496\ \text{A}$$

This is nearly 5 times the rated current, which is clearly unacceptable.

In order to obtain the desired $E_d = 8$ V, we must increase the firing angle such that

$$E_d = 1.35E\,(1 - \cos(120 - \alpha)) \quad \text{Eq. 22-1}$$
$$8 = 1.35 \times 184\,(1 - \cos(120 - \alpha))$$
$$\cos(120 - \alpha) = 0.9678$$

therefore

$$120 - \alpha = \arccos 0.9678 = 14.6°$$

consequently,

$$\alpha = 120 - 14.6° = 105.4°$$

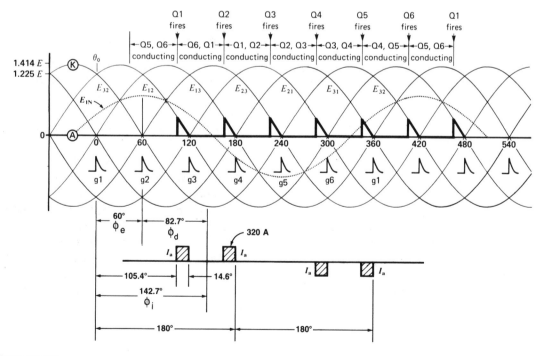

Figure 22-15b
Conventional rectifier and freewheeling diode supplying current to a dc armature. Firing angle is 105.4°.

With this firing angle, the waveshape of E_{KA} and I_a is as shown in Fig. 22-15b. The positive half cycle of I_a is composed of two current pulses. Each has an amplitude of 320 A and a duration of $(120° - 105.4°) = 14.6°$. The effective value of I_a is therefore:

$$I_a(\text{eff}) = 320\sqrt{(4 \times 14.6)/360} = 128.9 \text{ A}$$

Thus, the 3-phase line currents are much lower than in Fig. 22-14a, even though the dc current in the armature is the same.

The positive half cycle of current I_a is centered midway between 105.4° and 180° (see Fig. 22-15b).

Thus, angle $\phi_i = (105.4 + 180)/2 = 142.7°$. On the other hand, the positive peak of E_{1N} occurs at $\phi_e = 60°$.

Consequently, the fundamental component of I_a lags by angle $\phi_d = (142.7° - 60°) = 82.7°$ behind E_{1N}. The displacement angle is therefore 82.7° and the displacement power factor is:

$$PF \text{ (displacement)} = \cos 82.7° = 0.127$$

The active power P drawn from the 3-phase line is again 320 A × 8 V = 2.56 kW. The reactive power Q is

$$Q = P \tan \text{ (displacement angle)}$$
$$= 2.56 \tan 82.7°$$
$$= 20 \text{ kvar}$$

Thus, with the freewheeling diode, the converter only draws 20 kvar from the line, compared to 79.26 kvar with no diode.

The reader should note that the freewheeling diode only begins to produce an effect when the firing angle lies between 60° and 120°. It is only then that E_{KA} starts to become negative, thus permitting the diode to exert its influence. The line current is not sinusoidal and pos-

sesses a strong harmonic content. It can be shown that the fundamental component of the line current lags behind the corresponding line-to-neutral voltage by an angle ϕ_d given by:

$$\phi_d = 30° + \alpha/2 \qquad (22-2)$$

where

ϕ_d = displacement phase angle [°]
α = firing angle (must lie between 60° and 120°) [°]
30° = a constant for this type of converter

The effective value of the line current is given by:

$$I = I_d \sqrt{(120 - \alpha)/90} \qquad (22-3)$$

The displacement power factor is given by:

$$PF \text{ (displacement)} = \cos \phi_d \qquad (22-4)$$

The displacement angle ϕ_d can be used to calculate the reactive power Q:

$$Q = P \tan \phi_d \qquad (22-5)$$

22.9 Half bridge converter

The half-bridge converter is another way whereby the reactive power can be reduced when the dc output voltage is low. Figure 22-16 shows a 3-phase, 6-pulse converter in which three thyristors have been replaced by 3 diodes D2, D4, D6. This *half bridge converter* has properties similar to those of a conventional thyristor bridge with freewheeling diode.

The firing angle α can be varied from zero to 180°. However, the freewheeling diode effect only begins for angles greater than 60°. In Fig. 22-16, the firing angle is assumed to be 135°. As a result, E_{KA} is positive during successive 45° periods. Note that the positive and negative current pulses in each 3-phase line also flow for 45°.

Figure 22-16
Half bridge rectifier. Firing angle is 135°.

The average value of E_{KA} is given by the equation:

$$E_d = 0.675 \, E \, (1 + \cos \alpha) \qquad (22\text{-}6)$$

in which

E_d = dc voltage across the load [V]

E = effective value of line voltage [V]

α = firing angle (must lie between 60° and 180°) [V]

0.675 = a constant that only applies for α between 60° and 180° [exact value is $(1.5 \sqrt{2})/\pi$]

Using the same example as in Sec. 22.8, the firing angle needed to produce $E_d = 8$ V is given by:

$$\begin{aligned}
E_d &= 0.675 \, E \, (1 + \cos \alpha) \\
8 &= 0.675 \times 184 \, (1 + \cos \alpha) \\
-0.936 &= \cos \alpha
\end{aligned}$$

whence $\alpha = 159.32°$

The line current pulses have an amplitude equal to the dc armature current, namely 320 A. The duration of the current pulses is $(180° - 159.32°) = 20.68°$. Based upon the example of Fig. 22-16, the positive current pulse starts at 159.32° and ends at 180°. The negative current pulse starts at 279.32° and ends at 300°. Thus, current I_a essentially passes through zero at an angle ϕ_o given by

$$\phi_o = 1/2(159.32° + 300°) = 229.66°$$

Consequently, the positive peak of I_a occurs essentially at an angle of $(229.66° - 90°) = 139.66°$. Thus, the displacement angle is:

$$\phi_d = (139.66° - 60) = 79.66°$$

The displacement power factor is:

$$\text{PF (displacement)} = \cos \phi_d = 0.179$$

As before, the active power supplied by the 3-phase line is:

$$P = 2.56 \text{ kW}$$

The reactive power absorbed is

$$\begin{aligned}
Q &= P \tan \phi_d = 2.56 \tan 79.66° \\
&= 14 \text{ kvar}
\end{aligned}$$

The reader will note that even less reactive power is needed with the half bridge rectifier than with the freewheeling diode circuit of Fig. 22-15b.

The effective value I of the line current is given by

$$I_d^2 \times 20.68° \times 2 = I^2 \times 360°$$

whence

$$I = 0.339 I_d = 0.339 \times 320 = 108.4 \text{ A}$$

Table 22A sums up the basic properties of the three types of rectifier converters we have discussed. In the case of converters B and C, the firing angle limits are those during which freewheeling operation takes place. The values of E_d, ϕ_d, I, etc., are only valid within the stated limits. Furthermore, the load is resistive.

22.10 DC traction

Electric trains and buses have for years been designed to run on direct current, principally because of the special properties of the dc series motor. Many are now being modified to make use of the advantages offered by thyristors. Existing trolley lines still operate on dc and, in most cases, dc series motors are still used. To modify such systems, high-power *electronic choppers* are installed on board the vehicle (See Sec. 21-24). Such choppers can drive motors rated at several hundred horsepower, with outstanding results. To appreciate the improvement that has taken place, let us review some of the features of the older systems.

A train equipped with, say, two dc motors, is started with both motors in series with an external resistor. As the speed picks up, the resistor is shorted out. The motors are then paralleled

TABLE 22A   PROPERTIES OF SOME RECTIFIER CONVERTERS (RESISTIVE LOAD)

ITEMS	Converter A 3ph, 6-pulse	Converter B 3 ph, 6-pulse + free-wheel diode	Converter C half bridge
Firing angle (α) limits	0 to 90°	60° to 120°	60° to 180°
dc output voltage (E_d)	$1.35\,E \cos \alpha$	$1.35E\,(1 - \cos(120 - \alpha))$	$0.675\,E\,(1 + \cos \alpha)$
displacement angle (ϕ_d)	α	$30 + \alpha/2$	$\alpha/2$
PF (displacement) = $\cos \phi_d$	$\cos \alpha$	$\cos (30 + \alpha/2)$	$\cos \alpha/2$
effective line current (I)	$0.816\,I_d$	$I_d \sqrt{(120 - \alpha)/90}$	$I_d \sqrt{(180 - \alpha)/180}$
Total apparent power (S)	$EI \sqrt{3}$	$EI \sqrt{3}$	$EI \sqrt{3}$
Total active power (P)	$E_d I_d$	$E_d I_d$	$E_d I_d$
Total reactive power (Q)	$P \tan \phi_d$	$P \tan \phi_d$	$P \tan \phi_d$
PF (total)	P/S	P/S	P/S
PF (distortion)	$P/S \cos \phi_d$	$P/S \cos \phi_d$	$P/S \cos \phi_d$

and connected in series with another resistor. Finally, the last resistor is shorted out, as the train reaches its nominal torque and speed. The switching sequence produces small jolts, which, of course, are repeated during the electric braking process. Although a jolt affects passenger comfort, it also produces slippage on the tracks, with consequent loss of traction. The dc chopper overcomes these problems because it permits smooth and continuous control of torque and speed. We now study some simple chopper circuits used in conjunction with series motors.

Figure 22-17 shows the armature and field of a series motor connected to the output of a chopper. Supply voltage E_s is picked off from two overhead trolley wires. The inductor-capacitor combination $L_1 C_1$ acts as a dc filter, preventing the sharp current pulses I_s from reaching the trolley line. The capacitor can readily furnish these high current pulses. The

presence of the inductor has a smoothing effect so that current I drawn from the line has a relatively small ripple.

As far as the motor is concerned, the total inductance of the armature and series field is large enough to store and release the energy needed during the chopper cycle. Consequently, no external inductor is required. When the motor starts up, a low chopper frequency is used, typically 50 Hz. The corresponding "on" time T_a is typically 500 μs. In many systems, T_a is kept constant while the switching frequency varies. The top frequency (about 2000 Hz) is limited by the switching and turn-off time of the thyristors.

Other choppers function at constant frequency, but with a variable "on" time T_a. In still more sophisticated controls, both the frequency and T_a are varied. In such cases, T_a may range from 20 μs to 800 μs. Nevertheless, the basic chopper operation remains the same,

Figure 22-17
Direct-current series motor driven by a chopper. The chopper is not a switch as shown, but a force-commutated SCR.

Figure 22-18a
See Example 22-5.

no matter how the on-off switching times are varied. Thus, the chopper output voltage E_0 is related to the input voltage E_s by the equation:

$$E_0 = E_s f T_a \qquad \text{Eq. 21-14}$$

where f is the chopper frequency and T_a the "on" time.

Example 22-5:
A trolleybus is driven by a 150 hp, 1500 r/min, 600 V series motor. The nominal full-load current is 200 A and the total resistance of the armature and field is 0.1 Ω. The bus is fed from a 700 V dc line.

A chopper controls the torque and speed. The chopper frequency varies from 50 Hz to 1600 Hz, but the "on" time T_a is fixed at 600 μs:
a. calculate the chopper frequency and the current drawn from the line when the motor is at standstill and drawing a current of 240 A;
b. calculate the chopper frequency when the motor delivers its rated output.

Solution:
a. Referring to Fig. 22-18a, the armature IR drop is 240 A × 0.1 Ω = 24 V, and the cemf is zero because the motor is at standstill.
    Consequently, $E_0 = 24$ V, and $E_s = 700$ V.
    We can find the frequency from:

Figure 22-18b
Current pulses I_s drawn by the chopper from the 700 V source when the motor is stalled.

$$E_0 = E_s f T_a \qquad \text{Eq. 21-14}$$
$$24 = 700 f \times 600 \times 10^{-6}$$
$$f = 57.14 \text{ Hz}$$
$$T_a + T_b = 1/f = 1/57.14$$
$$= 17\,500 \text{ μs (Fig. 22-18b)}$$

dc current drawn from line is:
$$I = I_s = P/E_s = 24 \times 240/700$$
$$= 8.23 \text{ A}$$

(Note the very low current drawn from the line during start-up)

b. At rated output, the voltage across the motor terminals is 600 V (Fig. 22-19a).
$$E_0 = E_s f T_a$$
$$600 = 700 f \times 600 \times 10^{-6}$$
$$f = 1429 \text{ Hz}$$
$$T_a + T_b = 1/f = 1/1429$$
$$= 700 \text{ μs (Fig. 22-19b)}$$

Line current I:

$$I = I_s = P/E_s$$
$$= 600 \times 200/700$$
$$= 171 \text{ A}$$

Example 22-6:

Referring to Example 22-5 and Fig. 22-18a, calculate the peak value of currents I_s and I when the motor is at standstill.

Solution:

a. Although the average value of I_s is 8.23 A, its peak value is 240 A. The current flows in a series of brief, sharp pulses. On the other hand, the armature current is steady at 240 A.

b. The average value of line current I is 8.23 A, but since it flows continuously, the peak value is also 8.23 A. In practice, the current will have a ripple because inductor L_1 does not have infinite inductance. Consequently, the peak value of I will be slightly greater than the average.

22.11 Current-fed dc motor

Some electronic drives involve direct current motors that do not look at all like dc machines. The reason is that the usual rotating commutator is replaced by a stationary electronic converter. We now discuss the theory behind these so-called "commutatorless" dc machines.

Consider a 2-pole dc motor having 3 independent armature coils, A,B, and C spaced at 120° to each other (Fig. 22-20). The two ends of each coil are connected to diametrically opposite segments of a 6-segment commutator. Two narrow brushes are connected to a *constant-current* source that successively feeds current into the coils as the armature rotates. A permanent magnet N, S creates the magnetic field.

With the armature in the position shown, current flows in coil A and the resulting torque causes the armature to turn counterclockwise. As soon as contact is broken with this coil, it is immediately established in the next coil. Con-

(a)

Figure 22-19a
Conditions when the motor is running at rated torque and speed.

(b)

Figure 22-19b
Corresponding current pulses drawn by the chopper from the 700 V source.

sequently, conductors facing the N pole always carry currents that flow into the page, while those facing the S pole carry currents that flow out of the page (towards the reader). The motor torque is therefore continuous and may be expressed by:

$$T = kIB \qquad (22-7)$$

where

T = motor torque [N·m]
I = current in the conductors [A]
B = average flux density surrounding the current-carrying conductors [T]
k = a constant, dependent upon the number of turns per coil, and the size of the armature

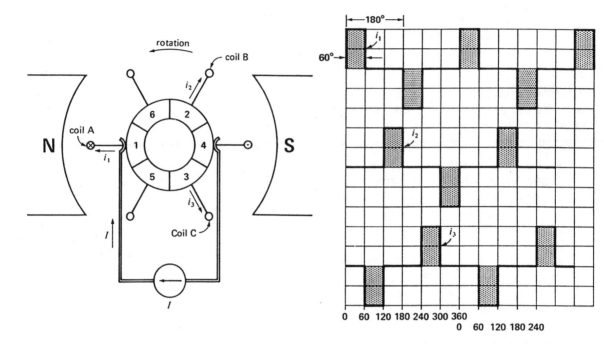

Figure 22-20
Special current-fed dc motor.

Figure 22-21
The dc current changes to ac current in the coils.

If the current and flux density are fixed, the resulting torque is also fixed, independent of motor speed.

The commutator segments are 60° wide; consequently, the current in each coil flows in 60° pulses. Furthermore, the current in the coil reverses every time the coil makes half a turn (Fig. 22-21). The alternating nature of the current is of crucial importance. If the current did *not* alternate, the torque developed by each coil would act first in one, then the opposite direction, as the armature rotates. The net torque would be zero, and so the motor would not develop any power.

Figure 22-21 shows that the ac currents in the 3 coils are out of phase by 120°. Consequently, the armature behaves as if it were excited by a 3-phase source. The only difference is that the current waveshapes are rectangular instead of sinusoidal. Basically, the commutator acts as a mechanical converter, changing the dc current from the dc source into ac current in the coils. The frequency is given by:

$$f = pn/120 \qquad (22\text{-}8)$$

where p is the number of poles and n the speed (r/min). The frequency in the coils is automatically related to the speed because the faster the machine rotates, the faster the commutator switches from one coil to the next. In effect, the commutator generates a frequency which at all times is appropriate to the instantaneous speed.

As the coils rotate, they cut across the magnetic field created by the N, S poles. An ac voltage is therefore induced in each coil, and its frequency is also given by Eq. 22-8. Furthermore, the voltages are mutually displaced at 120° owing to the way the coils are mounted on the armature. The induced ac voltages appear as a dc voltage between the brushes. The reason is that the brushes are always in contact with coils that are moving in the same direction through the magnetic field; consequently, the polarity is always the same (see Sec. 4-2).

If the brushes were connected to a dc *voltage source E*, the armature would accelerate until the induced voltage E_0 is about equal to E (Sec. 5.2). What determines the speed when the armature is fed from a *current source*, as it is in our case? The speed will increase until the load torque is equal to the torque developed by the motor. Thus, while the speed of a voltage-fed armature depends upon equilibrium between induced voltage and applied voltage, the speed of a *current-fed* armature depends upon equilibrium between motor torque and load torque. The torque of a mechanical load always rises with increasing speed. Consequently, for a given motor torque, a state of torque equilibrium is always reached, provided the speed is high enough. Care must be taken so that current-fed motors do not run away when the load torque is removed.

22.12 Commutator replaced by reversing switches

Recognizing that each coil in Fig. 22-20 carries an alternating current, we can eliminate the commutator by connecting each coil to a pair of slip rings and bringing the leads out to a set of mechanical reversing switches (Fig. 22-22). Each switch has 4 normally open contacts. Considering coil A, for example, switch contacts 7 and 8 are closed during the 60° interval when coil side 1 faces the N pole (Fig. 22-23). The contacts are then open for 120° until coil side 4 faces the N pole, whereupon contacts 9 and 10 close for 60°. Consequently, by synchronizing the switch with the *position* of coil A, we obtain the same result as if we used a commutator.

Figure 22-22
The commutator can be replaced by an array of mechanical switches and a set of slip rings.

Figure 22-23
Circuit showing how current is controlled in coil A.

Figure 22-24
The armature is now the stator, and the switches have been replaced by thyristors.

Coils B and C operate the same way, but they are energized at different times. Figure 22-22 shows how the array of 12 contacts and 6 slip rings are connected to the current source. The reversing switches really act as a 3-phase mechanical inverter, changing dc power into ac power. The slip rings merely provide electrical contact between the revolving armature and the stationary switches and power supply.

Clearly, the switching arrangement of Fig. 22-22 is more complex than the original commutator. However, we can simplify matters by making the armature stationary and letting the permanent magnets rotate. By thus literally turning the machine inside out, we can eliminate 6 slip rings. Then, as a final step, we can replace each contact by a thyristor (Fig. 22-24). The 12 thyristors are triggered by gate signals that depend upon the instantaneous *position* of the revolving rotor.

The dc motor in Fig. 22-24 looks so different from the one in Fig. 22-20 that we would never suspect they have the same properties. And yet they do.

For example:
1. If we increase the dc current I or the field strength of poles N, S, the torque increases, and consequently, the speed will increase.
2. If we shift the brushes against the direction of rotation in Fig. 22-20, current will start flowing in each coil a little earlier than before. Consequently, the ac current in each coil will *lead* the ac voltage induced across its terminals. We can produce exactly the same effect by firing the thyristors a little earlier in Fig. 22-24. Under these circumstances, the machine furnishes reactive power to the three thyristor bridges, at the same time as it absorbs active power from them.

3. If we shift the brushes by 180°, the current in each *coil* flows in the opposite direction to that shown in Fig. 22-20. However, the induced voltage in each coil remains unchanged because it depends only on the speed and direction of rotation. Consequently, the machine becomes a generator, feeding dc power back into the current source.

The same result occurs if we fire the thyristors 180° later in Fig. 22-24. The thyristors then behave as inverters feeding power back to the dc current source.

It is now clear that the machines in Figs. 22-20 and 22-24 behave the same way. The only difference between them is that one is equipped with a rotating mechanical commutator, while the other has a stationary electronic commutator composed of 12 thyristors. By firing the thyristors earlier or later, we produce the same effect as shifting the brushes.

22.13 Synchronous motor as a commutatorless dc machine

The revolving-field motor in Fig. 22-24 is built like a 3-phase synchronous motor. However, because of the way it receives its ac power, it behaves like a "commutatorless" dc machine. This has a profound effect upon its performance.

First, the "synchronous motor" can never pull out of step because the stator frequency is not fixed, but changes automatically with speed. The reason is that the gates of the SCRs are triggered by a signal that depends upon the instantaneous position of the rotor. For the same reason, the machine has no tendency to oscillate or hunt under sudden load changes.

Second, the phase angle between the ac current in a winding and the ac voltage across it can be modified by altering the timing of the gate pulses. This enables the synchronous motor to operate at leading, lagging, or unity power factor.

Third, because the phase angle between the respective voltages and currents can be fully controlled, the machine can even function as a generator, feeding power back to the dc current source. The thyristor bridges then operate as rectifiers.

Currents i_1, i_2, i_3 in Fig. 22-24 flow only during 60 degree intervals, as they did in the original dc machine. In practice, the conduction

Figure 22-25
Commutatorless dc motor being driven by a converter.

Figure 22-26
This elementary dc motor is equivalent to the entire circuit of Fig. 22-25.

period can be doubled to 120°, by connecting the coils in wye and exciting them by a 3-phase, 6-pulse converter (Fig. 22-25). This reduces the number of thyristors by half. Furthermore, it improves the current-carrying capacity of the windings because the duration of current flow is doubled. Gate triggering is again dependent upon the position of the rotor. The phase angle between line voltage E_s and line current I is modified by firing the gates earlier or later. In the circuit of Fig. 22-25, the power factor of the motor has to be slightly leading to provide the reactive power absorbed by the converter.

As a matter of interest, the converter and motor of Fig. 22-25 could be replaced by the dc motor shown in Fig. 22-26. The armature coils are connected in wye and the 3 leads are soldered to a 3-segment commutator. The respective voltages and currents are identical in the two figures.

22.14 Standard synchronous motor and commutatorless dc machine

The machine shown in Fig. 22-25 can be made to function as a conventional synchronous motor by applying a *fixed* frequency to the SCR gates. Under these conditions, the input to the gate triggering processor no longer depends on rotor position or rotor speed.

Obviously then, the behavior of the machine as a commutatorless dc motor or synchronous motor depends upon the way the gates are fired. If the triggering frequency is constant, the machine acts as a synchronous motor. On the other hand, if the triggering frequency depends on the speed of the rotor, it behaves like a commutatorless dc motor*.

22.15 Synchronous motor drive using current-fed dc link

Figure 22-27 shows a typical commutatorless dc motor circuit. It consists of two converters

---

* Readers familiar with feedback theory will recognize that the basic distinction between the two machines is that one functions on open loop while the other operates on closed loop.

connected between a 3-phase source and the "synchronous" motor. Converter 1 acts as a controlled rectifier, feeding dc power to converter 2. The latter behaves as a naturally-commutated inverter whose ac voltage and frequency are established by the motor.

A smoothing inductor L maintains a ripple-free current in the so-called *dc link* between the two converters. Current I is controlled by converter 1, which acts as a current source. A smaller bridge rectifier (converter 3) supplies the field excitation for the rotor.

Converter 2 is naturally-commutated by voltage E_s induced across the terminals of the motor. This voltage is created by the revolving magnetic flux in the air gap. The flux depends upon the stator currents and the exciting current I_f. The flux is usually kept fixed; consequently, the induced voltage E_s is proportional to the motor speed.

Gate triggering of converter 1 is done at line frequency (60 Hz) while that of converter 2 is done at motor frequency.

With regard to controls, information picked off from various points is assimilated in the gate triggering processors which then send out appropriate gate firing pulses to converters 1 and 2. Thus, the processors receive information as to desired speed of rotation, actual speed, instantaneous rotor position, stator voltage, stator current, field current, etc. They interpret whether these inputs represent normal or abnormal conditions, and emit appropriate gate pulses to correct the situation or to meet a specific command.

The gate pulses of converter 2 are controlled by the position of the rotor. This is accomplished by a set of transducers mounted on the stator next to the air gap. Other methods employ position transducers mounted on the end of the shaft. Owing to this method of gate control, the synchronous motor acts as a commutatorless dc machine. The motor speed may be increased by raising either the dc link current I or the field current I_f.

Figure 22-27
Commutatorless dc motor driven by a converter with a dc link. The output frequency can be considerably greater than 60 Hz, thus permitting high speeds.

Stator voltage E_s produces a dc emf E_2 given by:

$$E_2 = 1.35 \, E_s \cos \alpha_2 \qquad \text{Eq. 21-17}$$

where

E_2 = dc voltage generated by converter 2 [V]

E_s = effective line-to-line stator voltage [V]

α_2 = firing angle of converter 2 [°]

Similarly, the voltage produced by converter 1 is given by:

$$E_1 = 1.35 \, E_L \cos \alpha_1$$

Link voltages E_1 and E_2 are almost equal, differing only by the negligible IR drop in the inductor. Firing angle α_1 is automatically controlled so that the magnitude of the link current is just sufficient to develop the required torque.

The stator line current I_s flows in 120° rectangular pulses, as shown in Fig. 22-28. These step-like currents produce a revolving MMF that moves in jerks around the armature. This produces torque pulsations, but they are almost completely damped out (except at low speeds) owing to the inertia of the rotor. The shaft therefore turns smoothly when running at rated speed.

The motor line-to-neutral voltage E_{LN} and line voltage E_s are essentially sinusoidal. The field current, line current, and triggering are adjusted so that line current I_s leads the line-to-neutral voltage (Fig. 22-28). The reason is that the synchronous motor must operate at leading power factor to supply the reactive power absorbed by converter 2.

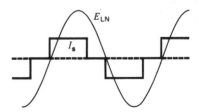

Figure 22-28
Typical voltage and current waveshapes in Fig. 22-27.

Converter 1 is designed so that under full-load conditions, firing angle α_1 is about 15° to minimize the reactive power drawn from the ac line.

Regenerative braking is accomplished by shifting the gate firing pulses so that converter 2 acts as a rectifier while converter 1 operates as an inverter. The polarity of E_2 and E_1 reverses, but the link current continues to flow in the same direction. Power is therefore pumped back into the ac line and the motor slows down. During this period, the machine functions as a commutatorless dc generator.

Starting the motor creates a problem, because the stator voltage E_s is zero at standstill. Consequently, no voltage is available to produce the natural commutation of converter 2. To get around this difficulty, the converters are fired in such a way that short current pulses flow successively in phases ab, bc, and ca. The successive pulses create N, S poles in the stator that are always just ahead of their opposite poles on the rotor. Like a dog chasing its tail, the rotor accelerates and, when it reaches about 10 percent of rated speed, converter 2 takes over and commutation takes place normally. This pulse mode of operation can also be used to continually brake the motor as it approaches zero speed.

The speed control of synchronous motors using a current-fed dc link is applied to motors ranging from 1 kW to several megawatts. Permanent-magnet synchronous motors for the textile industry and brushless synchronous motors for nuclear reactor circulating pumps are two examples. Pumped-storage hydropower plants also use this method to bring the huge synchronous machines up to speed so that they may be smoothly synchronized with the line.

22.16 Synchronous motor and cycloconverter

We have seen that cycloconverters can directly convert ac power from a higher frequency to a lower frequency (Sec. 21-22). These converters are sometimes used to drive slow-speed commutatorless dc motors rated up to several megawatts. If a 60 Hz source is used, the cycloconverter output frequency is typically variable from zero to 10 Hz. Such a low frequency permits excellent control of the waveshape of the output voltage by computer-controlled firing of the thyristor gates. As in drives with a dc link, no forced commutation is needed, with the result that the complexity of the electronics surrounding each SCR is considerably reduced.

Figure 22-29 shows three cycloconverters connected to the wye-connected stator of a 3-phase synchronous motor. Each cycloconverter produces a single-phase output, based upon the principle explained in Sec. 21-22. Referring to phase A, the associated cycloconverter is composed of two 3-phase bridges, +A and −A, each fed by the same 3-phase 60 Hz line.

Bridge +A generates the positive half cycle for line 1, while bridge −A generates the negative half. The two bridges are prevented from operating at the same time so as to eliminate circulating currents between them. The resulting low-frequency wave is composed of segments of the 60 Hz voltage between lines 1, 2, 3. By appropriate gate firing, the low frequency voltage can be made to approach a sine wave quite closely (Fig. 22-30). However, to reduce the reactive power absorbed from the 60 Hz line, the output voltage is usually designed to have a trapezoidal, flat-topped wave-shape.

Both the cycloconverter and the 3-phase controlled rectifier supplying field current I_f function as current sources. The magnitudes of the stator currents and of I_f are controlled so as to keep a constant flux in the air gap. Furthermore, the gate pulses are timed and the excitation is adjusted so that the motor operates at unity power factor at the low frequency. However, even at unity power factor (I_a in

Converter	Thyristors
+ A	Q1 to Q6
- A	Q7 to Q12
+ B	Q13 to Q18
- B	Q19 to Q24
+ C	Q25 to Q30
- C	Q31 to Q36

Typical inputs to gate triggering processor

a. rotor position

b. motor speed

c. cycloconverter input current

d. cycloconverter output current

e. input and output voltages

f. desired speed

Figure 22-29
Cycloconverter driving a commutatorless dc machine. The output voltage associated with phase A is a slowly changing sine wave having a frequency of 6 Hz, which is 10 times less than the supply frequency. Thyristors Q1 to Q12 are triggered so as to track the desired sine wave as closely as possible. This produces the saw-tooth output voltage shown in Fig.22-30. The power factor at the input to the motor is assumed to be unity. The corresponding power factor at the input to the cycloconverter is less than unity, owing to the delayed firing angles.

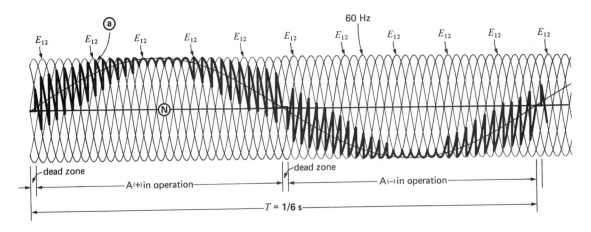

Figure 22-30
Voltage between lines a and N of Fig.22-29.

phase with E_{aN}), the cycloconverter absorbs reactive power from the 60 Hz line. The reason is that delayed triggering is needed on the 60 Hz line to generate the sinusoidal low-frequency voltage. The input power factor is typically 85 percent when the motor runs at rated power and speed.

Figure 22-31 shows a large slow-speed synchronous motor that is driven by a cycloconverter. The speed can be continuously varied from zero to 15 r/min. The low speed permits direct drive of the ball-mill without using a gear reducer. The motor is stopped by altering the gate firing so as to feed power back into the ac line.

Similar high-power low-speed systems are being introduced as propeller drives on board ships.

22.17 Cycloconverter voltage and frequency control

Returning to Fig. 22-30, we can see that the low-frequency output voltage is composed of selected segments of the ac line voltage. The segments are determined by the gate firing of the SCRs. The triggering is identical to that of a conventional 6-pulse rectifier, except that the

firing angle is continually varied so as to obtain an output voltage that approaches the desired sine wave. During the positive half cycle, thyristors Q1 to Q6 are triggered in sequence, followed by thyristors Q7 to Q12 for the negative half cycle. In our example, the low-frequency output voltage has the same peak amplitude as the 3-phase line voltage; consequently, it has the same effective value. The frequency is 1/10 of the line frequency, or 6 Hz on a 60 Hz system.

We can gain a better understanding of the triggering process by referring to Fig. 22-32. In this case, the output frequency is 20 Hz on a 60 Hz system. The 60 Hz line voltages are indicated, as well as the firing sequence for the various SCRs. Although the resulting waveshape is very jagged, it does follow the general shape of the desired sine wave (shown dotted). The gate triggering times are quite irregular (not evenly spaced) to obtain the desired output voltage. That is why the firing program has to be under computer control.

If this 20 Hz voltage is applied to the motor of Fig. 22-29, the resulting current will be a reasonably good sine wave. In effect, the inductance of the windings smooths out the ragged edges that would otherwise be produced by the

Figure 22-31a
Stator of a 3-phase synchronous motor rated 6400 kW (8576 hp), 15 r/min, 5.5 Hz, 80°C used to drive a
ball-mill in a cement factory. The stator is connected to a 50 Hz cycloconverter whose output frequency is
variable from zero to 5.5 Hz. Internal diameter of stator: 8000 mm; active length of stacking: 950 mm; slots:
456. The synchronous motor actually operates as a commutatorless dc machine *(courtesy* ABB*)*

Figure 22-31b
The 44 rotor poles are directly mounted on the ball-mill so as to eliminate the need of a gear box. The two slip rings on the right-hand side of the poles bring the dc current into the windings.

Figure 22-31c
End view of the ball-mill showing the enclosed stator frame in the background. The mill contains 470 tons of steel balls and 80 tons of crushable material. The motor is cooled by blowing 40 000 m³ of fresh air over the windings, per hour. (courtesy ABB)

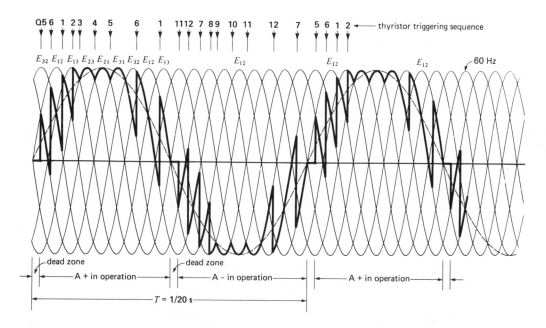

Figure 22-32
Waveshape of the output voltage E_{aN} in Fig.22-29, at a frequency of 20 Hz. The effective output voltage has the same value as the effective input voltage between the 3-phase lines.

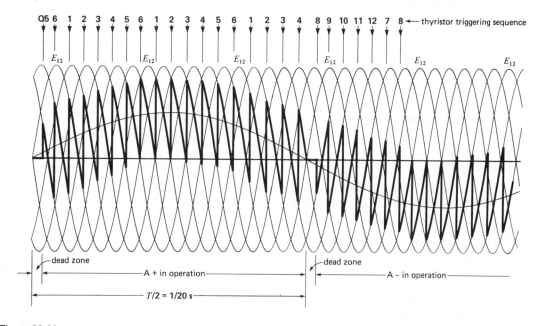

Figure 22-33
Waveshape of E_{aN} in Fig.22-29 at a frequency of 10 Hz. The thyristors are triggered later in the cycle so that the effective value of the output voltage is only half that between the 3-phase lines. As a result, the flux in the air gap is the same in this figure as it is in Fig.22-32.

saw-tooth voltage wave.

To reduce the speed, both the frequency and voltage have to be reduced in the same proportion. Thus, in Fig. 22-33, the frequency is now 10 Hz and the amplitude of the output voltage is also reduced by one half. The gate pulses are altered accordingly and, as we can see, a very jagged voltage is produced. Nevertheless, the current flowing in the windings will still be quite sinusoidal. A low output voltage requires a large firing angle delay, which in turn, produces a very low power factor on the 60 Hz line.

Although we have only discussed the behavior of phase A, the same remarks apply to phases B and C (Fig. 22-29). The gate firing is timed so that the low-frequency line currents I_a, I_b, I_c are mutually out of phase by 120°.

The cycloconverter drive is excellent when high starting torque and relatively low speeds are needed. However, it is not suitable for frequencies exceeding one half the system frequency. In such cases we use drives with a dc link, as previously discussed.

QUESTIONS AND PROBLEMS

Practical level

22-1 State in which quadrants a dc machine operates:
 a. as a motor;
 b. as a generator.

22-2 A dc machine is turning clockwise in quadrant 3. Does it develop a clockwise or counter-clockwise torque?

22-3 A 2-pole dc motor runs at 5460 r/min. What is the frequency of the voltage induced in the coils?

22-4 Referring to Fig. 22-1, the converter is connected to a 3-phase 480 V, 60 Hz line and the delay angle is set at 15°. Switch S is closed and the armature current is 270 A.

Calculate:
 a. the dc voltage across the armature;
 b. the power supplied to the motor
 c. the average current in each diode;
 d. the power output [hp] if the armature circuit has a resistance of 0.07 Ω.

22-5 Explain why the field or armature has to be reversed in order that the converter in Fig. 22-1 may feed power from the rotating armature back into the ac line.

22-6 Compare the basic behavior of the power drive of Fig. 22-5 with that of Fig. 22-8.

22-7 a. What is meant by the term commutatorless dc machine? Describe its construction and principle of operation.
 b. What is meant by a half-bridge converter? What advantages does it have over a conventional 3-phase bridge converter?

Intermediate level

22-8 The motor shown in Fig. 22-1 has a shunt field rated at 180 V, 2 A.
 a. calculate the effective value of the 60 Hz ac voltage that should be applied to the single-phase bridge circuit;
 b. what is the peak-to-peak voltage ripple across the field terminals?
 c. does the field current contain a substantial ripple? Explain.
 d. draw the waveshape of the current in the ac line.
 e. what is the effective value of the ac line current?

22-9 A 10 hp, 240 V, 1800 r/min permanent-magnet dc motor has an armature resistance of 0.4 Ω and a rated armature current of 35 A. It is energized by the converter shown in Fig. 22-1. If the ac line voltage is 208 V, 60 Hz and the motor operates at full-load, calculate:

a. the delay angle required so that the motor operates at its rated voltage;

b. the reactive power absorbed by the converter;

c. the effective value of the line currents;

d. the induced voltage E_0 at 900 r/min.

22-10 The motor in Problem 22-9 is started at reduced voltage, and the starting current is limited to 60 A. Calculate:

a. the delay angle required;

b. the reactive power absorbed by the inverter;

c. does inductor L absorb reactive power from the ac line?

22-11 Referring to Fig. 22-5, an ac ammeter inserted in series with line 1 gives a reading of 280 A. Furthermore, a 3-phase power factor meter indicates a lagging displacement power factor of 0.83. Calculate:

a. the value of the dc load current I_d;

b. the approximate delay angle if converter 1 is operating alone as an inverter.

22-12 The hoist motor shown in Fig. 22-9 is lifting a mass of 5000 lb at a constant speed of 400 ft/min.

a. neglecting gear losses, calculate the value of E_0 if the armature current I_d is 150 A.

b. knowing that $R_a = 0.1 \ \Omega$, calculate the value of converter voltage E_d.

22-13 In Problem 22-12 (and referring to Fig. 22-9), if the same mass is lowered at a constant speed of 100 ft/min, calculate:

a. the armature current and its direction;

b. the value of E_0 and its polarity;

c. the value of E_d and its polarity;

d. in which direction does the active power flow in the ac line?

22-14 In Problem 22-12, if the mass is simply held still in midair, calculate:

a. the value of E_0;

b. the armature current I_d;

c. the value and polarity of E_d.

22-15 If the 3-phase line voltage is 240 V, 60 Hz, calculate the delay angle required:

a. in Problem 22-13;

b. in Problem 22-14.

22-16 a. Referring to Fig. 22-18a, calculate the average current and also the peak current carried by the freewheeling diode;

b. what is the PIV across the diode?

22-17 An electronic chopper is placed between a 600 V dc trolley wire and the armature of a series motor. The switching frequency is 800 Hz and each power pulse lasts for 400 μs. If the dc current in the trolley wire is 80 A, calculate:

a. the armature voltage;

b. the armature current;

c. draw the waveshape of the current in the freewheeling diode, assuming the armature inductance is high.

Advanced level

22-18 Referring to Fig. 22-8, the cemf E_0 has the polarity shown when the armature turns clockwise. Furthermore, when the armature current actually flows in the direction shown and the machine is turning clockwise, it operates in quadrant 1. State whether converters 1 and 2 are acting as rectifiers or inverters when the machine operates:

a. in quadrant 2;

b. in quadrant 3;

c. in quadrant 4.

d. make a sketch of the actual direction of current flow and the actual polarity of E_0 in each case.

22-19 A 50 hp dc motor is required to drive a centrifuge at a speed ranging between 18 000 and 30 000 r/min. Owing to commutation problems associated with a standard commutator at these speeds, it is decided to use a commutatorless dc motor driven by two converters with a dc link (Fig. 22-27). A 2-pole motor is selected having a nominal rating of 50 hp, 30 000 r/min, 460 V, 60 A, 90 percent power factor leading. When the motor delivers its rated output, the delay angle for converter 2 is 155°. If the available 60 Hz line voltage is 575 V, calculate:

a. the triggering frequency of converter 2;
b. the dc link voltage;
c. the delay angle of converter 1;
d. the dc link current if the motor draws an input power of 41.5 kW;
e. the fundamental ripple frequency in E_1;
f. the fundamental ripple frequency in E_2;
g. the power factor of the 60 Hz line;
h. the effective value of I_s;
i. show the flow of active and reactive power in the converters.

22-20 The following specifications are given for the motor shown in Fig. 22-20:
armature diameter: 100 mm
armature axial length: 50 mm
turns per coil: 200
rotational speed: 840 r/min
flux density in air gap: 0.5 T
armature current: 5 A

Using this information, calculate:
a. the voltage induced in each coil;
b. the dc voltage between the brushes;
c. the frequency of the voltage in each coil;
d. the frequency of the current in each coil;
e. the power developed by the motor;
f. the torque exerted by the motor.

22-21 a. Referring to Fig. 22-17, what would be the effect if capacitor C_1 were removed?
b. In Fig. 22-18a, calculate the approximate value of capacitor C_1 [µF] so that the voltage across it does not drop by more than 50 V during the time of a current pulse.

22-22 A 3-phase 6-pulse rectifier is equipped with a freewheeling diode. The 3-phase feeder has a line voltage of 240 V and the dc load is composed of an armature having a resistance of 0.4 Ω. The rated armature current is 40 A. Calculate
a. the voltage needed to cause 60 A to flow through the armature when it is at standstill.
b. the firing angle needed to attain this current.
c. the reactive power absorbed by the converter.

23

ELECTRONIC CONTROL OF ALTERNATING CURRENT MOTORS

We saw in Chapter 22 that the electronic control of dc motors enables us to obtain high efficiency at all torques and speeds. Full 4-quadrant control is possible to meet precise high-speed industrial standards. The same remarks apply to the electronic control of ac motors. Thus, we find that squirrel-cage and wound-rotor induction motors, as well as synchronous motors, lend themselves well to electronic control. Whereas dc machines are controlled by varying the voltage and current, ac machines are often controlled by varying the voltage and frequency. Now we may ask, if dc machines do such an outstanding job, why do we also use ac machines? There are several reasons:

1. AC machines have no commutators; consequently, they require less maintenance;
2. AC machines cost less (and weigh less) than dc machines;
3. AC machines are more rugged and work better in hostile environments.

In this chapter, we cover 3-phase motor controls, in keeping with the power emphasis of the book. However, the reader should first review the basic principles covered in Chapters 21 and 22.

Furthermore, to better understand the basic principles of ac motor control, the reader should also review Secs. 20.18 to 20.20. They explain how variable frequency affects the behavior of a squirrel-cage induction motor.

23.1 Types of ac drives

Although there are many kinds of electronic ac drives, the majority can be grouped under the following broad classes:
1. Static frequency changers
2. Variable voltage controllers
3. Rectifier – inverter systems with natural commutation.
4. Rectifier – inverter systems with self-commutation

Static frequency changers convert the

543

incoming line frequency directly into the desired load frequency. Cycloconverters fall into this category, and they are used to drive both synchronous and squirrel-cage induction motors.

Variable voltage controllers enable speed and torque control by varying the ac voltage. They are used in squirrel-cage and wound-rotor induction motors. This method of speed control is the least expensive of all, and provides a satisfactory solution for small and medium-power machines used on fans, centrifugal pumps and electric hoists.

Rectifier – inverter systems rectify the incoming line frequency to dc, and the dc is reconverted to ac by an inverter. The inverter may be self-commutated, generating its own frequency, or it may be naturally-commutated by the very motor it drives. Such rectifier-inverter systems with a dc link are used to control squirrel-cage and wound-rotor induction motors and, in some cases, synchronous motors.

23.2 Squirrel-cage induction motor with cycloconverter

Figure 23-1 shows a 3-phase squirrel-cage induction motor connected to the output of a 3-phase cycloconverter. The circuit arrangement is similar to that of Fig. 22-29, except that the windings are fed from a common 3-phase source. Consequently, the windings cannot be connected in wye or delta, but must be isolated from each other. Motor speed is varied by applying appropriate gate pulses to the thyristors, to vary the output voltage and frequency. For example, the speed of a 2-pole induction motor can be varied from zero to 1500 r/min, on a 60 Hz line, by varying the output frequency of the cycloconverter from zero to 25 Hz.

Good torque-speed characteristics in all 4 quadrants can be obtained. Consequently, the motor can be started, stopped, reversed and decelerated by regenerative braking.

Standard 60 Hz motors can be used. The stator voltage is adjusted in proportion to the frequency to maintain a constant flux in the machine. Consequently, the torque-speed curves are similar to those shown in Fig. 23-2.

To understand the operation of the cycloconverter, consider phase A in Fig. 23-1. The voltage across the winding is E_a and the ac current through it is I_a. Because of the inductive nature of a squirrel-cage motor, I_a lags behind E_a by an angle of about 30° (Fig. 23-3). Suppose the cycloconverter produces a frequency of 15 Hz.

Referring to Fig. 23-1, a positive current I_a can only be furnished by converter 1 because only thyristors Q1, Q3, Q5 "point" in the proper direction. This converter can act either as a rectifier or inverter. When E_a is positive, it acts as a rectifier, and delivers power to the phase A winding. Conversely when E_a is negative, the converter acts as an inverter delivering power from the phase A winding to the 3-phase line.

Similarly, a negative current I_a can only be furnished by converter 2. This converter acts as a rectifier when E_a is negative and during this period the converter supplies power to the winding. Conversely, when E_a is positive, converter 2 transfers power from the winding to the 3-phase line.

It is important to note that only one converter operates at a time. Thus, when converter 1 is in operation, converter 2 is blocked, and vice versa. The rectifier/inverter behavior of the converters is easy to follow in Fig. 23-3.

The converters in phases B and C function the same way except that thyristors similar to those in phase A are fired respectively 120° and 240° later. At 15 Hz, an angle of 120° corresponds to a delay of (1/15) × (120/360) = 22.2 ms.

The smooth current and voltage waveshapes shown in Fig. 23-3 are actually jagged sine waves owing to the constant switching between output and input.

Figure 23-1
Squirrel-cage induction motor fed from a 3-phase cycloconverter.

Figure 23-2
Typical torque-speed curves of a 2-pole induction motor driven by a cycloconverter. The cycloconverter is connected to a 460 V, 3-phase, 60 Hz line.

Consequently, cycloconverter-fed motors run about 10°C hotter than normal, and adequate cooling must be provided. A separate ventilating fan may be needed at low speeds.

The cycloconverter can furnish the reactive power absorbed by the induction motor. However, a lot of reactive power is drawn from the 60 Hz line; the power factor is therefore poor. Indeed, with sinusoidal outputs the displacement power factor is always less than 84 percent, even for resistive loads. Cycloconverter drives are feasible on small and medium-power induction motors operating at top speeds of about 2000 r/min on a 60 Hz line.

Example 23-1

A 3-phase squirrel cage induction motor has a full-load rating of 25 hp, 480 V, 1760 r/min, 60 Hz. The three independent windings each carry a rated current of 20 A.

This motor is connected as shown in Fig. 23-1. The cycloconverter is connected to a 3-phase, 60 Hz line and generates a frequency of 8 Hz. Calculate the approximate value of:

a. the effective voltage across each winding
b. the no-load speed
c. the speed at rated torque
d. the effective current in the windings at rated torque
e. the effective voltage of the 60 Hz line

Solution:

a. The flux in the motor should remain the same at all frequencies. Consequently, for a frequency of 8 Hz the voltage across the windings must be reduced in proportion. Thus, the voltage is:

$$E = \frac{8 \text{ Hz}}{60 \text{ Hz}} \times 480 \text{ V} = 64 \text{ V}$$

b. The full-load speed at 60 Hz is 1760 r/min. Consequently, this is a 4-pole motor whose synchronous speed is 1800 r/min. The no-load speed at 8 Hz is therefore:

$$n = \frac{8 \text{ Hz}}{60 \text{ Hz}} \times 1800 = 240 \text{ r/min}$$

c. The slip at rated torque is $(1800 - 1760) = 40$ r/min when the motor operates on 60 Hz. Consequently, the slip is again 40 r/min when the motor develops rated torque at 8 Hz. The speed at rated torque is therefore:

$$n = 240 - 40 = 200 \text{ r/min}$$

d. Because the flux in the motor is the same at 8 Hz as it was at 60 Hz, it follows that rated torque will be developed when the current in the stator windings is the same, namely 20 A.

e. The effective value of the 60 Hz line voltage should be the same as the voltage output of the cycloconverter. Consequently, the line voltage should be about 64 V. However, a higher line voltage could be used but this would require a greater lag in the firing angle. The converter would draw more reactive power from the line and the power factor would be poorer.

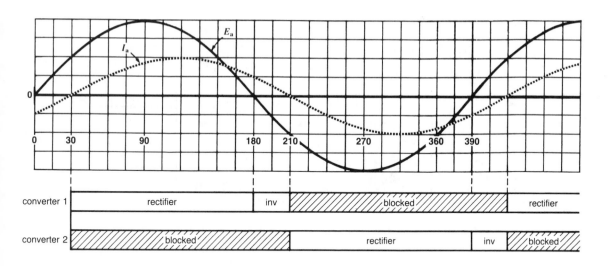

Figure 23-3
Operating mode of converter 1 and converter 2 when current I_a lags 30° behind E_a.

23.3 Squirrel-cage motor and variable voltage controller

We can vary the speed of a 3-phase squirrel-cage induction motor by simply varying the stator voltage. This is particularly useful for a motor driving a blower or centrifugal pump. Suppose the stator is connected to a variable-voltage 3-phase autotransformer (Fig. 23-4).

At rated voltage, the torque-speed characteristic of the motor is given by curve 1 of Fig. 23-5. To simplify the drawing, the curve is shown as two straight lines. If we apply half the rated voltage, we obtain curve 2. Because torque is proportional to the square of the applied voltage, the torques in curve 2 are only 1/4 of the corresponding torques in curve 1. Thus, the breakdown torque drops from 200 % to 50 %. Similarly, the torque at 60 percent speed drops from 175% to 43.75%.

The load torque of a blower varies nearly as the square of the speed. This typical characteristic, shown by curve 3, is superimposed on the motor torque-speed curves. Thus, at rated voltage, the intersection of curves 1 and 3 shows

that the blower runs at 90 percent of synchronous speed. On the other hand, at half rated voltage, the blower rotates at only 60 percent of synchronous speed. By reducing the voltage still more, we can make the speed as low as we please.

We can replace the variable-voltage autotransformer by 3 sets of thyristors connected back-to-back, as shown in Fig. 23-6. To produce rated voltage across the motor, the respective thyristors are fired with a delay θ equal to the phase angle lag that *would* exist if the motor were directly connected to the line.

Figure 23-4
Variable-speed blower motor.

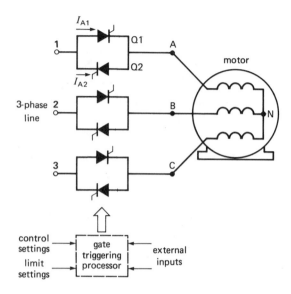

Figure 23-5
Torque-speed curve of blower motor at rated voltage (1) and 50% rated voltage (2). Curve (3) is the torque-speed characteristic of the fan.

Figure 23-6
Variable-voltage speed control of a squirrel-cage induction motor.

Figure 23-7 shows the resulting current and line-to-neutral voltage for phase A. The thyristors in phases B and C are triggered the same way, except for an additional delay of 120° and 240°, respectively.

To reduce the motor voltage, we delay the firing angle θ still more. For example, to obtain 50 percent rated voltage, all the pulses are delayed by about 100°. The resulting voltage and current waveshapes for phase A are given very approximately in Fig. 23-8. In actual fact, both the voltage and current are distorted, and the current lags considerably behind the voltage. The distortion increases the losses in the motor compared to the autotransformer method. Furthermore, the power factor is considerably lower, because of the large phase angle lag θ. Nevertheless, to a first approximation, the torque-speed characteristics shown in Fig. 23-5 still apply.

Owing to the considerable I^2R losses and low power factor, this type of electronic speed control is only feasible for motors rated below

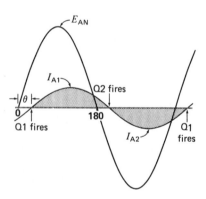

Figure 23-7
Waveshapes at rated voltage.

20 hp. Small hoists are well suited to this type of control, because they operate intermittently. Consequently, they can cool off during the idle and light-load periods.

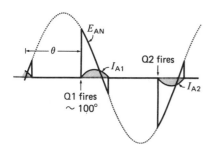

Figure 23-8
Waveshapes (very approximately) at 50% rated voltage.

23.4 Self-commutated inverters for squirrel-cage motors

In Section 22-15, we saw that a synchronous motor can be driven by a naturally-commutated inverter. This is possible because the synchronous motor can provide the reactive power needed by the inverter. Unfortunately, if we replace the synchronous motor by an induction motor, the frequency conversion system breaks down, because an induction motor cannot deliver reactive power. Worse still, it actually absorbs reactive power.

Nevertheless, we can drive an induction motor using a *self-commutated* inverter. Such an inverter operates quite differently from a naturally-commutated inverter. First, it can generate its own frequency, determined by the frequency of the pulses applied to the gates. Second, it can either absorb or deliver reactive power. The reactive power generated or absorbed depends upon the switching action of the power thyristors.

Self-commutated inverters convert dc power into ac power. The power thyristors are arranged in a conventional 3-phase bridge circuit. However, each thyristor is surrounded by an array of capacitors, inductors, diodes and auxiliary thyristors. The purpose of these auxiliary components is to force some power thyristors to conduct when normally they would not, and to force other power thyristors to stop

conducting before their "natural" time. It is precisely this special switching action that enables these converters to generate reactive power.

Owing to the complexity and variety of the switching circuits used, we show the self-commutated inverter as a simple 5-terminal device having a dc input and a 3-phase output (Figs. 23-9 and 23-12). This simple representation also helps us understand the basic features of all self-commutated inverters:

1. The inverter power loss is relatively small; consequently, we can assume that the dc input power is equal to the active ac output power.
2. The reactive power generated by the inverter is not produced by the commutating capacitors included in the circuitry. The reactive power is due to the switching action alone.
3. The reactive power output requires no dc power input.
4. The thyristors connect the dc input terminals to the ac output terminals in a controlled sequence, with negligible voltage drop. It follows that:
   a. in a *voltage-fed* inverter, the ac line voltages are successively equal to the dc input voltage;
   b. in a *current-fed* inverter, the ac line currents are successively equal to the dc current.

To control the speed of a squirrel-cage motor, we use a rectifier-inverter system. The rectifier is connected to the 3-phase supply line and the inverter is connected to the stator. The dc output of the rectifier is connected to the dc input to the inverter by a so-called dc link. Two types of dc links are used – constant current and constant voltage. This gives rise to the current-fed and voltage-fed inverters mentioned above.

The constant current link supplies a constant current to the inverter, which is then fed sequentially into the three phases of the motor (Fig. 23-9a). Similarly, the constant-voltage

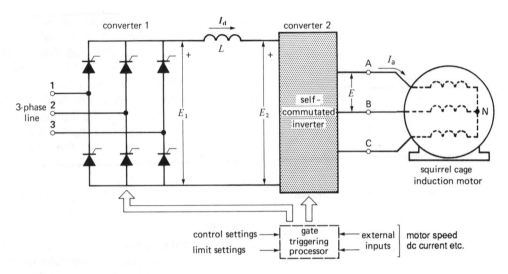

Figure 23-9a
Current-fed frequency converter.

Figure 23-9b
Motor voltage and current.

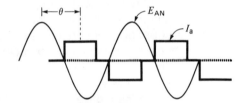

Figure 23-9c
Asynchronous generator voltage and current.

link furnishes a constant voltage to the inverter which is switched from one phase to the next of the induction motor (Fig. 23-12a). The switching means may be simple or complex, depending on the design of the converter. In the following sections, we describe some of the more common switching schemes that are used.

23.5 Current-fed self-commutated frequency converter

The current-fed frequency converter shown in Fig. 23-9a is used to control the speed of *individual* squirrel-cage motors. A large inductor L

helps to keep the current constant. The switching action of the inverter is such that the current in each phase is a rectangular pulse which flows for 120°. The resulting voltage between lines A, B, C is nearly sinusoidal. The reason is that the 3-phase rectangular current pulses together produce a revolving magnetic field that is almost sinusoidal in shape. Figure 23-9b shows the line current I_a in one phase, and the associated line-to-neutral voltage E_{AN}. Phase angle θ corresponds to the operating power factor of the motor. It depends upon the properties of the motor itself and not upon the switching action of the inverter. In

Figure 23-10
Current-source variable frequency electronic drive for a conventional 500hp, 460 V, 1780 r/min, 3-phase , 60 Hz induction motor. The output frequency can be varied from zero to 72 Hz and the efficiency at rated load and speed is 95%. The design is a current-source, 6-step output that can operate in all 4 quadrants (*courtesy Robicon Corporation*).

effect, although instant t_1 coincides with the firing of the thyristor connected to phase A, the *timing* of the pulse is not determined by the zero crossing point of voltage E_{AN}. The voltage finds its own place, so to speak, depending upon the particular speed, torque, and direction of rotation the motor happens to have.

We can obtain regenerative braking (generator action) by changing the firing angle of converter 2. This reverses the polarity of E_2. However, the dc current continues to flow in the same direction and so converter 2 feeds power into the dc link. By simultaneously retarding the triggering of the thyristors in converter 1, we also reverse the polarity of E_1. Consequently, converter 1 now acts as an inverter, feeding power back into the 3-phase line. The new phase relationship between stator voltage and stator current is shown in Fig. 23-9c. Note that converter 2 continues to supply reactive power to the motor during this regenerative braking period.

The direction of rotation is easily changed by altering the phase sequence of the pulses that trigger the gates of converter 2. Consequently, this static frequency converter can operate in all four quadrants with high efficiency. The torque-speed curve can be manipulated in any way, as already shown in Fig. 23-2. High inertia loads can be quickly brought up to speed by designing the control system so that the full breakdown torque is developed as the motor accelerates.

In practice, the output frequency of an inverter may be varied over a range of 20:1, with top frequencies of the order to 1000 Hz. At rated torque, the ac voltage changes in proportion to the frequency and so a constant stator flux is maintained. Consequently, the dc link voltage E_1 must be reduced as the speed decreases. The dc link voltage must be further reduced when the motor develops less than rated torque. This voltage reduction is achieved by increasing the firing angle of the thyristors in converter 1. Unfortunately, this tends to increase the reactive power drawn from the

3-phase line.

Various control systems have been developed to improve the dynamic performance of the motor. The inputs to the gate triggering processor are selected so that the system approaches the behavior of a dc drive. Some of these control schemes are very elaborate, involving the monitoring of speed, slip, current and even the instantaneous position of the revolving flux inside the motor. However, these sophisticated controls do not change the basic operation of the current-fed frequency converter. Figure 23-10 shows the physical size of a current-source variable frequency drive.

Example 23-2:

A 50 hp, 1165 r/min, 440 V, 78 A, 60 Hz, 3-phase motor is driven by a current-fed frequency converter. The efficiency of the motor is 82% and that of the inverter, 94 %. Referring to Fig. 23-9, calculate the approximate value of

a. the dc power input to converter 2
b. the current in the dc link
c. the dc voltage E_1 produced by converter 1.

Solution:

a. the active power absorbed by the motor is:

$$P = \frac{50 \times 746}{0.82} = 45.5 \text{ kW}$$

the active power absorbed by converter 2 is

$$P_2 = \frac{45.5}{0.94} = 48.4 \text{ kW}$$

b. the effective value of the full-load current is 78 A. Consequently, the dc current has an approximate value of

$$I_d = \frac{78}{0.816} = 95.6 \text{ A} \qquad \text{Eq. 21-6}$$

c. the dc value of E_1 (and of E_2) is:

$$E_1 = \frac{P_2}{I_d} = \frac{48.4 \times 10^3}{95.6} = 506 \text{ V}$$

Thyristors have the peculiarity of turning on whenever the voltage rises very rapidly between the anode and cathode, even when the gate is negative. To prevent such an unwanted turn-on, the rate of rise of the voltage must be limited. This is accomplished by connecting a capacitor and resistor in series across the anode, cathode terminals. The R-C circuit is called a snubber. Typical values are C = 0.1μF, R = 10 Ω. Snubbers introduce additional losses in the converter. In addition to snubbers, voltage-limiting surge arresters are often connected between the anode and cathode (Fig. 23-11).

23.6 Voltage-fed self-commutated frequency converter

In some industrial applications, such as in textile mills, the speeds of several motors have to move up and down together. These motors must be connected to a common bus in order to function at the same voltage. The current-fed frequency converter is not feasible in this case because it tends to supply a constant current to the total ac load, irrespective of the mechanical loading of individual machines. Under these circumstances, we use a *voltage-fed* frequency converter. It consists of a rectifier and a self-commutated inverter connected by a dc link (Fig. 23-12a).

Figure 23-11
R-C snubber to limit the rate of rise of E_{AK} and surge arrester to limit its peak value.

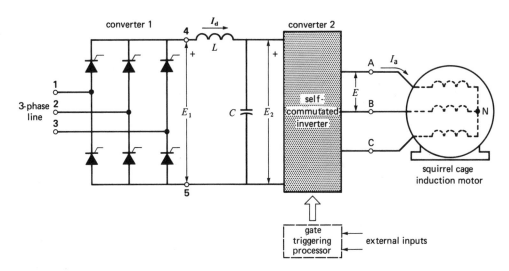

Figure 23-12a
Voltage-fed frequency converter.

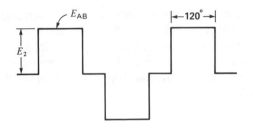

Figure 23-12b
Motor line-to-line voltages.

A 3-phase bridge rectifier produces a dc voltage E_1. An LC filter ensures a "stiff" dc voltage at the input to the inverter. The inverter successively switches this voltage across the lines of the 3-phase load. The switching produces positive and negative voltage pulses of 120° duration (Fig. 23-12b).

The amplitude of the inverter output voltage E is varied in proportion to the frequency so as to maintain a constant flux in the motor (or motors). Because the peak ac voltage is equal to the dc voltage E_2, it follows that rectifier voltage E_1 must be varied as the frequency varies. The speed of the motor can be controlled from

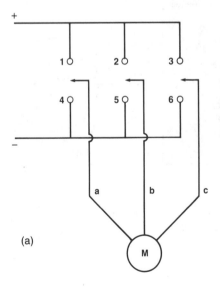

(a)

zero to maximum while developing full torque.

Regenerative braking is possible but, owing to the special arrangement of the auxiliary components in converter 2, the link current I_d reverses when the motor acts as a generator. Voltage E_2 does not change polarity as it does in a current-fed inverter. Because converter 1

Time intervals

contacts	T1	T2	T3	T4	T5	T6	T7	T8	T9	T10	T11
1		x	x	x				x	x	x	
2				x	x	x				x	x
3	x	x				x	x	x			
4	x					x	x	x			x
5	x	x	x				x	x	x		
6			x	x	x				x	x	x

(b)

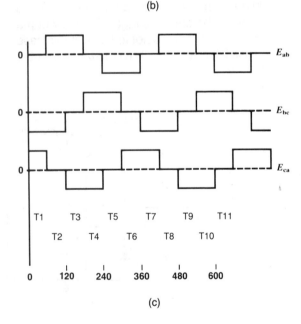

(c)

Figure 23-13
a. Three mechanical switches could produce the same voltage pulses as a voltage-fed inverter.
b. Table showing the switching sequence of the switches.
c. Voltages produced across the motor terminals.

cannot accept reverse current flow, a third converter (not shown) has to be installed in reverse parallel with converter 1 to permit regenerative braking. The third converter functions as an inverter and, while it operates, converter 1 is blocked. As a result, voltage-fed drives tend to be more expensive than current-fed drives, thus making them less attractive.

The switching action of converter 2 can be represented by the three mechanical switches shown in Fig. 23-13a. The opening and closing sequence is given in the table (Fig. 23-13b) together with the resulting ac line voltages. An X indicates that the switch contact is closed. This again illustrates that the thyristors and other electronic devices in converter 2 merely act as high-speed switches. The switching action is called 6-step because the switching sequence repeats after every 6th step. This is evident from the table.

23.7 Chopper speed control of a wound-rotor induction motor

We have already seen that the speed of a wound-rotor induction motor can be controlled by placing three variable resistors in the rotor circuit (Sec. 13.16). Another way to control the speed is to connect a 3-phase bridge rectifier across the rotor terminals and feed the rectified power into a *single* variable resistor. The resulting torque-speed characteristic is identical to that obtained with a 3-phase rheostat. Unfortunately, the single rheostat still has to be varied mechanically in order to change the speed.

We can make an all-electronic control (Fig. 23-14) by adding a chopper and a fixed resistor R_0 to the secondary circuit. In this circuit, capacitor C supplies the high current pulses drawn by the chopper. The purpose of inductor L and freewheeling diode D has already been explained in Sec. 21.24. By varying the chopper on-time T_a, the apparent resistance across the bridge rectifier can be made either high or low. The relationship is given by:

$$R_d = R_0/f^2 T_a^2 \qquad \text{Eq. 21-15}$$

where R_d is the apparent resistance between terminals A1, A2 and the other terms have the meaning previously explained in Sec. 21.24.

Example 23-3:
The wound-rotor motor shown in Fig. 23-14 is rated at 30 kW (40 hp), 1170 r/min, 460 V, 60 Hz. The open-circuit rotor line voltage is 400 V and the load resistor R_0 is 0.5 Ω. If the chopper frequency is 200 Hz, calculate time T_a so that the motor develops a torque of 200 N·m at 900 r/min.

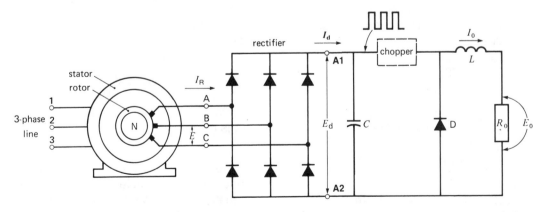

Figure 23-14
Speed control of a wound-rotor induction motor using a load resistor and chopper.

Solution:
This problem can be solved by applying the principles covered in Chapters 13 and 21.

a. The rated synchronous speed is clearly 1200 r/min. The slip at 900 r/min is:
$$s = (n_s - n)/n_s \qquad \text{Eq. 13-2}$$
$$= (1200 - 900)/1200$$
$$= 0.25$$

The rotor line voltage at 900 r/min is:
$$E = sE_{0c} \qquad \text{Eq. 13-4}$$
$$= 0.25 \times 400$$
$$= 100 \text{ V}$$

The dc voltage developed by the bridge rectifier is:
$$E_d = 1.35 E \qquad \text{Eq. 21-4}$$
$$= 1.35 \times 100$$
$$= 135 \text{ V}$$

Knowing the torque is 200 N·m, we can calculate the power P_r delivered to the rotor:
$$T = 9.55 P_r/n_s \qquad \text{Eq. 13-9}$$
$$200 = 9.55 P_r/1200$$
$$P_r = 25\ 130 \text{ W}$$

Part of P_r is dissipated as heat in the rotor circuit:
$$P_{jr} = sP_r \qquad \text{Eq. 13-7}$$
$$= 0.25 \times 25\ 130$$
$$= 6282 \text{ W}$$

The power of 6282 W is actually dissipated in resistor R_0, but it is obviously equal to the rectifier output E_dI_d. Thus,
$$E_dI_d = P_{jr}$$
$$135I_d = 6282$$
$$I_d = 46.5 \text{ A}$$

The apparent resistance at the input to the chopper circuit is therefore:
$$R_d = E_d/I_d$$
$$= 135/46.5$$
$$= 2.9 \ \Omega$$

Applying Eq. 21-15, we have;
$$R_0 = R_d f^2 T_a^2$$

$$0.5 = 2.9 \times 200^2 \times T_a^2$$
$$T_a = 2.08 \text{ ms}$$

The chopper on-time is therefore 2.08 ms. This may be compared with the time T of one chopper cycle:
$$T = 1/200 = 5 \text{ ms}$$

Example 23-4:
In Example 23-3, calculate the magnitude of the current pulses drawn from the capacitor.

a. The current I_0 in R_0 is a steady current given by:
$$I_0{}^2R_0 = P_{jr}$$
$$I_0{}^2 \times 0.5 = 6282$$
$$I_0 = 112 \text{ A}$$

The capacitor therefore delivers current pulses having an amplitude of 112 A. The pulse width is 2.08 ms and the repetition rate is 200 pulses per second. On the other hand, the rectifier continuously charges the capacitor with a current I_d of 46.5 A.

23.8 Recovering power in a wound-rotor induction motor

Instead of dissipating the rotor power in a resistor, we could use it to charge a large dc battery (Fig. 23-15). Assuming the battery voltage E_2 can be varied from zero to some arbitrary maximum value, let us analyze the behavior of the circuit.

The ac voltage E across the rotor terminals is given by:
$$E = sE_{0c} \qquad \text{Eq. 13-4}$$

where s is the slip and E_{0c} is the open-circuit rotor voltage at standstill (Sec. 13.10).

On the other hand, rectified output voltage E_d is given by:
$$E_d = 1.35E \qquad \text{Eq. 21-4}$$

Because the IR drop in the smoothing inductor is negligible, $E_d = E_2$. Combining equations 13-4 and 21-4, we obtain:

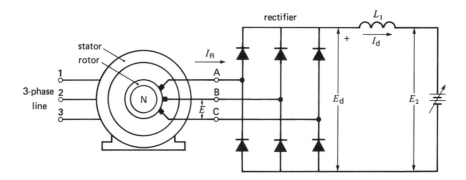

Figure 23-15
Speed control using a variable-voltage battery.

$$s = \frac{E_2}{1.35E_{0c}} \qquad (23\text{-}1)$$

Recognizing that E_{0c} is a fixed quantity, Eq. 23-1 shows that the slip depends exclusively upon the battery voltage E_2. Consequently, we can vary the speed from essentially synchronous speed ($s = 0$) to zero ($s = 1$) by varying the battery voltage from zero to $1.35E_{0c}$.

In practice, instead of charging a battery to absorb the rotor power, we use a 3-phase inverter which returns the power to the ac source. The naturally-commutated inverter is connected to the same line that supplies power to the stator (Fig. 23-16). A transformer T is usually added, so that the effective value of E_T lies between 80 and 90 percent of E_2. This ensures that the firing angle is reasonably close to 165° which reduces the reactive power absorbed by the inverter. As usual, the voltages are related by the equation:

$$E_2 = 1.35E_T \cos \alpha \qquad \text{Eq. 21-17}$$

where

E_2 = dc voltage developed by the inverter [V]

E_T = secondary line voltage of transformer T [V]

α = firing angle [°]

This method of speed control is very efficient because the rotor power is not dissipated in a group of resistors, but is returned to the line. Another advantage is that for any given setting of E_T and α the speed is practically constant from no-load to full-load.

Figure 23-17a shows the torque-speed curves for two settings of E_T. When $E_T = 0$ (curve 1), both E_2 and E_d are zero. Consequently, the slip-rings are short-circuited. When E_T is adjusted to equal $0.4 E_{0c}$ the torque-speed characteristic (curve 2) has the same shape but is shifted to the left. The slip increases only slightly with increasing torque.

The rotor current I_R is rectangular and flows during 120° intervals (Fig. 23-17b). It is symmetrical with respect to the respective line-to-neutral rotor voltage E_{AN}. Consequently, the power factor of the "load" across the rotor is always unity.

This method of speed control is economical because the rectifier and inverter only have to carry the *slip* power of the rotor, which is considerably less than the input power to the stator. For example, if the lowest motor speed is 80% of synchronous speed, the power handled by the converters (at rated torque) is only 20% of the input power to the stator. It follows that the converters are much smaller than if they were placed in the stator circuit where the full stator power would have to be controlled.

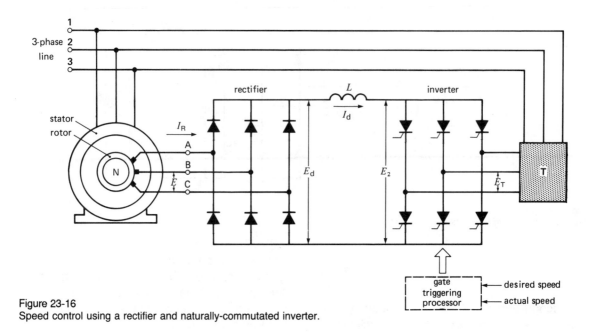

Figure 23-16
Speed control using a rectifier and naturally-commutated inverter.

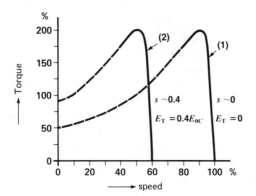

Figure 23-17a
Torque-speed characteristics of a wound-rotor motor for two settings of voltage E_T.

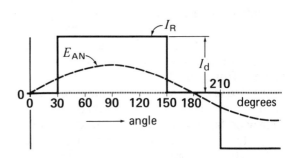

Figure 23-17b
Rotor voltage and current in Fig. 23-14.

Example 23-5:

A 3-phase, 3000 hp, 4000 V, 60 Hz, 8-pole wound-rotor induction motor drives a variable-speed centrifugal pump. When the motor is connected to a 4160 V line, the open-circuit rotor line voltage is 1800 V. A 3-phase 4160 V/480 V transformer is connected between the inverter and the line (Fig. 23-18). If

the motor has to develop 800 kW at a speed of 700 r/min, calculate:

a. the power output of the rotor;
b. rotor voltage and link voltage;
c. link current I_d and rotor current;
d. firing angle of the inverter;
e. current in the primary and secondary lines of transformer T.

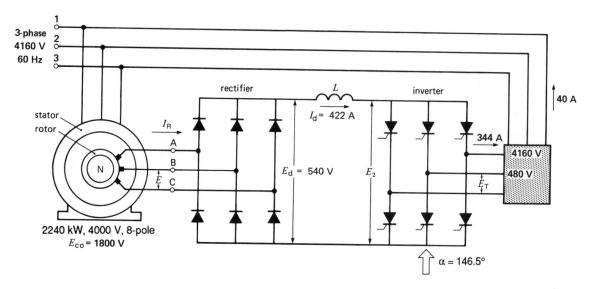

Figure 23-18
See Example 23-5.

Solution:

a. Synchronous speed is:

$$n_s = 120f/p \qquad \text{Eq. 13-1}$$
$$= 120 \times 60/8$$
$$= 900 \text{ r/min}$$

Slip is:

$$s = (n_s - n)/n_s \qquad \text{Eq. 13-2}$$
$$= (900 - 700)/900$$
$$= 0.222$$

Mechanical power is:

$$P_m = 800 \text{ kW} \ (= 1072 \text{ hp})$$

but

$$P_m = P_r (1 - s) \qquad \text{Eq. 13-8}$$
$$800 = P_r (1 - 0.222)$$

Power supplied to rotor is:

$$P_r = 1028 \text{ kW}$$

Power "dissipated" by rotor:

$$P_{jr} = sP_r \qquad \text{Eq. 13-7}$$
$$= 0.222 \times 1028$$
$$= 228 \text{ kW}$$

Consequently, 228 kW are fed back to the ac line.

b. Rotor voltage

$$E = sE_{0c} \qquad \text{Eq. 13-4}$$
$$E = 0.222 \times 1800$$
$$= 400 \text{ V}$$

Link voltage

$$E_d = 1.35 E$$
$$= 1.35 \times 400 = 540 \text{ V}$$

c. Link current

$$I_d = P_{jr}/E_d = 228\ 000/540$$
$$= 422 \text{ A}$$

The effective value of the rotor current is:

$$I_R = 0.816I_d \qquad \text{Eq. 21-6}$$
$$= 0.817 \times 422$$
$$= 344 \text{ A}$$

d. $$E_2 = 1.35 E_T \cos \alpha$$
$$540 = 1.35 \times 480 \cos \alpha$$
$$\alpha = 33.5°$$

The firing angle is actually $(180 - 33.5) = 146.5°$ because the converter acts as an inverter.

e. The current in each phase of the 480 V line flows during 120° intervals and has a peak

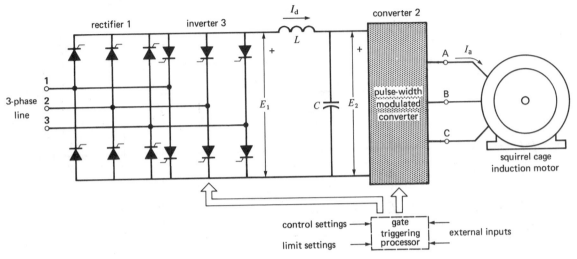

Figure 23-19
Speed control by pulse width modulation.

value of 422 A. The effective value is given by Eq. 21-6:

$$I = 344 \text{ A}$$

Effective line current on the 4160 V side is:

$$I = (480/4160) \times 344 = 40 \text{ A}$$

23.9 Pulse width modulation

The frequency converters discussed so far create substantial harmonic voltages and currents. When these harmonics flow in the windings, they produce torque pulsations that are superimposed on the main driving torque. The pulsations are damped out at moderate and at high speeds owing to mechanical inertia. However, at low speeds, they may produce considerable vibration. Such torque fluctuations are unacceptable in some industrial applications, where fine speed control down to zero speed is required. Under these circumstances, the motor can be driven by *pulse width modulation* techniques.

To understand the technique, consider the voltage-fed frequency converter system shown in Fig. 23-19. A 3-phase bridge rectifier 1 produces a fixed voltage E_1 which appears essentially undiminished as E_2 at the input to the self-commutated inverter 2. The inverter is triggered in a special way so that the output voltage is composed of a series of short positive pulses of constant amplitude, followed by an equal number of short negative pulses (Fig. 23-20a). The pulse widths and pulse spacings are arranged so that their weighted average approaches a sine wave. The pulses as shown all have the same width, but in practice, the ones near the middle of the sine wave are made broader than those near the edges. By increasing the number of pulses per half cycle, we can make the output frequency as low as we please. Thus, to reduce the output frequency of Fig. 23-20a by a factor of 10, we increase the pulses per half-cycle from 5 to 50.

The pulse widths and pulse spacings are specially designed so as to eliminate the low-frequency voltage harmonics, such as the 3rd, 5th, and 7th harmonics. The higher harmonics, such as the 17th, 19th, etc., are unimportant because they are damped out, both mechanically and electrically. Such pulse width modulation produces output currents having very low harmonic distorsion. Consequently, torque

Figure 23-20a
Voltage waveform across one phase.

Figure 23-20b
Waveform yielding the same frequency but half the voltage.

vibrations at low speeds are greatly reduced.

In some cases, the output voltage has to be reduced while maintaining the same output frequency. This is done by reducing all the pulse widths in proportion to the desired reduction in output voltage. Thus, in Fig. 23-20b, the pulses are half as wide as in Fig. 23-20a, yielding an output voltage half as great, but having the same frequency. We can therefore vary both the output frequency and output voltage using a fixed dc input voltage. As a result, a simple diode bridge rectifier can be used to supply the fixed dc link voltage. The power factor of the 3-phase supply line is therefore high.

Regenerative braking can be achieved, but during such power reversal, current I_d reverses while the polarity of E_2 remains the same. Consequently, an extra inverter 3 has to be placed in reverse parallel with rectifier 1 in order to feed power back to the line (Fig. 23-19). Rectifier 1 is automatically blocked while inverter 3 is in operation, and vice versa.

Pulse-width modulation is effected by computer control of the gate triggering. It can

be used to control induction motors up to several hundred horsepower.

QUESTIONS AND PROBLEMS

Practical level

23-1   Name 3 types of drives used to power squirrel-cage induction motors.

23-2   Why are two converters needed for each phase winding in Fig. 23-1?

23-3   A standard 3-phase, 16-pole squirrel-cage induction motor is rated at 460 V, 60 Hz. We want the motor to run at a no-load speed of about 225 r/min while maintaining the same flux in the air gap. Calculate the voltage and frequency to be applied to the stator.

23-4   The blower motor shown in Fig. 23-4 has 4 poles and a full-load rating of 2 hp, 240 V, 3-phase, 60 Hz. It has the torque-speed characteristic shown in Fig. 23-5. Calculate:
a. the full-load rated speed [r/min]
b. the full-load rated torque [N·m]

23-5   A 3-phase, 6-pole induction motor is driven by a cycloconverter that is fed from a 60 Hz line. What is the approximate maximum speed that can be attained with this arrangement?

23-6   What is the basic difference between a naturally-commutated and a self-commutated inverter?

23-7   A naturally-commutated inverter can be used to drive a 3-phase synchronous motor but not a 3-phase induction motor. Explain why.

23-8   In comparing the physical arrangement of the bridge rectifiers in one phase of the cycloconverter of Fig. 23-1, is there any difference with the bridge rectifier arrangement of Fig. 23-19?

23-9 A large squirrel-cage induction motor has to run at a very low, steady speed. If electronic control is required, what type of control would be most appropriate?

Intermediate level

23-10 The induction motor in Fig. 23-1 has 6 poles and runs at a no-load speed of 160 r/min. The effective voltage across the windings is 42 V. Calculate:
a. the frequency generated by the cycloconverter [Hz]
b. the smallest possible effective line voltage of the 60 Hz source. [V]

23-11 When the induction motor in Problem 23-10 operates at full load (with the frequency unchanged) its power factor is 80 percent. Calculate:
a. the time during which converter 1 acts as a rectifier [ms]
b. the time during which it acts as an inverter [ms]

23-12 The motor and blower shown in Fig. 23-4 have the properties shown in Fig. 23-5. The synchronous speed is 1200 r/min, the rated torque is 8 N·m, and the rated voltage is 240 V. Calculate:

a. the torque, speed and horsepower when the voltage is 240 V
b. the torque, speed and horsepower when the voltage is 120 V.

23-13 Referring to Fig. 23-9a, the dc link voltage and current are respectively 250 V and 60 A. The 3-phase line voltage is 240 V, 60 Hz. The motor has an efficiency of 82%. Calculate:

a. the approximate mechanical power developed by the motor [hp]
b. the firing angle of converter 1
c. the reactive power absorbed by converter 1.

23-14 Why is it easier to achieve regenerative braking with a current-fed drive than with a voltage-fed drive?

23-15 Referring to Fig. 23-18, calculate:
a. the average current in each diode
b. the peak current in each diode
c. the peak inverse voltage across each diode
d. the frequency of rotor current I_R.

23-16 In Fig. 23-19 the value of E_2 remains fixed, but in Fig. 23-12 it varies linearly with the frequency generated by converter 2. Can you explain why?

23-17 The blower motor in Fig. 23-4 has a nominal rating of 1/4 hp, 1620 r/min, 3-phase, 460 V. The respective torque-speed characteristics of the motor and blower are given in Fig. 23-5. Calculate the rotor I^2R losses (with the motor coupled to the blower):
a. when the motor runs at rated voltage;
b. when the stator voltage is reduced to 230 V;
c. Is the rotor hotter in (a) or (b)?

23-18 In Problem 23-17, calculate the stator voltage required so that the blower runs at a speed of 810 r/min.

23-19 A 30 hp, 208 V, 3-phase, 3500 r/min, 60 Hz, wound-rotor induction motor produces an open-circuit rotor line voltage of 250 V. We wish to limit the locked-rotor torque to a maximum value of 40 N·m so as to ensure a small starting current. A 3-phase bridge rectifier composed of six diodes is connected to the three slip rings. A single manual rheostat is connected across the dc output of the rectifier. Calculate:
a. the synchronous speed of the motor;
b. the power that is dissipated in the rotor circuit under locked-rotor conditions;

c. the approximate dc output voltage;

d. the approximate resistance of the rheostat and its power-handling capacity.

23-20 In Problem 23-19, a dc chopper is connected between the dc output of the rectifier and a 0.2 Ω resistor. If the chopper operates at a fixed frequency of 500 Hz, calculate the duration of the on-time T_a under locked-rotor conditions.

Advanced level

23-21 a. The squirrel-cage induction motor shown in Fig. 23-1 has a nominal rating of 50 hp, 460 V per phase, 60 Hz, 1100 r/min. The 3-phase line voltage is 208 V, 60 Hz. If we want the motor to run at a speed of about 200 r/min, while developing full-load torque, calculate the approximate voltage and frequency to be applied to the stator windings.

b. If current I_a has an effective value of 60 A, calculate the approximate value of the peak current carried by each thyristor.

23-22 The self-commutated inverter in Fig. 23-9a furnishes a motor current having an effective value of 26 A. What is the value of the dc link current?

23-23 A standard 50 hp, 1750 r/min, 3-phase, 200 V, 60 Hz squirrel-cage induction motor is driven by a current-fed self-commutated inverter shown in Fig. 23-9a. Calculate the voltage and frequency to be applied to the stator so that the motor develops its rated torque at 400 r/min. Assume the flux in the machine is constant.

PART IV

ELECTRIC UTILITY POWER SYSTEMS

24

GENERATION
OF ELECTRICAL
ENERGY

Now that we are familiar with the principal machines and power devices, we are in a position to see how they are used in a large electrical system. Such a system comprises all the apparatus used in the generation, transmission and distribution of electrical energy, starting from the generating station and ending up in the most remote summer home in the country. The next four chapters are therefore devoted to the following major topics:

- the generation of electrical energy
- the transmission of electrical energy
- the distribution of electrical energy
- the cost of electricity

24.1 Demand of an electrical system

The total power drawn by the customers of a large utility system fluctuates between wide limits, depending on the seasons and time of day. Figure 24-1 shows how the system demand (power) varies during a typical summer and winter day. The pattern of the daily

demand is remarkably similar. During the winter the peak demand of 15 GW (= 15 000 MW) is higher than the summer peak of 10 GW. Nevertheless, both peaks occur around 17:00 (5 p.m.) because increased domestic activity at this time coin-

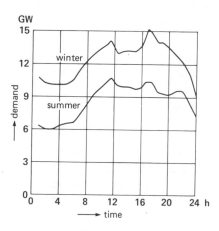

Figure 24-1
Demand curve of a larger system during a summer and winter day.

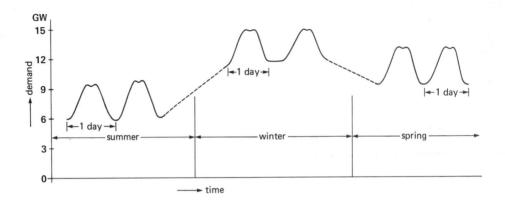

Figure 24-2
Demand curve of a large system during one year.

cides with industrial and commercial centers that are still operating at full capacity.

The load curve of Fig. 24-2 shows the *seasonal* variations for the same system. Note that the peak demand during winter (15 GW) is more than twice the minimum demand during summer (6 GW).

In examining the curve, we note that the demand throughout the year never falls below 6 GW. This is the *base load* of the system. We also see that the annual *peak load* is 15 GW. The base load has to be fed 100 percent of the time but the peak load may be on for only 0.1 percent of the time. Between these two extremes, we have *intermediate loads* that have to be fed for less than 100 percent of the time.

If we plot the duration of each demand on an annual base, we obtain the *load duration curve* of Fig. 24-3. For example, the curve shows that a demand of 9 GW lasts 70 percent of the time while a demand of 12 GW lasts for only 15 percent of the time. The graph is divided into base, intermediate and peak load sections. The peak load portion usually includes demands that last for less than 15 percent of the time. On this basis, the system has to deliver 6 GW of base power, another 6 GW of intermediate power and 3 GW of peak power.

These power blocks give rise to three types of generating stations:

a. *Base power stations* that deliver full power at all times. Nuclear stations and coal-fired stations are particularly well adapted to furnish base demand;

b. *Intermediate power stations* that can respond relatively quickly to changes in demand, usually by adding or removing one or more generating units. Hydropower stations are well adapted for this purpose;

c. *Peak generating stations* that deliver power for brief intervals through the day. Such stations must be put in service very quickly. Consequently, they are equipped with prime movers such as diesel engines, gas turbines, compressed-air motors or pumped-storage turbines that can be started up in a few minutes. In this regard, it is worth mentioning that thermal generating stations, using gas or coal, take from 4 to 8 hours to start up, while nuclear stations may take several days. Obviously, such generating stations cannot be used to supply short-term peak power.

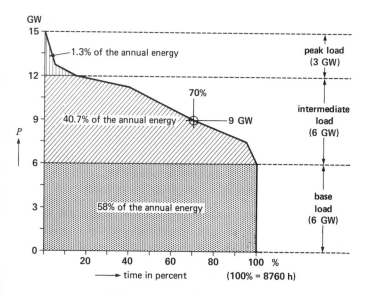

Figure 24-3
Load duration curve of a large system.

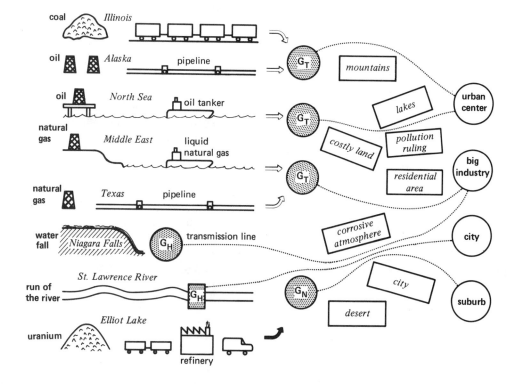

Figure 24-4
Extracting, hauling and transforming the primary sources of energy is done in different ways. The dotted transmission lines connecting the generating stations G with the consumers must go around various obstacles.

Returning to Fig. 24-3, it so happens that the areas of the dotted and cross-hatched parts are proportional to the relative amount of energy (kW·h) associated with the base, intermediate and peak loads. Thus, the base power stations supply 58 percent of the total annual energy requirements, while the peak load stations contribute only 1.3 percent. The peak load stations are in service for an average of only 1 hour per day. Consequently, peak power is very expensive because the stations that produce it are idle most of the time.

24.2 Location of the generating station

In planning an electric utility system, the physical location of the generating station, transmission lines and substations must be carefully planned to arrive at an acceptable, economic solution. We can sometimes locate a generating station next to the primary source of energy (such as a coal mine) and use transmission lines to carry the electrical energy to where it is needed. When this is neither practical or economical, we have to transport the primary energy (coal, gas, oil) by ship, train, or pipeline to the generating station (Fig. 24-4). The generating station may therefore be near to, or far away from, the user of electrical energy. Figure 24-4 also shows some of the obstacles that prevent transmission lines from following the shortest route. Owing to these obstacles, both physical and legal, transmission lines often follow a zig-zag path between the generating station and the ultimate user.

24.3 Types of generating stations

There are three main types of generating stations:

    a. thermal generating stations;

    b. hydropower generating stations;

    c. nuclear generating stations.

Thermal generating stations produce most of the electrical energy in the United States.

Nevertheless, important hydropower stations and nuclear generating stations produce about 20 percent of the total requirements.

Some of the largest hydropower stations are located in Quebec and British Columbia, in Canada.

Although we can harness the wind, tides and solar energy, these energy sources represent a tiny part of the total energy we need.

24.4 Controlling the power balance between generator and load

The electrical energy consumed by the thousands of customers must immediately be supplied by the ac generators because electrical energy cannot be stored. How do we maintain this almost instantaneous balance between customer requirements and generated power? To answer the question, let us consider a single hydropower station supplying a regional load R_1 (Fig. 24-5). Water behind the dam flows through the turbine, causing the turbine and generator to rotate.

The mechanical power P_T developed by the turbine depends exclusively on the opening of the wicket gates that control the water flow. The greater the opening, the more water is admitted to the turbine and the increased power is immediately transmitted to the generator.

On the other hand, the electric power P_L drawn from the generator depends exclusively on the load. When the mechanical power P_T supplied to the rotor is equal to the electrical power P_L consumed by the load, the generator is in dynamic equilibrium and its speed remains constant. The electrical system is said to be *stable*.

However, we have just seen that the system demand fluctuates continually, so that P_L is sometimes greater and sometimes less than P_T. If P_L is greater than P_T, the generating unit (turbine and generator) begins to slow down. Conversely, if P_L is less than P_T, the unit speeds up.

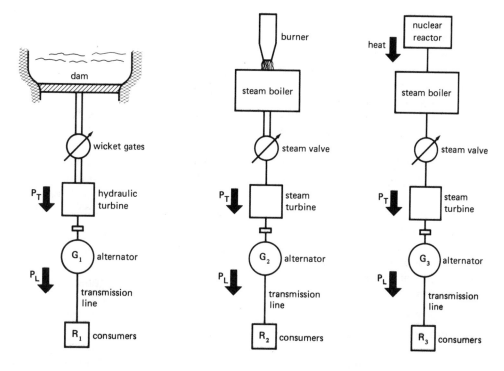

Figure 24-5
Power supplied to three independent regions.

The speed variation of the generator is therefore an excellent indicator of the equilibrium between P_L and P_T, and hence of the stability of the system. If the speed falls, the wicket gates must open, and if it rises, they must close so as to maintain a continuous state of equilibrium between P_T and P_L. Although we could adjust the gates manually by observing the speed, we always use an automatic speed regulator.

Speed regulators, or governors, are extremely sensitive devices. They can detect speed changes as small as 0.02 percent. Thus, if the speed of a generator increases from 1800 r/min to 1800.36 r/min, the governor begins to act on the wicket gate mechanism. If the load should suddenly increase, the speed will drop momentarily, but the governor will quickly bring it back to rated speed. The same

corrective action takes place when the load is suddenly removed.

Clearly, any speed change produces a corresponding change in the system frequency. The frequency is therefore an excellent indicator of the stability of a system. *The system is stable so long as the frequency is constant.*

The governors of thermal and nuclear stations operate the same way, except that they regulate the steam valves, allowing more or less steam to flow through the turbines (Fig. 24-5). The resulting change in steam flow has to be accompanied by a change in the rate of combustion. Thus, in the case of a coal-burning boiler, we have to reduce combustion as soon as the valves are closed off, otherwise the boiler pressure will quickly exceed the safety limits.

24.5 Advantage of interconnected systems

Consider the three generating stations of Fig. 24-5, connected to their respective regional loads R_1, R_2, and R_3. Because the three systems are not connected, each can operate at its own frequency, and a disturbance on one does not affect the others. However, it is preferable to interconnect the systems because 1) it improves the overall stability, 2) it provides better continuity of service and 3) it is more economical. Figure 24-6 shows four interconnecting transmission lines, tying together both the generating stations and the regions being serviced. High-speed circuit breakers d_1 to d_{10} are installed to automatically interrupt power in case of a fault, and to reroute the flow of electric power. We now discuss the advantages of such a network.

1. **Stability**. Systems that are interconnected have greater reserve power than a system working alone. In effect, a large system is better able to withstand a large disturbance and, consequently, it is inherently more stable. For example, if the load suddenly increases in region R_1, energy immediately flows from stations G_2 and G_3 and over the interconnecting tie lines. The heavy load is therefore shared by all three stations instead of being carried by one alone.

2. **Continuity of service**. If a generating station should break down, or if it has to be shut down for annual inspection and repair, the customers it serves can temporarily be supplied by the two remaining stations. Energy flowing over the tie lines is automatically metered, and credited to the station that supplies it.

3. **Economy.** When several regions are interconnected, the load can be shared between the various generating stations so that the overall operating cost is minimized. For example, instead of operating all three stations at reduced capacity during the night when demand is low, we can shut down one station completely and let the others carry the load. In this way, we reduce the

Figure 24-6
Three networks connected by 4 tie lines.

Figure 24-7
Technicians in the control rooms of two generating stations communicate with each other or with a central dispatching office, while supervising the operation of their respective generating units.

operating cost of one station to "zero" while improving the efficiency of the other stations, because they now run closer to their rated capacity.

Electrical utility companies are therefore interested in grouping their resources by a grid of interconnecting transmission lines. A central dispatching office (control center) distributes the load among the various companies and generating stations so as to minimize the costs (Fig. 24-7). Owing to the complexity of some systems, control decisions are often made with the aid of a computer. The dispatching office also has to predict daily and seasonal load changes and direct the start-up and shut-down of generating units so as to maintain good stability of the immense and complicated network.

For example, the New England Power Exchange (NEPEX) coordinates the resources of 13 electrical utility companies serving Connecticut, Rhode Island, Maine and New Hampshire. It also supervises power flow between this huge network and the State of New York and Canada.

Although such interconnected systems must necessarily operate at the same frequency, the load can still be allocated among the individual generating units, according to a specific program. Thus, if a generating unit has to deliver more power, its governor setting is changed slightly so that more power is delivered to the generator. The increased electrical output from this unit produces a corresponding decrease in the total power supplied by all the other generating units.

24.6 Conditions during an outage

A major disturbance on a system (called "contingency") creates a state of emergency and immediate steps must be taken so that it does not spread to other regions. The sudden loss of an important load or a permanent short-circuit of a transmission line constitutes a major contingency.

If a big load is suddenly lost, the turbines begin to speed up and the frequency increases everywhere on the system. On the other hand, if a generator is disconnected, the speed of the remaining generators decreases because they suddenly have to carry the entire load. The frequency starts to decrease – sometimes at the rate of 5 Hz per second. Under these conditions, no time must be lost and, if conventional methods are unable to bring the frequency back to normal, one or more loads must be dropped. Such *load shedding* is done by frequency-sensitive relays that open selected circuit breakers as the frequency falls. For example, on a 60 Hz system the relays may be set to shed 15 percent of the system load when the frequency reaches 59.3 Hz, another 15 percent when it reaches 58.9 Hz and a final 30 percent when the frequency is 58 Hz. Load shedding must be done in less than one second to "save" the loads judged to be of prime importance. As far as the disconnected customers are concerned, such an outage creates serious problems. Elevators stop between floors, arc furnaces start to cool down, paper tears as it moves through a paper mill, traffic lights stop functioning, and so forth. Clearly, it is in everyone's interest to provide uninterrupted service.

Experience over many years has shown that most system short-circuits are very brief. They may be caused by lightning, by polluted insulators, by falling trees, or by overvoltages created when circuit breakers open and close. Such disturbances usually produce a short-circuit between two phases or between one phase and ground. Three-phase short-circuits are very rare.

Statistical evidence also reveals that line short-circuits are, in general, very brief. Consequently, a major outage can usually be prevented by simply opening a short-circuited line and reclosing it very quickly. Naturally, such fast switching of circuit breakers is done automatically because it all happens in a matter of a few cycles.

24.7 Frequency and electric clocks

The frequency of a system fluctuates as the load varies, but the turbine governors always bring it back to 60 Hz. Owing to these fluctuations, the system gains or loses a few cycles throughout the day. When the accumulated loss or gain is about 180 cycles, the error is corrected by making all the generators turn either faster or slower for a brief interval. The frequency correction is effected according to instructions from the dispatching center. In this way, a 60 Hz network generates exactly 5 184 000 cycles in a 24-hour period. Electric clocks connected to the network are therefore very accurate, because the position of the second hand is directly related to the number of elapsed cycles.

HYDROPOWER GENERATING STATIONS

Hydropower generating stations convert the energy of moving water into electrical energy by means of a hydraulic turbine coupled to a synchronous generator.

24.8 Available hydro power

The power that can be extracted from a waterfall depends upon its height and rate of flow. The size and physical location of a hydropower station depends therefore on these two factors. The available hydro power can be calculated by the equation:

$$P = 9.8 \, qh \tag{24-1}$$

where

P = available water power [kW]
q = water flow [m³/s]
h = head of water [m]

Owing to friction losses in the water conduits, turbine casing, and the turbine itself, the mechanical power output of the turbine is somewhat less than that calculated by Eq. 24-1. However, the efficiency of large hydraulic turbines is between 90 and 94 percent. The generator efficiency is even higher, ranging from 95 to 99 percent, depending on the size.

Example 24-1:
A large hydropower station has a head of 324 m and an average flow of 1370 m³/s. The reservoir of water behind the dams and dikes is composed of a series of lakes covering an area

of 6 400 km². Calculate:

a. the available hydraulic power;

b. the number of days this power could be sustained if the level of the impounded water were allowed to drop by 1 m (assume no precipitation or evaporation and neglect water brought in by surrounding rivers and streams).

Solution:

a. The available hydropower is:

$$P = 9.8\ qh$$
$$= 9.8 \times 1370 \times 324$$
$$= 4\ 350\ 000\ kW = 4350\ MW$$

b. A drop of 1 m in the water level corresponds to 6 400 × 10^6 m³ of water. Because the flow is 1370 m³/s, the time for all this water to flow through the turbines is:

$$t = 6400 \times 10^6/1370$$
$$= 4.67 \times 10^6\ s$$
$$= 1298\ h = 54\ days$$

As a matter of interest, a flow of 1370 m³/s is about 10 times the amount of water used by the city of New York, including all its suburbs.

24.9 Types of hydropower stations

Hydropower stations are divided into three groups depending on the head of water:

1. high-head development;

2. medium-head development;

3. low-head development.

High-head developments have heads in excess of 300 m, and high-speed Pelton turbines are used. Such generating stations are found in the Alps and other mountainous regions. The amount of impounded water is usually small.

Medium-head developments have heads between 30 m and 300 m and medium-speed Francis turbines are used. The generating sta-

tion is fed by a huge reservoir of water retained by dikes and a dam. The dam is usually built across a river bed in a relatively mountainous region. A great deal of water is impounded behind the dam (Fig. 24-8).

Low-head developments have heads under 30 m and low-speed Kaplan or Francis turbines are used. These generating stations often extract the energy from flowing rivers. The turbines are designed to handle large volumes of water at low pressure. No reservoir is provided (Fig. 24-9).

24.10 Makeup of a hydropower plant

A hydropower installation consists of dams, waterways and conduits that form a reservoir and channel the water towards the turbines. These, and other items described below, enable us to understand some of the basic features and components of a hydropower plant (see Fig. 24-10).

1. **Dams.** Dams made of earth or concrete are built across river beds to create storage reservoirs. Reservoirs can compensate for the reduced precipitation during dry seasons, and for the abnormal flows that accompany heavy rains and melting snow. In effect, dams permit us to regulate the water flow throughout the year, so that the powerhouse may run at close to full capacity.

Spillways adjacent to the dam are provided to discharge water whenever the reservoir level is too high. In effect, we have seen that the demand for electricity varies considerably throughout the day and from season to season. Consequently, we cannot always use the available water to supply energy to the system. If the water reservoir is small or almost nonexistent (such as in run-of-river stations), we unfortunately have to let the water through the spillway, without using it.

Dams often serve a dual purpose, providing irrigation and navigation facilities, in addition to

Figure 24-8
Grand Coulee Dam on the Columbia River, in the State of Washington, is 108 m high and 1270 m wide. It is the largest hydropower plant in the world, having 18 generating units of 125 MW each and 12 generating units of 600 MW each, for a total of 9450 MW of installed capacity. The spillway can be seen in the middle of the dam. (*General Electric*)

their power-generating role. The integrated system of the Tennessee Valley Authority is a good example.

2. **Conduits penstocks and scroll-case**. In large installations, conduits lead the water from the dam site to the generating plant. They may be open canals, or tunnels carved through rock. The conduits feed one or more penstocks (huge steel pipes) which bring the water to the individual turbines. Enormous valves, sometimes several meters in diameter, enable the water supply to be shut off in the conduits.

The penstocks channel the water into a scroll-case that surrounds the runner (turbine) so that water is evenly distributed around its circumference. Guide vanes and wicket gates control the water so that it flows smoothly into the runner blades (see Figs. 24-11, 24-12 and 24-13). The wicket gates open and close in response to a powerful hydraulic mechanism that is controlled by the respective turbine governors.

3. **Draft tube and tailrace**. Water that has passed through the runner moves next through a carefully designed vertical channel, called draft tube. The draft tube improves the hydraulic efficiency of the turbine. It leads out to the tailrace which channels the water into the downstream river bed.

4. Powerhouse. The powerhouse contains the synchronous generators, transformers, circuit breakers, etc., and associated control apparatus. Instruments, relays, and meters are contained in a central control room where the entire station can be monitored and controlled. Finally, many other devices, too numerous to mention here, make up the complete hydropower station.

24.11 Pumped storage installations

We have already seen that peak power stations are needed to meet the variable system demand. To understand the different types of peaking systems used, consider a network (electric system) in which the daily demand varies between 100 MW and 160 MW, as shown in Fig. 24-14. One obvious solution is to install a 100 MW base-power station and a peak-power unit of 60 MW, driven intermittently by a gas turbine.

However, another solution is to install a larger base-power unit of 130 MW and a smaller peaking station of 30 MW. The peaking station must be able to both deliver and absorb

Figure 24-9
The Beauharnois generating station, on the St-Lawrence River, contains 26 3-phase alternators rated 50 MVA, 13.2 kV, 75 r/min, 60 Hz at a power factor of 0.8 lagging. An additional 10 units rated 65 MVA, 95.7 r/min, make up the complete installation. The output ranges between 1000 MW and 1575 MW depending upon the water flow. (*Hydro-Quebec*)

30 MW of electric power. During lightly loaded periods (indicated by a minus sign in Fig. 24-15), the peaking station receives and stores energy provided by the base-power generating plant. Then, during periods of heavy

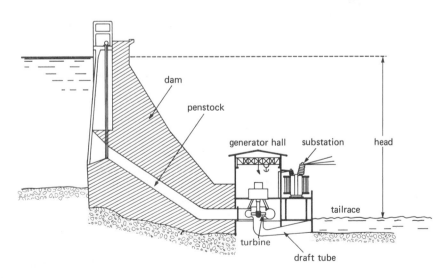

Figure 24-10
Cross section view of a medium-head hydropower plant.

Figure 24-11
Spiral case feeds water around the circumference of a 483 MW turbine. (*Marine Industrie*)

Figure 24-12
Inside the spiral case a set of adjustable wicket gates control the amount of water flowing into the turbine. (*Marine Industrie*)

Figure 24-13
Runner of a Francis-type turbine being lowered into position at the Grand Coulee Dam. The turbine is rated at 620 MW, 72 r/min and operates on a nominal head of 87 m. Other details: runner diameter: 10 m; runner mass: 500 t; maximum head: 108 m; minimum head: 67 m; turbine efficiency: 93 percent; number of wicket gates: 32; mass per wicket gate: 6.3 t; turbine shaft length: 6.7 m; mass of shaft: 175 t. (*Les Ateliers d'Ingénierie Dominion*)

demand (shown by a plus), the peaking station returns the energy it had previously stored.

This second solution has two advantages:

1. the base-power station is larger and, consequently, more efficient;

2. the peak-power station is much smaller and, therefore, less costly.

Large blocks of energy can only be stored mechanically, and that is why we often resort to a hydraulic pumped-storage station. Such a peak-power generating station consists of an upper and a lower reservoir of water connected by a penstock and an associated generating/pumping unit. During system peaks, the station acts like an ordinary hydropower generating station, delivering electrical energy as water flows from the upper to the lower reservoir. However, during light load periods, the process is reversed. The generator then operates as a synchronous motor, driving the turbine as an enormous pump. Water now flows from the lower to the upper reservoir, thereby storing energy in preparation for the next system peak (Fig. 24-16).

Figure 24-14
A 100 MW base power station and a 60 MW peak power station can supply the network demand.

Figure 24-15
A 130 MW base power station and a 30 MW pumped storage unit can also supply the network demand.

This generating/pumping cycle is repeated once or twice per day, depending on the nature of the system load. Peak-power generators have ratings between 50 MW and 500 MW. They are reversible because the direction of rotation has to be changed when the turbine operates as a pump. Starting such big synchronous motors puts a heavy load on the transmission line, and special methods must be used to bring them up to speed. Pony motors are often used, but static electronic frequency converters are also gaining ground.

Pumped storage installations operating in conjunction with nuclear plants make a very attractive combination because the latter give best efficiency when operating at constant load.

Figure 24-16
This pumped storage station, in Tennessee, pumps water from Lake Nickajack, to the top of Raccoon Mountain, where it is stored in a 2 km^2 (~ 500 acres) reservoir, giving a 316 m head. The four alternator/pump units can each deliver 425 MVA, during the system peaks. The units can be changed over from generators to pumps in a few minutes. (*Tennessee Valley Authority*)

THERMAL GENERATING STATIONS

The hydraulic resources of most modern countries are already fully developed. Consequently, we have to rely on thermal and nuclear stations to supply the growing need for electrical energy.

Thermal generating stations produce electricity from the heat released by the combustion of coal, oil, or natural gas. Most stations have ratings between 200 MW and 1500 MW so as to attain the high efficiency and economy of a large installation. Such a station has to be seen to appreciate its enormous complexity and size.

Thermal stations are usually located near a river or lake because large quantities of cooling water are needed to condense the steam as it exhausts from the turbines.

The efficiency of thermal generating stations is always low because of the inherent low efficiency of the turbines. The maximum efficiency of any machine that converts heat energy into mechanical energy is given by the equation:

$$\eta = (1 - T_2/T_1)\ 100 \qquad (24\text{-}2)$$

where

η = efficiency of the machine [%]

T_1 = temperature of the gas entering the turbine [K]

T_2 = temperature of the gas leaving the turbine [K]

In most thermal generating stations the gas is steam. In order to obtain a high efficiency, the quotient T_2/T_1 should be as small as possible. However, temperature T_2 cannot be lower than the ambient temperature, which is usually about 20°C. As a result, T_2 cannot be less than

$$T_2 = 20° + 273° = 293\ K$$

This means that to obtain high efficiency, T_1 should be as high as possible. The problem is that we cannot use temperatures above those that steel and other metals can safely withstand, bearing in mind the corresponding high steam pressures. It turns out that the highest feasible temperature T_1 is about 550°C. As a result,

$$T_1 = 550° + 273° = 823\ K$$

It follows that the maximum possible efficiency of a turbine driven by steam that enters at 823 K and exists at 293 K is

$$\eta = (1 - 293/823)100 = 64.4\ \%$$

Owing to other losses, some of the most efficient steam turbines have efficiencies of 45%. This means that 65% of the thermal energy is lost during the thermal-to-mechanical conversion process. The enormous loss of heat and how to dispose of it represents one of the major aspects of a thermal generating station.

24.12 Makeup of a thermal generating station

The basic structure and principal components of a thermal generating station are shown in Fig. 24-17. They are itemized and described below.

- A huge boiler (1) acts as a furnace, transferring heat from the burning fuel to row upon row of water tubes S_1, which entirely surround the flames. Water is kept circulating through the tubes by a pump P_1.

- A drum (2) containing water and steam under high pressure, produces the steam required by the turbines. It also receives the water delivered by boiler feed pump P_3. Steam races towards the high-pressure turbine HP after having passed through superheater S_2. The superheater, composed of a series of tubes surrounding the flames, raises the steam temperature by about 200°C. This ensures that the steam is absolutely dry and raises the overall efficiency of the station.

- High-pressure (HP) turbine (3) converts thermal energy into mechanical energy by

Figure 24-17
Principal components of a thermal power plant.

letting the steam expand as it moves through the turbine blades. The temperature and pressure at the output of the turbine are therefore less than at the input. In order to raise the thermal efficiency and to prevent premature condensation, the steam passes through a reheater S_3, composed of a third set of heated tubes.

- Medium-pressure (MP) turbine (4) is similar to the high-pressure turbine, except that it is bigger so that the steam may expand still more.

- Low pressure (LP) turbine (5), is composed of two identical left-hand and right-hand sections. The turbine sections remove the remaining available energy from the steam (Fig. 24-18). The steam flowing out of LP

expands into an almost perfect vacuum created by the condenser (6).

- Condenser (6) causes the steam to condense by letting it flow over cooling pipes S_4. Cold water from an outside source, such as a river or lake, flows through the pipes, thus carrying away the heat. It is the condensing steam that creates the vacuum.

  A condensate pump P_2 removes the luke-warm condensed steam and drives it through a reheater (7) towards feedwater pump (8).

- Reheater (7) is a heat exchanger. It receives hot steam, bled off from high-pressure turbine HP, to raise the temperature of the feedwater. Thermodynamic studies show that the overall thermal efficiency is

Figure 24-18
Low-pressure section of a 375 MW, 3600 r/min steam-turbine generator set, showing the radial blades. (*General Electric*)

improved when some steam is bled off this way, rather than letting it follow its normal course through all three turbines.

- The burners (9) supply and control the amount of gas, oil or coal injected into the boiler. Coal is pulverized before it is injected. Similarly, heavy bunker oil is preheated and injected as an atomized jet to improve surface contact (and combustion) with the surrounding air.

- A forced-draft fan (10) furnishes the enormous quantities of air needed for combustion (Fig. 24-19).

- An induced-draft fan (11) carries the gases and other products of combustion towards cleansing apparatus and from there to the stack and the outside air.

- Generator G, directly coupled to all three turbines, converts the mechanical energy into electrical energy.

In practice, a steam station has hundreds of other components and accessories to ensure high efficiency, safety, and economy. For example, control valves regulate the amount of steam flowing to the turbines; complex water purifiers maintain the required cleanliness and chemical composition of the feedwater; oil pumps keep the bearings properly lubricated. However, the basic components we have just described enable us to understand the operation and some of the basic problems of a thermal station.

24.13 Turbines

The low-, medium-, and high-pressure turbines possess a series of blades mounted on the drive shaft (Fig. 24-18). The steam is deflected by the blades, producing a powerful torque. The blades are made of special steel to withstand the high temperature and intense centrifugal forces.

Figure 24-19
This forced-draft fan provides 455 m³/s of air at a pressure difference of 5.8 kPa for a thermal power station. It is driven by a 3-phase induction motor rated 12 000 hp (8955 kW), 60 Hz, 890 r/min. (*Novenco Inc.*)

The HP, MP and LP turbines are coupled together to drive a common generator. However, in some large installations, the HP turbine drives one generator while the MP and LP turbines drive another having the same rating.

24.14 Condenser

About one-half the energy produced in the boiler has to be removed from the steam when it exhausts into the condenser. Consequently, enormous quantities of cooling water are needed to carry away the heat. The temperature of the cooling water increases typically by 5°C to 10°C as it flows through the condenser tubes. The condensed steam (condensate) usually has a temperature between 27°C and 33°C and the corresponding absolute pressure is a near-vacuum of about 5 kPa. The cooling water temperature is only a few degrees below the condensate temperature (see Fig. 24-20).

24.15 Cooling towers

If the thermal station is located in a dry region, or far away from a river or lake, we still have to cool the condenser, in one way or another. We often use *evaporation* to produce the cooling effect. To understand the principle, consider a lake which exposes a large surface to the surrounding air. A lake evaporates continually, even at low temperatures, and it is known that for every kilogram of water that evaporates, the lake loses 2.4 MJ of heat. Consequently, evaporation causes the lake to cool down.

Consider now a tub containing 100 kg of water at a certain temperature. If we can somehow cause 1 kg of water to evaporate, the temperature of the remaining 99 kg will drop by 5.8°C. We conclude that whenever 1 percent of a body of water evaporates, the temperature of the remaining water drops by 5.8°C. Evaporation is therefore a very effective cooling process.

But how can we produce evaporation? Surprisingly, all that is needed is to expose a large surface of water to the surrounding air. The

Figure 24-20
Condenser rated at 220 MW. Note the large pipes feeding cooling water into and out of the condenser. The condenser is as important as the boiler in thermal and nuclear power stations. (*Foster-Wheeler Energy Corporation*)

simplest way to do this is to break up the water into small droplets, and blow air through this artificial rain.

In the case of a thermal station, the warm cooling water flowing out of the condenser is piped to the top of a cooling tower where it is broken up into small droplets. (Fig. 24-21). As the droplets fall towards the open reservoir below, evaporation takes place, and the droplets are chilled. The cool water is pumped from the reservoir and recirculated through the condenser where it again removes heat from the condensing steam. The cycle then repeats. Approximately 2 percent of the cooling water that flows through the condenser is lost by evaporation. This loss can be made up by a stream, or small lake.

24.16 Boiler feed pump

The boiler feed pump drives the feedwater into the high-pressure drum. The high pressure together with the large volume of water flowing through the pump requires a very powerful motor to drive it. In modern steam stations, the

Figure 24-21
Cooling tower installed in a nuclear power station in Oregon. The generator output is 1280 MVA at a power factor of 0.88. Tower characteristics: height: 152 m; diameter at the base: 117 m; diameter at the top: 76 m; cooling water: 27 m^3/s; water loss by evaporation: 0.7 m^3/s. The temperature of the cooling water drops from 44.5° to 24° as it passes through the tower. (*Portland General Electric Company*)

pumping power represents about 1 percent of the generator output. Although this appears to be a significant loss, we must remember that the energy expended in the pump is later recovered when the high-pressure steam flows through the turbines. Consequently, the energy supplied to the feed pump motor is not really lost, except for the small portion consumed by the losses in the motor and pump.

24.17 Energy flow diagram for a steam plant

Modern thermal generating stations are very similar throughout the world because all designers strive for high efficiency at lowest cost. This means that materials are strained to the limits of safety as far as temperature, pressure, and centrifugal forces are concerned. Because the same materials are available to all, the resulting steam plants are necessarily

similar. Figure 24-22 shows a typical 540 MW turbine-generator set and Fig. 24-23 is a view of the control room.

Most modern boilers furnish steam at a temperature of 550°C and a pressure of 16.5 MPa. The overall efficiency (electrical output/thermal input) is then about 40 percent. The relative amounts of energy, steam flow, losses, and so forth, do not change very much, provided the temperature and pressure have the approximate values indicated above. This enables us to draw a diagram showing the energy flow, steam flow, water flow, etc., in a reduced-scale model of a typical thermal generating station. Figure 24-24 shows such a model, producing 12 MW of electrical power.

For example, a 480 MW station (40 times more powerful than the model) has the following approximate characteristics:

Electric power output:	40 × 12 MW	480 MW
Coal consumption:	40 × 1 kg/s	40 kg/s
Air intake:	40 × 10 kg/s	400 kg/s
Boiler thermal power:	40 × 30 MW	1 200 MW
Steam output:	40 × 8 kg/s	320 kg/s
Cooling water:	40 × 360 kg/s	14 400 kg/s
Heat carried away by the cooling water:	40 × 15 MW	600 MW

Figure 24-22
This 540 MW steam-turbine generator set runs at 3600 r/min, generating a frequency of 60 Hz. The low-pressure turbine and alternator are in the background. (*General Electric*)

Figure 24-23
Control room of the 540 MW generator set. (*General Electric*)

If a large river or lake is not available and a cooling tower is required, it would have to evaporate:

$$q = 2\% \times 14\,400 = 288 \text{ kg/s}$$

of cooling water. This loss by evaporation has to be made up by a local source of water.

24.18 Thermal stations and the environment

The products of combustion of thermal generating stations are an increasing subject of concern, owing to their impact on the environment.

Carbon dioxide (CO_2), sulfur dioxide (SO_2), and water are the main products of combustion when oil, coal, or gas are burned. Carbon dioxide and water produce no immediate environmental effects, but sulfur dioxide creates substances that give rise to acid rain. Dust and fly ash are other pollutants that may reach the atmosphere. Natural gas produces only water and CO_2. This explains why gas is used (rather than coal or oil), when atmospheric pollution must be reduced to a minimum.

A good example of pollution control is the large Eraring generating station located in Newcastle, Australia, about 100 km north of Sydney (Fig. 24-25). It is equipped with a special fabric filter flue gas cleaning system (Fig. 24-26). The fabric filters act like huge vacuum cleaners to remove particles from the

Figure 24-24
Scale model of a typical thermal generating station.

boiler gas flue stream. The fabric filter for each boiler is composed of 48 000 filter bags, each 5 m long and 16 cm in diameter (Fig. 24-27). When a boiler operates at full capacity, they capture dust particles at the rate of 28 kg/s. A substantial proportion of this material is later mixed with concrete for road-building projects. The following technical specifications enable us to appreciate the size of this station.

Electrical data

number of generators: 4
power per generator: 660 MW
speed: 3000 r/min
voltage: 23 000 V
frequency: 50 Hz phases: 3

Thermal and mechanical data

number of steam turbines: 4
number of condensers: 4
number of boilers: 4
steam flow per turbine: 560 kg/s
steam temperature: 540° C
steam pressure: 16.55 MPa
cooling water per condenser: 21 000 kg/s
coal consumption per boiler: 51.5 kg/s
dust captured by cleaning system: 28 kg/s

length of one turbine-generator unit: 50 m
weight of one turbine-generator unit: 1342 tonnes

number of emission stacks: 2
height of emission stack: 200 m
outside diameter at bottom: 20 m
outside diameter at top: 11.6 m

Figure 24-25
View of the Eraring Power Station situated in Newcastle, Australia. The large building on the left is the turbine-generator hall: 27 m wide × 38 high × 418 m long. To the right can be seen the four structures that house the steam boilers. A portion of the flue gas cleaning system can be seen between the emission stack in the foreground and the boiler structures. (*Courtesy of The Electricity Commission of New South Wales*)

Another interesting feature is that coal for the station is brought in by conveyor belts from two mines that are only 1.5 km and 4.5 km away. Thus, the station is ideally located near its source of fuel and near its source of cooling water, on the shore of Lake Macquarie.

Figure 24-26
General construction of the flue gas cleaning system showing the filter bags that capture the dust which then falls into the hoppers below. (*Electricity Commission of New South Wales*)

Figure 24-27
Installation of the fabric filter bags. Each bag is 15 m long and 16 cm in diameter. (*Electricity Commission of New South Wales*)

NUCLEAR GENERATING STATIONS

Nuclear stations produce electricity from the heat released by a *nuclear reaction*. When the nucleus of an atom splits in two (a process called atomic fission), a considerable amount of energy is released. Note that a chemical reaction, such as the combustion of coal, produces only a rearrangement of the atoms, without in any way affecting their nuclei.

A nuclear station is identical to a thermal station except that the boiler is replaced by a nuclear reactor. The reactor contains the fissile material that generates the heat. A nuclear station therefore contains a synchronous generator, steam turbine, condenser, etc., similar to those found in a conventional thermal station. The overall efficiency is also similar (between 30 and 40 percent) and a cooling system must be provided for. Consequently, nuclear stations are also located close to rivers and lakes. In dry areas, cooling towers are installed. Owing to these similarities, we will only examine the operating principle of the reactor itself.

24.19 Composition of an atomic nucleus; isotopes

The nucleus of an atom contains two types of particles – protons and neutrons. The proton carries a positive charge, equal to the negative charge on an electron. The neutron, as its name implies, has no electric charge. Neutrons are therefore neither attracted to nor repelled by protons and electrons.

Protons and neutrons have about the same mass and both weigh 1840 times as much as electrons do. The mass of an atom is therefore concentrated in its nucleus.

The number of protons and neutrons in the nucleus depends upon the element. Furthermore, because an atom is electrically neutral, the number of electrons is equal to the number of protons. Table 24A gives the atomic structure of a few important elements used in nuclear reactors. For example, there are two types of hydrogen atoms that can be distinguished from each other only by the makeup of the nucleus. First, there is ordinary hydrogen (H), whose nucleus contains 1 proton and no neutrons. Next, there is a rare form, deuterium (D) whose nuclei contain 1 neutron, in addition to the usual proton. This rare form is called an *isotope* of hydrogen.

When two atoms of ordinary hydrogen unite with one atom of oxygen, we obtain ordinary water (H_2O) called *light* water. On the other hand, if 2 atoms of deuterium unite with 1 atom of oxygen, we obtain a molecule of *heavy* water (D_2O). The oceans contain about 1 kg of heavy water for every 7000 kg of sea water.

In the same way, two isotopes of uranium are found in nature: uranium 238 (^{238}U) and uranium 235 (^{235}U). Each contains 92 protons, but ^{238}U has 146 neutrons while ^{235}U has 143. Uranium 238 is very common, but the isotope ^{235}U is rare.

Uranium 235 and heavy water deserve our attention because both are essential to the operation of the nuclear reactors we are about to describe.

TABLE 24A		ATOMIC STRUCTURE OF SOME ELEMENTS			
Element	**Symbol**	**Protons**	**Electrons**	**Neutrons**	**mass number (neutrons + protons)**
hydrogen	H	1	1	0	1
deuterium	D	1	1	1	2
tritium	^3H	1	1	2	3
helium	He	2	2	2	4
carbon	C	6	6	6	12
iron	Fe	26	26	30	56
uranium 235	^{235}U	92	92	143	235
uranium 238	^{238}U	92	92	146	238

24.20 The source of uranium

Where does uranium come from? It is obtained from the ore found in uranium mines. This ore contains the compound U_3O_8 (3 atoms of uranium and 8 atoms of oxygen). It so happens that U_3O_8 is actually composed of $^{238}UO_8$ and $^{235}UO_8$ in the relatively precise ratio of 1398 to 10.

In other words, the ore contains 1398 parts of the less interesting ^{238}U for every 10 parts of the isotope ^{235}U. It is very difficult to separate $^{238}UO_8$ from $^{235}UO_8$ because they possess identical chemical properties.

In order to use these substances in nuclear reactors, they are processed into uranium dioxide (UO_2). The natural UO_2 again contains $^{238}UO_2$ and $^{235}UO_2$ in the ratio of 1398 to 10.

Some nuclear reactors require UO_2 that has more of the isotope ^{235}U than natural UO_2 does. This is produced by an enrichment process whereby the ratio of $^{235}UO_2$ to $^{238}UO_2$ is raised to 50 to 1398 rather than the natural ratio of 10 to 1398. In this enrichment process, a lot of $^{238}UO_2$ is obtained as a byproduct that must

be stored. As we shall see, this byproduct also has useful applications.

The process of converting uranium ore into these uranium derivatives is shown in highly simplified form in Fig. 24-28.

24.21 Energy released by atomic fission

When the nucleus of an atom fissions, it splits in two. The total mass of the two atoms formed in this way is usually less than that of the original atom. If there is a loss in mass, energy is released according to Einstein's equation:

$$E = mc^2 \qquad (24\text{-}3)$$

where
E = energy released [J]
m = loss of mass [kg]
c = speed of light [3×10^8 m/s]

An enormous amount of energy is released because a loss in mass of only 1 μg produces 9×10^{10} J, which is equivalent to the heat given off by burning 3 tons of coal. Uranium is one of those elements that loses mass when it fissions. However, uranium 235 is "fissionable"

Figure 24-28
Various steps in the manufacture of nuclear fuel for heavy-water and light-water reactors. This extremely simplified diagram shows that in the process of enriching uranium dioxide, it is inevitable that large amounts of ^{238}U remain as a byproduct.

whereas uranium 238 is not, and so large separating plants have been built to isolate molecules containing ^{235}U from those containing ^{238}U.

24.22 Chain reaction

How can we provoke the fission of a uranium atom? One way is to bombard its nucleus with neutrons. A neutron makes an excellent projectile because it is not repelled as it approaches the nucleus and, if its speed is not too great, it has a good chance of scoring a hit. If the impact is strong enough, the nucleus will split in two, releasing energy. The fission of one atom of ^{235}U releases 218 MeV of energy, mainly in the form of heat. Fission is a very violent reaction on an atomic scale, and it produces a second important effect: it ejects 2 or 3 neutrons that move at high speed away from the broken nucleus. These neutrons collide with other uranium atoms and so a chain reaction quickly takes place, releasing a tremendous amount of heat.

This is the principle that causes atomic bombs to explode. Although a uranium mine also releases neutrons, the concentration of ^{235}U atoms is too low to produce a chain reaction.

In the case of a nuclear reactor, we have to slow down the neutrons to increase their chance of striking other uranium nuclei. Towards this end, small fissionable masses of uranium fuel (UO_2) are immersed in a *moderator*. The moderator may be ordinary water, heavy water, graphite, or any other material that can slow down neutrons without absorbing them. By using an appropriate geometrical distribution of the uranium fuel within the moderator, the speed of the neutrons can be reduced so they have the required velocity to initiate other fusions. Only then will a chain reaction take place, causing the reactor to "go critical".

As soon as the chain reaction starts, the temperature rises rapidly. To keep it at an acceptable level, a liquid or gas has to flow rapidly through the reactor to carry away the heat. This so-called *coolant* may be heavy water, ordinary water, liquid sodium or a gas like helium or carbon dioxide. The hot coolant moves in a closed circuit which includes a heat exchanger. The latter transfers the heat to a steam generator that drives the turbines (Fig. 24-29). Thus, contrary to what its name would lead us to believe, the coolant is not "cool" but searingly hot.

24.23 Types of nuclear reactors

There are several types of reactors, but the following are the most important:

1. **Pressure water reactor (PWR).** Water is used as a coolant and it is kept under such high pressure that it cannot boil off into steam. We can use ordinary water, as in light-water reactors, or heavy water as in CANDU* reactors.

2. **Boiling water reactors (BWR).** The coolant in this reactor is ordinary water boiling under high pressure and releasing steam. This eliminates the need for a heat exchanger, because the steam circulates directly through the turbines. However, as in all light-water reactors, we must use enriched uranium dioxide containing about 3 percent ^{235}U.

3. **High temperature gas reactor (HTGR).** This reactor uses an inert gas coolant such as helium or carbon dioxide. Owing to the high operating temperature (typically 750°C), graphite is used as a moderator. The steam created by the heat exchanger is as hot as that produced in a conventional coal-fired steam boiler. Consequently, the overall efficiency of HTGR stations is about 40 percent.

---

* CANDU: Canada Deuterium Uranium, developed by the Atomic Energy Commission of Canada.

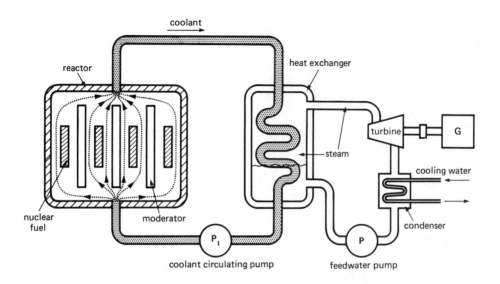

coolant

reactor

heat exchanger

turbine

G

steam

cooling water

nuclear
fuel

moderator

condenser

P_1

P

coolant circulating pump

feedwater pump

Figure 24-29
Schematic diagram of a nuclear power station.

4. Fast breeder reactor (FBR).

This reactor has the remarkable ability to both generate heat and create additional nuclear fuel while it is operating.

24.24 Example of a light-water reactor

Reactors that use ordinary water as a moderator are similar to those using heavy water, but the uranium-dioxide fuel has to be enriched. Enrichment means that the fuel bundles contain between 2 and 4 percent of ^{235}U, the remainder being ^{238}U. This enables us to reduce the size of the reactor for a given power output. On the other hand, the reactor has to be shut down about once a year to replace the expended fuel.

The generated heat, created mainly by the fission of uranium 235, is carried away by a coolant such as ordinary water, liquid sodium or a gas such as CO_2. As it flows through the heat exchanger, the coolant creates the steam that drives the turbine.

A typical nuclear power station located in Connecticut (Figs. 24-30 and 24-31) possesses a light-water reactor that is composed of a massive vertical steel tank having an external diameter of 4.5 m and a height of 12.5 m. The tank contains 157 vertical tubes which can lodge 157 large fuel assemblies. Each assembly is 3 m long and groups 204 fuel rods containing a total of 477 kg of enriched UO_2. The nuclear reaction is kept under control by 45 special-alloy control rods. When these rods are gradually lowered into the moderator, they absorb more and more neutrons. Consequently, they control the rate of the nuclear reaction and hence the amount of heat released by the reactor.

The nuclear station drives a 3-phase, 667 MVA, 90 percent power factor, 19 kV, 60 Hz, 1800 r/min alternator.

24.25 Example of a heavy-water reactor

The CANDU reactor uses heavy water, both as

Figure 24-30
Aerial view of a light-water nuclear generating station. The large rectangular building in the foreground houses a 667 MVA, 90 percent power factor, 19 kV, 60 Hz, 1800 r/min turbogenerator set; the circular building surrounds the reactor. (*Connecticut Yankee Atomic Power Company; photo by Georges Betancourt*)

moderator and as a coolant. It differs from all other reactors in that it uses *natural* uranium dioxide as a fuel. One of the biggest installations of its kind is located at Pickering, a few kilometers east of Toronto, Canada. The nuclear station has 4 reactors. Each reactor is coupled to 12 heat exchangers that provide the interface between the heavy water coolant and the ordinary steam that drives the turbines (Fig. 24-32).

Each reactor is enclosed in a large horizontal vessel (calandria) having a diameter of 8 m and a length of 8.25 m. The calandria possesses 390 horizontal tubes each housing 12 fuel bundles containing 22.2 kg of UO_2. Each bundle releases about 372.5 kW while it is in operation. Because there is a total of 4680 bundles, the reactor develops 1740 MW of thermal power.

Twelve pumps, each driven by an 1100 kW motor, push the heavy water coolant through the reactor and the heat exchangers, in a closed loop. The heat exchangers produce the steam to drive the 4 turbines. The steam exhausts into a condenser which is cooled by water drawn from Lake Ontario.

Each turbine drives a 3-phase, 635 MVA, 85 percent power factor, 24 kV, 1800 r/min, 60 Hz alternator.

The fuel bundles are inserted at one end of the calandria and, after a 19-month stay in the tubes, they are withdrawn from the other end. The bundles are inserted and removed on a

Figure 24-31
Looking down into the water-filled refuelling cavity of the reactor. (*Connecticut Yankee Atomic Power Company; photo by Georges Betancourt*)

Figure 24-32
Simplified schematic diagram of a CANDU nuclear generating unit composed of one heavy-water reactor driving one alternator. (*Atomic Energy of Canada*)

continuous basis - an average of 9 bundles per day.

Table 24B gives typical characteristics of a light-water and heavy-water reactor.

24.26 Principle of the fast breeder reactor

A fast breeder reactor differs from other reactors because it can extract more of the available energy in the nuclear fuel. It possesses a central core containing fissionable plutonium 239 (^{239}Pu). The core is surrounded by a blanket composed of substances containing nonfissionable uranium 238 (^{238}U). No moderator is used; consequently, the high-speed ("fast") neutrons generated by the fissioning ^{239}Pu bombard the nonfissionable atoms of ^{238}U. This nuclear reaction produces two important results:

a. the heat released by the fissioning core can be used to drive a steam-turbine;

b. some atoms of ^{238}U, in the surrounding blanket, capture the flying neutrons, thereby becoming fissionable ^{239}Pu. In other words, the passive atoms of uranium 238 are transmuted into fissionable atoms of plutonium 239.

As time goes by, the blanket of nonfissionable ^{238}U is gradually transmuted to fissionable ^{239}Pu and waste products. The blanket is periodically removed and the materials are processed to recover the substances containing ^{239}Pu. The nuclear fuel recovered is placed in the central core to generate heat and to produce still more fuel in a newly relined blanket of substances containing uranium 238.

This process can be repeated until nearly 80 percent of the available energy in the uranium is extracted. This is much more efficient that the 2 percent now being extracted by conventional reactors.

TABLE 24B	TYPICAL LIGHT-WATER AND HEAVY-WATER REACTORS	
	Light-Water Reactor	**Heavy-Water Reactor**
Reactor Vessel		
external diameter	4.5 m	8 m
length	12.5 m	8.25 m
vessel thickness	274 mm	25.4 mm
weight empty	416 t	604 t
position	vertical	horizontal
number of fuel canals	157	390
type of fuel	enriched UO_2 (3.3%)	natural UO_2
total mass of fuel	75 t	104 t
Moderator		
type	light-water	heavy-water
volume	13.3 m^3	242 m^3
Reactor Cooling		
Heat produced in reactor	1825 MW	1661 MW
Coolant	light-water	heavy-water
Volume	249 m^3	130 m^3
Flow rate	12.8 t/s	7.73 t/s
Coolant temperature entering the reactor	285°C	249°C
Coolant temperature leaving the reactor	306°C	294°C
Coolant pumps	4	12
Total pump power	12 MW	14 MW
Electrical output		
3-phase, 1800 r/min, 60 Hz alternator	600 MW	540 MW

The breeder reactor is particularly well adapted to complement existing light water reactors. The reason is that a great deal of ^{238}U is available as a byproduct in the manufacture of enriched ^{235}U (see Fig. 24-28). This otherwise useless material (now being stored) could be used to surround the core of a fast breeder reactor. By capturing fast neutrons, it could be rejuvenated as explained above, until most of the potential energy in the uranium is used up.

24.27 Nuclear fusion

We have seen that splitting the nucleus of a heavy element such as uranium results in a decrease in mass and the release of energy. We can also produce energy by *combining* the nuclei of two light elements in a process called *nuclear fusion*. For example, energy is released by the fusion of an atom of deuterium with an atom of tritium. However, owing to the strong repulsion between the two nuclei (both are positive), they only unite (fuse) when they approach each other at high speed. The required velocity is close to the speed of light and corresponds to a thermodynamic temperature of several million degrees. If both the atomic concentration and speed are high enough, a self-sustaining chain reaction will result.

We can therefore produce heat by the fusion of two elements, and the hydrogen bomb is a good example of this principle. Unfortunately, we run into almost insurmountable problems when we try to *control* the fusion reaction, as we must do in a nuclear reactor. Basically, scientists have not yet succeeded in confining and controlling high-speed particles without at the same time slowing them down.

A major world-wide research effort is being devoted to solve this problem. If scientists succeed in domesticating nuclear fusion, it could mean the end of the energy shortage because hydrogen is the most common element on earth.

QUESTIONS AND PROBLEMS

Practical level

24-1 Explain the difference between a base load and peak load generating plant.

24-2 Why are nuclear power stations not suited to supply peak loads?

24-3 Referring to the coal-mine in Fig. 24-4, we have the choice of hauling the coal to a generating plant or installing the generating plant next to the mine mouth. What factors come into play in determining the best solution?

24-4 What is the best indicator of stability (or instability) of an electric utility system?

24-5 What is meant by the term network?

24-6 Give two reasons why electric utility systems are interconnected.

24-7 The river flow in Fig. 24-9 is 5000 m^3/s at a head of 24 m. Calculate the available hydraulic power.

24-8 Explain the operating principle of a thermal plant; a hydropower plant; a nuclear plant.

24-9 Name two basic differences between a light-water reactor and a heavy-water reactor.

24-10 Explain what is meant by: moderator, fission, fusion, neutron, heavy water.

Intermediate level

24-11 The Zaïre River, in Africa, discharges at a constant rate, 1300 km^3 of water per year. It has been proposed to build a series of dams in the region of Inga, where the river drops by 100 m. Calculate:
a. the water flow [m^3/s];
b. the power that could be harnessed [MW];
c. the discharge in cubic miles per year.

24-12 For how long does a 1500 MW generator have to run to produce the same quantity of energy as that released by a 20 kiloton atomic bomb? (See conversion charts in Appendix).

24-13 The demand of a municipality regularly varies between 60 MW and 110 MW in the course of one day, the average power being 80 MW. To produce the required energy, we have the following options:

a. install a base power generating unit and a diesel engine peaking plant;

b. install a base power generating unit and a pumped storage unit.

What are the respective capacities of the base power and peaking power plants in each case?

24-14 On a particular day, the head of Grand Coulee dam is 280 feet and the generators deliver 6000 MVA at a power factor of 0.9 lagging. Assuming the average turbine efficiency is 0.92 and the average generator efficiency is 0.98, calculate:

a. the active power output [MW];

b. the reactive power supplied to the system [Mvar];

c. the amount of water flowing through the turbines [yd³/s].

24-15 Explain the principle of operation of a cooling tower.

24-16 A modern coal-burning thermal station produces an electrical output of 720 MW. Calculate the approximate value of:

a. the amount of coal consumed [tons (not tonnes) per day ;

b. the amount of smoke, gas, and fly ash released [tons per day];

c. the cooling water flowing through the condenser, assuming a temperature rise of 10°C [m³/s].

24-17 In Problem 24-16, if a cooling tower is required, how much water must be drawn from a local stream [m³/s]? Can this water be recycled?

24-18 A fuel bundle of natural uranium dioxide has a mass of 22.2 kg when first inserted into a heavy-water reactor. If it releases an average of 372.5 kW of thermal energy during its 19-month stay in the reactor, calculate:

a. the total amount of heat released [J] and [Btu];

b. the reduction in weight of the bundle, due to the energy released [g].

Advanced level

24-19 a. Calculate the annual energy consumption [TW·h] of the electric utility system having the load duration curve given in Fig. 24-3.

b. If this energy were consumed at an absolutely uniform rate, what would the peak load be [GW]?

24-20 Referring to Fig. 24-32, the temperature of the heavy water coolant drops from 294°C to 249°C in passing through the heat exchangers. Knowing that the reactor is cooled at the rate of 7.7 t/s of heavy-water, calculate the heat [MW] transmitted to the heat exchangers (specific heat of heavy-water is 4560 J/kg).

25

TRANSMISSION
OF ELECTRICAL
ENERGY

The transmission of electrical energy does not usually raise as much interest as does its generation and utilization: consequently, we sometimes tend to neglect this important subject. This is unfortunate because the human and material resources involved in transmission are much greater than those employed in generation.

Electrical energy is carried by conductors such as overhead transmission lines and underground cable. Although these conductors appear very ordinary, they possess important electrical properties that greatly affect the transmission of electrical energy. In this chapter, we shall study these properties for several types of transmission lines: high-voltage, low-voltage, high power, low power, aerial lines and underground lines.

25.1 Principal components of a power distribution system

In order to provide electrical energy to consumers in usable form, a transmission and distribu-

tion system must satisfy some basic requirements. Thus, the system must:
1. Provide, at all times, the power that consumers need;
2. Maintain a stable, nominal voltage that does not vary by more than ±10%;
3. Maintain a stable frequency that does not vary by more than ±0.1 Hz;
4. Supply energy at an acceptable price;
5. Meet standards of safety;
6. Respect the standards of environment.

Figure 25-1 shows an elementary diagram of a transmission and distribution system. It consists of two generating stations G_1 and G_2, a few substations, an interconnecting substation and several commercial, residential and industrial loads. The energy is carried over lines designated *extra high* voltage (EHV), *high* voltage (HV), *medium* voltage (MV) and *low* voltage (LV). This voltage classification is made according to a scale of standardized voltages whose limiting values are given in Table 25A.

Figure 25-1
Single-line diagram of a generation, transmission and distribution system.

Substations (Fig. 25-1) serve to change the line voltage by means of step-up and step-down transformers and to regulate it by means of synchronous condensers, static var compensators, or transformers with variable taps. Substations also contain circuit breakers, fuses and lightning arrestors, to protect expensive apparatus and to provide for quick isolation of faulted lines from the system.

Interconnecting substations serve to tie different power systems together, to enable power exchanges between them, and to increase the stability of the overall network*.

Electrical power utilities divide their power distribution systems into two major categories:

1. **transmission** systems in which the line voltage is roughly between 115 kV and 800 kV;

2. **distribution** systems in which the voltage generally lies between 120 V and 69 kV. Distribution systems, in turn, are divided into *medium-voltage* distribution systems

---

* A network is "an aggregation of interconnected conductors consisting of feeders, mains and services" (ref. IEEE Standard Dictionary of Electrical and Electronic Terms).

(2.4 kV to 69 kV) and *low-voltage* distribution systems (120 V to 600 V).

25.2 Types of power lines

The design of a power line depends upon:
1. the amount of active power it has to transmit;
2. the distance over which the power must be carried;
3. the cost;
4. esthetic considerations, urban congestion, ease of installation, and expected load growth.

As mentioned previously, we distinguish four types of power lines, according to their voltage class:

1. Low-voltage (LV) lines are installed inside buildings, factories and houses to supply power to motors, electric stoves, lamps, heaters and air conditioners. The service entrance panel constitutes the source and the lines are made of insulated cable or bus-bars operating at voltages below 600 V.

In some metropolitan areas, the distribution system feeding the factories, homes and commercial buildings consists of a grid of underground cables operating at 600 V or less. Such

TABLE 25A	VOLTAGE CLASSES AS APPLIED TO INDUSTRIAL AND COMMERCIAL POWER		
voltage class	**nominal system voltage**		
	two-wire	**three-wire**	**four-wire**
low voltage LV	120 single phase	120/240 □ single phase 480 V □ 600 V	— 120/208 □ 277/480 □ 347/600
medium voltage MV		2 400 4 160 □ 4 800 6 900 13 800 □ 23 000 34 500 46 000 69 000 □	 7 200/12 470 □ 7 620/13 200 □ 7 970/13 800 14 400/24 940 □ 19 920/34 500 □
high voltage HV		115 000 □ 138 000 □ 161 000 230 000 □	
extra high voltage EHV		345 000 □ 500 000 □ 735 000 - 765 000 □	

All voltages are 3-phase unless indicated otherwise.
Voltages designated by the symbol □ are preferred voltages.

Note: Voltage class designations were approved for use by IEEE Standards Board
(September 4, 1975)

a network provides dependable service, because even the outage of one or several cables will not interrupt customer service.

2. Medium-voltage (MV) lines tie the load centers to the main substation of the utility company. The voltage is usually between 2.4 kV and 69 kV. Such medium-voltage radial distribution systems are preferred in the larger cities. In radial systems, the transmission lines spread out like fingers from one or more substations to feed power to various load centers, such as high-rise buildings, shopping centers and universities.

3. High-voltage (HV) lines connect the main substations to the generating stations. The lines are composed of aerial wire or underground cable operating at voltages below 230 kV. In this category, we also find lines that transmit energy between two power systems, to increase the stability of the network.

4. Extra high voltage (EHV) lines are used when generating stations are very far from the load centers. We put these lines in a separate class because of their special electrical properties. Such lines operate at voltages up to 1000 kV and may be as long as 1000 km.

25.3 Standard voltages

To reduce the cost of distribution apparatus and to facilitate its protection, standards-setting organizations have established a number of standard voltages for transmission lines. These standards, given in Table 25A, reflect the various voltages presently used in North America. Voltages that bear the symbol ☐ are *preferred* voltages. Unless otherwise indicated, all voltages are three-phase.

25.4 Components of a transmission line

A transmission line is composed of conductors, insulators and supporting structures.

1. Conductors. Conductors for high-voltage lines are always bare. We use stranded copper conductors or steel-reinforced aluminum cable (ACSR). ACSR conductors are usually preferred, because they result in a lighter and more economical line. Conductors have to be spliced when a line is very long. Special care must be taken so that the joints have low resistance and great mechanical strength.

2. Insulators. Insulators serve to support and anchor the conductors and to insulate them from ground. Insulators are usually made of porcelain, but glass and other synthetic insulating materials are also used.

From an electric standpoint, insulators must offer a high resistance to surface leakage currents and they must be sufficiently thick to prevent breakdown under the high voltage stresses they have to withstand. To increase the leakage path (and hence the leakage resistance), the insulators are molded with wave-like folds. From a mechanical standpoint, they must be strong enough to resist the pull due to the weight of the conductors.

There are two main types of insulators: *pin-type* insulators and *suspension-type* insulators (Figs. 25-2 and 25-3). The pin-type insulator has several porcelain skirts (folds) and the conductor is fixed at the top. A steel pin screws into the insulator so it can be bolted to a support.

For voltages above 70 kV, we always use suspension-type insulators, strung together by their cap and pin metallic parts. The number of insulators depends upon the voltage: for 110 kV, we generally use from 4 to 7; for 230 kV, from 13 to 16. Figure 25-4 shows an insulator arrangement for a 735 kV line. It is composed of 4 strings of 35 insulators each, to provide both mechanical and electrical strength.

3. Supporting structures. The supporting structure must keep the conductors at a safe height from the ground and at an adequate distance from each other. For voltages below 70 kV, we can use single wooden poles

Figure 25-2
Sectional view of a 69 kV pin-type insulator. BIL: 270 kV; 60 Hz flashover voltage under wet conditions: 125 kV. (*Canadian Ohio Brass Co. Ltd.*)

Figure 25-3
Sectional view of a suspension insulator. Diameter: 254 mm; BIL: 125 kV, 60 Hz flashover voltage, under wet conditions: 50 kV (*Canadian Ohio Brass Co. Ltd.*)

Figure 25-4
Lineman working "bare-handed" on a 735 kV line. He is wearing a special conductive suit so that his body is not subjected to high differences of potential. In the position shown, his potential with respect to ground is about 200 kV. (*Hydro-Québec*)

equipped with cross-arms, but for higher voltages, a wooden H-frame must be used. The wood is treated with creosote or special metallic salts to prevent it from rotting. For very high voltage lines, we always use steel towers made of galvanized angle-iron pieces that are bolted together.

The spacing between conductors must be sufficient to prevent arc-over under gusty wind conditions. The spacing has to be increased as we increase the distance between towers and as the line voltages becomes higher.

25.5 Construction of a line

Once we know the conductor size, the height of the poles and the distance between the poles (span), we can direct our attention to stringing the conductors. A wire supported between two points (Fig. 25-5) does not remain horizontal, but loops down at the middle. The vertical distance between the straight line joining the points of support and the lowest point of the conductor is called *sag*. The tighter the wire, the smaller the sag will be.

Before undertaking the actual construction of a line, it is important to calculate the permissible sag and the corresponding mechanical pull. Among other things, the summer to winter temperature range must be taken into account because the length of the conductor varies with the temperature. Thus, if the line is strung in the winter, the sag must not be too great, otherwise the wire will stretch even more during the summer heat, with the result that the clearance to ground may no longer be safe. On the other hand, if the line is installed in the summer, the sag must not be too small otherwise the wire, contracting in winter, may become so dangerously tight as to snap. Wind and sleet add even more to the tractive force, which may also cause the wire to break (Fig. 25-6)

25.6 Galloping lines

If a coating of sleet is deposited on a line during windy conditions, the line may begin to oscil-

Figure 25-5
Span and sag of a line.

late. Under certain conditions, the oscillations may become so large that the line is seen to actually "gallop". Galloping lines can produce short-circuits between phases or snap the conductors. To eliminate the problem we sometimes load the line with special mechanical weights, to dampen the oscillations or to prevent them from building up.

Figure 25-6
During winter, steel towers must carry the combined weight of conductors and accumulated ice. (*Hydro-Québec*)

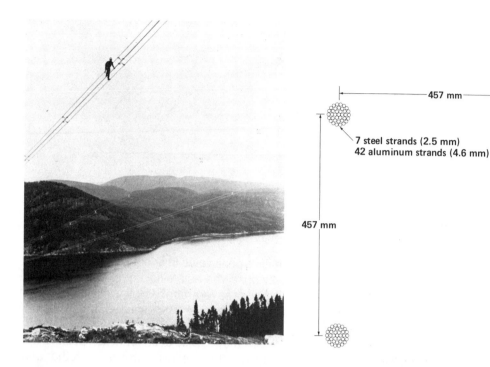

Figure 25-7a
Four bundled conductors make up this phase of a 3-phase, 735 kV line.

Figure 25-7b
Details of the bundled conductors.

25.7 Corona effect – radio interference

The very high voltages in use today produce a continual electrical discharge around the conductors, owing to local ionization of the air. This discharge, or corona effect, produces losses over the entire length of the transmission line. In addition, corona emits high frequency noise which interferes with nearby radio receivers and TV sets. To diminish corona, we must reduce the electric field (V/m) around the conductors, either by increasing their diameter or by arranging them in sets of two, three or more bundled conductors per phase (see Figs. 25-7a and 25-7b). This *bundling* arrangement also reduces the inductance of the line, enabling it to carry more power. This constitutes an important additional benefit.

25.8 Pollution

Dust, acids, salts, and other pollutants in the atmosphere settle on insulators and reduce their insulating properties. Insulator pollution may produce short-circuits during storms or under momentary overvoltage conditions. The possibility of service interruption and the necessity to clean insulators periodically is therefore a constant concern to the utility company.

And added to pollution there is the problem of lightning, which we now discuss.

25.9 Lightning strokes

During stormy weather, by a process not yet fully understood, a charge separation takes place inside certain clouds, so that positive

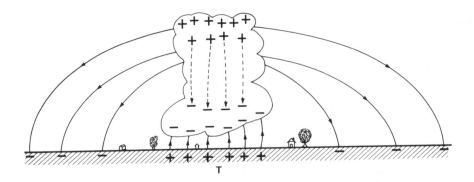

Figure 25-8
Electric fields created by a thundercloud.

charges move to the upper part of the cloud while negative charges stay below (Fig. 25-8). This transfer of electric charge sets up an electric field within the cloud. Furthermore, the negative charge at the base of the cloud repels the free electrons on the ground below. Consequently, region T becomes positively charged, by induction. It follows that an electric field and a difference of potential will also be established between the base of the cloud and the earth. Finally, yet another electric field exists between the electrons repelled from region T, and the positive charge at the top of the cloud.

As more and more positive charges move upward within the cloud, the electric field below the cloud becomes more and more intense. Ultimately, it reaches the critical ionization level where air begins to break down. Ionization takes place first at the tips of church spires and the top of high trees, and may sometimes give rise to a bluish light. Mariners of old observed this light around the masts of their ships and called it St. Elmo's fire.

When the electric field becomes sufficiently intense, lightning will suddenly strike from cloud to earth. A single stroke may involve a charge transfer of from 0.2 to 20 coulombs, under a difference of potential of several hundred million volts. The current per stroke rises to peak in one or two microseconds and falls to

half its peak value in about 50 μs. What is visually observed as a single stroke, is often composed of several strokes following each other in rapid succession. The total discharge time may last as long as 200 ms.

Discharges also occur between positive and negative charges inside the cloud, rather than between the base of the cloud and ground.

The thunder we hear is produced by a supersonic pressure wave. It is created by the sudden expansion of air around the intensely-hot lightning stroke.

25.10 Lightning arresters on buildings

The simplest lightning arresters are metallic rods which rise above the highest point of a building, channeling the lightning towards a ground electrode by means of a conducting wire. This prevents the high current from passing through the building itself, which might cause a fire, or endanger its occupants. A lightning arrester and anything connected to it can be dangerous to the touch; during a discharge, a very high voltage can exist between the protective system and ground.

Much more sophisticated lightning arresters are used on electrical utility systems. They divert lightning and high-voltage switching surges to ground before they damage costly and critical electrical equipment.

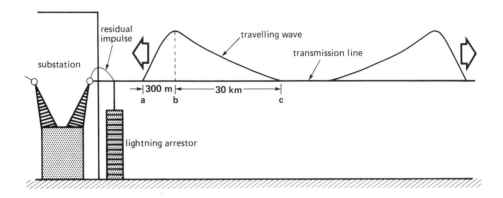

Figure 25-9
Flow of electric charge along a transmission line.

25.11 Lightning and transmission lines

When lightning makes a direct hit on a transmission line, it deposits a large electric charge, producing an enormous overvoltage between the line and ground. The dielectric strength of air is immediately exceeded and a flashover occurs. The line discharges itself and the overvoltage disappears in typically less than 50 μs.

Unfortunately, the arc between the line and ground, (initiated by the lightning stroke) produces a highly ionized path which behaves like a conducting short-circuit. Consequently, the normal ac line voltage immediately delivers a large ac current which follows the ionized path. This *follow-through* current sustains the arc until the circuit-breakers open at the end of the line. The fastest circuit-breakers will trip in about 1/15th of a second, which is almost 1000 times longer than the duration of the lightning stroke itself.

Direct hits on a transmission line are rare; more often, lightning will strike the overhead ground wire that shields the line. In the latter case, a local charge still accumulates on the line, producing a very high local overvoltage. This concentrated charge immediately divides into two waves that swiftly move in opposite directions at close to the speed of light (300 m/μs). The height of the impulse wave

represents the magnitude of the so-called surge-voltage that exists from point to point, between the line and ground (Fig. 25-9). The peak voltage (corresponding to the crest of the wave) may attain one or two million volts. Wave front **ab** is concentrated over a distance of about 300 m, while tail **bc** may stretch out over several kilometers.

The wave also represents the point-to-point value of the current flowing in the line. For most aerial lines, the ratio between surge voltage and surge current corresponds to a resistance of about 400 Ω. A surge voltage of 800 000 V at a given point is therefore accompanied by a surge current of 800 000/400 = 2000 A.

As the wave travels along the line, the I^2R and corona losses gradually cause it to flatten out, and the peak of the surge voltage decreases.

Should the wave encounter a line insulator, the latter will be briefly subjected to a violent overvoltage. The overvoltage period is equal to the time it takes for the wave to sweep past the insulator. The voltage rises from its nominal value to several hundred kilovolts in about 1 μs, corresponding to the length of wavefront **ab**. If the insulator cannot withstand this overvoltage, it will flash over, and the resulting follow-through current will cause the circuit-breakers to trip. On the other hand, if the insulator does not fail, the wave will continue to travel along the line until it eventually encounters a substation. It is here that the impulse wave can produce real havoc. The windings of transformers, synchronous condensers, reactors, etc., are seriously damaged when they flash over to ground. Expensive repairs and even more costly shut-downs are incurred while the apparatus is out of service. The overvoltage may also damage circuit-breakers, switches, insulators, relays, etc., which make up a substation. To reduce the impulse voltage on station apparatus, lightning arresters are installed on all incoming lines.

Lightning arresters are designed to clip off all voltage peaks that exceed a specified level, say 400 kV. In turn, the apparatus within the substation is designed to withstand an impulse voltage considerably higher than the arrester "clipping voltage", say 550 kV. Consequently, if a 1000 kV surge voltage enters a substation, the station arrester diverts a substantial part of the surge energy to ground. The residual impulse wave which travels beyond the arrester then has a peak of only 400 kV. This impulse can easily be borne by station apparatus built to withstand an impulse of 550 kV.

25.12 Basic impulse insulation level (BIL)

How do insulating materials react to impulse voltages? Tests have shown that the withstand capability increases substantially when voltages are applied for very brief periods. To illustrate, suppose we wish to carry out an insulation test on a transformer, by applying a 60 Hz sinusoidal voltage between the windings and ground. As we slowly raise the voltage, a point will be reached where breakdown occurs. Let us assume that the breakdown voltage is 46 kV (RMS) or 65 kV crest.

On the other hand, if we apply a dc impulse of short duration between the windings and ground, we discover that it takes about twice the peak voltage (or 130 kV) before the insula-

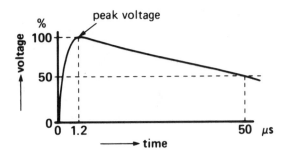

Figure 25-10
Standard shape of impulse voltage used to determine the BIL rating of electrical apparatus.

| TABLE 25B | TYPICAL PEAK VOLTAGES FOR |
| | 1.2 x 50 μs BIL TESTS |

Values are in kilovolts		
1550	825	250
1425	750	200
1300	650	150
1175	550	110
1050	450	90
900	350	30

tion breaks down. The same is true of suspension insulators, bushings, spark gaps, etc., except that the ratio between impulse voltage and crest ac voltage is closer to 1.5.

In the interest of standardization, and to enable a comparison between the impulse withstand capability of similar devices, standards organizations have precisely defined the shape and crest values of impulse waves. Figure 25-10 shows such a standard impulse wave. It attains its peak after 1.2 μs and falls to one-half the peak in 50 μs. The peak voltage has a defined set of values that range from 30 kV to 1550 kV (see Table 25B).

Figure 25-11
A 4 000 000 V impulse causes a flashover across an insulator string rated at 500 kV, 60 Hz. Such impulse tests
increase the reliability of equipment in the field. The powerful impulse generator in the center of the photo is 24 m high
and can deliver 400 kJ of energy at a potential of 6.5 mV
(*IREQ*)

The peak voltage is used to specify the basic impulse insulation levels (BIL) of equipment. Thus, a piece of equipment (transformer, insulator, capacitor, resistor, bushing, etc.) that can withstand a 1.2 × 50 microsecond wave of 900 kV, is said to possess a basic impulse insulation level (or BIL) of 900 kV. Figure 25-11 shows an insulator string being subjected to a BIL impulse test.

The BIL of a device is sometimes several times higher than its nominal ac operating voltage. For example, a 69 kV distribution transformer must have a BIL of 350 kV. However, there is no special relationship between BIL and nominal voltage. As the BIL rises, we must increase the amount of insulation which, in turn, increases the size and cost of equipment.

In conclusion, the peak voltage across an arrester must never exceed the BIL of the apparatus it is intended to protect.

25.13 Ground wires

In Figure 25-6, we see two bare conductors supported at the very top of the transmission-line towers. These conductors, called *ground wires*, serve to shield the line and intercept lightning strokes before they hit the current-carrying conductors below. Grounding wires normally do not carry current; consequently, they are often made of steel. They are solidly connected to ground at each tower.

25.14 Tower grounding

Transmission-line towers are always solidly connected to ground. Great care is taken to ensure that the ground resistance is low. In effect, when lightning hits a line, it creates a sudden voltage rise across the insulators as the lightning current discharges to ground. Such a voltage rise may produce a flashover across the insulators and a consequent line outage, as shown by the following example.

Example 25-1
A 3-phase 69 kV transmission line having a BIL of 350 kV is supported on steel towers and protected by a circuit-breaker (Fig. 25-12). The ground resistance at each tower is 20 Ω whereas the neutral of the transmission line is *solidly* grounded at the transformer just ahead of the circuit-breaker. During an electric storm, one of the towers is hit by a lightning stroke of 20 kA.
a. Calculate the voltage across each insulator string under normal conditions;
b. Describe the sequence of events during and after the lightning stroke.

Solution:
a. Under normal conditions, the line-to-neutral voltage is 69 kV/$\sqrt{3}$ = 40 kV and the current flowing in the tower ground resistance is zero. The steel tower is therefore at the same potential as the ground. It follows that the voltage across each insulator string (line to tower) is 40 kV (RMS).
b. When lightning strikes the tower, the voltage across the ground resistance suddenly leaps to 20 kA × 20 Ω = 400 kV. The voltage between the tower and *solid* ground is therefore 400 kV, and so the potential difference across all three insulator strings jumps to the same value. Because this impulse exceeds the insulator BIL of 350 kV, a flashover immediately occurs across the insulators, short-circuiting all three lines to the steel cross-arm. The resulting 3-phase short-circuit initiated by the lightning stroke will continue to be fed and sustained by a heavy follow-through current from the 3-phase source. This short-circuit current I_{sc} will trip the circuit breaker, producing a line outage.

In view of the many customers affected by such a load interruption, we try to limit the number of outages by ensuring a low resistance between the towers and ground. In the prece-

Figure 25-12
Flashover produced by a lightning as it flows to ground.

ding example, if the tower resistance had been 10 Ω instead of 20 Ω, the impulse voltage across the insulators would have risen to 200 kV and no flashover would have occurred.

Note that lightning currents of 20 kA are quite frequent, and they last for only a few microseconds.

25.15 Fundamental objectives of a transmission line

The *fundamental* purpose of a transmission or distribution line is to carry *active* power from one point to another. If it also has to carry reactive power, the latter should be relatively small. In addition, a transmission line should possess the following basic characteristics:

1. The voltage should remain as constant as possible over the entire length of the line, from source to load, and for all loads between zero and rated load;
2. The line losses must be small so as to attain a high transmission efficiency;
3. The I^2R losses must not overheat the conductors.

If the line alone cannot satisfy the above requirements, supplementary equipment, such as capacitors and inductors, must be added until they are met.

25.16 Equivalent circuit of a line

In spite of their great differences in power rating, voltage levels, lengths and mechanical construction, transmission lines possess similar electrical properties. In effect, an ac line possesses a resistance R, an inductive reactance X_L, and a capacitive reactance X_C. These impedances are uniformly distributed over the entire length of the line; consequently, we can represent the line by a series of identical sections, as shown in Figure 25-13. Each section represents a portion of the line (1 km, for example), and elements r, x_L, x_C represent the impedances corresponding to this unit length.

We can simplify the circuit of Figure 25-13 by lumping the individual resistances r together to yield a total resistance R. In the same way, we obtain a total inductive reactance X_L equal to the sum of the individual reactances x_L. On the other hand, the total capacitive reactance X_C is equal to the sum of the x_C reactances connected in parallel. It is convenient to assume

Figure 25-13
Distributed impedance of a transmission line

that the total capacitive reactance X_C of the line is composed of two parts, each having a value $2X_C$ located at each end of the line. The resulting equivalent circuit of Figure 25-14 is a good approximation of any 50 Hz or 60 Hz power line, provided its length is less than 250 km. Note that R and X_L increase as the length of the line increases, whereas X_C decreases with increasing length.

The equivalent circuit of Figure 25-14 can also be used to represent one phase of a 3-phase line. Current I corresponds to the actual current flowing in one conductor and E is the voltage between the same conductor and neutral.

25.17 Typical impedance values

Table 25C gives typical values of the inductive and capacitive reactances per kilometer for practical transmission lines operating at 60 Hz. Surprisingly, the respective impedances per unit length are reasonably constant for all aerial lines. Thus, x_L is about 0.5 Ω/km and x_C is about 300 000 Ω/km whether the transmission line voltage is high or low, or whether the power is great or small.

The same can be said of underground cables except that the inductive and capacitive reactances of three-phase cables are much smaller. Thus, x_C is about one hundred times smaller than that of aerial lines whereas x_L is about five times smaller. This fact has a direct bearing on the maximum distance that ac power can be transmitted by cable.

The resistance per unit length varies greatly with conductor size and so it is impossible to

Figure 25-14
Equivalent lumped circuit of a transmission line.

TABLE 25C	TYPICAL IMPEDANCE VALUES PER KILOMETER FOR 3-PHASE, 60 Hz LINES	
type of line	x_L Ω	x_C Ω
aerial line;	0.5	300 000
underground cable	0.1	3 000

give a "typical" value. Table 25D gives the resistance as well as the ampacity for several aerial conductors.

Example 25-2:
A 3-phase 230 kV transmission line having a length of 50 km is composed of three ACSR conductors having a cross section of 1000 MCM. The voltage of the source is 230 kV (line-to-line) and that at the load is 220 kV.
a. Determine the equivalent circuit, per phase;
b. Draw the complete equivalent circuit of the 3-phase line.

TABLE 25D		RESISTANCE AND AMPACITY OF SOME BARE AERIAL CONDUCTORS			
conductor size		**resistance per conductor at 75°C**		**ampacity in free air***	
AWG	cross section mm^2	copper Ω/km	ACSR Ω/km	copper A	ACSR A
10	5.3	3.9	6.7	70	—
7	10.6	2.0	3.3	110	—
4	21.1	0.91	1.7	180	140
1	42.4	0.50	0.90	270	200
3/0	85	0.25	0.47	420	300
300 MCM	152	0.14	0.22	600	500
600 MCM	304	0.072	0.11	950	750
1000 MCM	507	0.045	0.065	1300	1050

* The ampacity indicated is the maximum that may be used without weakening the conductor by overheating. In practice, the actual line current may be only 25% of the indicated value.

Solution:

a. Referring to Tables 25C and 25D, the line impedances per unit length are about:

$r = 0.065$ Ω/km;
$X_L = 0.5$ Ω/km;
$X_C = 300$ kΩ/km

The line impedances *per phase* are:

$R = 0.065 \times 50 = 3.25$ Ω
$X_L = 0.5 \times 50 = 25$ Ω
$X_C = 300\ 000/50 = 6000$ Ω

The capacitive reactance at each end of the line is:

$2X_C = 2 \times 6000 = 12$ kΩ

The voltage per phase is $230/\sqrt{3} = 133$ kV at the source and $220/\sqrt{3} = 127$ kV at the load.

The equivalent circuit per phase is given in Figure 25-15.

b. The complete equivalent circuit of the 3-phase line is shown in Fig.25-16. Note that the line capacitance acts as if it were composed of six capacitors connected between the lines and ground. This circuit arrangement holds true even when the neutrals of the source and load are not grounded. The result is that the voltage across each capacitor is given by $E_c = E/\sqrt{3}$ where E is the respective line voltage at the source or load. Thus, knowing that the line voltage is 230 kV at the source and 220 kV at the load, the voltages across the respective capacitors are:

$E_{CS} = 230/\sqrt{3} = 133$ kV
$E_{CL} = 220/\sqrt{3} = 127$ kV

Figure 25-15
Equivalent circuit of one phase (Example 25-2).

Figure 25-16
Equivalent circuit of a 3-phase line.

25.18 Simplifying the equivalent circuit

We can sometimes simplify the equivalent circuit of a transmission line by eliminating one, two or all the elements shown in Figure 25-14. The validity of this simplification depends upon the relative magnitude of the active and reactive powers P_J, Q_L, Q_C associated with the line, compared to the active power P which it delivers to the load. Referring to Figure 25-17, these powers are:

P = active power absorbed by the load;
P_J = I^2R, active power dissipated in the line;

Q_L = I^2X_L, reactive power absorbed by the line;
Q_C = E^2/X_C, reactive power generated by the line. (We assume the source and load voltages have the same magnitude)

If one of these powers P_J, Q_L, Q_C is negligible compared with the active power P, we can neglect the corresponding circuit element that produces it. For example, low-voltage lines are always short; consequently, X_C is high and so E^2/X_C is negligible. This permits us to represent low-voltage lines by the circuit shown in Figure 25-18. On the other hand, long, high-voltage lines can be represented by the circuit of Figure 25-19 because the I^2R losses are relatively small compared to P whereas the reactive powers Q_L and Q_C are not.

Example 25-3:
The transmission line shown in Figure 25-16 delivers 300 MW to the 3-phase load. If the line voltage at both the sending end (source) and receiving end (load) is 230 kV, determine:
a. the active and reactive powers associated with the line;
b. the approximate equivalent circuit, per phase.

Solution:
Referring to Figure 25-20a, we have:
a. The line-to-neutral voltage at the load is:
E = $230/\sqrt{3}$ = 133 kV

Figure 25-17
Active and reactive powers of a transmission line.

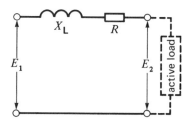

Figure 25-18
Equivalent circuit of a short LV line.

Figure 25-19
Equivalent circuit of a long HV line.

The active power transmitted to the load per phase is:

$$P = 300 \text{ MW}/3 = 100 \text{ MW}$$

The load current is:

$$I = 100 \text{ MW}/133 \text{ kV} = 750 \text{ A}$$

If we temporarily neglect the presence of the 12 kΩ capacitor in parallel with the load, the line current is equal to the load current (Fig. 25-20b).

I^2R losses in the line are:

$$P_J = I^2R = 3.25 \times 750^2$$
$$= 1.83 \text{ MW (1.8 percent of } P)$$

Reactive power absorbed by the line is:

$$Q_L = I^2X_L = 25 \times 750^2$$
$$= 14.1 \text{ Mvar (14 percent of } P)$$

Reactive power generated by the line at each end is:

$$Q = E^2/X_C = 133\,000^2/12\,000$$
$$= 1.47 \text{ Mvar}$$

Total reactive power generated by the line is:

$$Q_C = 2 \times 1.47$$
$$= 3 \text{ Mvar (3 percent of } P)$$

b. Comparing the relative values of P_J, Q_L, Q_C and P, it is clear that we can neglect the resistance and the capacitance of the line. The resulting equivalent circuit is a simple inductive reactance of 25 Ω (Fig. 25-20c).

(a)

(b)

(c)

Figure 25-20
Progressive simplification of a 735 kV line (Example 25-3).

25.19 Voltage regulation and power transmission capability of transmission lines

Voltage regulation and power-handling capacity are two important features of a transmission line. Thus, the voltage of a transmission line should remain as constant as possible even under variable load conditions. Ordinarily, the voltage regulation from zero to full load should not exceed ± 5% of the nominal voltage (though we can sometimes accept a regulation as high as ± 10%).

As regards power-handling capacity, it may come as a surprise that a transmission line can deliver only so much power and no more. The power that can be transported from source to load depends upon the impedance of the line. We are mainly interested in transmitting active power because only it can do useful work. In order to determine the voltage regulation and to establish their power transmission capability, we now examine four types of lines:

1. resistive line;
2. inductive line;
3. inductive line with compensation;
4. inductive line connecting two large systems.

In our analysis, the lines connect a load (or receiver) R to a source (or sender) S. The load can have all possible impedance values, ranging from no-load to a short-circuit. However, we are only interested in the *active* power the line can transmit. Consequently, the load can be represented by a variable resistance absorbing a power P. The sender voltage E_s is fixed but the receiver voltage E_R depends upon the resistance of the load.

25.20 Resistive line

The transmission line of Figure 25-21a possesses a resistance R. Starting from an open circuit, we gradually reduce the load resistance until it becomes zero. During this process, we observe the receiver voltage E_R across the load, as well as the active power P it absorbs. If

(a)

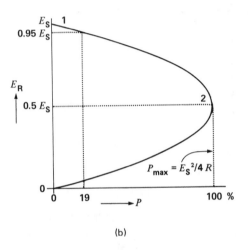

(b)

Figure 25-21
Characteristics of a resistive line.

numerical values were given, a few simple calculations would enable us to draw a graph of E_R as a function of P. However, we prefer to use a generalized curve that shows the relationship between E_R and P for *any* transmission line having an arbitrary resistance R.

The generalized shape of this graph is given in Figure 25-21b. It reveals the following information:

a. There is an upper limit to the power the line can transmit to the load. In effect,
$$P_{max} = E_S^2/4R \qquad (25\text{-}1)$$
and this maximum is reached when the receiver voltage is
$$E_R = 0.5E_S.$$

b. The power delivered to the load is maximum

when the impedance of the load is equal to the resistance of the line.

c. If we permit a maximum regulation of 5 percent ($E_R = 0.95E_S$), the line can carry a load that is only 19% of P_{max}. The line could transmit more power, but the customer voltage E_R would then be too low.

Note that the sender must furnish the power P absorbed by the load, plus the I^2R losses in the line.

Example 25-4:
A single-phase transmission line having a resistance of 10 Ω is connected to a fixed sender voltage of 1000 V.

Calculate:
a. the maximum power the line can transmit to the load;
b. the receiver power for a receiver voltage of 950 V.

Solution:
a. The maximum power that can be transmitted is:
$$P_{max} = E_S^2/4R$$
$$= 1000^2/(4 \times 10)$$
$$= 25 \text{ kW}$$

b. when $E_R = 950$ V, the voltage drop in the line is:
$$E_S - E_R = 1000 - 950 = 50 \text{ V}$$

The line current is therefore:
$$I = (E_S - E_R)/R = 50/10$$
$$= 5 \text{ A}$$

The receiver power is:
$$P = E_S I = 950 \times 5 = 4750 \text{ W}$$
$$= 4.75 \text{ kW}$$

Note that 4.75 kW/25 kW = 0.19 or 19%, which is the percentage predicted by the curve of Figure 25-21b.

25.21 Inductive line

Let us now consider a line having negligible

(a)

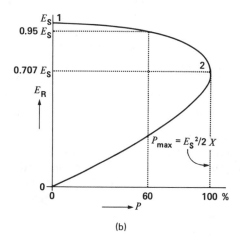

(b)

Figure 25-22
Characteristics of an inductive line.

resistance but possessing an inductive reactance X (Fig. 25-22a). The receiver again operates at unity power factor, and so it can be represented by a variable resistance absorbing a power P. As in the case of a resistive line, voltage E_R diminishes as the load increases, but the regulation curve has a different shape (Fig. 25-22b). In effect, the generalized graph of E_R as a function of P reveals the following information:

a. The line can transmit a maximum power to the load given by
$$P_{max} = E_S^2/2X. \tag{25-2}$$

The corresponding receiver voltage is,
$$E_R = 0.707E_S.$$

Thus, for a given line impedance and sender

voltage, the reactive line can deliver twice as much power as a resistive line can (compare $P = E_S^2/2X$ and $P = E_S^2/4R$).

b. The power delivered to the load is maximum when the resistance of the load is equal to the reactance of the line.

c. If we again allow a maximum regulation of 5 percent, we discover the line can carry a load that is 60% of P_{max}. Thus, for a given line impedance, and a regulation of 5 percent, the inductive line can transmit *six* times as much active power as a resistive line can.

The sender has to supply the active power P consumed by the load plus the reactive power I^2X absorbed by the line.

Example 25-5:
A single-phase transmission line having an inductive reactance of 10 Ω is connected to a fixed sender voltage of 1000 V. Calculate:

a. the maximum active power that the line can deliver to a resistive load.
b. the corresponding receiver voltage.
c. the receiver power when the receiver voltage is 950 V.

Solution:

a. The maximum power that can be transmitted to the load is:

$$P_{max} = E_s^2/2X$$
$$= 1000^2/2 \times 10$$
$$= 50 \text{ kW}$$

b. According Figure 25-22b, the receiver voltage at maximum power is:
$$E_R = 0.707E_S = 0.707 \times 1000$$
$$= 707 \text{ V}$$

c. In order to calculate the receiver power when $E_R = 950$ V, we first calculate the value of the current I. Taking $E_R = 950$ V as the reference phasor, we draw the phasor diagram for the circuit of Figure 25-23.

Current I is in phase with E_R because the

Figure 25-23
See Example 25-5.

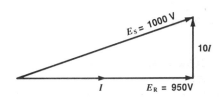

Figure 25-24
See Example 25-5.

load is resistive. Furthermore, we can write:
$$E_S = E_R + jIX$$
$$E_S = 950 + 10jI$$

This equation corresponds to the phasor diagram of Figure 25-24. From this diagram we can write, by inspection:

$$E_S^2 = E_R^2 + (10I)^2$$
$$1000^2 = 950^2 + 100I^2$$

from which

$$I = \sqrt{975} = 31.22 \text{ A}$$

The power to the receiver is therefore:

$$P = E_sI = 950 \times 31.22$$
$$= 29.66 \text{ kW}$$

Note that 29.66 kW is equal to 60 percent of P_{max} (50 kW) as predicted by the curve of Figure 25-22.

25.22 Compensated inductive line

We can improve the regulation and power-handling capacity of an inductive line by adding a variable capacitive reactance X_C across the load (Figure 25-25a). All we have to do is to adjust the value of X_C so that the reactive power E_S^2/X_C supplied by the capacitor is at all times equal to *one half* the reactive power I^2X absorbed by the line. For such a compensated line the value of the receiver voltage E_R will always be equal to the sender voltage E_S, irrespective of the active power P absorbed by the load.

However, there still is an upper limit to the power the line can transmit. A detailed analysis shows that we can maintain a constant load voltage ($E_R = E_S$) up to a maximum of

$$P_{max} = E_S^2/X \qquad (25\text{-}3)$$

Beyond this limit, E_R gradually decreases to zero in a diagonal line, as shown by the graph of Figure 25-25b. Note that:

a. The voltage regulation is perfect until the load power reaches the limiting value $P_{max} = E_S^2/X$.

b. The compensated inductive line can deliver twice as much power (P_{max}) as an uncompensated line can. Moreover, it has the advantage of maintaining a constant load voltage.

Capacitor X_C supplies one-half the reactive power I^2X_L absorbed by the line; the remaining half is supplied by the sender E_S. If necessary, we can add a second capacitor X_C (shown dotted in Fig. 25-25) at the input to the line. The source has then only to supply the active power P, while the reactive power is supplied by the capacitors at both ends.

Example 25-6:
A single phase line possesses an inductive reactance X of 10 Ω and is connected to a fixed sender voltage of 1000 V. If it is fully compensated, calculate:

(a)

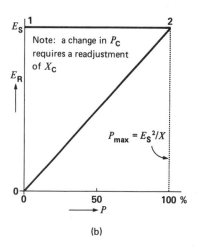

(b)

Figure 25-25
Characteristics of a compensated inductive line.

a. the maximum active power that the line can deliver to a resistive load.

b. the capacitive reactance that must be installed on the receiver side.

c. the capacitive reactance that must be installed on the receiver side when the active power is 40 kW.

Solution:

a. The maximum power that can be transmitted to the receiver is:

$$P_{max} = E_S^2/X = 1000^2/10$$
$$= 100 \text{ kW}$$

b. Figure 25-26 shows the compensated line with the capacitive reactance carrying a current I_C. The phasor diagram of Figure 25-27

Figure 25-26
See Example 25-6.

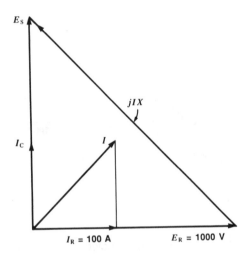

Figure 25-27
See Example 25-6.

gives us the key to solving the value of X_C that is required. Using $E_R = 1000$ V as the reference phasor we reason as follows:

The current in the resistive load is:
$$I_R = P_{max}/E_R = 100\ 000/1000$$
$$= 100\ A.$$

The current I_C leads E_R by 90°.

The reactive power generated by X_C is:
$$Q_C = E_R I_C$$
$$= 1000 I_C$$

The current I in the line is:
$$I = \sqrt{I_C^2 + I_R^2}$$
$$= \sqrt{I_C^2 + 100^2}$$

The reactive power absorbed by the line is:
$$Q_L = I^2 X = (I_C^2 + 100^2)10$$

In a fully compensated line we have:
$$Q_C = 0.5 Q_L$$

Consequently:
$$1000 I_C = 5(I_C^2 + 100^2)$$
$$5I_C^2 - 1000 I_C + 5 \times 10^4 = 0$$

Solving this quadratic equation we find:
$$I_C = 100\ A$$

The value of X_C is given by:
$$X_C = E_R/I_C$$
$$= 1000/100$$
$$= 10\ \Omega$$

c. When the load is 40 kW, the load current is:
$$I_R = P/E_R = 40\ 000/1000$$
$$= 40\ A$$

Letting the current in the capacitor be I_C, the resulting line current is:
$$I = \sqrt{I_C^2 + I_R^2}$$
$$= \sqrt{I_C^2 + 40^2}$$

The reactive power generated by X_C is:
$$Q_C = I_C E_R$$
$$= 1000 I_C$$

The reactive power absorbed by the line is:
$$Q_L = I^2 X = (40^2 + I_C^2)10$$

Full compensation requires that
$$Q_C = 0.5 Q_L$$
$$1000 I_C = 5(40^2 + I_C^2)$$

Solving this equation, we find:
$$I_C = 8.35\ A$$

The value of X_C is:
$$X_C = \frac{E_R}{I_C} = \frac{1000}{8.35} = 119.8\ \Omega$$

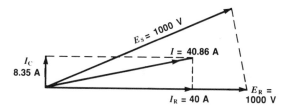

Figure 25-28
See Example 25-6.

The circuit and phasor diagram for this compensated line is given in Figure 25-28.

We note that although the values of E_R and E_S are the same, E_R lags considerably behind E_S.

25.23 Inductive line connecting two systems

Large cities and other regional users of electrical energy are always interconnected by one or more transmission lines. Such a network improves the stability of the system and enables it to better endure momentary short-circuits and other disturbances. Interconnecting lines also permit energy exchanges between electrical utility companies.

On such lines, the voltages at each end remain essentially independent of each other, both in value and in phase. In effect, because of their enormous power the regional consumers at each end act as independent, infinite buses.* Figure 25-29 shows the equivalent circuit of an inductive line connecting two such regional consumers S and R. We assume the terminal voltages E_S and E_R are fixed, each possessing

* See Section 16-23.

the same magnitude E. Regarding the exchange of active power between the two regional consumers, we examine three distinct cases:

1. E_S and E_R in phase;
2. E_S leading E_R by an angle δ
3. E_S lagging E_R by an angle δ.

1. **E_S and E_R in phase.** In this case, the line current is zero and no power is transmitted.

2. **E_S leads E_R by an angle δ** (Fig. 25-29). Region S supplies power to region R and, from the phasor diagram, we can prove (Sec. 16-23) that the active power transmitted is given by:

$$P = \frac{E^2}{X} \sin \delta \qquad (25\text{-}4)$$

where

P = active power transmitted per phase [MW]†

E = line-to-neutral voltage [kV]

X = inductive reactance of the line, per phase [Ω]

δ = phase angle between the voltages at each end of the line [°]

Figure 25-29
E_S leads E_R.

† In this equation, if E represents the line voltage, P is the *total* power transmitted by the three phases.

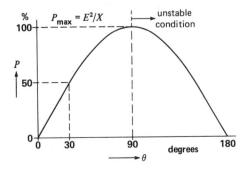

Figure 25-30a
Power versus angle characteristic.

Figure 25-30b
Voltage versus power characteristic.

Figure 25-30a shows the active power transmitted as a function of the voltage phase angle between the two regions. Note that the power increases progressively and attains a maximum value of E^2/X when the phase angle is 90°. In effect, just as in the other transmission lines we have studied, a line connecting two power centers can transmit only so much power and no more. The power limit is the same as that of a compensated inductive line. Although we can still transmit power when the phase angle exceeds 90°, we avoid this condition because it corresponds to an unstable mode of operation. When δ approaches 90°, the two regions are at the point of "pulling apart" and the line circuit-breakers will trip.

Figure 25-30b shows the load voltage E_R as a function of the active power transmitted. It is simply a horizontal line that stretches to a maximum value $P_{max} = E^2/X$ before falling back again to zero (dotted line). This regulation curve should be compared with that of Figure 25-25b.

Note that the line voltage drop E_x is quite large, even though the terminal voltages E_S and E_R are equal. Referring to Figure 25-29, it is clear that the line voltage "drop" increases as the phase angle between E_S and E_R increases.

3. **E_S lags behind E_R by an angle** δ (Fig. 25-31). The active power has the same value as before, but it now flows in the opposite direction, from region R towards region S. The graph of active power versus phase angle is identical to that shown in Figure 25-30a.

If we compare Figures 25-29 and 25-31, we note that the direction of power flow does not depend upon the relative magnitudes of E_S and E_R (they are equal), but *only upon the phase angle between them*. On inductive lines, active

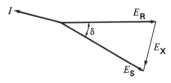

Figure 25-31
E_R leads E_S.

Figure 25-32
Comparison of the power-handling properties of various transmission lines.

power always flows from the leading to the lagging voltage side.

25.24 Review of power transmission

In summary, there is always a limit to the amount of power a line can transmit. The maximum power is proportional to the square of the sender voltage and inversely proportional to the impedance of the line. Figure 25-32 enables us to compare actual values of power and voltage for the four types of transmission lines we have studied. Each line possesses an impedance of 10 Ω and the sender furnishes a voltage E_S of 1000 V. It is clear that the E_R versus P curves become flatter and flatter as we progress from a resistive to an inductive to a compensated line.

The table next to the graph shows the maximum power that can be transmitted assuming a regulation of 5% or better. Thus, the resistive

line can transmit 4.75 kW, whereas the inductive line can transmit 30 kW.

Because all lines possess some resistance, we also show the voltage-power curve of a compensated line having a reactance of 9.8 Ω and a resistance of 2 Ω (curve 5). This line also has an impedance of 10 Ω, but the maximum power it can transmit drops to 80 kW, compared to 100 kW for a line possessing no resistance.

In practice, the voltages and powers are much higher than those given in these examples. Nevertheless, the method of analysis is the same.

25.25 Choosing the line voltage

We have seen that for a given transmission line and for a given voltage regulation, the power P that can be transmitted is proportional to E^2/Z, where E is the voltage of the line and Z, its impedance. Because Z is proportional to the length of the line, we deduce that the line voltage is given by:

$$E^2 = k'Pl$$

that is:

$$E = k\sqrt{Pl} \qquad (25\text{-}5)$$

where

E = 3-phase line voltage [kV]
P = power to be transmitted [kW]
l = length of the transmission line [km]
k = coefficient that depends on the type of line and the allowable voltage regulation. Typical values are:
$\quad k$ = 0.1 for an uncompensated line having a regulation of 5 percent
$\quad k$ = 0.06 for a compensated line

Equation 25-5 is very approximate, but it does give an idea of the magnitude of the line voltage E. The value finally chosen depends upon economic factors as well as technical considerations; in general, the actual voltage selected will lie between 0.6 E and 1.5 E.

Example 25-7:
Power has to be carried over a distance of 20 km to feed a 10 MW unity power factor load. If the line is uncompensated:
a. Determine the line voltage;
b. Select an appropriate conductor size;
c. Calculate the voltage regulation.

Solution:
a. Because the line is not compensated, we assume k = 0.1. A power of 10 MW is equal to 10 000 kW. Consequently:
$$E = k\sqrt{Pl} \qquad \text{Eq. 25-5}$$
$$= 0.1 \sqrt{10\ 000 \times 20}$$
$$= 44.7 \text{ kV}$$

Any voltage between 0.6 × 44.7 kV (=27 kV) and 1.5 × 44.7 kV (=67 kV) is feasible. We shall use a standard line voltage of 34.5 kV.

The line-to-neutral voltage is:
$$E = 34.5/\sqrt{3} = 19.9 \text{ kV}$$

b. The conductor size depends, first, upon the current to be carried. The line current is:
$$I = S/(\sqrt{3}E) \qquad \text{Eq. 8-9}$$
$$= 10 \times 10^6/(1.73 \times 34\ 500)$$
$$= 167 \text{ A}$$

According to Table 25D, we can use a No. 1 ACSR conductor:
$$\text{ampacity} = 200 \text{ A}$$
$$R = 0.9 \text{ }\Omega/\text{km} \times 20 \text{ km} = 18 \text{ }\Omega$$

From Table 25C we find:
$$X_L = 0.5 \text{ }\Omega/\text{km} \times 20 \text{ km} = 10 \text{ }\Omega$$

c. The IR drop in the line is:
$$IR = 167 \times 18 = 3006 \text{ V}$$

The IX_L drop in the line is:
$$IX_L = 167 \times 10 = 1670 \text{ V}$$

The line-to-neutral load voltage is 19 900 V. The complete circuit diagram per phase is given in Figure 25-33a. The corresponding phasor diagram is given in Figure 25-33b. The sender voltage may be calculated as follows:
$$E_S = \sqrt{(19\ 900 + 3006)^2 + 1670^2}$$
$$= 22\ 967 \text{ V}$$

If the load were removed, E_R would rise to 22 967 V. The voltage regulation is therefore:
$$\text{regulation} = (22\ 967 - 19\ 000)/22\ 967$$
$$= 3967/22\ 967$$
$$= 0.172 \text{ or } 17.3 \text{ percent}$$

25.26 Methods of increasing the power capacity

High-voltage lines are mainly inductive and they possess a reactance of about 0.5 Ω/km. This creates problems when we have to transmit large blocks of power over great distances. Suppose, for example, that we have to transmit 4000 MW over a distance of 400 km. The react-

Figure 25-33a
Transmission line under load.

Figure 25-33b
Corresponding phasor diagram.

ance of the line is 400 km × 0.5 Ω/km = 200 Ω per phase. Since the highest practical voltage is about 800 kV, the line can transmit no more than:

$$P_{max} = E^2/X$$
$$= 800^2/200 \qquad \text{Eq. 25-3}$$
$$= 3200 \text{ MW}$$

To transmit 4000 MW, the only solution is to use *two lines in parallel,* one beside the other. Note that doubling the size of the conductors would not help, because for such a line it is the *reactance* and not the resistance of the conductors that determines the maximum power that can be transmitted.

Additional lines are also useful to provide system security in the event that a parallel line trips out, due to a disturbance. Thus, if one line is lost, the scheduled power can still be carried by the remaining line.

To carry large blocks of power, we sometimes erect two, three and even four transmission lines in parallel, which follow the same corridor across the countryside (Fig. 25-34). In addition to high cost, the use of parallel lines often creates serious problems of land expropriation. Consequently, we sometimes use special methods to increase the maximum power of a line. In effect, when we can no longer increase the line voltage, we try to reduce the line reactance X_L by greatly increasing the effective size of the conductors. This is done by using two or more conductors per phase, kept apart by spacers. Such *bundled conductors*

can reduce the reactance by as much as 40 percent permitting an increase of 67 percent in the power-handling capability of the line.

Another method uses capacitors in series with the three lines to artificially reduce the value of X_L. With this arrangement, the maximum power is:

$$P_{max} = E^2/(X_L - X_{cs}) \qquad (25\text{-}6)$$

where X_{cs} is the reactance of the series capacitors per phase. Such series compensation is

Figure 25-34
Two 735 kV transmission lines in parallel carrying electrical energy to a large city. Each phase is composed of 4 bundled conductors (See Fig. 25-7). (*Hydro-Québec*)

also used to regulate the voltage of medium-voltage lines when the load fluctuates rapidly.

25.27 Extra high voltage lines

When electrical energy is transmitted at extra high voltages, special problems arise that require the installation of large compensating devices to regulate the voltage and to guarantee stability. Among these devices are synchronous capacitors, inductive reactors, static var compensators, and shunt and series capacitors.

To understand the need for such devices, and to appreciate the magnitude of the powers involved, consider a three-phase, 735 kV, 60 Hz line, having a length of 600 km. The line operates at 727 kV and the inductive and capacitive reactances are respectively 0.5 Ω and 300 kΩ for each kilometer of length. We shall first determine the equivalent circuit of the transmission line per phase:

Sender voltage per phase (line-to-neutral) is:

$$E_S = 727/\sqrt{3} = 420 \text{ kV}$$

Inductive reactance per phase is:
$$X_L = 0.5 \times 600 = 300 \ \Omega$$

Capacitive reactance per phase is:
$$X_C = 300 \text{ k}\Omega/600 = 500 \ \Omega$$

Equivalent capacitive reactance at each end of the line is:
$$X_{C1} = X_{C2} = 2 \times 500 \ \Omega$$
$$= 1000 \ \Omega$$

The equivalent circuit per phase is shown in Figure 25-35. Let us now study the behavior of the line under no-load and full-load conditions.

No-load operation. At no-load, the circuit formed by X_L in series with X_{C2} produces a partial resonance and the terminal voltage E_R rises to 600 kV. This represents an increase of 43 percent above the nominal voltage of 420 kV (Fig. 25-35). Such an abnormally high voltage is unacceptable. The only feasible way to reduce it is to connect an inductive reactance X_{L2} at the end of the line (Fig. 25-36). If we make X_{L2} equal to X_{C2}, the resulting parallel resonance brings voltage E_R back to 420 kV. In effect, the reactive power generated by X_{C2} ($420^2/1000 = 176$ Mvar) is entirely absorbed by X_{L2}. The latter must, therefore, have a capacity of 176 Mvar *per phase*.

Despite this inductive compensation, we still have a reactive power of 176 Mvar, generated by X_{C1}, which has to be absorbed by synchronous generator G. However, a capacitive load at the terminals of a generator creates overvoltages, unless we reduce the alternator exciting current (Sec. 16-13). But underexcitation is not recommended because it leads to instability. Consequently, we must install a second inductive reactance of 176 Mvar close to the generating station. In the case of very long transmission lines, several inductive reactances are installed along the line to distribute the inductive compensation evenly over its length.

Figure 25-35
EHV transmission line at no-load.

Figure 25-36
EHV reactor compensation.

Figure 25-37
Three large 110 Mvar, single-phase reactors, installed in a substation to compensate the line capacitance of a very long 3-phase 735 kV transmission line. (*Hydro-Québec*)

Inductive reactors (fixed or variable) are composed of a large coil placed inside a tank and immersed in oil (Fig. 25-37). A laminated steel core split up into a series of short air-gaps carries the magnetic flux. Intense magnetic forces are developed across the air gaps. On a 60 Hz system, these forces continuously oscillate between zero and several tons, at a mechanical frequency of 120 Hz. The core laminations and all metallic parts must therefore be firmly secured to reduce vibration and to limit noise to an acceptable level.

Operation under load, characteristic impedance. Returning again to the original uncompensated line on open-circuit

Figure 25-38
Surge impedance loading of a line.

(Fig. 25-35), let us connect a variable unity power factor load across the receiver terminals. If we progressively increase the megawatt load, receiver voltage E_R will gradually decrease from its open-circuit value of 600 kV and, for one particular load, it will become exactly equal to the sender voltage E_S (Fig. 25-38). This particular load is called the *surge-impedance load*. For most aerial lines, the impedance of this load corresponds to a line-to-neutral load resistance of about 400 Ω per phase. This particular load resistance (called surge impedance) is independent of the system frequency. The surge impedance load (SIL) of a 3-phase transmission line is therefore given by the approximate equation:

$$SIL = E^2/400 \qquad (25\text{-}7)$$

where

   SIL = surge impedance load [MW]
   E = 3-phase line voltage [kV]

In our example, the total surge-impedance load is approximately $727^2/400 = 1320$ MW.

When a transmission line delivers active power corresponding to its surge-impedance load, the reactive power generated by the

Figure 25-39
Static var compensator for HV line. (*General Electric*)

capacitance of the line is equal to that absorbed by its inductance. The line, in effect, compensates itself. If the load exceeds the surge-impedance load, we can keep E_R at 420 kV by adding extra capacitors at the receiver end of the line. However, the maximum power is still limited to the value given by Eq. 25-3, namely:

$$P_{max} = 3 \times (E^2/X)$$
$$= 3 \times (420^2/300)$$
$$= 1764 \text{ MW.}$$

On the other hand, if the load is less than the surge-impedance load, we must add inductive reactance at the receiver end of the line to maintain a constant voltage. Because the load continually changes throughout the day, we must continually vary the magnitude of the capacitive and inductive reactance to keep a steady voltage. This is done by means of static var compensators (Fig. 25-39), or rotating synchronous machines. The latter can deliver or absorb reactive power according to whether they are over- or under-excited (Sec. 17.15).

25.28 Power exchange between power centers

We sometimes have to install an additional

transmission line on systems that are already tightly interconnected. Such a line may be required to meet the energy needs of a rapidly-growing area or to improve the overall stability of the network. In such cases, we use special methods so that the additional line will transmit the required power.

Consider, for example, two major power centers A and B that are already interconnected by a grid of transmission lines (not shown) (Fig. 25-40). The respective voltages E_a and E_b are equal, but E_a leads E_b by an angle δ. If we decide to connect the two centers by an extra transmission line having a reactance X, the active power P will automatically flow from A to B because E_a leads E_b (see Sec. 25-23). Furthermore, phase angle δ and reactance X will completely dictate the magnitude of the power transmitted because $P = (E^2/X) \sin δ$.

However, the magnitude and direction of P may not correspond at all to what we want to achieve. For example, if we wish to transmit energy from region B to region A, the installation of a simple line will not do, for reasons we have just explained.

However, we can *force* a power exchange in one direction or the other by artificially modifying the phase angle between the two regions. All we have to do is to introduce a phase-shift autotransformer (Sec. 12.11) at one end of the line; by varying the phase angle of this transformer, we can completely control the active power flow between the two centers.

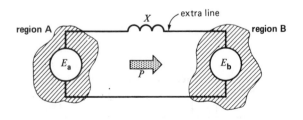

Figure 25-40
Power flow between two regions.

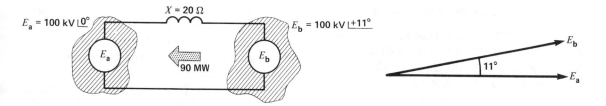

Figure 25-41a
An ordinary transmission line causes power to flow in the wrong direction.

Example 25-8:
Figure 25-41a shows the voltages and phase angle between two regions A and B that are already interconnected by a network (not shown). Voltage E_b is known to lead E_a by 11° and both voltages have a value of 100 kV. A new tie line having a reactance $X = 20\ \Omega$ connects the two regions. Calculate:
a. the power transmitted by the line and the direction of power flow, if no phase-shift transformer is employed;
b. the required phase-shift of the transformer so that the line will transmit 70 MW from A to B.

Solution:
a. The power transmitted is given by:

$$P = (E^2/X) \sin \delta \qquad \text{Eq. 25-4}$$
$$= (100^2/20) \sin 11° = 95\ \text{MW}$$

Because E_b leads E_a, the 95 MW will flow from B to A.

b. Let us first calculate the phase angle δ_1 required between opposite ends of the line, so that it will transmit 70 MW. We have:

$$P = (E^2/X) \sin \delta_1 \qquad \text{Eq. 25-4}$$
$$70 = (100^2/20) \sin \delta_1$$
$$\sin \delta_1 = 0.14$$

from which
$$\delta_1 = 8°$$

Consequently, voltage E_d produced by the phase-shift transformer must *lead* E_b by 8° in order that 70 MW may flow from A to B (Fig. 25-41b). Referring to the phasor diagram, and noting that E_b already leads E_a by 11°, it follows that E_d must be 11° + 8° = 19° ahead of E_a. The autotransformer T must therefore produce a phase-shift of 19° between its primary and secondary windings and the secondary voltage E_d must lead the primary voltage E_a. We can put the autotransformer at either end of the line or even in the middle.

Figure 25-41b
A phase-shift autotransformer can force power to flow in the desired direction (Example 25-8).

25.29 Practical example of power exchange

We now consider a practical application of a phase-shift transformer. To facilitate power exchange between the State of Connecticut and Long Island, New York, six 138 kV submarine cables (2 per phase) were installed between Norwalk and Northport, at the bottom of Long Island Sound (Fig. 25-42). Because the

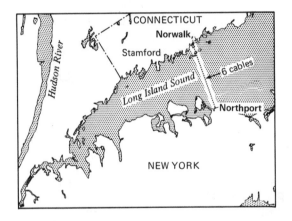

Figure 25-42
Six single-phase submarine cables join the State of Connecticut to Long Island.

two regions were already interconnected by a grid of transmission lines above ground, it was decided to install a phase-shift autotransformer at Northport to control a maximum power flow of 300 MW. Variable taps enable a maximum phase-shift of ± 25 degrees. On the other side, at Norwalk, a 300 MVA variable-voltage autotransformer was installed to provide voltage control of up to 10 percent, without phase-shift. By varying the phase-shift and the voltage at either end of the 19 km line, it is possible to control the power flow between the two regions, in one direction or the other, depending on the need. The following technical details point out the magnitude of the powers involved in such a cable transmission system.

Each of the six single-phase cables (Fig. 25-44) possesses a resistance of 1.3 Ω, an inductive reactance of 1.1 Ω and a capacitive reactance of 375 Ω. The latter can be represented by two reactances of 750 Ω at each end of the cable. The full-load current per cable is 630 A and the line-to-neutral voltage is 80 kV. Referring to the equivalent circuit diagram for one cable (Fig. 25.43), we can readily calculate the value of the inductive, capacitive and resistive powers associated with it. These powers Q_L, Q_C, P_J are listed in Table 25E.

In comparing them with the active power transmitted (300 MW), we can see that the capacitance of the cable is far more important

Figure 25-43
Equivalent circuit of each submarine cable.

TABLE 25E   POWERS IN A SUBMARINE CABLE

Submarine cable installation	Power per cable	Total power
I^2R losses: $P_J = 630^2 \times 1.3$ = 0.516 MW	0.516 MW	3.1 MW
Reactive power generated: $Q_C = (80\ 000)^2/375$ = 17.06 Mvar	17 Mvar	102 Mvar
Reactive power absorbed: $Q_L = 630^2 \times 1.1$ = 0.436 Mvar	0.44 Mvar	2.6 Mvar
Active power transmitted: $P = 630 \times 80\ 000$ = 50 MW	50 MW	300 MW

than are its resistance or inductance. A cable behaves like an enormous capacitor, contrary to an aerial line which is mainly inductive. It is precisely the inherent large capacitance that prevents us from using cables to transmit energy over long distances. This restriction does not apply when direct current is used because capacitance then has no effect.

High-voltage dc transmission lines are covered in Chapter 28.

QUESTIONS AND PROBLEMS

Practical level

25-1   Standard voltages are grouped into four main classes. Name them and state the approximate voltage range of each.

- oil duct
- copper conductor
- carbon black screen
- paper insulation
- metallized paper screen
- lead alloy sheath
- reinforcement
- polyethylene sheath
- treated jute bedding
- aldrey wire armoring

Figure 25-44
Cross section view of submarine cable (138 kV, 630 A). This cable is one of seven cables submerged in Long Island Sound between Northport (Long Island) and Norwalk (Connecticut). (*Pirelli Cables Limited*)

25-2 Explain what is meant by:
- suspension-type insulator
- ground wire
- corona effect
- sag of a transmission line
- reactance of a line

25-3 Why must transmission line towers be solidly connected to ground?

25-4 A 735 kV transmission line, 745 miles long, transmits a power of 800 MW.

a. Is there an appreciable voltage difference between the two ends of the line, measured line to neutral?

b. Is there an appreciable phase angle between corresponding line to neutral voltages?

25-5 In some areas, two identical three-phase lines are installed side by side, supported by separate towers. Could we replace these two lines by a single line by simply doubling the size of the conductors? Explain.

25-6 Why do we seldom install underground cable (instead of aerial transmission lines) between generating stations and distant load centers?

25-7 In Problem 25-4, the average line span is 480 m. How many towers are needed between the source and the load?

25-8 A 20 km transmission line operating at 13.2 kV has just been disconnected from the source. A lineman could receive a fatal shock if he does not first connect the line to ground before touching it. Explain.

25-9 What is the ampacity of a 600 MCM ACSR cable suspended in free air? Why can a copper conductor having the same cross-section carry a current that is considerably greater?

Intermediate level

25-10 Each phase of the two 735 kV lines shown in Fig. 25-34 is composed of 4 bundled subconductors. The current per phase is 2000 A and the resistance of each subconductor is 0.045 Ω/km. Calculate:
a. the total power transmitted by both lines at unity power factor load;
b. the total I^2R loss, knowing the lines are 350 miles long;
c. the I^2R loss as a percent of the total power transmitted.

25-11 A single-phase transmission line possesses a resistance R of 15 Ω (Fig. 25-21a). The source E_s is 6000 V and the impedance of the unity power factor load varies between 285 Ω and 5 Ω.
a. Calculate the terminal voltage E_R and the power P absorbed by the load when the impedance is successively 285 Ω, 45 Ω, 15 Ω, and 5 Ω.
b. Draw the graph of the terminal voltage E_R as a function of the power P.

25-12 In Problem 25-11, what is the phase angle between E_R and E_S when the load is 45 Ω?

25-13 The transmission line in Problem 25-11 is replaced by another having an inductive reactance of 15 Ω (Fig. 25-22a).
a. Calculate the voltage E_R at the terminals of the load and the power P it absorbs for the same impedance values.
b. Draw the graph of E_R as a function of P.

25-14 In Problem 25-13, what is the phase angle between E_R and E_S when the load impedance is 45 Ω? Does E_R lead or lag behind E_S?

25-15 A single-phase transmission line possesses an inductive reactance of 15 Ω. It is supplied by a source E_S of 6000 V.
  a. Calculate the voltage E_R at the end of the line for the following capacitive loads: 285 Ω, 45 Ω.
  b. Calculate the phase angle between E_R and E_S when the load is 45 Ω.

25-16 The line in Problem 25-15 possesses a resistance of 15 Ω (instead of a reactance of 15 Ω).
  a. Calculate the voltage E_R at the end of the line for a capacitive load of 45 Ω.

  b. If a line is purely resistive, can we raise the voltage at the end of the line by connecting a capacitor across it?

25-17 The following information is given for Fig. 25-25a:

  – terminal voltage E_R: 6000 V
  – equivalent load resistance: 45 Ω
  – inductive reactance X_L of the line: 15 Ω
  – capacitive reactance X_C in parallel with the load: 150 Ω

  Neglecting the dotted reactance X_C in parallel with the source, calculate:
  a. the reactive power supplied by the capacitor;
  b. the line current I;
  c. the reactive power absorbed by the line;
  d. the reactive power supplied by the sender;
  e. the apparent power supplied by the sender;
  f. the voltage E_S of the sender;
  g. the voltage E_R and the power P when E_S is 6 kV.

25-18 In Problem 25-17, what is the phase angle between E_R and E_S?

25-19 Referring to Fig. 25.13, each section of the circuit represents a transmission line length of 1 km in which the impedances are
  $x_L = 0.5\ \Omega;\ r = 0.25\ \Omega;$
  $x_c = 300\ k\Omega.$
  Calculate the values of X_L, R and X_C if the circuit is reduced to that shown in Fig. 25-14.

25-20 What is meant by the term surge impedance load?

Advanced level

25-21 A three-phase 230 kV transmission line having a reactance of 43 Ω per phase connects two regions that are 50 miles apart. The phase angle between the voltages at the two ends of the line is 20°. Calculate:
  a. the active power transmitted by the line;
  b. the current in each conductor;
  c. the total reactive power absorbed by the line;
  d. the reactive power supplied to the line by each region.

25-22 A three-phase aerial line connected to a 115 kV 3-phase source has a length of 200 km. It is composed of three 600 MCM type ACSR conductors. Referring to Fig. 25-17 and Tables 25C and 25D, if there is no load on the line, calculate:
  a. the value of R, X_L and X_C, per phase;
  b. the voltage between conductors at the "load" (open end);
  c. the current drawn from the source, per phase;
  d. the total reactive power received by the source;
  e. the total I^2R loss in the line.

25-23 a. In Problem 25-22 calculate the current drawn from the source if a 3-phase short occurs across the end of the line.

   b. Compare this current with the ampacity of the conductors.

26

DISTRIBUTION OF ELECTRICAL ENERGY

In Chapter 25, we mentioned that an electrical power system is composed of high-voltage transmission lines that feed power to a medium-voltage (MV) network by means of substations. In North America, these MV networks generally operate at voltages between 2.4 kV and 69 kV. In turn, they supply millions of independent low-voltage systems that function between 120 V and 600 V.

In this chapter, we cover the following main topics:

1. Substations;
2. Protection of medium-voltage distribution systems;
3. Low-voltage distribution;
4. Electrical installation in buildings.

SUBSTATIONS

Substations are used throughout an electrical system. Starting with the generating station, a substation raises the medium-voltage generated by the synchronous generators to the high-voltage needed to transmit the energy economically.

The high transmission-line voltage is then reduced in those substations located close to the power-consuming centers. The electrical equipment in such distribution substations is similar to that found in substations associated with generating plants.

26.1 Substation equipment

A medium-voltage substation usually contains the following major apparatus:

Transformers
Circuit-breakers
Horn-gap switches
Disconnect switches
Grounding switches

Surge arresters
Current-limiting reactors
Instrument transformers
Relays and protective devices

In the description which follows, we study the basic principles of this equipment. Furthermore, to understand how it all fits together, we conclude our study with a typical substation that provides power to a large suburb.

26.2 Circuit-breakers

Circuit-breakers are designed to interrupt either normal or short-circuit currents. They behave like big switches that may be opened or closed by local push-buttons or by distant telecommunication signals emitted by the system of protection. Thus, circuit-breakers will automatically open a circuit whenever the line current, line voltage, frequency, etc., departs from a preset limit.

The most important types of circuit-breakers are:

1. Oil circuit-breakers (OCB's);
2. Air-blast circuit-breakers;
3. SF$_6$ circuit-breakers;
4. Vacuum circuit-breakers.

The nameplate on a circuit breaker usually indicates (1) the maximum steady-state current it can carry, (2) the maximum interrupting current, (3) the maximum line voltage, and (4) the interrupting time in cycles. The interrupting time may last from 3 to 8 cycles on a 60 Hz system. To interrupt large currents this quickly, we have to ensure rapid deionization of the arc, combined with rapid cooling. High-speed interruption limits the damage to transmission lines and equipment and, equally important, it helps to maintain the stability of the system when a contingency occurs.

The triggering action which causes a circuit-breaker to open is usually produced by means of an overload relay that can detect abnormal line conditions. For example, the relay coil in Figure 26-1 is connected to the secondary of a current transformer. The primary carries the line current of the phase that has to be protected. If the line current exceeds a preset limit, the secondary current will cause relay contacts C_1C_2 to close. As soon as they close, the tripping coil is energized by an auxiliary dc source. This causes the main line contacts to open, thus interrupting the circuit.

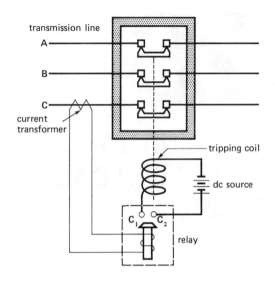

Figure 26-1
Elementary tripping circuit for a circuit-breaker.

1. Oil circuit-breakers. Oil circuit-breakers are composed of a steel tank filled with insulating oil. In one version (Fig. 26-2), three porcelain bushings channel the 3-phase line currents to a set of fixed contacts. Three movable contacts, actuated simultaneously by an insulated rod, open and close the circuit. When the circuit-breaker is closed, the line current for each phase penetrates the tank by way of one porcelain bushing, flows through the first fixed contact, the movable contact, the second fixed contact, and then out by a second bushing.

If an overload occurs, the tripping coil releases a powerful spring which pulls on the insulated rod, causing the contacts to open. As soon as the contacts separate, a violent arc is created, which volatilizes the surrounding oil. The pressure of the hot gases creates turbulence around the contacts. This causes cool oil to swirl around the arc, thus extinguishing it.

In modern high-power breakers, the arc is confined to an explosion chamber so that the pressure of the hot gases produces a powerful

current
transformer

movable
contact

actuating rod

fixed contact

steel tank

bushing

heater

Figure 26-2
Cross section of an oil circuit-breaker. The diagram shows four of the six bushings; the heater keeps the oil at a satisfactory temperature during cold weather. (*Canadian General Electric*)

Figure 26-3
Three-phase oil circuit-breaker rated at 1200 A at 115 kV. It can interrupt a current of 50 kA in 3 cycles on a 60 Hz system. Other characteristics: height: 3660 mm; diameter: 3050 mm; mass: 21 t; BIL 550 kV. (*General Electric*)

Figure 26-4
Minimum oil circuit-breaker installed in a 420 kV, 50 Hz substation. Rated current: 2000 A; rupturing capacity: 25 kA; height (less support): 5400 mm; length: 6200 mm; 4 circuit breaking modules in series per circuit-breaker. *(courtesy* ABB*)*

Figure 26-5
Air blast circuit-breaker rated 2000 A at 362 kV. It can interrupt a current of 40 kA in 3 cycles on a 60 Hz system. It consists of 3 identical modules connected in series, each rated for a nominal voltage of 121 kV. The compressed-air reservoir can be seen at the left. Other characteristics: height: 5640 mm; overall length: 9150 mm; BIL 1300 kV. (*General Electric*)

jet of oil. The jet is made to flow across the path of the arc, to extinguish it. Other types of circuit-breakers are designed so that the arc is deflected and lengthened by a self-created magnetic field. The arc is blown against a series of insulating plates that break up the arc and cool it down. Figures 26-3 and 26-4 show the appearance of two typical OCB's.

2. Air-blast circuit-breakers. These circuit-breakers interrupt the circuit by blowing compressed air at supersonic speed across the opening contacts. Compressed air is stored in reservoirs at a pressure of about 3 MPa (~ 435 psi) and is replenished by a com-

pressor located in the substation. The most powerful circuit-breakers can typically open short-circuit currents of 40 kA at a line voltage of 765 kV in a matter of 3 to 6 cycles on a 60 Hz line. The noise accompanying the air blast is so loud that noise-suppression methods must be used when the circuit-breakers are installed near residential areas. Figure 26-5 shows a typical 3-phase air-blast circuit-breaker. Each phase is composed of three contact modules connected in series. Figure 26-6 shows a cross section of the contact module.

3. SF$_6$ circuit-breakers. These totally enclosed circuit-breakers, insulated with SF$_6$ gas*, are used whenever space is at a pre-

* Sulfur hexafluoride.

Figure 26-6
Cross section of one module of an air blast circuit-breaker. When the circuit-breaker trips, the rod is driven upwards, separating the fixed and movable contacts. The intense arc is immediately blown out by a jet of compressed air coming from the orifice. The resistor dampens the overvoltages that occur when the breaker opens. (*General Electric*)

Figure 26-8
Three-phase vacuum circuit-breaker having a rating of 1200 A at 25.8 kV. It can interrupt a current of 25 kA in 3 cycles on a 60 Hz system. Other characteristics: height: 2515 mm; mass: 645 kg; BIL: 125 kV. (*General Electric*)

mium, such as in downtown substations (Fig. 26-7). They are much smaller than any other type of circuit-breaker of equivalent power and are far less noisy than air circuit-breakers.

4. Vacuum circuit breakers. These circuit-breakers operate on a different principle from other breakers because there is no gas to ionize when the contacts open. They are hermetically sealed; consequently, they are silent and never become polluted (Fig. 26-8). Their interrupting capacity is limited to about 30 kV. For higher voltages, several circuit-breakers are connected in series.

Vacuum circuit-breakers are often used in underground systems.

Figure 26-7
Group of 15 totally enclosed SF$_6$ circuit breakers installed in an underground substation of a large city. Rated current: 1600 A; rupturing current: 34 kA; normal operating pressure: 265 kPa (38 psi); pressure during arc extinction: 1250 kPa (180 psi). These SF$_6$ circuit-breakers take up only 1/16 of the volume of conventional circuit-breakers having the same interrupting capacity *(courtesy ABB)*

Figure 26-9
One pole of a horn-gap disconnecting switch rated 600 A, 27 kV, 60 Hz; (left) in the open position, (right) in the closed position. (*Dominion Cutout*)

Figure 26-10
One pole of a 3-phase 3000 A, 735 kV, 60 Hz horn gap disconnecting switch in the open position (left); in the closed position (right). The switch can be operated manually by turning a hand wheel or remotely by means of a motorized drive located immediately below the hand wheel. Other characteristics: height when closed: 12 400 mm; length: 7560 mm; mass: 3 t; maximum current-carrying capacity for 10 cycles: 120 kA; BIL: 2200 kV. (*Kearney*)

26.3 Air-break switches

Air-break switches can interrupt the exciting currents of transformers, or the moderate capacitive currents of unloaded transmission lines. They cannot interrupt normal load currents.

Air-break switches are composed of a movable blade that engages a fixed contact; both are mounted on insulating supports (Figs. 26-9, 26-10). Two arcing horns are attached to the fixed and movable contacts, so that when the main contact is broken, an arc is set up between the arcing horns. The arc moves upward due to the combined action of the hot air currents it produces and the magnetic field. As the arc rises, it becomes longer until it eventually blows out (Fig. 26-11). Although the arcing horns become pitted and gradually wear out, they can easily be replaced.

26.4 Disconnecting switches

Unlike air-break switches, disconnecting switches are unable to interrupt any current at all. They must only be opened and closed when the current is zero. They are basically isolating switches, enabling us to isolate oil circuit-breakers, transformers, transmission lines, and

Figure 26-12
This hookstick-operated disconnecting switch is rated 2000 A, 15 kV and has a BIL of 95 kV. (*Dominion Cutout*)

Figure 26-11
The arc produced between the horns of a disconnecting switch as it cuts the exciting current of a HV transformer provides the light to take this night picture. (*Hydro-Québec*)

Figure 26-13
Disconnecting switch rated 10 kA, 1 kV for indoor use. (*Montel/Sprecher and Schuh*)

Figure 26-14
Disconnecting switch rated 600 A, 46 kV for sidewise operation. (*Kearney*)

so forth, from a live network. Disconnecting switches are essential to carry out maintenance work and to reroute power flow.

Figure 26-12 shows a 2000 A, 15 kV disconnecting switch. It is equipped with a latch to prevent the switch from opening under the strong electromagnetic forces that accompany short-circuits. The latch is disengaged by inserting a hookstick into the ring and pulling the movable blade out of the fixed contact.

Figure 26-13 shows another so-called "disconnect" that carries a larger current, but at a much lower voltage. It, too, is opened by means of a manual hookstick. Figure 26-14 shows another type of disconnecting switch and Figure 26-15 shows how the fixed and movable contacts engage. Figure 26-16 shows maintenance personnel working on a large disconnecting switch in a HV substation.

26.5 Grounding switches

Grounding switches are safety switches that ensure a transmission line is definitely grounded while repairs are being carried out. Figure 26-17 shows such a 3-phase switch with the blades in the open (horizontal) position. To

Figure 26-15
The blade of a vertical motion disconnecting switch is in tight contact with two fixed contacts, due to the pressure exerted by two powerful springs. When the switch opens, the blade twists on its axis as it moves upward. During switch closure, the reverse rotary movement exerts a wiping action against the fixed contacts, thus ensuring excellent contact. (*Kearney*)

short-circuit the line to ground, all three grounding blades swing up to engage the stationary contact connected to each phase. Grounding switches are opened and closed only after the lines are de-energized.

26.6 Surge arresters

The purpose of a surge arrester* is to limit the overvoltages that may occur across transformers and other electrical apparatus due either to lightning or switching surges. The upper end of the arrester is connected to the line or terminal that has to be protected, while the lower end is solidly connected to ground.

* Also called lightning arrester or surge diverter.

Figure 26-16
Like all electric equipment, disconnecting switches have to be overhauled and inspected at regular intervals. During such operations, the current has to be diverted by way of auxiliary tie lines within the substation. The picture shows one pole of a 3-phase disconnect rated 2000 A, 345 kV. (*Hydro-Québec*)

Figure 26-17
Combined disconnecting switch and grounding switch rated at 115 kV. The grounding switch blades are shown in the open, horizontal position. These blades pivot upwards to engage three fixed contacts at the same time as the line is opened. (*Kearney*)

The arrester is composed of an external porcelain tube containing an ingenious arrangement of stacked discs, air gaps, ionizers, and coils. The discs (or valve blocks) are composed of a silicon carbide material known by trade names such as thyrite®, autovalve®, etc. This material has a resistance that decreases dramatically with increasing voltage. The typical E-I characteristic of a surge arrester is given in Figure 26-18.

Under normal voltage conditions, spark gaps prevent any current from flowing through the tubular column. If an overvoltage occurs, the spark gaps break down and the surge discharges to ground. The 60 Hz follow-through current is limited by the resistance of the valve blocks and the arc is simultaneously stretched and cooled in a series of arc chambers. The arc is quickly snuffed out and the arrester is then ready to protect the line against the next voltage surge. The discharge period is very short, rarely lasting more than a fraction of a millisecond.

Zinc-oxide arresters are also used. They are composed of stacked circular discs enclosed in a porcelain tube. The E-I characteristics are similar to those of silicon-carbide arresters. However, ZnO arresters do not require air gaps or other auxiliary devices.

Lightning arresters also enable us to reduce the BIL requirements of apparatus installed in substations (Sec. 25-12). On HV and EHV systems, the reduction in BIL significantly reduces the cost of the installed apparatus. Figure 26-19 shows a lightning arrester installed in an EHV substation.

26.7 Current-limiting reactors

The MV bus in a substation usually energizes several feeders which carry power to various load centers around the substation. It so happens that the output impedance of the MV bus is usually very low. Consequently, if a short-circuit should occur on one of the feeders, the

Figure 26-18
Voltage-current characteristic of a surge arrester having a nominal rating of 30 kV (42.4 kV peak), used on a 34.5 kV line.

Figure 26-19
Surge arresters protect this EHV transformer. (*General Electric*)

Figure 26-20
MV bus bar feeding 8 lines, each protected by a circuit-breaker.

Figure 26-21
Current-limiting reactors reduce the short-circuit current.

resulting short-circuit current can be disastrous.

Consider, for example, a 3-phase 69 MVA, 220 kV/24.9 kV transformer having an impedance of 8% and a nominal secondary current of 1600 A. It supplies power to eight 200 A feeders connected to the common MV bus (Fig. 26-20). Each feeder is protected by a 24.9 kV, 200 A circuit-breaker having an interrupting capacity of 4000 A. Because the transformer impedance is 8%, it can deliver a secondary short-circuit current of:

$$I = 1600 \times (1/0.08)$$
$$= 20\ 000\ A$$

This creates a problem because if a feeder becomes short-circuited, the resulting current flow could be as high as 20 000 A, which is 5 times greater than the interrupting capacity of the circuit-breaker. The circuit-breaker could be destroyed in attempting to interrupt the circuit. Furthermore, the feeder might burn over its entire length, from the circuit-breaker to the fault. Finally, a violent explosion would take place at the fault itself owing to the tremendous amount of thermal energy released by the burning arc.

To prevent this from happening, a current-limiting reactor is connected in series with each phase of the feeder (Fig. 26-21). The reactance

Figure 26-22
Three 2.2 Ω reactors rated 500 A are connected in series with a 120 kV, 3-phase, 60 Hz line. They are insulated from ground by four insulating columns and each is protected by a surge arrester. (*Hydro-Québec*)

must be high enough to keep the current below the interrupting capacity of the circuit-breaker, but not so high as to produce a large voltage drop under normal full-load conditions. Figure 26-22 shows three current-limiting reactors in series with a HV line.

26.8 Grounding transformer

We sometimes have to create a neutral on a 3-phase, 3-wire system, to change it into a 3-phase, 4-wire system. This can be done by means of a *grounding* transformer. It is basically a 3-phase autotransformer in which identical primary and secondary windings are connected in series in zig-zag fashion on a 3-legged core (Fig. 26-23).

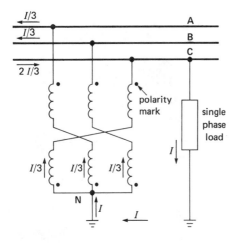

Figure 26-23
Grounding transformer to create a neutral.

If we connect a single-phase load between one line and neutral, load current I divides into three equal currents $I/3$ in each winding. Because the currents are equal, the neutral point stays fixed and the line-to-neutral voltages remain balanced as they would be on a regular 4-wire system. In practice, the single-phase loads are distributed as evenly as possible between the three phases and neutral so that the unbalanced load current I remains relatively small.

26.9 Example of a substation

Figure 26-24 shows the principal elements of a typical modern substation providing power to a large suburb. Power is fed into the substation at 220 kV and is distributed at 24.9 kV to various load centers within about a 5 km radius.

The substation is fed by 3 different lines, all operating at 220 kV. It contains six 3-phase transformers rated at 36/48/60 MVA, 220 kV/24.9 kV. The windings are connected in wye-delta and automatic tap-changers regulate the secondary voltage.

A neutral is established on the MV side by means of 3-phase grounding transformers. Consequently, single-phase power can be provided at $24.9/\sqrt{3} = 14.4$ kV.

Minimum-oil circuit-breakers having an interrupting capacity of 32 kA protect the HV side. Conventional oil circuit-breakers having an interrupting capacity of 25 kA are used on the MV side. Furthermore, all the outgoing feeders are protected by circuit-breakers having an interrupting capacity of 12 kA.

This completely automatic and unattended substation covers an area of 235 m × 170 m. However, line switching and other operations can be carried out by telecommunications from a dispatching center.

The substation provides service to hundreds of single-family homes, dozens of apartment buildings, several business and shopping centers, a large university and some industries. Figures 26-25 and 26-26 show the basic layout and components of the substation. We now study the distribution system that branches out from the substation.

26.10 Medium-voltage distribution

Thirty-six 3-phase feeders (30 active and 6 spares), rated at 24.9 kV, 400 A lead outwards from the substation. Each feeder is equipped with three current-limiting reactors

Figure 26-24
Aerial view of a substation serving a large suburb. The 220 kV lines (1) enter the substation and move through disconnecting switches (2) and circuit-breakers (3) to energize the primaries of the transformers (4). The secondaries are connected to a MV bus (5) operating at 24.9 kV. Grounding transformers (6) and MV circuit-breakers (7) feed power through current limiting reactors (8). The power is carried away by 36 aerial and underground feeders to energize the suburb (9).

that limit the line to ground short-circuit currents to a maximum of 12 kA. Some feeders are underground, others overhead and still others are underground/overhead.

Underground feeders are composed of three single-phase stranded aluminium cables insulated with polyethylene. The insulation is in turn surrounded by a spiral wrapping of tinned copper conductors which act as the ground. The cable is pulled through underground concrete duct (Fig. 26-27) or simply buried in the ground. Spare cables are invariably buried along with active cables to provide alternate service in case of a fault.

The 24.9 kV aerial lines are supported on wooden poles. The latter also carry the LV circuits, and telephone cable. The 24.9 kV lines are tapped off at various points to supply

1

2

3

4

5

6

7

8

9

10

11

12

3-phase and single-phase power to residences, commercial establishments and recreation centers (Fig. 26-28).

For nearby areas, the 24.9 kV line voltage is regulated within acceptable limits by the tap-changing transformers at the substation. In more remote districts, special measures have to be taken to keep the voltage reasonably stable with changing load. Thus, self-regulating autotransformers (Fig. 26-29) are often installed.

26.11 Low-voltage distribution

Two systems are provided on the low-voltage side of this typical suburban network:
1. Single-phase 120/240 V with grounded neutral;
2. Three-phase 600/347 V with grounded neutral.

The first system is mainly used in individual dwellings and for single-phase power up to 150 kVA. The second is used in industry, large buildings and commercial centers where the power requirement is under 2000 kVA.

For single-phase service, the transformers are usually rated between 10 kVA and 167 kVA and they are pole-mounted. The voltage rating is 14 400 V/240-120 V. The transformers possess a single high-voltage bushing connected to one side of the HV winding. The other side of the winding is connected to the steel enclosure which, in turn, is connected to the neutral conductor and also to ground (Fig. 26-30).

In the case of 3-phase installations, 3 single-phase transformers rated at 14 400 V/347 V are used. The units are connected in wye-wye and the neutral on the primary side is solidly grounded.

The secondary side provides a line voltage of 600 V, and it may or may not be grounded. Such standard distribution transformers have no taps, and no circuit-breakers or fuses are used on the secondary side. The primary HV terminal is, however, protected by a cut-out in order to prevent excessive damage to equipment in case of a fault (see Figs. 26-30 and 26-31).

Figure 26-25
This sequence of photos enables us to see how energy flows through the substation, starting from the 220 kV lines until it leaves by the 24.9 kV feeders.

1. 220 kV incoming line.
2. The line passes through 3 CT's (left) and the substation apparatus is protected by 3 lightning arresters (right).
3. Three HV disconnecting switches are placed ahead of the circuit-breakers.
4. Minimum volume oil circuit-breakers composed of 3 modules in series permit the line to be opened and closed under load.
5. Three-phase transformer bank steps down the voltage from 220 kV to 24.9 kV. Lightning arresters on the right protect the HV windings.
6. MV line from the transformer feeds the 24.9 kV bus.
7. Grounding transformer and its associated oil circuit-breaker having an interrupting capacity of 25 kA.
8. Current-limiting reactors.
9. Three-phase circuit-breaker having an interrupting capacity of 12 kA.
10, 11. MV underground feeder rated 400 A, 24.9 kV/14.4 kV leads into the ground towards a load center in the suburb.
12. All steel supports are solidly grounded by bare copper conductors to prevent overvoltages across equipment due to lightning strokes and other disturbances. Typical station ground resistance: 0.1 Ω.

Figure 26-26
Schematic diagram of the substation in Fig. 26-24.

Figure 26-27
MV underground feeders in concrete duct.

Figure 26-28
MV aerial feeder serving a residential district.

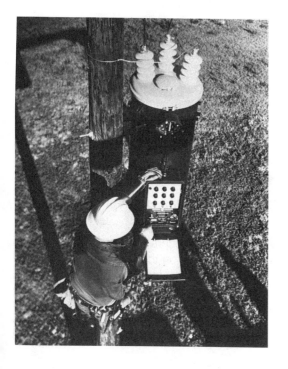

Figure 26-29
Automatic tap-changing autotransformer maintains steady
voltages on long rural lines. (*General Electric*)

Figure 26-30
A fused cutout (top left) and lightning arrester (top right) protect a single-phase transformer rated 25 kVA, 14.4 kV/240 V – 120 V, 60 Hz.

Figure 26-31
Expulsion type fused cutout rated 7.5 kV, 300 A. (*Dominion Cutout*)

Protection of Medium-Voltage Distribution Systems

Medium-voltage lines must be adequately protected against short-circuits so as to limit damage to equipment and to restrict the outage to as small an area as possible. Such line faults can occur in various ways: falling branches, icing, defective equipment, lines that touch, and so forth. According to statistics, 85 percent of the short-circuits are temporary, lasting only a fraction of a second. The same studies reveal that 70 percent take place between one line and ground. Finally, short-circuits involving all 3 phases of a transmission line are rare. The methods of protection are based upon these statistics, and on the necessity to provide continuity of service to the customers.

26.12 Coordination of the protective devices

When a fault occurs, the current increases sharply, not only on the faulted line, but on all lines that directly or indirectly lead to the short-circuit. To prevent the overload current from simultaneously tripping all the associated protective devices, we must design the system so that the devices trip selectively.

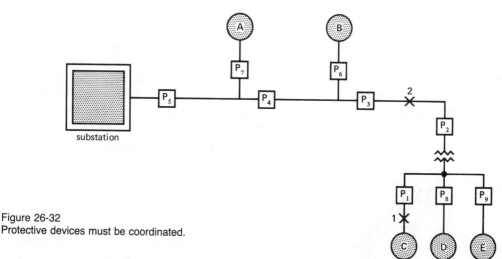

Figure 26-32
Protective devices must be coordinated.

A well-coordinated system will cause only those devices next to the short-circuit to open, leaving the others intact. To achieve this, the tripping current and tripping time of each device is set to protect the line and associated apparatus, while restricting the outage to the smallest number of customers.

Consider, for example, the simple distribution system of Figure 26-32 composed of a main feeder from a substation, supplying a group of subfeeders. The subfeeders deliver power to loads A, B, C, D and E. A protective device is installed at the input to each subfeeder so that if a short-circuit occurs, it alone will be disconnected from the system. For instance, a short-circuit at point (1) will trip device P_1 but not P_2. Similarly, a fault at point (2) will open P_3 but not P_4, as so on. A short-circuit must be cleared in a matter of a few cycles. Consequently, the coordination between the protective devices involves delays that are measured in milliseconds. We must also know the tripping characteristics of the fuses and circuit-breakers throughout the system. The most important protective devices used on MV lines are:

1. Fused cutouts;
2. Reclosers;
3. Sectionalizers.

26.13 Fused cutouts

A *fused cutout* is essentially a fused disconnecting switch. The fuse pivots about one end and the circuit can be opened by pulling on the other end with a hookstick (Fig. 26-31). Cutouts are relatively inexpensive and are used to protect transformers and small single-phase feeders against overloads. They are designed so that when the fuse blows, it automatically swings down, indicating that a fault has occurred on the line.

Fused cutouts possess a fuse link that is kept taut by a spring. The fuse link assembly is placed inside a porcelain or glass tube filled with boracic acid, oil, or carbon tetrachloride. The fuse link must be replaced every time it blows, which often produces a relatively long outage. To ensure good coordination, the current/time characteristics are selected very carefully for each cutout. A burned-out fuse link must always be replaced by another having exactly the same rating.

26.14 Reclosers

A *recloser* is a circuit-breaker that opens on short-circuit and automatically recloses after a brief time delay. The delay may range from a fraction of a second to several seconds. The open/close sequence may be repeated 2 or 3 times, depending on the internal control setting of the recloser. If the short-circuit does not clear itself after 2 or 3 attempts to reclose the line, the recloser opens the circuit permanently. A repair crew must then locate the fault, remove, it, and reset the recloser.

Figure 26-33
Automatic recloser protecting a 3-phase feeder.

Reclosers rated at 24.9 kV can interrupt fault currents up to 12 000 A. They are made for either single-phase or 3-phase lines, and are usually pole-mounted (Fig. 26-33). Reclosers are self-powered, drawing their energy from the line and storing it in powerful actuating springs and electromagnets.

26.15 Sectionalizers

When a main feeder is protected by several fuses spotted over the length of the line, it is often difficult to obtain satisfactory coordination between them, based on fuse-blowing time alone. Under these circumstances, we resort to *sectionalizers*. A sectionalizer is a special circuit-breaker that trips depending on the number of times a recloser has tripped further up the line. In other words, a sectionalizer works according to the "instructions" of a recloser.

For example, consider a recloser R and a sectionalizer S protecting an important main feeder (Fig. 26-34). If a fault occurs at the point shown, the recloser will automatically open and reclose the circuit, according to the predetermined program. A recorder inside the *sectionalizer* counts the number of times the recloser has tripped and, just before it trips for the last time, the sectionalizer itself trips – but permanently. In so doing, it deprives customers C and D of power but it also isolates the fault. Consequently, when the recloser closes for the last time, it will stay closed and customers A and B will continue to receive service. Unlike reclosers, sectionalizers are not designed to interrupt line currents. Consequently, they must trip during the interval when the line current is zero, which coincides with the time when the recloser itself is open.

Sectionalizers are available for single-phase and 3-phase transmission lines. They offer several advantages over fused cutouts. They can be reclosed on a dead short-circuit without fear of exploding, and there is no delay in searching for a fuse link having the correct caliber.

26.16 Review of MV protection

If we examine the single line diagram of a typical distribution system, we find it contains dozens of automatic reclosers, sectionalizers and fused cutouts. The reclosing of circuit-breakers at the substation may be coordinated with reclosers and sectionalizers of the system. The variety of devices available makes it possible to provide adequate protection of MV lines by using combinations such as:

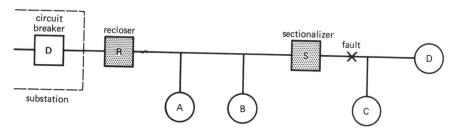

Figure 26-34
Recloser/sectionalizer protective system.

1. circuit-breaker – fuse;
2. circuit breaker – fuse – fuse;
3. circuit breaker – recloser – fuse;
4. circuit breaker – recloser – sectionalizer;
5. circuit breaker – sectionalizer – recloser – sectionalizer – fuse, etc.

In urban areas, the lines are relatively short and the possibility of faults is rather small. Such lines are subdivided along their length into 3 or 4 sections, each protected by single-pole fused cutout. Reclosers and sectionalizers are not required. On the other hand, in outlying districts, a 24.9 kV line may be quite long and consequently more exposed to faults. In such cases, the line is subdivided into sections and protected by reclosers and sectionalizers to provide satisfactory service.

LOW-VOLTAGE DISTRIBUTION

We have seen that electrical energy is delivered to the customer via HV substations through MV networks and finally by LV circuits. In this section, we briefly cover the organization of a LV distribution system.

26.17 LV distribution system

The most common LV systems used in North America are:

1. Single-phase, 2-wire, 120 V;
2. Single-phase, 3-wire, 240/120 V;
3. Three-phase, 4-wire, 208/120 V;
4. Three-phase, 3-wire, 480 V;
5. Three-phase, 4-wire, 480/277 V;
6. Three-phase, 3-wire, 600 V;
7. Three-phase, 4-wire, 600/347 V

In Europe and other countries of the world, 3-phase 380/220 V, 50 Hz systems are widely used. Despite the different voltages employed, the basic principles of LV distribution are everywhere the same.

Single-phase, 2-wire, 120 V system. This simple distribution system is only used for very small loads. When heavier loads have to be serviced, the 120 V system is not satisfactory because large conductors are required. Furthermore, the line voltage drop under load becomes significant even over short distances.

Single-phase, 3-wire, 240/120 V system. In order to reduce the current and hence conductor size, the voltage is raised to 240 V. However, because the 120 V level is still very useful, the 240 V/120 V 3-wire system was developed (Fig. 26-35). This type of distribution system is widely used. The dual voltage is produced by a distribution transformer having a double secondary winding (Sec. 11-1). The common wire, called neutral, is solidly connected to ground. When the "live" lines A and B are equally loaded, the current in the neutral is zero. When the loading is unequal, the neutral

Figure 26-35
Single-phase 240 V/120 V distribution system.

current is equal to the difference between the line currents $I_A - I_B$ (Fig. 26-35). We try to distribute the 120 V loads as equally as possible between the two live wires.

What are the advantages of such a 3-wire system?

1. The line to ground voltage is only 120 V, which is reasonably safe for use in a home;
2. Lighting and small motor loads can be energized at 120 V, while larger loads such as electric stoves and large motors can be fed from the 240 V line.

Both live lines are protected by fuses or circuit-breakers. However, such protective devices must *never* be placed in series with the neutral conductor. The reason is that if the device trips, the line-to-neutral voltages become unbalanced. The voltage across the lightly loaded 120 V line goes up, while that across the more heavily loaded side goes down. This means that some lights are dimmer than others and, moreover, the intensity will vary as refrigerator motors, electric stove elements, and so forth, are switched on and off.

Three-phase, 4-wire, 208/120 V system. We can create a 3-phase, 4-wire system by using 3 single-phase transformers connected in delta-wye. The neutral of the secondary is grounded, as shown in Figure 26-36.

This distribution system is used in commercial buildings and small industries because the 208 V line voltage can be used for electric motors or other large loads, while the 120 V lines can be used for lighting circuits and convenience outlets. The single-phase loads between the three respective "live" lines and neutral are arranged to be about equal. When the loads are perfectly balanced, the currents in the neutral wire is zero.

Three-phase, 3-wire, 600 V system. A 600 V, 3-phase, 3-wire system is used in factories where fairly large motors are installed, ranging up to 300 hp (Fig. 26-37). Separate 600 V/240-120 V step-down transformers, spotted throughout the premises, are used to service the lighting loads and convenience outlets.

Three-phase, 4-wire, 600/347 V system. In large buildings and commercial centers, a 600 V, 4-wire distribution system is used because it enables motors to run at 600 V while fluorescent lights operate at 347 V. For 120 V convenience outlets, separate transformers are required, usually fed from the 600 V line.

The same remarks apply to 480/277 V, 4-wire systems.

26.18 Grounding electrical installations

The grounding of electrical systems is probably one of the less well understood aspects of electricity. Nevertheless, it is a very important sub-

Figure 26-36
Three-phase 4-wire 208 V/120 V distribution system.

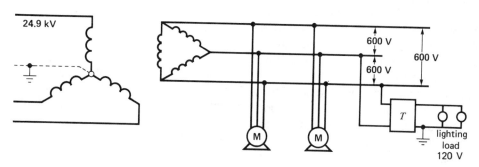

Figure 26-37
Three-phase 3-wire 600 V distribution system.

ject and an effective way of preventing accidents.

As we have seen, most electrical distribution systems in buildings are grounded, usually by connecting the neutral to a water pipe or the massive steel structure. On low-voltage systems, the purpose of grounding is mainly to reduce the danger of electric shock. In addition, for reasons of safety, the metal housing of electrical equipment is systematically grounded on HV, MV and LV systems.

26.19 Electric shock

It is difficult to specify whether a voltage is dangerous or not because electric shock is actually caused by *the current* that flows through the human body. The current depends mainly upon the skin contact resistance because, by comparison, the resistance of the body itself is negligible. The contact resistance varies with the thickness, wetness, and resistivity of the skin.

It is generally claimed that currents below 5 mA are not dangerous. Between 10 mA and 20 mA, the current is potentially dangerous because the victim loses muscular control and so may not be able to let go; above 50 mA, the consequences can be fatal.

The resistance of the human body, taken between two hands, or between one hand and a leg, ranges from 500 Ω to 50 kΩ. If the resist-

Figure 26-38
Transformer capacitance may produce high voltages on the LV side of a transformer.

ance of a dry hand is, say 50 kΩ, then momentary contact with a 600 V line may not be fatal (I = 600 V/50 kΩ = 12 mA). But the resistance of a clammy hand is much lower, so that an ac voltage as low as 25 V may prove fatal (I = 25 V/500 Ω = 50 mA).

When ac current flows through the body, the muscular contractions prevent the victim from letting go. The current is particularly dangerous when it flows in the region of the heart. It induces temporary paralysis and, if it flows long enough, fibrillation may result. Fibrillation is a rapid and uncoordinated heart beat that is not synchronized with the pulse beat. In such cases, the person can be revived by applying artificial respiration.

Statistical investigations have shown that a current may cause death if it satisfies the following equation:

$$I = 116/\sqrt{t} \qquad (26\text{-}1)$$

where

I = current flow through the body [mA]
t = time of current flow [s]
116 = an empirical constant, expressing the probability of a fatal outcome

---

* Daziel, C.F., 1968 "Reevaluation of Lethal Electric Currents", *IEEE, Transactions on Industry and General Application, Vol. IGA-4, No. 5, pages 467 to 475.*

For example, a current of 116 mA flowing for 1 s could be fatal.

26.20 Grounding of 120 V and 240 V/120 V systems

Suppose the primary winding of a distribution transformer is connected between the line and neutral of a 14.4 kV line (Fig. 26-38a). If the secondary conductors are ungrounded, it would appear that a person could touch either one without harm because there is no ground return. However, this is not true.

First, the capacitive coupling C_1, C_2 between the primary, secondary, and ground, can produce a high voltage between the secondary lines and ground. Depending upon the relative magnitude of C_1 and C_2, it may be as high as 20 to 40 percent of the primary voltage. If a person touches either one of the secondary wires, the resulting capacitive current I_c could be dangerous. For example, if I_c is only 20 mA, the person may no longer be able to let go (Fig. 26-38b).

Even more serious, suppose that a high-voltage wire accidentally touches a 120 V conductor. This could be caused by an internal fault in the transformer, or by a branch or tree falling across the MV and LV lines. Under these circumstances, the much higher than normal

Figure 26-39
A HV to LV fault is not dangerous if the secondary is solidly grounded.

14.4 kV between the secondary conductors and ground would immediately puncture the 120 V insulation, causing a massive flashover. The flashover could occur anywhere on the secondary network, possible inside a home or factory. Consequently, an undergrounded secondary system is a potential fire hazard and may produce serious accidents under abnormal conditions.

On the other hand, if one of the secondary lines is grounded, the accidental contact between a HV and a LV conductor produces a short-circuit. The short-circuit current follows the dotted path shown in Figure 26-39. The high current will blow the fuse on the MV side, thus effectively disconnecting the transformer and secondary distribution system from the MV line. In conclusion, if the neutral on the 120 V system is *solidly* grounded, the potential difference between the ground and live conductor 1 will only slightly exceed 120 V. However, if the ground electrode has a high resistance (say 50 Ω), the voltage on conductor 1 produced by a MV-LV short-circuit current may still be large, and potentially dangerous.

26.21 Equipment grounding

The consumer of electricity is constantly touching electrical equipment of all kinds, ranging from domestic appliances and hand-held tools to industrial motors, switchgear and heating equipment. As we have seen, the voltages and currents associated with this equipment far exceeds those the human body can stand. Consequently, special precautions are taken to ensure that the equipment is safe to the touch.

In order to understand the safety features of modern distribution systems, let us begin with a simple single-phase circuit composed of a 120 V source connected to a motor M (Fig. 26-40). The neutral is solidly grounded at

Figure 26-40
Undergrounded metallic enclosures are potentially dangerous.

the service entrance. Suppose the motor is part of an appliance, such as a refrigerator, and that the motor frame is connected to the ungrounded metal enclosure. If a person touches the enclosure, nothing will happen if the equipment is functioning properly. But should the motor winding insulation become faulty, resistance R_e between the windings and the motor frame may drop from several megohms to only a few hundred ohms or less. A person having a body resistance R_b would complete the current path to ground as shown in Figure 26-40. If R_e is small, the leakage current I_L flowing through the person's body could be dangerously high. This system is unsafe.

As a first approach, we could remedy the situation by bonding the enclosure to the grounded neutral wire (Fig. 26-41). The leakage current now flows from the motor windings,

Figure 26-41
Bonding the enclosure to the neutral wire appears safe.

through the motor frame and the enclosure, and straight back to the neutral wire. But the enclosure remains at ground potential and so a person touching the enclosure would not experience any shock.

The trouble with this solution is that the neutral wire may become open, either accidentally or due to a faulty installation. For example, if the switch controlling the motor is inadvertently connected in series with the neutral rather than the live wire, the motor can still be turned on and off. However, if a person touches the enclosure while the motor is *off*, he could receive a bad shock (Fig. 26-42). The reason is that when the motor with the defective windings is off, the potential of the motor frame and the enclosure rises to that of the live conductor.

To get around this problem, we install a third wire, called *ground* wire. It permanently connects the enclosure to the system ground at the service entrance panel (Fig. 26-43). With this arrangement, the neutral wire must never be connected to the enclosure, except for special cases permitted by the National Electrical Code.

Figure 26-42
Enclosure-to-neutral bonding can still be dangerous.

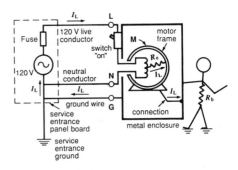

Figure 26-43
A separate ground conductor bonded to the enclosure is safe.

The ground wire may be bare, or, if insulated, it is colored green. In armored-cable and conduit installations, the armor and conduit serve as the ground conductor. The locknuts, squeeze-connectors, threads and bushings must be screwed on tight so as to make a good electrical connection between the service entrance ground and the hundreds of outlets that sometimes make up a large installation.

Most electrical outlets are now provided with receptacles having three contacts – one live, one neutral, and one ground (Fig. 26-44). Consequently, electrical appliances and portable hand-tools, such as electric drills, are equipped with a 3-conductor cord and a 3-prong plug.

Figure 26-44
The metal housing of hand tools must be grounded.

One exception is the so-called double-insulated devices that are completely enclosed in plastic enclosures. They are exempt from the ground wire requirement and so they are equipped with two-prong plugs.

26.22 Ground-fault circuit-breaker

The grounding methods we have covered so far are usually adequate, but further safety measures are needed in some cases. Suppose for example, that a person sticks his finger into a lamp socket (Fig. 26-45). Although the metal enclosure is securely grounded, the person will still receive a painful shock. Again, suppose a 120 V electric toaster tumbles into a swimming pool. The heating elements and contacts will produce a hazardous leakage current throughout the pool, even if the frame of the toaster is securely grounded. Devices have been developed that will cut the source of power as soon as such accidents occur. These *ground-fault*

circuit-breakers will typically trip in 25 ms if the leakage current exceeds 5 mA. How do these protective devices operate?

A small current transformer surrounds the live and neutral wires as shown in Figure 26-46. The secondary winding is connected to a sensitive electronic detector that can trigger a circuit-breaker CB that is in series with the 120 V line. Under normal conditions, the current I_W in the line conductor is exactly equal to the current I_N in the neutral, and so the net current $(I_W - I_N)$ flowing through the hole in the toroidal core is zero. Consequently, no flux is produced in the core, the induced voltage E_F is zero, and breaker CB does not trip.

Suppose now that a fault current I_F leaks directly from the live wire to ground. This could happen if someone touched a live terminal (Fig. 26-46). A fault current would also be produced if the insulation broke down between a motor and its grounded enclosure. Under any of these conditions, the net current flowing through the hole of the CT is no longer zero but equal to I_F. A flux is set up and a voltage E_F is induced which trips CB. Because an unbalance of only 5 mA has to be detected, the core of the transformer must be very permeable at low flux densities. The relative permeability is typically 40 000 at a density of 4 mT and Supermalloy is often used for this purpose.

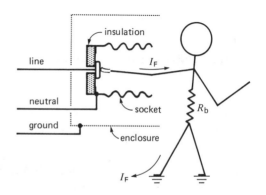

Figure 26-45
Special case where a grounding wire does not afford protection.

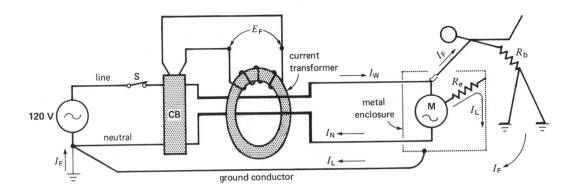

Figure 26-46
Ground fault circuit-breaker trips when leakage currents I_L or I_F exceed about 5 mA.

26.23 Electrical installation in buildings

The electrical distribution system in a building is the final link between the consumer and the original source of electrical energy. All such "in-house" distribution systems, be they large or small, must meet certain basic requirements. They relate to:

1. Safety:
   a. protection against electric shock;
   b. protection of conductors against physical damage;
   c. protection against overloads;
   d. protection against hostile environments.

2. Conductor voltage drop:
   It should not exceed 1 or 2 percent

3. Life expectancy:
   The distribution system should last a minimum of 50 years.

4. Economy:
   The cost of the installation should be minimized while observing the pertinent standards.

Standards are set by the National Electrical Code* and every electrical installation must be approved by an inspector before it can be put into service.

26.24 Principal components of an electrical installation

Many components are used in the makeup of an electrical installation. The block diagrams of Figs. 26-47 and 26-48, together with the following definitions †, will help the reader understand the purpose of some of the more important items.

1. Service conductors. These are the conductors that extend from the street main feeder or from a transformer, to the service equipment on the consumer premises.

2. Service equipment. The necessary equipment, usually consisting of a circuit-breaker or switch and fuses, and their accessories, located near the point of entrance of service conductors to a building or other structure, or an otherwise defined area, and intended to constitute the main control and means of cutoff of the supply.

3. Metering equipment. Various meters and recorders to indicate the electrical energy consumed on the premises.

* In Canada, by the Canadian Electrical Code.

† Some taken from the National Electrical Code.

4. Panel board. A single panel or group of panel units designed for assembly in the form of a single panel; including buses, automatic overcurrent devices, and with or without switches for the control of light, heat or power circuits; designed to be placed in a cabinet or cutout box placed in or against a wall or partition and accessible only from the front.

5. Switchboard. A large single panel, frame, or assembly of panels on which are mounted, on the face, or back, or both, switches, overcurrent and other protective devices buses, and usually instruments. Switchboards are generally accessible from the rear as well as from the front and are not intended to be installed in cabinets.

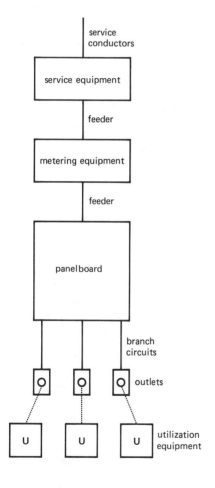

Figure 26-47
Block diagram of the electrical system in a residence.

Figure 26-48
Block diagram of the electrical system in an industrial or commercial establishment.

6. Feeder. All circuit conductors between the service equipment, or the generator switchboard of an isolated plant, and the final branch-circuit overcurrent device.

7. Branch circuit. The circuit conductors between the final overcurrent device protecting the circuit and the outlet(s).

8. Outlet. A point on the wiring circuit at which current is taken to supply utilization equipment.

9. Receptacle. A contact device installed at the outlet for the connection of a single attachment plug.

10. Utilization equipment. Equipment which utilizes electrical energy for mechanical, chemical, heating, lighting, or similar services.

The greatly simplified diagrams of Figures 26-47 and 26-48 indicate the type of distribution systems used respectively in a home and in an industrial or commercial establishment.

QUESTIONS AND PROBLEMS

Practical level

26-1   What is the difference between a circuit-breaker and a disconnecting switch?

26-2   Name four types of circuit breakers.

26-3   The disconnecting switch shown in Fig. 26-13 can dissipate a rated maximum of 200 W. Calculate the maximum allowable value of the contact resistance.

26-4   What is the purpose of a grounding switch?

26-5   Name some of the main components of a substation.

26-6   The ground resistance of a substation is 0.35 Ω. Calculate the rise in potential of the steel structure if the station is hit by a 50 kA lightning stroke.

26-7   What is the purpose of the following equipment:

fuse cutout	receptacle
recloser	ground wire
current-limiting	surge
reactor	arrester

Intermediate level

26-8   A lightning arrester having the characteristics shown in Fig. 26-18 is connected to a line having a line-to-neutral voltage of 34.5 kV. Calculate:
   a. the peak voltage between line and neutral;
   b. the current flowing in the arrester under these conditions.

26-9   In Problem 26-8, an 80 kV surge appears between line and neutral. Calculate:
   a. the peak current in the arrester;
   b. the peak power dissipated in the arrester;
   c. the energy dissipated in the arrester if the surge effectively lasts for 5 μs.

26-10   Figure 26-26 is a schematic diagram of a substation and Fig. 26-25 shows the actual equipment. Can you correlate the symbols of the schematic diagram with the equipment?

26-11   Repairs have to be carried out on HV circuit breaker No. 6 shown in Fig. 26-26. If the three 220 kV lines must be kept in service, which disconnecting switches must be kept open?

26-12   The current-limiting reactors (8) shown in Fig. 26-25 limit the short-circuit current to 12 kA on the 24.9 kV feeders. Calculate the reactance and inductance of each coil.

26-13   In Fig. 26-35, resistive loads, 1, 2 and 3

respectively absorb 1200 W, 2400 W and 3600 W. Calculate the current:

a. in lines A and B;

b. in the neutral conductors;

c. in the HV line.

26-14 In Fig. 26-37, the lighting circuit is off and the two motors together draw 420 kVA from the 600 V line. Calculate the current in the MV lines.

26-15 Draw a graph (I versus t) of Equation 26-1 for currents between 10 mA and 2 A. Crosshatch the potentially mortal regions. State whether the following conditions are dangerous:

a. 300 mA for 10 ms;

b. 30 mA for 2 min.

26-16 Explain the operation of a ground fault circuit-breaker.

Advanced level

26-17 The following loads are connected to the 240 V/120 V line shown in Fig. 26-35.

load 1: 6 kW, cos θ = 1.0
load 2: 4.8 kW, cos θ = 0.8 lagging
load 3: 18 kVA, cos θ = 0.7 lagging

a. Calculate the currents in lines A and B, and the neutral;

b. What is the current in the MV line?

c. What is the power factor on the MV side?

26-18 Referring to Fig. 26-36, the connected loads have the following ratings:
single phase loads: 30 kW each
motor M1: 50 kVA, cos θ = 0.5 lagging
motor M2: 160 kVA, cos θ = 0.80 lagging

a. Calculate the currents in the secondary windings.

b. Calculate the line currents and power factor on the 2400 V side.

26-19 In Fig. 26-37, a sensitive voltmeter reads 300 V between one 600 V line and ground, even though the 600 V system is not grounded. Can you explain this phenomenon?

26-20 The oil in the big power transformer shown on the extreme left-hand side of Fig. 26-26 has to be filtered and cleaned. Without interrupting the power flow from the three 220 kV incoming lines, state which circuit breakers and which disconnecting switches have to be opened, and in what order?

26-21 Referring to Fig. 26-25, the three aluminium conductors that make up the 3-phase 24.9 kV feeder (item 10,11.) each have a cross-section of 500 MCM. The cable possesses the following characteristics, per phase, and per kilometer of length:

resistance: 0.13 Ω
inductive reactance: 0.1 Ω
capacitive reactance: 3000 Ω

a. Draw the equivalent circuit of one phase if the line length is 5 km.

b. If no current-limiting line reactors are used, calculate the short-circuit current if a fault occurs at the end of the line.

c. Given the 12 kA rating of the circuit breakers, is a line reactor needed in this special case?

27

THE COST OF ELECTRICITY

In 1987, the electric power utilities in the United States supplied approximately 2345 TW·h of electrical energy to their industrial, commercial and residential customers (Table 27A). This enormous amount of energy represents 1.1 kW of power continually at the service of every man, woman and child, 24 hours per day. The production, transmission and distribution of this energy involves important costs that may be divided into two main categories – fixed costs and operating costs.

Fixed costs comprise the depreciation charges against buildings, dams, turbines, alternators, circuit breakers, transformers, transmission lines and other equipment used in the production, transmission and distribution of electrical energy. These investments represent enormous sums: in 1987, they were about $520 000 per employee, compared to $90 000 per employee in the automobile industry.

Operating costs include salaries, fuel costs, administration, and any other daily or weekly expense.

Bearing in mind the relative importance of these two types of costs, utility companies have established rate structures that attempt to be as equitable as possible for their customers. The rates are based upon the following guidelines:

1. the amount of energy consumed [kW·h];

2. the *demand* or rate at which energy is consumed [kW];

3. the power factor of the load.

27.1 Tariff based upon energy

The cost of electricity depends, first, upon the amount of energy consumed. However, even if a customer uses no energy at all, he has to pay a minimum service charge, because it costs money to keep him connected to the line.

As consumption increases, the cost per kilowatthour drops, usually on a sliding scale. Thus, the domestic tariff may start at 15 cents per kilowatthour for the first one hundred kW·h, fall to 8 cent/kW·h for the next two hundred

Table 27A   CONSUMPTION OF ELECTRICAL ENERGY IN THE UNITED STATES (1987)*

Type of customer	Number of customers	Total annual consumption [TW·h]	Monthly consumption per customer [kW·h]
Industrial	494 000	843	142 000
Commercial	11 300 000	657	4 850
Residential	92 300 000	845	765

* These statistics for 1987 were drawn from information supplied by the Edison Electric Institute. It is expected that the yearly increase in energy consumption will be about 2%. Thus, in 1990, the total energy will amount to 2510 TW·h.

kW·h, and bottom out at 6 cent/kW·h for the rest of the energy consumed. The same general principle applies to medium-power and large-power users of electrical energy.

27.2 Tariff based upon demand

The monthly cost of electricity supplied to a large customer depends not only upon the energy he consumes, but also upon the rate at which he uses it. In other words, the cost also depends upon the active *power* drawn from the line. To understand the reason for this dual rating consider the following example.

Two factories A and B are respectively connected to a high-tension line by transformers T_A and T_B (Fig.27-1). Factory A operates at full load, night and day, Saturdays and Sundays, constantly drawing 1000 kW of active power. At the end of the month (720 h), it has consumed a total of 1000 kW × 720 = 720 000 kW·h of energy.

Factory B consumes the same amount of energy, but its load is continually changing. Thus, power fluctuates between 50 kW and 3000 kW, as shown in the figure. Obviously, the capacity of the transformer and the transmission line supplying factory B must be greater than that supplying factory A. The electric utility must therefore invest more capital to service factory B; consequently, it is reasonable that factory B should pay more for its energy.

It is advantageous, both to the customer and the utility company that energy be consumed at a constant rate. The more regular the power flow, the less the cost.

27.3 Demand meter

The graphs of Fig. 27-1 show a number of power spikes that last for a very short time. These spikes correspond to the high power drawn by motors when they are started up. However, the high start-up power does not last long enough to warrant the installation of correspondingly large equipment by the utility company. The question then arises: How long does the power surge have to last, in order to be considered significant? The answer depends upon several factors, but the period is usually taken to be 10, 15 or 30 minutes. For very large power users, such as municipalities, the averaging period may be as long as 60 minutes. It is called the *demand* interval.

To monitor the power flowing into the plant, a special meter is installed at the customer's service entrance. It automatically measures the average power during successive demand intervals (15 minutes, say). The average power measured during each interval is called the *demand*. As time goes by, the meter faithfully records the demand every 15 minutes and a pointer moves up and down a calibrated scale as the demand changes. In order to record the

Figure 27-1
Comparison between two factories consuming the same energy but having different demands.

maximum demand, the meter carries a second pointer that is pushed upscale by the first. The second pointer simply sits at the highest position to which it is pushed. At the end of the month, a meterman takes the maximum demand reading and resets the pointer to zero.

This special meter is called a *demand meter*, and it is installed at the service entrance

Figure 27-2
Combined energy and demand meter. (*Sangamo*)

Figure 27-3
Printing demand meter. (*General Electric*)

of most industries and commercial establishments (Fig. 27-2). Figure 27-3 shows a printing demand meter for metering large industrial loads. The printout constitutes a permanent record of the demand and is used for both diagnostic and billing purposes.

Example 27-1:

The graph in Fig. 27-4 represents the active power drawn by a large factory between 7:00 and 9:00 in the morning. The demand meter has a 30 min demand interval. Let us assume that at 7:00 the first pointer reads 2 MW while the second (pushed) pointer indicates 3 MW. What are the meter readings at the following times:

a.	7:30;	b.	8:00;
c.	8:30;	d.	9:00.

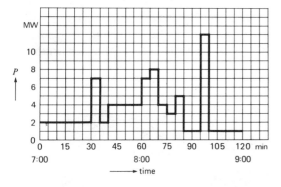

Figure 27-4
Instantaneous record of the active power absorbed by a plant.

Solution:

a. According to the graph, the average power (or demand) between 7:00 and 7:30 is 2 MW. Consequently, pointer 1 continues to indicate 2 MW at 7:30 and pointer 2 stays where it was at 3 MW.

b. The average power (or demand), between 7:30 and 8:00 is equal to the energy divided by time:

$$P_d = (7 \text{ MW} \times 5 \text{ min} + 2 \text{ MW} \times 5 \text{ min} + 4 \text{ MW} \times 20 \text{ min})/30 \text{ min}$$
$$= 4.17 \text{ MW}$$

During this 30-minute interval, pointer 1 gradually moves from 2 MW (at 7:30) to 4.17 MW (at 8:00), pushing pointer 2 to 4.17 MW. Consequently, at 8:00, both pointers indicate 4.17 MW. Note that the demand reading is considerably less than the 7 MW peak which occurred during this interval.

c. The demand between 8:00 and 8:30 is:
$$P_d = (7 \times 5 + 8 \times 5 + 4 \times 5 + 3 \times 5 + 5 \times 5 + 1 \times 5)/30$$
$$= 4.67 \text{ MW}$$

Thus, at 8:30, both pointers have moved up to 4.67 MW.

d. The demand between 8:30 and 9:00 is:
$$P_d = (1 \times 5 + 12 \times 5 + 1 \times 20)/30$$
$$= 2.83 \text{ MW}$$

During this 30-minute interval, pointer 1 drops from 4.67 MW to 2.83 MW, but pointer 2 sits at 4.67 MW, the previous maximum demand.

27.4 Tariff based upon power factor

Many alternating current machines, such as induction motors and transformers, absorb reactive power to produce their magnetic fields. The power factor of these machines is therefore less than unity and so, too, is the power factor of the factory where they are installed. A low power factor increases the cost of electrical energy, as the following example shows.

Consider two factories X and Y that consume the same amount of energy and, in addition, have the same maximum demand. However, the power factor of X is unity while that of Y is 50 percent (Fig. 27-5).

The energy and demand being the same, the watthourmeters and demand meters will

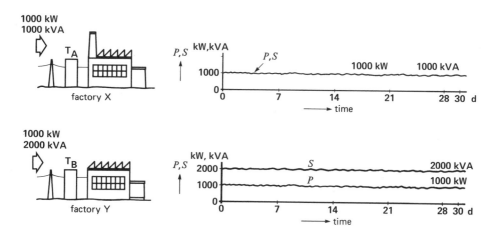

Figure 27-5
A low plant power factor requires larger utility company lines and equipment.

show the same reading at the end of the month. At first glance, it would appear that both users should pay the same bill. However, we must not overlook the *apparent power* drawn by each plant:

Apparent power drawn by factory X:

$$S = P/\cos\theta \qquad \text{Eq. 7-7}$$
$$= 1000/1.0$$
$$= 1000 \text{ kVA}$$

Apparent power drawn by factory Y;

$$S = P/\cos\theta = 1000/0.5$$
$$= 2000 \text{ kVA}$$

Because the line current is proportional to apparent power, factory Y draws twice as much current as factory X. The line conductors feeding Y must therefore be twice as big. Worse still, the transformers, circuit breakers, disconnect switches, and other devices supplying energy to Y, must have twice the rating of those supplying X.

The utility company must therefore invest more capital to service factory Y; consequently, it is logical that it should pay more for its energy, even though it consumes the same amount. In practice, the rate structure is designed to automatically increase the billing whenever the

power factor is low. Most electrical utilities require that the power factor of their industrial clients be 90 percent or more, in order to benefit from the minimum rate. When the power factor is too low, it is usually to the customer's advantage to improve it, rather than pay the higher monthly bill. This is usually done by installing capacitors at the service entrance to the plant, on the load side of the metering equipment. These capacitors may supply part of, or all the reactive power required by the plant. Industrial capacitors for power factor correction are made in single-phase and three-phase units rated from 5 kvar to 200 kvar.

27.5 Typical rate structures

Electrical power utility rates vary greatly from one area to another, and so we can only give a general overview of the subject. Most companies divide their customers into categories, according to their power demand. For example, one utility company distinguishes four power categories:

1. *Domestic power* – power corresponding to the needs of houses and rented apartments;

2. *Small power* – power of less than 100 kilowatts;

TABLE 27B	**TYPICAL RATE STRUCTURES**

Residential Rate Structure

Typical clauses:

1. " . . . This rate shall apply to electric service in a single private dwelling . . .

2. . . . This rate applies to single-phase alternating current at 60 Hz . . ."

Rate Schedule

Minimum monthly charge - $5.00 plus
First 100 kW·h per month at 5 cent/kW·h
Next 200 kW·h per month at 3 cent/kW·h
Excess over 200 kW·h per month at 2 cent/kW·h

General Power Rate Structure (medium power)

Typical contract clauses:

1. " . . . The customer's maximum demand for the month, or its contract demand, is at least 50 kW but not more than 5000 kW . . .

2. . . . Utility Company shall make available to the customer 1000 kW of firm power during the term of this contract . . .

3. . . . The power shall be delivered at a nominal 3-phase line voltage of 480 V, 60 Hz . . .

4. . . . The power taken under this contract shall not be used to cause unusual disturbances on the Utility Company's system. In the event that unreasonable disturbances, including harmonic currents, produce interference with communication systems, the customer shall at his expense correct such disturbances . . .

5. . . . Voltage variations shall not exceed 7 percent up or down from the nominal line voltage . . .

6. . . . Utility Company shall make periodic tests of its metering equipment so as to maintain a high standard of accuracy . . .

7. . . . Customer shall use power so that current is reasonably balanced on the three phases. Customer agrees to take corrective measures if the current on the more heavily loaded phase exceeds the current in either of the two other phases by more than 20 percent. If said unbalance is not corrected, Utility Company may meter the load on individual phases and compute the billing demand as being equal to 3 times the maximum demand on any phase . . .

TABLE 27B   **TYPICAL RATE STRUCTURES**

8. . . . The maximum demand for any month shall be the greatest of the demands measured in kilowatts during any 30 minute period of the month . . .

9. . . . If 90 percent of the highest average kVA measured during any 30 minute period is higher than the maximum demand, such amount will be used as the billing demand . . ."

Rate Schedule

Demand charge: $3.00 per month per kW of billing demand
Energy charge: 4 cent/kW·h for the first 100 hours of billing demand
                 2 cent/kW·h for the next 50 000 kW·h per month
                 1.2 cent/kW·h for the remaining energy

General Power Rate Structure (large power)

Typical contract clauses:

1. ". . . Customer's maximum demand for the month or its contract demand is greater than 5000 kW . . ."

2. Clauses similar to clauses 2 to 8 listed in the General Power (medium power) contract given above.

3. . . . Contract shall be for a duration of 10 years . . .

4. . . . The minimum bill for any one month shall equal 70 percent of the highest maximum demand charge during the previous 36 months . . .

5. . . . Utility Company shall not be obligated to furnish power in greater amounts than the customer's contract demand . . ."

Rate Schedule

Demand charge: First 75 000 kW of demand per month at $2.50 per kW
                Excess over 75 000 kW of demand per month at $2.00 per kW
                Additional charge for any demand in excess of customer's contract demand at $2.20 per month per kW
Energy charge: First 20 million kW·h per month at 6.1 mills per kW·h*
                 Next 30 million kW·h per month at 6.0 mills per kW·h
                 Additional energy at 5.9 mills per kW·h

* 1 mill = one thousandth of a dollar or one-tenth of a cent.

3. *Medium power* – power of 100 kilowatts and more, but less than 5000 kilowatts;

4. *Large power* – power in excess of 5000 kilowatts.

Table 27B shows typical rate structures that apply to each of these categories. In addition, a contract is usually drawn up between the electrical utility and the medium- or large-power customer. The contract may stipulate a minimum monthly demand, a minimum power factor, the voltage regulation, and various other clauses concerning firm power, growth rate, liability, off-peak energy, seasonal energy, price increases, and so forth. However, in the case of domestic tariffs, the rate schedule is quite straightforward (see Table 27B).

27.6 Demand controllers

For industrial and commercial consumers the maximum demand plays an important role in compiling the electricity bill. Substantial savings can be made by keeping the maximum

Figure 27-6
Automatic demand controller that sheds nonessential loads whenever the demand reaches a preset level or according to a set schedule. This model can control up to 96 loads. (*Gentec Inc.*)

Figure 27-7
All-electric home that consumes a maximum of 9400 kW·h in January, and a minimum of 2100 kW·h in July.

demand as low as possible. Thus, an alarm can be installed to sound a warning whenever the demand is about to exceed a pre-established maximum. Loads that are not absolutely essential can then be temporarily switched off until the peak has passed. This procedure can be carried out automatically by a demand controller that connects and disconnects individual loads so as to stay within the prescribed maximum demand (Fig. 27-6). Such a device can easily save thousands of dollars per year for a medium-power customer.

Example 27-2:

Billing of a domestic customer

A homeowner consumes 900 kW·h during the month of August. Calculate the electricity bill using the residential rate schedule given in Table 27B.

Solution:

Minimum charge	= $ 5.00
First 100 kW·h @ 5 cent/kW·h	= 5.00
Next 200 kW·h @ 3 cent/kW·h	= 6.00
Remaining energy consumed	
(900 − 300) = 600 kW·h	
600 kW·h @ 2 cent/kW·h	= 12.00
TOTAL bill for the month:	$ 28.00

This represents an average cost of 2800/900 = 3.11 cent/kW·h.

Demand meters are not usually installed in homes because the maximum demand seldom exceeds 10 kW.

Table 27C shows the energy consumed by various electrical appliances found in a home. Figure 27-7 is an example of an all-electric home, heated by baseboard heaters.

Example 27-3:

Billing for a medium-power customer

A small industry operating night and day, 7 days a week, consumes 260 000 kW·h per month. The maximum demand is 1200 kW, and the maximum kVA demand is 1700 kVA. Note that the demand in this factory is measured for both active power (kW) and apparent power (kVA). Calculate the electricity bill using the medium-power rate schedule (Table 27B).

Solution:

Clause 9 is important here because 90% of the kVA demand is equal to 90% × 1700 = 1530 kVA which is greater than the maximum demand of 1200 kW. Consequently, the demand for billing purposes is 1530 kW and not 1200 kW. In effect, the power factor of the plant is low; consequently, the "billing demand" (1530) is higher than the metered demand (1200).

Applying the rate schedule, the demand charge is:

1530 kW @ $3.00/kW = $ 4 590

The energy charge for the first 100 h is:
1530 kW × 100 hours
= 153 000 kW·h @ 4 cent/kW·h
= 153 000 × 0.04 = 6 120

The energy charge for the next 50 000 kW·h is:
50 000 kW·h @ 2 cent/kW·h = 1 000

TABLE 27C **AVERAGE MONTHLY CONSUMPTION OF HOUSEHOLD APPLIANCES**

Average monthly consumption of a family of 5 persons in a modern house equipped with an automatic washing machine and a diswasher

appliance	kW·h consumed	appliance	kW·h consumed
Hot water heater (2000 gallons/month)	500	Automatic washing machine	100
Freezer	100	Coffee maker	9
Stove	100	Stereo system	9
Lighting	100	Radio	7
Dryer	70	Lawn mower	7
Dishwasher	30	Vacuum cleaner	4
Electric kettle	20	Toaster	4
Electric skillet	15	Clock	2
Electric iron	12		

The remainder of the energy is:
(260 000 − 153 000 − 50 000) = 57 000 kW·h

The energy charge for the remainder of the energy is:
57 000 kW·h @ 1.2 cent/kW·h   =   684
TOTAL bill for the month: $12 394

The average cost of energy is:
  unit cost = 12 394/260 000
         = 4.77 cent/kW·h

Figure 27-8 shows a plant equipped with both a demand controller and power-factor correcting capacitors.

Example 27-4:

Billing of a large-power customer

A paper mill consumes 28 million kilowatthours of energy per month. The demand meter registers a peak demand of 43 000 kW. Calculate the monthly bill using the large-power rate schedule given in Table 27B.

Solution:
a. The demand charge is:
    43 000 kW @ $2.50/kW   = $ 107 500

b. The energy charge for the first
    20 million kW·h is:
        20 million kW·h @ 6.1 mill/kW·h
    20 × 10^6 × 6.1/1000   = 122 000
    The energy charge for the next
    (28 − 20) = 8 million kW·h is:
    8 million kW·h @ 6.0 mill/kW·h
    8 × 10^6 × 6.0/1000   =   48 000
    TOTAL bill for the month: $ 277 500

Average cost per kW·h =
  277 500/28 × 10^6 = 9.9 × 10^{-3} = 9.9 mill

  or about 1 cent/kW·h

A monthly bill of nearly $300 000 may appear high, but we must remember that it probably represents less than 5 percent of the selling price of the finished product.

Figure 27-9 gives an idea of the power and energy consumed by a large city.

Figure 27-8
All-electric industry covering an area of 1300 m^2. It is heated by passing current through the reinforcing wire mesh embedded in the concrete floor. A load controller connects and disconnects the heating sections (non-priority loads) depending on the level of production (priority loads). The demand is thereby kept below the desired preset level. Annual energy consumption: 375 000 kW·h; maximum demand in winter: 92 kW; maximum demand during the summer: 87 kW. (*Lab-Volt*)

27.7 Power factor correction

Power factor correction (or improvement) is economically feasible whenever the decrease in the annual cost of electricity exceeds the amortized cost of installing the required capacitors. In some cases, the customer has no choice, but must comply with the minimum power factor specified by the utility company.

The power factor may be improved by installing capacitors at the service entrance to the factory or commercial enterprise. In other cases, it may be desirable to correct the power factor of an individual device, if its power factor is particularly low.

Example 27-5:
A factory draws an apparent power of 300 kVA at a power factor of 65% (lagging). What capac-

ity in kvar must be installed at the service entrance to bring the overall power factor to:
a. unity;
b. 85 percent lagging.

Solution:
a. Apparent power absorbed by the plant is:
$$S = 300 \text{ kVA}$$

Active power absorbed by the plant is:
$$P = S \cos \theta \qquad \text{Eq. 7-7}$$
$$= 300 \times 0.65 = 195 \text{ kW}$$

Reactive power absorbed by the factory is:
$$Q = \sqrt{S^2 - P^2} \qquad \text{Eq. 7-4}$$
$$= \sqrt{300^2 - 195^2} = 228 \text{ kvar}$$

To raise the power factor to unity, we have to supply all the reactive power absorbed by the load (228 kvar). The 3-phase capacitors must therefore have a rating 228 kvar. Figure 27-10a shows the active and reactive power flow.

b. The factory always draws the same amount of active power (195 kW) because the mechanical and thermal loads are fixed. Consequently, because the new overall power factor is to be 0.85 lagging, the apparent power drawn from the line must be:
$$S = P/\cos\theta$$
$$= 195/0.85 = 230 \text{ kVA}$$

The new reactive power supplied by the line is:
$$Q = \sqrt{230^2 - 195^2} = 121 \text{ kvar}$$

Because the plant still draws 228 kvar and the line furnishes only 121 kvar, the difference must come from the capacitors. The rating of these units is:
$$Q = (228 - 121) = 107 \text{ kvar}$$

Thus, if we can accept a power factor of 0.85 (instead of unity), we can install a smaller capacitor bank, and hence reduce

Figure 27-9
In 1988, the City of Montreal with about 2 million inhabitants consumed 26 500 GW·h of electric energy. Maximum demand during the winter, 5600 MW; during the summer, 2700 MW. (*Photo courtesy of Service de la C.I.D.E.M., Ville de Montréal*)

(a)

(b)

Figure 27-10
a. Overall power factor corrected to unity (Example 27-5)
b. Overall power factor corrected to 0.85.

the cost. Figure 27-10b shows the new power flow in the transmission line and the factory. Note that the factory draws the same active and reactive power, irrespective of the size of the capacitor installation.

The demand of commercial and industrial customers varies greatly throughout the day. As a result, it is common practice to install a variable capacitor unit at the service entrance. An automatic controller switches capacitor units in and out so that the power factor always lies between 90% and 95%.

Example 27-6:
A 600 kW induction furnace, connected to a 800 V single-phase line, operates at a power

factor of 0.6 lagging. It is supplied by a 4 kV line and a step-down transformer (Fig. 27-11).

a. Calculate the current in the 4000 V line.

b. If a 500 kvar capacitor is installed on the HV side of the transformer, calculate the new power factor and the new line current.

Solution:

This is an interesting example where individual power factor correction must be applied. The reason is that the induction furnace is a single-phase device whereas the plant is certainly energized by a 3-phase line. We cannot correct the power factor of single-phase equipment by adding balanced 3-phase capacitors at the service entrance.

a. Active power absorbed by the furnace is:
    $P = 600$ kW

    Apparent power absorbed by the furnace is:
    $S = P/\cos\theta = 600/0.6$
    $= 1000$ kVA

    Current in the 4 kV line is:
    $I = S/E = 1000/4$
    $= 250$ A

b. Reactive power absorbed by the furnace is:

    $Q = \sqrt{S^2 - P^2}$

    $= \sqrt{1000^2 - 600^2}$

    $= 800$ kvar

    Reactive power supplied by the capacitor is:
    $Q_C = 500$ kvar

    Reactive power which the line must supply is:
    $Q_L = Q - Q_c$
    $= 800 - 500 = 300$ kvar

    Active power drawn from the line is:
    $P_L = 600$ kW

Apparent power drawn from the line is:

$$S_L = \sqrt{P_L^2 + Q_L^2}$$

$$= \sqrt{600^2 + 300^2}$$

$$= 671 \text{ kVA}$$

New power factor is of the line:
$\cos\theta = P_L/S_L = 600/671 = 0.89$

The new line current is:
$I = S_L/E = 671/4$
$= 167$ A

By installing a single-phase capacitor bank, the line current drops from 250 A to 167 A, which represents a decrease of 33%. It follows that the I^2R loss, and voltage drop on the supply line will be greatly reduced. Furthermore, the power factor rises from 60% to 89% which will significantly reduce the monthly power bill. Finally, the 3-phase line currents are more likely to be reasonably balanced at the service entrance despite the presence of this large single-phase load.

Figure 27-11
Individual power factor correction (Example 27-8).

27.8 Measuring electrical energy, the watthourmeter

We have already seen that the SI unit of energy is the joule. However, for many years, power utilities have been using the kilowatthour to measure the energy supplied to industry and private homes. One kilowatthour (kW·h) is exactly equal to 3.6 MJ.

Meters which measure industrial and residential energy are called *watthourmeters*; they are designed to multiply power by time. The electricity bill is usually based upon the number of kilowatthours consumed during one month. Watthourmeters must therefore be very precise. *Induction* watthourmeters are practically the only types employed for residential metering.

Figure 27-12 shows the principal parts of such a meter: a potential coil B_p wound with many turns of fine wire; a current coil B_c; an aluminum disc D supported on a vertical spindle; a permanent magnet A; and a gear mechanism that registers the number of turns made by the disc. When the meter is connected to a single-phase line, the disc is subjected to a torque which causes it to turn, like a high-precision motor.

27.9 Operation of the watthourmeter

The operation of a watthourmeter can be understood by referring to Fig. 27-13. Load current I produces an alternating flux Φ_c which crosses the aluminum disc, inducing in it a voltage and, consequently, eddy currents I_f. On the other hand, potential coil B_p produces an alternating flux Φ_p which intercepts current I_f. The disc is therefore subjected to a torque which causes it to rotate. The torque is proportional to flux Φ_p and current I_f and the cosine of the angle between them. Because these two quantities depend respectively upon voltage E and load current I, the torque is proportional to $EI \cos \theta$, which is the active power delivered to the load. Thus, the average torque on the disc is zero when the E and I are 90° out of phase.

Figure 27-12a
Complete watthourmeter. (*General Electric*).

This, however, is only part of the story.

As the disc moves between the poles of permanent magnet A, a braking torque is produced whose value is proportional to the *speed* of the disc. Because the motor torque is always equal to the braking torque (see Sec. 3.11), it follows that the speed is proportional to the motor torque. The latter, we have seen, is proportional to the active power supplied to the load. Consequently, the number of turns per second is proportional to the number of joules per second. It follows that the number of turns of the disc is proportional to the number of joules (energy) supplied to the load.

27.10 Meter readout

In addition to other details, the nameplate of a watthourmeter lists the rated voltage, current and frequency, and a metering constant K_h. Constant K_h is the amount of energy, in watthours, which flows through the meter for each turn of the disc. Consequently, we can

Figure 27-12b
Components making up the meter. (*General Electric*).

Figure 27-13
Principle of operation of watthourmeter.

calculate the amount of energy which flows through a meter by counting the number of turns. Then, dividing energy by time, we can calculate the active power supplied to the load.

Example 27-7:
The nameplate of a watthourmeter shows K_h = 3.0. If the disc makes 17 turns in two minutes, calculate the energy consumed by the load during this interval, and the average power of the load.

Solution:
Each turn corresponds to an energy consumption of 3.0 W·h. Energy consumed during the 2-minute interval is:

$E_h = K_h \times$ number of turns
$\qquad = 3.0 \times 17$
$\qquad = 51$ W·h

Average power absorbed by the load during this interval is:

$P = E_h/t = 51/2$ min
$\qquad = 51/(1/30 \text{ h})$
$\qquad = 51 \times 30$
$\qquad = 1530$ W

Most watthourmeters have four dials to indicate the amount of energy consumed. The dials are read from left to right and the number so obtained is the number of kilowatthours consumed since the meter was first put in service. In reading the individual dials, we always take the number which the pointer swept over

last. For example, in Fig. 27-14, the reading is 1-5-9-0, or 1590 kW·h. Obviously, to measure the energy consumed during one month, we must subtract the readings at the beginning and end of the month. Some modern watthourmeters give a digital readout which, of course, is much easier to read.

27.11 Measuring three-phase energy and power

The energy consumed by a 3-phase load (three-wire system) can be measured with two single-phase watthourmeters. The two meters are often combined into one by mounting the two discs on the same spindle and using a single register (Fig. 27-15). The current and potential coils are connected to the line in the same way as those of two wattmeters.

Figure 27-16 is a 3-phase solid-state watthour-meter having a precision exceeding that of induction-type watthourmeters.

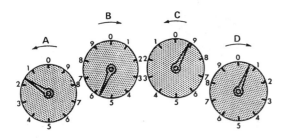

Figure 27-14
Reading the dials of a watthourmeter.

Figure 27-15
Watthourmeter for a 3-phase, 3-wire circuit (*General Electric*).

Figure 27-16
This high-precision electronic watthour-meter gives a numerical readout of the energy delivered by a 3-phase transmission line. It has an accuracy of 0.2 percent, which compares favorably with the 0.5 percent accuracy of some of the best induction-type watthourmeters. This meter is used on high-power lines where the monthly consumption exceeds 10 GW·h. (*Siemens*)

QUESTIONS AND PROBLEMS

Practical level

27-1 Explain what is meant by the following terms:

demand   billing demand
maximum demand   fixed cost
mill   demand interval

27-2 Using the rate schedule given in Table 27B calculate the power bill for a homeowner who consumes 920 kW·h in one month.

27-3 Explain why a low power factor in a factory results in a higher power bill.

27-4 Explain the behavior of a demand meter.

27-5 Using the rate schedule given in Table 27B for a medium power customer, calculate the monthly power bill under the following conditions:
demand meter reading = 120 kW
billing demand = 150 kW
energy consumed = 36 000 kW·h

Intermediate level

27-6 The demand meter in a factory registers a maximum demand of 4300 kW during the month of May. The power factor is known to be less than 70 percent.
a. If capacitors had been installed so as to raise the power factor to 0.9, would the maximum demand have been affected?
b. Would the billing demand have been affected?

27-7 According to Example 27-1, the maximum demand registered at 8:00 is 4.17 MW. If the demand meter were replaced by another one having a demand interval of 15 min, calculate the new value of the maximum demand at 8:00.

27-8 a. Give an estimate of the energy consumed in one year by a modern city of 300 000 inhabitants in North America (refer to Fig. 27-9).
b. If the average rate is 40 mill/kW·h, calculate the annual cost of servicing the city.

27-9 A motor draws 75 kW from a 3-phase line, at a cos θ = 0.72 lagging.
a. Calculate the value of Q and S absorbed by the motor.

b. If a 20 kvar, 3-phase capacitor is connected in parallel with the motor, what is the new value of P and Q supplied by the line?

c. Calculate the percent drop in line current after the capacitor is installed.

27-10 A plant draws 160 kW at a lagging power factor of 0.55.

a. Calculate the capacitors [kvar] required to raise the power factor to unity.

b. If the power factor is only raised to 0.9 lagging, how much less would the capacitor bank cost (in percent)?

27-11 a. Assuming that power in a large industry can be purchased at 15 mill/kW·h, estimate the hourly cost of running a 4000 hp motor having an efficiency of 96 percent.

b. If the motor runs night and day, 365 days per year, what would the annual saving be if the motor were redesigned to have an efficiency of 97%.

27-12 a. Referring to the residential rate schedule given in Table 27B, calculate the cost per kW·h if only 20 kW·h are consumed during a given month.

b. The heating element on an electric stove is rated at 1200 W. Using the same rate schedule, what is the least possible cost of running it for one hour?

27-13 A barrel of oil costing 32 dollars contains 42 gal (U.S.) having a heating value of 115 000 Btu/gal. When the fuel is burned in a thermal generating station to produce electricity, the overall efficiency is typically 35%. Calculate the minimum cost per kilowatthour, considering only the price of the fuel.

27-14 Describe the construction of a watthourmeter. Explain why the disc rotates.

27-15 The disc in Fig. 27-13 turns at 10 r/min for a load of 10 kW. If a 5 kvar capacitor is connected in parallel with the load, what is the new rate of rotation?

27-16 We want to determine the power of an electric heater, installed in a home, by means of a watthourmeter. All other loads are shut off and it is found that the disc makes ten complete turns in one minute. If $K_h = 3.0$, calculate the power of the heater.

27-17 a. The flux created by the permanent magnet in Fig. 27-13 decreases by 0.5% in 10 years. What is the effect on the speed of rotation and the precision of the meter?

b. The resistance of coil B_c changes with temperature. Does this affect the speed of rotation if the active load remains fixed?

27-18 A domestic watthourmeter has a precision of 0.7 percent. Calculate the maximum possible error if the monthly consumption is 800 kW·h.

28

DIRECT CURRENT TRANSMISSION

The development of high-power, high voltage converters has made it possible to transmit and control large blocks of power using direct current. Direct current transmission offers unique features that complement the characteristics of existing ac networks. We cover here some of the various ways it is being adapted and used, both in North America and throughout the world. However, before undertaking this chapter, the reader should first review the principles of power electronics covered in Chapter 21.

28.1 Features of dc transmission

What are the advantages of transmitting power by dc rather than by ac? They may be listed as follows:

1. DC power can be controlled much more quickly. For example, power in the megawatt range can be reversed in a dc line in less than one second. This feature makes it useful to operate dc transmission lines in parallel with existing ac networks. When instability is about to occur (due to a disturbance on the ac system), the dc power can be changed in amplitude to counteract and dampen out the power oscillations. Quick power control also means that dc short-circuit currents can be limited to much smaller values than those encountered on ac networks.

2. DC power can be transmitted in cables over great distances. We have seen that the capacitance of a cable limits ac power transmission to a few tens of kilometers (Sec. 25.29). Beyond this limit, the reactive power generated by cable capacitance exceeds the rating of the cable itself. Because capacitance does not come into play under steady-state dc conditions, there is theoretically no limit to the distance that power may be carried this way. As a result, power can be transmitted by cable under large bodies of water, where the use of ac cables is unthinkable. Furthermore, underground dc cable may be used to deliver power into large urban centers. Unlike over-

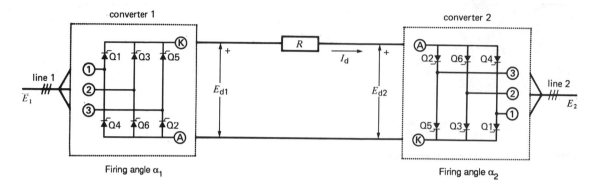

Figure 28-1
Elementary dc transmission system connecting 3-phase line 1 to 3-phase line 2.

head lines, underground cable is invisible, free from atmospheric pollution, and solves the problem of securing rights of way.

3. We have seen that ac power can only be transmitted between centers operating at the same frequency. Furthermore, the power transmitted depends upon line reactance and the phase angle between the voltages at each end of the line (Sec. 25.23). But when power is transmitted by dc, frequencies and phase angles make no difference, and line reactance does not limit the steady-state power flow. If anything, it is only the resistance of the line that limits the flow. This also means that power can be transmitted over greater distances by using dc. However, this is a marginal benefit because large blocks of ac power are already being carried over distances exceeding 1000 km.

4. Overhead dc transmission lines become economically competitive with ac lines when the length of the line exceeds several hundred kilometers. The width of the power corridor is less, and experience to date has shown that outages due to lightning are somewhat reduced. Consequently, dc transmission lines are being used to carry bulk power directly from a generating station located near a coal mine or waterfall, to the load center.

5. At the opposite extreme are so-called "back-to-back" converters which interconnect large adjacent ac systems with a dc transmission line that is only a few meters long. Back-to-back converters enable the two systems to operate at their respective frequencies and phase angles. As a result, disturbances on one system do not tend to destabilize the other system. Furthermore, the power flow between the systems can be modified and even reversed in a matter of milliseconds – much faster than could be achieved on an ac system.

Unlike ac transmission lines, it is not easy to tap power off at different points along a dc line. In effect, dc lines are usually point-to-point systems, tying one large generating station to one large power-consuming center. Electronic converters are installed at each end of the transmission line, but none in between. However, a multiterminal dc line is being planned to supply power to New England and points in Quebec.

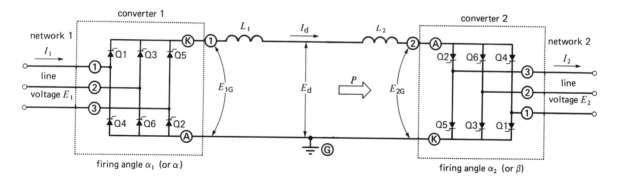

Figure 28-2
Smoothing inductors L_1 and L_2 are required between fluctuating dc voltages E_{1G}, E_{2G} and ripple-free voltage E_d.

28.2 Basic dc transmission system

A dc transmission system consists basically of a dc transmission line connecting two ac systems. A converter at one end of the line converts ac power into dc power while a similar converter at the other end reconverts the dc power into ac power. One converter acts therefore as a rectifier, the other as an inverter.

Stripped of everything but the bare essentials, the transmission system may be represented by the circuit of Fig. 28-1. Converter 1 is a 3-phase 6-pulse rectifier which converts the ac power of line 1 into dc power. The dc power is carried over a 2-conductor transmission line and reconverted to ac power by means of converter 2, acting as an inverter. Both the rectifier and inverter are naturally-commutated by the respective line voltages to which they are connected (Secs. 21.9, 21.20, 21.28, and 21.29). Consequently, the networks can function at entirely different frequencies, without affecting power transmission between them.

Power flow may be reversed by changing the firing angles α_1 and α_2, so that converter 1 becomes an inverter and converter 2 a rectifier. Changing the angles reverses the polarity of the conductors, but the direction of current flow remains the same. The reason is that thyristors can only conduct in one direction.

The dc voltages E_{d1} and E_{d2} at each converter station are identical, except for the IR drop in the line. The drop is usually so small that we can neglect it, except insofar as it affects losses, efficiency and conductor heating.

Owing to the high voltages encountered in transmission lines, each thyristor shown in Fig. 28-1 is actually composed of several thyristors connected in series. Such a group of thyristors is often called a valve. Thus, a valve for a 50 kV, 1000 A converter would typically be composed of 50 thyristors connected in series. Each converter in Fig. 28-1 would therefore contain 300 thyristors. The 50 thyristors in each bridge arm are triggered simultaneously so together they act like a super-thyristor.

28.3 Voltage, current and power relationships

In a practical transmission line, smoothing inductors L_1 and L_2 (Fig. 28-2) must be used as a buffer between the ripple-free dc voltage E_d and the undulating output of the converters (Secs. 21.10, 21.26, and 21.31). Thus, the potential difference between E_{1G} and E_d appears across inductor L_1. Similarly, the difference between E_{2G} and E_d appears across L_2. The inductors also reduce the ac harmonic currents flowing in the transmission line to an acceptable level.

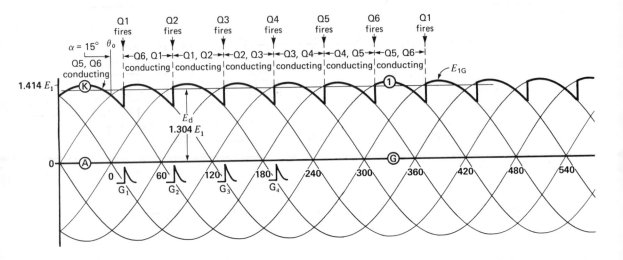

Figure 28-3a
Rectifier voltage waveshape E_{1G} for $\alpha = 15°$ (neglecting commutation overlap).

If we neglect commutation overlap, the waveshape of E_{1G} on the dc side of the rectifier is as shown in Fig. 28-3a. Similarly, the waveshape of E_{2G} is as shown in Fig. 28-3b. We have assumed firing angles of $\alpha_1 = 15°$ and $\alpha_2 = 150°$, respectively, for the rectifier and inverter. Consequently, the dc line voltage is given by:

$$E_d = 1.35\ E_1 \cos \alpha_1$$
$$= 1.35\ E_1 \cos 15° = 1.304\ E_1$$

similarly,

$$E_d = 1.35\ E_2 \cos \alpha_2$$

In the case of dc transmission lines, the rectifier angle α_1 is simply designated α. Furthermore, the inverter firing angle is not considered as a delay (α_2) with respect to the rectifier zero firing point, but as an angle of advance β. Thus, as regards the inverter, instead of stating that $\alpha_2 = 150°$ (as we have done in all previous inverter circuits, Chapters 21, 22 and 23), we now refer to it as an angle of advance $\beta = 30°$ (Fig. 28-3b). The value of β is related to α_2 by

the simple equation:

$$\beta = 180 - \alpha_2$$

Note that the inverter voltage is zero when β is 90°, and maximum when $\beta = 0°$.

The voltage, current, and power relationships of a dc transmission system are the same as those for any circuit containing ac/dc power converters. Referring to Fig. 28-2, and based upon equations we have already seen, the relationships may be stated as follows:

$$P = E_d I_d$$
$$E_d = 1.35\ E_1 \cos \alpha \qquad \text{Eq. 21-17}$$
$$E_d = 1.35\ E_2 \cos \beta$$
$$I_1 = I_2 = 0.816 I_d \qquad \text{Eq. 21-6}$$
$$Q_1 = P \tan \alpha \qquad \text{Eq. 21-18}$$
$$Q_2 = P \tan \beta$$

where

P = active power transmitted [W]
E_d = dc line voltage [V]
I_d = dc line current [A]
I_1, I_2 = effective values of the currents in ac lines 1 and 2 [A]

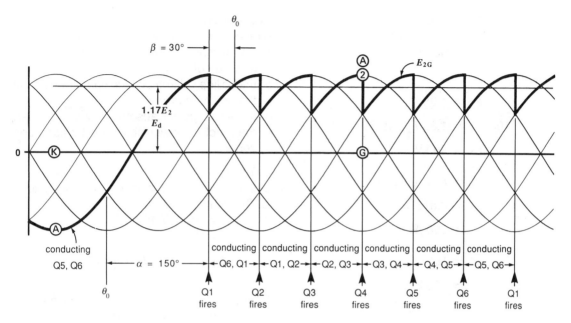

Figure 28-3b
Inverter voltage waveshape E_{2G} for $\beta = 30°$ (neglecting commutation overlap).

E_1, E_2 = effective values of the respective ac line voltages [V]

Q_1, Q_2 = reactive powers absorbed by converters 1 and 2 [var]

α = rectifier angle of delay [°]

β = inverter angle of advance [°]

In order to keep the reactive powers Q_1 and Q_2 as low as possible, we attempt to make α

and β approach 0°. However, for practical reasons, and taking account of commutation overlap, the "effective" value of α is about 25°, while that of β is about 35°**. Using these values, we can calculate the relative magnitudes of the voltages and currents in a scale model of a transmission line. Let the line deliver 1000 A at a potential of 100 kV (Fig. 28-4).

Thus, we have:

Figure 28-4
Scale model of a simple dc transmission system.

$$E_d = 1.35 \, E_1 \cos \alpha$$
$$100 \text{ kV} = 1.35 \, E_1 \cos 25°$$
$$E_1 = 82 \text{ kV}$$

Furthermore,

$$E_d = 1.35 \, E_2 \cos \beta$$
$$100 \text{ kV} = 1.35 \, E_2 \cos 35°$$
$$E_2 = 90 \text{ kV}$$

also

$$I_1 = I_2 = 0.816 \, I_d$$
$$= 0.816 \times 1000$$
$$= 820 \text{ A}$$

$$P = E_d I_d$$
$$= 100 \text{ kV} \times 1000 \text{ A}$$
$$= 100 \text{ MW}$$

$$Q_1 = P \tan \alpha$$
$$= 100 \tan 25°$$
$$= 47 \text{ Mvar}$$

$$Q_2 = P \tan \beta$$
$$= 100 \tan 35°$$
$$= 70 \text{ Mvar}$$

Figure 28-4 may thus be used as a scale model to determine the order of magnitude of the voltages, currents, and power in *any* dc transmission system.

** The "effective" firing angles α^* and β^* depend upon:
1. the actual firing angles α and β;
2. the commutation overlap angle μ;
3. the extinction angle γ.
The approximate relationship between these quantities is:
$\cos \alpha^* = 1/2(\cos \alpha + \cos (\alpha + \mu))$
$\cos \beta^* = 1/2(\cos \beta + \cos (\beta - \mu))$
$\beta = \gamma + \mu$

Example 28-1:
A dc transmission line operating at 150 kV carries a current of 400 A. Calculate the approximate value of:
a. the ac line voltage at each converter station;
b. the ac line current;
c. the active power absorbed by the rectifier;
d. the reactive power absorbed by each converter;

Solution:
Using the scale model of Fig. 28-4 we find:
a. The ac line voltages are:
$$E_1 = (150 \text{ kV}/100 \text{ kV}) \times 82$$
$$= 123 \text{ kV}$$

$$E_2 = (150 \text{ kV}/100 \text{ kV}) \times 90$$
$$= 135 \text{ kV}$$

b. The effective value of the line current is:
$$I_1 = I_2 = (400/1000) \times 820$$
$$= 328 \text{ A}$$

c. The active power absorbed by the rectifier:
$$P_1 = 150 \text{ kV} \times 400 \text{ A}$$
$$= 60 \, 000 \text{ kW}$$
$$= 60 \text{ MW}$$

d. The reactive power absorbed by each converter:
$$Q_1 = (60/100) \times 47$$
$$= 28 \text{ Mvar}$$

$$Q_2 = (60/100) \times 70$$
$$= 42 \text{ Mvar}$$

28.4 Power fluctuations on a dc line

In order to ensure stability in transmitting dc power, the rectifier and inverter must have special voltage-current characteristics. These characteristics are shaped by computer-controlled gate firing circuits. We can best understand the need for such controls by studying the behavior of the system when the controls are absent.

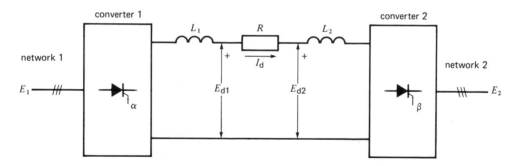

Figure 28-5
A small change in either E_{d1} or E_{d2} produces a very big change in I_d.

Figure 28-5 shows a dc transmission line having a resistance R. The converters produce voltages E_{d1} and E_{d2}, and the resulting dc line current is obviously given by:

$$I_d = (E_{d1} - E_{d2})/R \qquad (28\text{-}1)$$

The line resistance is always small; consequently, a very small difference between E_{d1} and E_{d2} can produce full-load current in the line. Furthermore, a small variation in either E_{d1} or E_{d2} can produce a very big change in I_d. For example, if E_{d1} increases by only a few percent, the line current can easily double. Conversely, if E_{d2} increases by only a few percent, the line current can fall to zero.

Unfortunately, both E_{d1} and E_{d2} are subject to sudden changes because the associated ac line voltages E_1 and E_2 may fluctuate. The fluctuations may be due to sudden load changes on the ac networks or to any number of other system disturbances that can occur. Owing to the almost instantaneous response of the converters and transmission line, the dc current could swing wildly under these conditions, producing erratic power swings between the two networks. Such power surges are unacceptable because they tend to destabilize the ac networks at each end, and because they produce misfiring of the SCRs.

It is true that the firing angles α and β could be modulated to counteract the ac line voltage fluctuations. However, it is preferable to design the system so that large, unpredictable dc power surges are inherently impossible. We now show how this is done.

28.5 Typical rectifier and inverter characteristic

In a practical dc transmission system, the computer-controlled rectifier circuit is designed to yield the E-I curve shown in Fig. 28-6a. Assuming a fixed ac line voltage E_1, the dc output voltage E_d is kept constant until the line current I_d reaches a value I_1. Beyond this point, E_d drops sharply as can be seen on the curve. This E-I characteristic is obtained by keeping α constant until current I_d approaches the desired value I_1. The firing angle then increases (automatically) at a very rapid rate, so that I_d is equal to I_1, when $E_{d1} = 0$. In other words, if a short-circuit were to occur across the dc side of the rectifier, the resulting dc current would be equal to I_1.

As regards the inverter, it is designed to give the E-I curve shown in Fig. 28-6b. Assuming a fixed ac line voltage E_2, voltage E_{d2} is maintained at zero until the dc line current reaches a value I_2. This means that from zero to I_2, the fir-

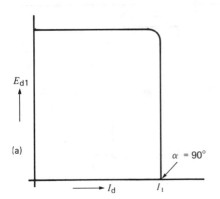

E_{d1}

(a)

$\alpha = 90°$

I_d I_1

Figure 28-6a
Rectifier $E\text{-}I$ characteristic.

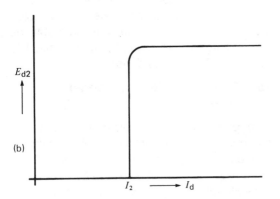

E_{d2}

(b)

I_2 I_d

Figure 28-6b
Inverter $E\text{-}I$ characteristic.

ing angle $\beta = 90°$. As soon as the dc current approaches the desired value I_2, the firing angle β decreases (automatically) to a limiting value of about $30°$.

Under normal operating conditions, the inverter voltage level is made to be slightly below the rectifier voltage level. Furthermore, limiting current I_2 is made slightly smaller than I_1. The effect of these constraints can best be seen by superposing the rectifier and inverter characteristics (Fig. 28-7a). The actual transmission-line voltage and current corresponds to the point of intersection of the two

curves. It is obvious that the line current I_d is equal to I_1 (determined by the rectifier characteristic) while the line voltage E_d is equal to E_{d2} (determined by the inverter characteristic). The difference between I_1 and I_2 is called the current margin ΔI. It is made to equal about 10 percent of the rated line current.

If the line has appreciable resistance, the IR drop modifies the effective rectifier characteristic so that it follows the dotted line shown in Fig. 28-7a. This, however, does not affect the operating point under normal conditions. The effective power input to the inverter is therefore given by the product $E_{d2}I_1$.

28.6 Power control

To vary the power flow over the dc line, the rectifier and inverter $E\text{-}I$ characteristics are modified simultaneously. Voltages E_{d1} and E_{d2} are kept constant but I_1 and I_2 are varied simultaneously while keeping the current margin fixed. Thus, Fig. 28-7b shows the new $E\text{-}I$ characteristics for a transmission line current I_d smaller than that in Fig. 28-7a. By thus shifting the $E\text{-}I$ characteristics back and forth, we can cause the dc power to vary from zero to maximum. Note that the line voltage E_{d2} is constant, and that it is always determined by the inverter. On the other hand, the magnitude of the line current is determined by the rectifier.

At this point, the reader may wonder why the $E\text{-}I$ characteristics have been given such odd shapes to attain such a simple result. The reason is that the dc system must be able to accommodate serious ac voltage fluctuations at either end of the line without affecting the dc power flow too much. It must also limit the magnitude of the fault currents, should a short-circuit occur on the dc line. We now explain how this is achieved.

28.7 Effect of voltage fluctuations

Referring to Fig. 28-7a, let us assume that the dc line carries full-load current I_1. If the ac voltage of line 1 increases suddenly, E_{d1} rises in

Figure 28-7a
Operating point when the transmission line delivers rated power.

Figure 28-7b
Operating point when the line delivers 20 percent of rated power.

proportion, but this has no effect on I_1 or E_{d2}. Consequently, the power flow over the line is unaffected.

On the other hand, if line voltage E_2 decreases, E_{d2} decreases in proportion. The dc line current is unaffected, but, because E_{d2} is smaller than before, the dc power is also less. However, the percent change in power cannot exceed the percent change in ac voltage E_2.

Next, if a large disturbance occurs on line 1,

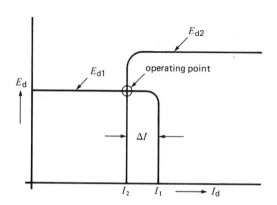

Figure 28-8
Change in operating point when E_1 falls drastically.

E_{d1} may fall drastically. This produces a new operating point, shown in Fig. 28-8. The dc line current decreases suddenly from I_1 to I_2, while the dc voltage decreases equally suddenly from E_{d2} to E_{d1}. With a current margin of 10 percent, the drop in current is not excessive. Consequently, the power flow is again not affected too much. As soon as the disturbance is cleared, the E-I characteristics return to the original curves given in Fig. 28-7a.

Finally, one of the worst conditions that can arise is a short-circuit on the dc line. Here again, the rectifier supplies a maximum current I_1 while the inverter draws a maximum current I_2. Consequently, the fault current

$$I_F = (I_1 - I_2)$$

is only 10 percent of the normal line current (Fig. 28-9). Fault currents are therefore much smaller than on ac transmission lines. In addition, because E_{d1} and E_{d2} are close to zero, the power delivered to the fault is small.

It is now clear that the special shape of the E-I characteristics prevents large power fluctuations on the line, and limits the short-circuit currents. In practice, the actual E-I characteris-

Figure 28-9
The short-circuit current at the fault cannot exceed 10 percent of the rated line current.

tics differ slightly from those shown in Fig. 28-7. However, the basic principle remains the same.

28.8 Bipolar transmission line

Most dc transmission lines are *bipolar*. They possess a positive line and a negative line and a common ground return (Fig. 28-10a). A converter is installed at the end of each line, and the line currents I_{d1} and I_{d2} flow in the directions shown. Converters 1 and 3 act as rectifiers while converters 2 and 4 are inverters. Power obviously flows over both lines from ac network 1 to ac network 2. The ground current is $I_{d1} - I_{d2}$. It is usually small because the converters automatically maintain equal currents in

the positive and negative lines.

The bipolar arrangement has three advantages. First, the ground current is small, under normal conditions. Consequently, corrosion of underground pipes, structures, and so forth is minimized. Second, the same transmission-line towers can carry two lines, thus doubling the power, with a relatively small increase in cost. Third, if power flow on one line is interrupted, the other can continue to function, delivering half the normal power between the ac networks.

28.9 Power reversal

To reverse power flow in a bipolar line, we change the firing angles, so that all the rectifiers become inverters and vice versa. This reverses the polarity of the transmission lines, but the line currents I_{d1} and I_{d2} continue to flow in the same direction as before (Fig. 28-10b).

28.10 Components of a dc transmission line

In order to function properly, a dc transmission system must have auxiliary components, in addition to the basic converters. Referring to Fig. 28-11, the most important components are:

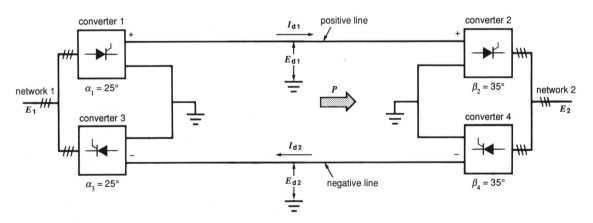

Figure 28-10a
Bipolar line transmitting power from network 1 to network 2.

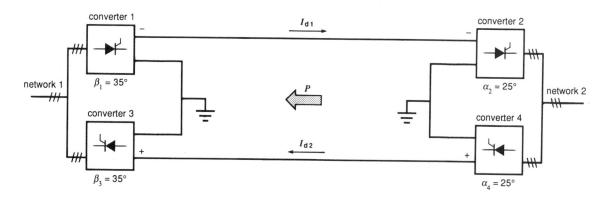

Figure 28-10b
Power reversal from network 2 to network 1 is obtained by reversing the line polarities.

1. dc line inductors (L)
2. harmonic filters on the dc side (F_{dc})
3. converter transformers (T)
4. reactive power source (Q)
5. harmonic filters on the ac side (F_{ac})
6. microwave communications link between the converter stations
7. ground electrodes (Gd)
   The need for these components is explained in the following sections.

28.11 Inductors and harmonic filters on the dc side (6-pulse converter)

Voltage harmonics are produced on the dc side of both the rectifier and inverter (Sec. 21.31). They give rise to 6th and 12th harmonic currents, and such currents, if allowed to flow over the dc line, could produce serious noise on neighboring telephone lines. Consequently, filters are required to prevent the currents from

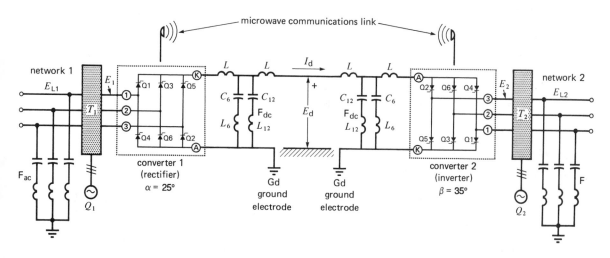

Figure 28-11
Schematic diagram showing some of the more important components of a HVDC transmission system.

flowing over the line. The filters consist of two inductors L and a shunt filter F_{dc}. The latter is composed of two series LC circuits, each tuned to respectively short-circuit the 6th and 12th harmonic currents to ground (Fig. 28-11).

The inductors L also prevent the dc line current from increasing too rapidly if a line fault should occur. This enables the thyristors to establish control before the current becomes too large to handle electronically.

28.12 Converter transformers

The basic purpose of the converter transformer on the rectifier side is to transform the network voltage E_{L1} to yield the ac voltage E_1 required by the converter. Three-phase transformers, connected in either wye-wye or wye-delta, are used. A tertiary winding (Sec. 12.5) is sometimes used for direct connection to a source of reactive power (Q_1).

As we have seen, the dc line voltage E_d is kept essentially constant from no-load to full-load. Furthermore, to reduce the reactive power absorbed by the converters, firing angles α and β should be kept small. This means that the ratio between ac voltage input and dc voltage output of the converters is essentially fixed. Because E_d is constant, it follows that E_1 must also be essentially constant.

Unfortunately, the network voltage E_{L1} may vary significantly throughout the day. Consequently, the converter transformers on the rectifier side must be equipped with taps so that the variable input voltage E_{L1} will give a reasonably constant output voltage E_1. The taps are switched automatically by a motorized tap changer whenever the network voltage E_{L1} changes for a significant length of time. For the same reasons, taps are needed on the converter transformers on the inverter side.

28.13 Reactive power source

The reactive power Q absorbed by the converters must be supplied by the ac network or by a local reactive power source. Because the

active power transmitted varies throughout the day, the reactive source must also be varied. Consequently, either variable static capacitors or a synchronous capacitor are required (Sec. 17.15).

28.14 Harmonic filters on the ac side

Three-phase 6-pulse converters produce 5th, 7th, 11th, 13th (and higher) current harmonics on the ac side. These harmonics are a direct result of the choppy current waveforms (Sec. 21.11). Again for reasons of telephone interference, these currents must not be allowed to flow over the ac lines. Consequently, the currents are bypassed through low-impedance filters F_{ac} connected between the three-phase lines and ground. The filters for each frequency are connected in wye, and the neutral point is grounded. On a 60 Hz network, each 3-phase filter is composed of a set of series-resonant LC circuits respectively tuned to 300, 420, 660, and 780 Hz.

At 60 Hz, these LC circuits are almost entirely capacitive. Consequently, they also furnish part of the reactive power Q absorbed by each converter.

28.15 Communications link

In order to control the converters at both ends of the line, a communications link between them is essential. For example, to maintain the current margin ΔI (Fig. 28-7), the inverter at one end of the line must "know" what the rectifier current setting I_1 is. This information must continually be relayed by a high-speed communications link between the two converters.

28.16 Ground electrode

Particular attention is paid to the ground electrode at each end of the dc line. Direct currents in the ground are much more corrosive than ac currents. Consequently, the actual ground electrode is usually situated several kilometers from the converter station, to ensure that dc leakage currents create no local problem. The dc ground wire between the station and the actual grounding site is either pole-mounted or enclosed in a shielding cable. At the grounding site, special means are used to minimize electrode resistance. This is particularly important when a bipolar system operates temporarily in the monopolar mode. Under these circumstances, the ground current may exceed 1000 A, and the heat generated may eventually dry out the grounding bed, causing the ground resistance to increase.

The best grounds are obtained next to, or in, large bodies of water. But even in this case, elaborate grounding methods must be used.

28.17 Example of a monopolar converter station

Figure 28-12 shows the elementary circuit diagram of a monopolar mercury-arc inverter station. The incoming dc line operates at 150 kV, and power is fed into a 230 kV, 3-phase, 60 Hz power line. Two smoothing inductors, each

having an inductance of 0.5 H, are connected in series with the dc line. The two LC filters effectively short-circuit the 6th and 12th harmonic voltages generated on the dc side of the converter. The 9 Ω and 11 Ω resistors make the filters less sensitive to slight frequency changes on the outgoing ac line.

Three single-phase transformers connected wye-wye (with a tertiary winding) are connected to the ac side of the converter. A 160 Mvar synchronous capacitor, connected to the tertiary winding, provides the reactive power for the converter.

Filters for the 5th, 7th, 11th, and 13th harmonic currents are connected between the three ac lines and neutral of the 230 kV system. As previously explained, the filters shunt the ac harmonic currents so that they do not enter the 230 kV line.

Figures 28-13 through 28-17 give us an idea of the size of these various components, and of the immense switchyard needed to accommodate them.

Figure 28-12
Simplified circuit of a 150 kV, 1800 A, 60 Hz mercury-arc inverter. See components illustrated in Figs. 28-13 to 28-17.

Figure 28-13
These 12 single-phase harmonic filters at an inverter station are tuned for 300 Hz, 420 Hz, 660 Hz and 780 Hz. They are connected between the three lines and neutral of the outgoing 230 kV, 60 Hz transmission line. The filter in the foreground is tuned to 720 Hz. It is a series circuit composed of a 2 Ω resistor, a group of capacitors having a total capacitance of 0.938 μF and an oil-filled inductor of 44.4 mH. The 720 Hz reactive power associated with the *LC* circuit amounts to 18.8 Mvar. (*GEC Power Engineering Limited, England*)

28.18 Thyristor converter station

Mercury-arc converters have been supplanted by thyristor converters and the design of the latter has become almost standardized. Thus, each pole is composed of two 6-pulse converters. Figure 28-18a shows how two 200 kV converters are connected to produce a 400 kV dc output. The dc sides are connected in series, while the ac sides are essentially connected in parallel across the 230 kV, 3-phase line. This means that converter 2 (and its transformer T2) functions at a dc potential of 200 kV. The windings must be especially well insulated to withstand these dc voltages, in addition to the 180 kV ac voltage.

The 180 kV windings of transformer T1 are connected in wye-delta, while those of transformer T2 are connected in wye-wye. This produces a 30° phase shift between the sec-ondary voltages of T1 and T2. Consequently, the thyristors in converter 1 and converter 2 do not fire at the same time. In effect, the two converters act as a 12-pulse converter.

One important result of the 30° shift is that the 5th and 7th harmonic currents generated by the two converters tend to cancel each other on the primary side of T1 and T2 and do not therefore appear in the 230 kV line. Consequently, the filtering equipment for these frequencies is substantially reduced. Furthermore, the 30° phase shift eliminates the 6th harmonic on the dc side, which reduces the filtering equipment needed for Fdc.

Figure 28-18b shows the three valves that make up a 12-pulse converter, together with the ac and dc connections. The valves are called *quadruple* valves because each is composed of four bridge arms. Figure 28-21b illustrates the impressive size of these valves.

Figure 28-14
Three-phase converter transformer bank rated
230 kV/127 kV/17 kV composed of 3 single-phase
transformers each rated 323 MVA. (*Manitoba Hydro*)

Figure 28-16
Portion of the three-phase 6-pulse mercury-arc inverter
rated 270 MW, 150 kV. (*Manitoba Hydro*)

Figure 28-15
View of one 0.5 H smoothing inductor on the 450 kV dc side
of the inverter station. The second inductor can be seen in
the distance (lower right - hand corner). The space between
the two units permits installing the filters on the dc side.
(*Manitoba Hydro*)

Figure 28-17
Partial view of the refrigeration unit needed to cool the
inverters. (*Manitoba Hydro*)

Figure 28-18a
Schematic diagram of one pole of a ± 400 kV converter station. It consists of two 200 kV converters connected in series on the dc side. The converters are 6-pulse units, respectively connected to 3-phase voltages that are 30° out of phase. The dc filter is tuned to the 12th harmonic. The ac filters prevent the 11th, 13th and higher harmonics from entering the 230 kV system. The static var compensator supplies the reactive power needed by the converter.

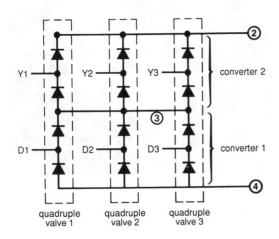

Figure 28-18b
Schematic diagram of the 12-pulse converter, showing the two 6-pulse converters and line connections. Three quadruple valves constitute the main components in one pole of the valve hall.

28.19 Typical installations

Power transmission by direct current is being used in many parts of the world. It will no doubt grow in importance as the reliability, speed and versatility of power converters is realized. The following installations give the reader an idea of the various types of systems that have been built, and the particular problem they were designed to solve.

1. **Schenectady**. Of historical interest is the 17 mile, 5.25 MW, 30 kV transmission line installed between Mechanicville and Schenectady, New York, in 1936. Using mercury arc converters, it tied together a 40 Hz and 60 Hz system.

2. **Gotland**. The first important dc transmission line was installed in Sweden, in 1954. It connected the Island of Gotland (in the middle of the Baltic Sea) to the mainland, by a 96 km submarine cable. The single-conductor cable operates at 100 kV and transmits 20 MW. The ground current returns by the sea.

3. **English Channel**. In 1961, a bipolar submarine link was laid in the English Channel between England and France. Two cables, one operating at + 100 kV and the other at − 100 kV, laid side by side, together carry 160 MW of power in one direction or the other. The power exchange between the two countries was found to be economical because the time zones are different, and consequently, the system peaks do not occur at the same time. Furthermore, France has excess hydro generating capacity during the spring, thus making the export of power attractive.

4. **Pacific Intertie**. In 1970, a bipolar link operating at ± 400 kV was installed between the Dalles, Oregon, and Los Angeles, California. The overhead line transmits a total of 1440 MW over a distance of 1370 km. Power can be made to flow in either direction, depending upon the requirements of the respective NW and SW regions. The dc link also helps stabilize the 3-phase ac transmission system connecting the two regions.

Figure 28-19
This converter station and switchyard at Eel River connects the ac networks of Quebec and New Brunswick by means of a dc link. The rectifier and inverter are both housed in the large building in the center. It pioneered the commercial use of solid-state thyristors in HVDC applications. (*New Brunswick Electric Power Commission*)

5. **Nelson River**. The hydropower generated by the Nelson River, situated 890 km north of Winnipeg, Canada, is transmitted by means of two bipolar lines operating at ± 450 kV. Each bipolar line carries 1620 MW, which is converted and fed into the ac system near Winnipeg. According to studies made, it was slightly more economical to transmit power by dc rather than by ac over this considerable distance.

6. **Eel River**. The converter station at Eel River, Canada, provides an asynchronous intertie between the 230 kV electrical systems of Quebec and New Brunswick. Although both systems operate at a nominal frequency of 60 Hz, it was not feasible to connect them directly, owing to stability considerations. In this application, the dc "transmission line" is only a few meters long, representing the length of the conductors needed to connect the rectifiers and inverters. Power may flow in either direction, up to a maximum of 320 MW (See Figs. 28-19 and 28-20).

7. **CU Project**. The power output of a generating station situated next to the lignite coal mines near Underwood, North Dakota, is converted to dc and transmitted 436 miles eastwards to a terminal near Minneapolis, Minnesota, where it is reconverted to ac. The bipolar line transmits 1000 MW at 1250 A, ± 400 kV. A metallic ground return is provided, in the event that one line should be out for a prolonged period (see Fig. 28-21a). Figure 28-21b shows the three quadruple valves that make up one 12-pulse converter.

8. **Châteauguay substation.** The Châteauguay substation, located near Montreal, is a back-to back converter station rated at 1000 MW (Fig. 28-22). In order to ensure high reliability, it is composed of two independent valve halls, each rated at 500 MW.

Power usually flows from the Hydro-Québec 735 kV ac system to the 765 kV ac system in the state of New York. Owing to the rectifier/ inverter link, frequency changes on one system do not affect the other system. Furthermore, the direction of power flow can be reversed, depending upon the circumstances.

The thyristors are water-cooled, using deionized water and an elaborate water/glycol/air heat exchanger.

Figure 28-20
View of one 6-pulse thyristor valve housed in its rectangular cubicle. It is fed by a 3-phase converter transformer and yields an output of 2000 A at 40 kV. The hundreds of individual thyristors it contains are triggered by a reliable, interference-free fiber-optic control system.
Eight such cubicles, together with three synchronous capacitors and four converter transformers make up the entire converter terminal. (*General Electric*)

Figure 28-21a

Simplified schematic diagram of the bipolar HVDC transmission system that links the Coal Creek Terminal in North Dakota to the Dickinson Terminal in Minnesota. The ±400 kV line delivers 1000 MW over a distance of 702 km. The output from two 500 MW turboalternators is stepped up to 230 kV and transmitted to the Coal Creek Terminal where the ac power is converted to dc. The on-load tap-changing converter transformers TCC are connected wye-wye and wye-delta for 12-pulse converter operation.

The 0.4 H smoothing inductors are in series with the grounded lines, thus significantly reducing the insulation requirements. The dc filters Fdc, each composed of a 48.8 mH inductor in series with a 1 μF capacitor bank, prevent the 12th harmonic voltage from reaching the dc lines.

The positive and negative transmission lines consist of two bundled conductors (2 × 1590 MCM, ACSR). The dc grounds are situated at 10.3 km and 20 km from the respective terminals. Under normal conditions the line currents are controlled automatically so that the ground current is 20 A or less. However, if one pole is out of service for short periods, the ground current can be as high as 1375 A.

The 12-pulse inverter station (Dickinson Terminal) feeds into a 345 kV, 60 Hz system and tap-changing converter transformers TD are used to regulate the inverter voltage level.

The control system is arranged to operate each terminal unmanned from a telecommunications control center located in Minnesota.

Figure 28-21b

View of three quadruple valves being installed in one of the Coal Creek Terminal valve halls. Together they constitute one pole of the 400 kV system. (*Courtesy United Power Association*)

Figure 28-21c
Aerial view of the Coal Creek Terminal, showing the two valve halls that respectively produce the + 400 kV and − 400kV dc voltage. The switchyard contains circuit breakers, transformers and filters.(*Courtesy United Power Association*)

Figure 28-22a
A quadruple valve, rated at 140 kV, 1200 A dc, being tested at IREQ, the Hydro-Québec research center. The valve is 12 m high, 6.9 m wide and 2.7 m deep. It is designed for the Châteauguay substation and contains a total of 400 thyristors.(*Courtesy Hydro-Québec*)

Figure 28-22b
View of one 500 MW, 140 kV, 3600 A dc back-to-back converter at the Châteauguay substation. It is composed of 6 quadruple valves. The three valves on the right usually operate as rectifiers while the three on the left usually function as inverters. The valve hall is 17.5 mm wide and 18 m high. (*Courtesy Hydro-Québec*)

QUESTIONS AND PROBLEMS

Practical level

28-1 Give three examples where dc power transmission is particularly useful.

28-2 Name the principal components making up a dc transmission system.

28-3 Which harmonics occur on the ac side of a converter? On the dc side?

28-4 What is the purpose of the large dc line inductors?

28-5 A dc transmission line operating at 50 kV, carries a current of 600 A. The terminal contains a single 3-phase, 6-pulse converter.
a. Calculate the approximate value of the secondary ac line voltage of the converter transformer.
b. What is the effective value of the secondary line current?

28-6 The bipolar line shown in Fig. 28-10a operates at a potential of ± 150 kV. If the dc line currents are respectively 600 A and 400 A, calculate:
a. the power transmitted between the two ac networks;
b. the value of the ground current.

28-7 The transmission line shown in Fig. 28-5 possesses a resistance of 10 Ω. The rectifier (converter 1) produces a dc voltage of 102 kV while the inverter generates 96 kV.
a. Calculate the dc line current and the power transmitted to network 2.
b. If the gates of the inverter are fired a little earlier in the cycle, will the dc line current increase or decrease? Explain.
c. If the inverter gates are fired so that the inverter generates 110 kV, will the power flow reverse? Explain.

28-8 When a short-circuit occurs on a dc line, the current in the fault itself is less than the full load current. Explain.

28-9 Why is a communications link needed between the converter stations of a dc line?

Intermediate level

28-10 The converters shown in Fig. 28-2 are identical 3-phase 6-pulse units, producing a voltage E_d of 50 kV and a current I_d of 1200 A.
a. Calculate the dc current per valve (bridge arm).
b. What is the approximate peak inverse voltage across each valve?

28-11 Referring to Section 28.19, part 4, calculate the line current per pole on the Pacific Intertie, and estimate the total reactive power absorbed by each converter station.

28-12 The ground electrode of a bipolar converter station is located 15 km from the station, and possesses a ground resistance of 0.5 Ω. If the line currents in each pole are respectively 1700 A and 1400 A, calculate the power loss at the electrode.

28-13 Referring to Fig. 28-11, it is given that E_d = 450 kV, I_d = 1800 A and the two dc smoothing inductors L each have an inductance of 0.5 H. If a short-circuit occurs between line and ground close to the rectifier station, calculate the magnitude of the rectifier current after 5 ms, assuming the gate triggering remains unaltered.

28-14 Each pole of the bipolar Nelson project (Fig. 28-18) is composed of two conductors (2-conductor bundle) of ACSR cable. Each conductor is composed of

72 strands of aluminum (diameter 0.16 in) and a central 7-strand (diameter 0.1067 in) steel core. Each 2-conductor bundle carries a nominal current of 1800 A over a distance of 550 miles. The voltage at the rectifier terminal is 450 kV. Neglecting the presence of the steel core, calculate:

a. the effective cross section of the 2-conductor bundle [in^2];
b. the line resistance of the 2-conductor bundle at a temperature of 20°C;
c. the corresponding I^2R loss;
d. the dc voltage at the inverter terminal;
e. the efficiency of the line (neglecting corona losses).

Advanced level

28-15 Referring to Fig. 28-12, calculate:
a. the resonant frequency of the two dc filters;

b. the value of the respective series impedances;
c. what is the dc voltage across the capacitors?

28-16 a. In Problem 28-15, if the sixth harmonic voltage generated on the dc side of the inverter is 20 kV, calculate the approximate value of the corresponding harmonic current.
b. Calculate the value of the 6th harmonic voltage across the 0.5 H inductor.
c. What is the value of the 360 Hz voltage at the input to the second 0.5 H line inductor?

REFERENCES

BOOKS

Allegheny, 1961. *Electrical materials handbook.* Pittsburgh: Allegheny Ludlum Steel Corp.

ANSI, 1971. *Schedules of preferred ratings and related required capabilities for ac high-voltage circuit breakers rated on a symmetrical current basis (ANSI C 37.06-1971).* New York: American National Standards Institute.

———, 1976. *Metric Practice* (ANSI Z210.1-1976)/(IEEE Std 268-1976)/(ASTM E380-76). New York: Institute of Electrical and Electronic Engineers.

Arillaga, J. 1983. *High Voltage Direct Current Transmission.* London: Peregrinus Ltd.

ASME, 1968. *Letter symbols for quantities used in electrical science and electrical engineering.* New York: American Soc. of Mech. Eng.

ASTM, 1979. *Annual Book of ASTM standards: Part 6 Copper and copper alloys.* Philadelphia: American Society for Testing and Materials.

Bean, R.L.; Chackan, N.; Moore, H.R. and Wentz, E.C. 1959. *Transformers for the electric power industry.* New York: McGraw.

Bedford, B.D. and Hoft, R.G. 1964. *Principles of inverter circuits.* New York: Wiley.

Beeman, D. 1955. *Industrial power systems handbook.* New York: McGraw.

Biddle, 1967. *"Getting down to earth . . . " A manual on earth resistance testing for the practical man.* Plymouth Meeting: James G. Biddle Co.

Bloomquist, W.C. (ed). 1950. *Capacitors for industry.* New York: Wiley.

Boll, R. 1978. *Soft magnetic materials.* Philadelphia: Heyden.

Bonneville Power Administration (1976). *Transmission line reference book HVDC to ± 600 kV.* Palo Alto: Electric Power Research Institute.

Bosich, J.F. 1970. *Corrosion prevention for practicing engineers.* New York: Barnes and Noble.

Boylestad, R.L. 1977. *Introductory circuit analysis.* Columbus: Merrill.

Bruins, P.F. 1968. *Plastics for electrical insulation.* New York: Wiley.

Cairns, E.J. and McBreen, J. 1975. *Advanced batteries: candidates for vehicle propulsion.* GMR-1871. Warren, Mich.: Research Labs, Gen. Motors.

Canadian Standards Association, 1979. *Canadian metric practice guide Z 234.1-79.* Toronto: CSA.

Considine, D.M. (ed). 1977. *Energy technology handbook.* New York: McGraw.

Dunning, J.S.; Bradley, T.G. and Zeitner, E.J. 1976. *Development of compact lithium/iron disulfide electrochemical cells.* GMR-2168. Warren, Mich.: Research Labs, Gen. Motors.

Edison Electric Institute, 1973. *Questions and answers about the electric utility industry.* New York: Edison Electric Institute.

Elgerd, O. 1971. *Electric energy systems theory: an introduction.* New York: McGraw.

Federal Power Commission, 1978. *Electric power statistics.* Washington: Superintendent of Documents.

Fink, D.G. and Beaty, H.W. (eds). 1978. *Standard handbook for electrical engineers.* New York: McGraw.

General Electric, 1975. *Transmission line reference book 345 kV and above.* New York: Fred Weidner and Son Printers Inc.

———, 1971. *Nickel cadmium battery application engineering handbook (GET 3148).* Gainesville: General Electric Company Battery Products Section.

Gingrich, H.W. 1979. *Electrical machinery, transformers, and controls.* Englewood Cliffs: Prentice-Hall.

Gyugyi, L. and Pelly, B.R. 1976. *Static power frequency changers.* New York: Wiley.

Hogerton, J.F. 1963. *Atomic fuel.* Oak Ridge: U.S. Atomic Energy Commission.

Holden, A. 1965. *The nature of solids.* New York: Columbia University Press.

IEEE, 1972. *IEEE standard dictionary of electrical and electronic terms.* New York: Wiley.

———, 1972. *IEEE recommended practice for grounding of industrial and commercial power systems.* New York: Inst. Electrical and Electronic Eng.

———, 1973. *IEEE standard and general requirements for distribution, power, and regulating transformers.* (IEEE Std 462-1973/ANSI C57. 12.00). New York: Inst. Electrical and Electronic Eng.

———, 1974. *IEEE Standard practices and requirements for thyristor converters for motor drives.* New York: Inst. Electrical and Electronic Eng.

———, 1979. *IEEE Standard metric practice.* New York: Inst. Electrical and Electronic Eng.

———, 1968. *IEEE guide for transformer impulse tests.* New York: Inst. Electrical and Electronic Eng.

———, 1971. *Graphic symbols for electrical and electronics diagrams.* New York: Inst. Electrical and Electronic Eng.

Indiana, 1963. *Design and application of permanent magnets, Manual 7.* Valparaiso: Indiana General Corp.

Jackson, H.W. 1976. *Introduction to electric circuits.* Englewood Cliffs: Prentice-Hall.

Johnson, R.C. 1971. *Electrical wiring.* Englewood Cliffs: Prentice-Hall.

Kimbark, E.W. 1971. *Direct current transmission.* New York: Wiley.

———, 1956. *Power system stability.* New York: Dover.

Kip, A.F. 1969. *Fundamentals of electricity and magnetism.* New York: McGraw.

Kosow, I.L. 1972. *Electric machinery and transformers.* Englewood Cliffs: Prentice-Hall.

———, 1973. *Control of electric machines.* Englewood Cliffs: Prentice-Hall.

Kurtz, E.B. and Showmaker, T.J. 1976. *The lineman's and cableman's handbook.* New York: McGraw.

Levine, J.N. 1961. *Selected papers on new techniques for energy conversion.* New York: Dover.

Liptai, R.G. and Pearson, J.W. 1975. *Metrication - managing the industrial transition.* Philadelphia: American Society for Testing and Materials.

Lye, R.W. (ed). 1976. *Power converter handbook.* Peterborough: Can. Gen. Electric Co. Ltd.

Marbury, R.E. 1949. *Power capacitors.* New York: McGraw.

Matsch, L.W. 1977. *Electromagnetic and electromechanical machines.* New York: Dun-Donnelley.

Morganite, 1961. *Carbon brushes for electrical machines.* London: Morganite Carbon Limited.

Moskowitz, L.R. 1976. *Permanent magnet design and application handbook.* Boston: Cahners.

NEMA, 1970. *Electrical insulation terms and definitions.* New York: Insulating materials division, National Elect. Manufacturers Ass'n.

———, 1978. *Motors and Generators* (ANSI/NEMA No. MG1-1978). Washington: National Electrical Manufacturers Association.

Nesbitt, E.A. 1962. *Ferromagnetic domains, a basic approach to the study of magnetism.* Baltimore: Bell Telephone Labs, Inc.

OECD, 1974. *Twenty-seventh survey of electric power equipment.* Paris: Organization for Economic Co-operation and Development.

Ogorkiewicz, R.M. 1970. *Engineering properties of thermoplastics.* London: Wiley.

Oklahoma State University, 1977. *Power factor improvement.* Stillwater: OSU Elect. Power Technology Dept.

Olson, E. 1966. *Applied Magnetism.* New York: Springer-Verlag.

Pearman, R.A. 1980. *Power electronics, solid state motor control.* Reston: Reston Publ. Co.

Pelly, B.R. 1971. *Thyristor phase-controlled converters and cycloconverters.* New York: Wiley.

Ramshaw, R.S. 1973. *Power electronics thyristor controlled power for electric motors.* London: Chapman and Hall/Halsted Pr.

Richardson, D.V. 1978. *Rotating electric machinery and transformer technology.* Reston: Reston Publ. Co.

Russel, C.R. 1967. *Elements of energy conversion.* New York: Pergamon Press.

Sanford, R.L. and Cooter, I.L. 1962. *Basic magnetic quantities and the measurement of the magnetic properties of materials.* Washington: Nat. Bur. Stds., U.S. Dept. Comm.

Schaefer, J. 1965. *Rectifier circuits.* New York: Wiley.

Schonland, B. 1964. *The flight of thunderbolts.* London: Clarendon Press.

Seel, F. 1963. *Atomic structure and chemical bonding.* New York: Wiley.

Stevenson, W.D. Jr. 1975. *Elements of power system analysis.* New York: McGraw.

Stigant, S.A. and Franklin, A.C. 1973. *The J & P transformer book.* Boston Newnes-Butterworths.

Tarnawecky, M.Z. 1971. *Manitoba power conference EHV-DC.* Winnipeg: University of Manitoba.

TVA, 1975. *Annual report of the Tennessee Valley Authority, Volume II - appendices.* Knoxville: Tennessee Valley Authority.

Uhlmann, E. 1975. *Power transmission by direct current.* New York: Springer-Verlag.

Uman, M.A. 1969. *Lightning.* New York: McGraw.

Underwriters Laboratories, 1978. *Systems of insulating materials* (UL 1446). Northbrook, Ill: Underwriters Laboratories Inc.

———, 1977. *Ground-fault circuit interrupters* (UL 943). Northbrook, Ill: Underwriters Laboratories Inc.

Veinott, C.G. 1972. *Computer-aided design of Electrical Machinery.* Cambridge: MIT Pr.

Vielstich, W. 1970. *Fuel cells.* London: Wiley.

Vinal, G.W. 1950. *Primary batteries.* New York: Wiley.

———, 1955. *Storage batteries.* New York: Wiley.

Werninck, E.H. (ed). 1978. *Electric motor handbook.* New York: McGraw.

Waddicor, H. 1935. *The principles of electric power transmission by alternating currents.* New York: Wiley.

Westinghouse, 1964. *Electrical transmission and distribution reference book.* Pittsburgh: Westinghouse Electric Corp.

——, 1965. *Electrical utility engineering reference book.* Pittsburgh: Westinghouse Electric Corp.

Wildi, T. 1973. *Understanding Units.* Québec: Sperika Enterprises Ltd.

——, 1988. *Units and Conversion Charts.* Québec: Sperika Enterprises Ltd.

REFERENCES

ARTICLES AND TECHNICAL REPORTS

Allan, R. (ed). 1975. Power semiconductors. *Spectrum.* Nov.: 37-44.

Allan, J.A. et al 1975. Electrical aspects of the 8750 hp gearless ball-mill drive at St. Lawrence Cement Company. *IEEE Trans. Ind. Gen. Appl.* 1A-11, No. 6: 681-687.

Apprill, M.R. 1978. Capacitors reduce voltage flicker. *Electrical World* 189, No. 4: 55-56.

American National Standard, 1979. Standard nominal diameters and cross sectional areas of AWG sizes of solid round wires used as electrical conductors. *ANSI/ASTM* B 258-65 (Part 6): 536-542.

Bachmann, K. 1972. Permanent magnets. *Brown Boveri Rev.* 9: 464-468.

Balzhiser, R.E. 1977. Capturing a star: controlled fusion power. *EPRI Journal* Dec.: 6-13.

Beaty, H.W. (ed). 1978. Underground distribution. *Electrical World* 189, No. 9: 51-66.

Beaty, H.W. 1978. Charts determine substation grounds. *Electrical World* 189, No. 2: 56-58.

Beutler, A.J.; Staats, G. 1969. A 100 000-Joule system for charging permanent magnets. *IEEE Trans. Ind. and Gen. Appl.* 1GA-5, No. 1: 95-100.

Creek, F.R.L. 1976. Large 1200 MW four-pole generators for nuclear power stations in the USA. *GEC J. Sc. and Tech.* 43, No. 2: 68-76.

Dalziel, C.F. 1968. Reevaluation of lethal electric currents. *IEEE Trans. Ind. and Gen. Appl.* 1GA-4, No. 5: 467-475.

Drake, W.D. and Sarris, A.E. 1968. Lightning protection for cement plants. *IEEE Trans. Ind. Gen. Appl.* 1GA-4, No. 1: pp. 57-67.

Duff, D.L. and Ludbrook, A. 1968. Semiconverter rectifier go high power. *IEEE Trans. Ind. Gen. Appl.* 1GA-4, No. 2: 185-192.

DuPont, 1972. "Tefzel" ETFE fluoropolymer: an exciting new resin. *Journal of "Teflon"* 13: No. 1.

Electrical World 1978. Statistical Report. *Electrical World* 189, No. 6: 75-106.

Elliott, T.C. (ed). 1976. Demand control of industry power cuts utility bills, points to energy savings. *Power* 120, No. 6: 19-26.

Engstrom, L.; Mutanda, N.M. Adams, N.G. and Flisberg, G. 1975. Refining copper with HVDC. *Spectrum* 12, No. 12: 40-45.

Fagenbaum, J. 1980. Cogeneration: an energy saver. *Spectrum* 17, No. 8: 30-34.

Fickett, A. 1976. Fuel cells - versatile power generators. *EPRI Journal*, April: 14-19.

Friedlander, G.D. 1977. UHV: onward and upward. *Spectrum* 14, No. 2: pp. 57-65.

Gazzana-Priaroggia, P.; Piscioneri, J.H. and Margolin, S.W. 1971. The Long Island Sound submarine cable interconnection. *Spectrum* 8, No. 10: pp. 63-71.

Goodbrand, W. and Ross, C.A. 1972. Load-rejection testing of steam turbine generators. *Ontario Hydro Rsrch Qtly.* 24, No. 2: 1-7.

Graf, K. 1973. AC pump drives with static-converter slip-power recovery system for the Lake Constance water supply scheme. *Siemens Rev.*, XL: 539-542.

Hand, C. (ed). 1977. Batteries. *Canadian Electronics Engineering*, May: 19-28.

Harris, S.W. 1960. Compensating dc motors for fast response. *Control Enginnering* Oct. 1960: 115-118.

Herbst, W.; Käuferle, F.P. and Reichert, K. 1974. Controllable static reactive power compensation. *Brown Boveri Rev.* 61: 433-439.

Hoffmann, A.H. 1969. Brushless synchronous motors for large industrial drives. *IEEE Trans. Ind. and Gen. Appl.* 1GA-5, No. 2: 158-162.

Holburn, W.W. 1970. Brushless excitation of 660 MW generators. *Journal of Science and Technology.* 37, No. 2: 85-90.

Hossli, W. (ed). 1976. Large steam turbines. *Brown Boveri Rev.* 63: 84-147.

Iwanusiw, O.W. 1970. Remanent flux in current transformers. *Ont. Hydro Rsrch Qtly.* 3rd Qtr: 18-21.

Jacobs, A.P. and Walsh, G.W. 1968. Application considerations for SCR dc drives and associated power systems. *IEEE Trans. Ind. Gen. Appl.* 1GA-4, No. 4: pp. 396-404.

Javetski, J. 1978. Cogeneration. *Power.* 122, No. 4: 35-40.

Joos, G. and Barton, H.B. 1975. Four-quadrant DC variable-speed drives - design considerations. *Proc. IEEE,* 63, No. 12: 1660-1668.

Joyce, J.S. 1974. Factors influencing reliability of large generators for nuclear power plants. *IEEE Trans. Pwr App. Syst.,* PAS-93, No. 1: 210-219.

Kolm, H.H. and Thornton, R.D. 1973. Electromagnetic flight. *Scientific American.* 229, No. 4: pp. 17-25.

Krick, N. and Noser, R. 1976. The growth of turbo-generators. *Brown Boveri Rev.* 63: 148-155.

Kronenberg, K.J. and Bohlmann, M.A. 1968. Stability of permanent magnets. *Applied Magnetics; Indiana Current Corp.* 15, No. 1.

Lee, R.H. 1975. The effect of color on temperature of electrical enclosures subject to solar radiation. *IEEE Trans. Ind. Appl.* IA-11, No. 6: 646-653.

Lengyel, G. 1962. Arrhenius theorem aids interpretation of accelerated life test results. *Can. Electronics Engineering.* Nov.: 35-39.

Lyman, J. (ed). 1975. Battery technology. *Electronics.* April: 75-82.

Machine Design, 1979. Electrical and electronics reference issue. *Machine Design Magazine* 51, No. 11: entire issue.

Manian, V.S. 1971. Electric water heaters for high-usage residences. *Ontario Hydro Rsrch Quarterly,* 3rd quarter: 7-13.

McCormick, L.S. and Hedding, R.A. 1974. Phase-angle regulators optimize transmission line power flows. *Westinghouse Engineer.* July: 87-91.

McLaughlin, M.H. and Vonzastrow, E.E. 1975. Power semiconductor equipment cooling methods and application criteria. *IEEE Trans. Ind. Appl.* IA-11, No. 5: 546-555.

Matta, U. (ed). 1975. Industrial electro-heat. *Brown Boveri Rev.* 62: 4-67.

Methé, M. 1971. The distribution of electrical power in large cities. *Canadian Engineering Journal.* October: 15-18.

Minnick, L. and Murphy, M. 1976. The breeder - when and why. *EPRI Journal* March: 6-11.

Moor, J.C. 1977. Electric drives for large compressors. *IEEE Trans. Ind. Gen. Appl.* 1A-13, No. 5: pp. 441-449.

Patterson, W.A. 1977. The Eel River HVDC scheme. *Can. Elec. Eng. J.* 2, No. 1: 9-16.

Quinn, G.C. 1977. Plant primary substation trends. *Power* 121, No. 3: 29-35.

Rieder, W. 1971. Circuit breakers: *Scientific American* 224, No. 1: 76-84.

Robinson, E.R. 1968. Redesign of dc motors for applications with thyristors power supplies. *IEEE Trans. Ind. Gen. Appl.* 1GA-4, No. 5: 508-514.

Robichaud, G.G. and Tulenko, J.S. 1978. Core design: fuel management as a balance between in-core efficiency and economics. *Power* 122, No. 5: 52-57.

Schwieger, R. (ed). 1978. Industrial Boilers - what's happening to-day. *Power* 122, No. 2: s1-s24.

Sebesta, D. 1978. Responsive ties avert system breakup. *Electrical World* 180, No. 7: 54-55.

Selzer, A. 1971. Switching in vacuum: a review. *Spectrum* 8, No. 6: pp. 26-37.

Sullivan, R.P. 1978. Uranium: key to the nuclear fuel cycle. *Power* 122, No. 4: 58-63.

Summers, C.M. 1971. The conversion of energy *Scientific American* 224, No. 3: 148-160.

Vendryes, G.A. 1977. Superphénix: a full-scale breeder reactor. *Scientific American* 236, No. 3: pp. 26-35.

Waldinger, H. 1971. Converter-fed drives. *Siemens Rev.* 38: 387-390.

Woll, R.F. 1964, High temperature insulation in ac motors. *Westinghouse Engineer* Mar. 1964: 1-5.

Woll, R.F. 1975. Effect of unbalanced voltage on operation of polyphase induction motors. *IEEE Trans. Ind. Appl.* IA-11, No. 1: 38-42.

Woll, R.F. 1977. Electric motor drives for oil well walking beam pumps. *IEEE Trans. Ind. Gen. Appl.* 1A-13, No. 5: 437-441.

Woodson, R.D. 1971. Cooling towers. *Scientific American* 224, No. 5: 70-78.

Zimmermann, J.A. 1969. Starting requirements and effects of large synchronous motors. *IEEE Trans. Ind. and Gen. Appl.* 1GA-5, No. 2: 169-175.

APPENDIX

AX0 CONVERSION CHARTS

The conversion charts listed in this Appendix make it very easy to convert from one unit to any other. Their use is explained in Chapter 1, Section 10.

Quantities such as AREA, MASS, VOLUME, and so forth, are listed in alphabetical order for quick reference. Multipliers between units are either exact, or accurate to ± 0.1 percent.

Examples at the bottom of each page are intended to assist the reader in applying the conversion rule which basically states:

WITH THE ARROW — MULTIPLY

AGAINST THE ARROW — DIVIDE

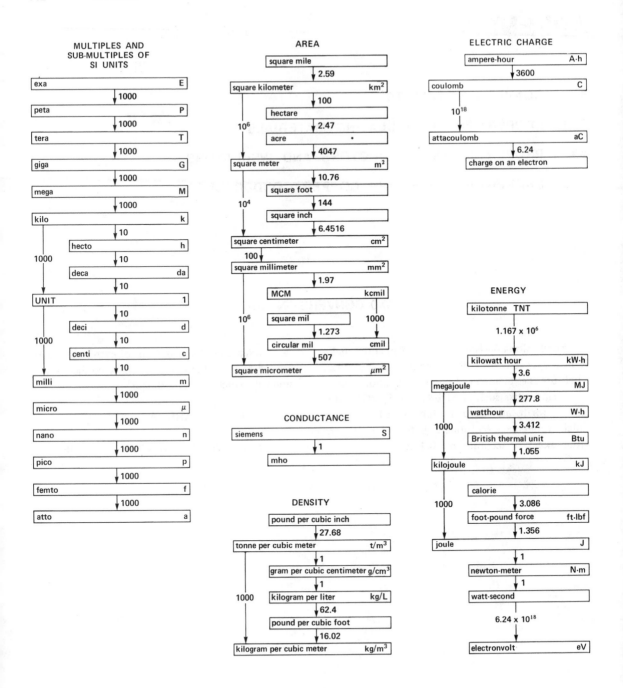

MULTIPLES AND SUB-MULTIPLES OF SI UNITS

exa	E
	↓ 1000
peta	P
	↓ 1000
tera	T
	↓ 1000
giga	G
	↓ 1000
mega	M
	↓ 1000
kilo	k
	↓ 10
hecto	h
	↓ 10
deca	da
	↓ 10
UNIT	1
	↓ 10
deci	d
	↓ 10
centi	c
	↓ 10
milli	m
	↓ 1000
micro	μ
	↓ 1000
nano	n
	↓ 1000
pico	p
	↓ 1000
femto	f
	↓ 1000
atto	a

(1000 from kilo to milli; 1000 from milli-side vertical arrows)

AREA

square mile
↓ 2.59
square kilometer   km²
↓ 100
hectare
↓ 2.47
acre   *
↓ 4047
square meter   m²
↓ 10.76
square foot
↓ 144
square inch
↓ 6.4516
square centimeter   cm²
↓ 100
square millimeter   mm²
↓ 1.97
MCM   kcmil
square mil   1000
↓ 1.273
circular mil   cmil
↓ 507
square micrometer   μm²

(10⁶, 10⁴, 10⁶ side multipliers)

CONDUCTANCE

siemens   S
↓ 1
mho

DENSITY

pound per cubic inch
↓ 27.68
tonne per cubic meter   t/m³
↓ 1
gram per cubic centimeter g/cm³
↓ 1
kilogram per liter   kg/L
↓ 62.4
pound per cubic foot
↓ 16.02
kilogram per cubic meter   kg/m³

(1000 side multiplier)

ELECTRIC CHARGE

ampere-hour   A·h
↓ 3600
coulomb   C
↓ 10¹⁸
attacoulomb   aC
↓ 6.24
charge on an electron

ENERGY

kilotonne TNT
↓ 1.167 x 10⁶
kilowatt hour   kW·h
↓ 3.6
megajoule   MJ
↓ 277.8
watthour   W·h
↓ 3.412
British thermal unit   Btu
↓ 1.055
kilojoule   kJ
calorie
↓ 3.086
foot-pound force   ft·lbf
↓ 1.356
joule   J
↓ 1
newton-meter   N·m
↓ 1
watt-second
↓ 6.24 x 10¹⁸
electronvolt   eV

(1000 side multipliers)

Example: Convert 1590 MCM to square inches.

Solution: 1590 MCM = 1590 (÷ 1.97) (÷ 100) (÷ 6.4516) in² = 1.25 in².

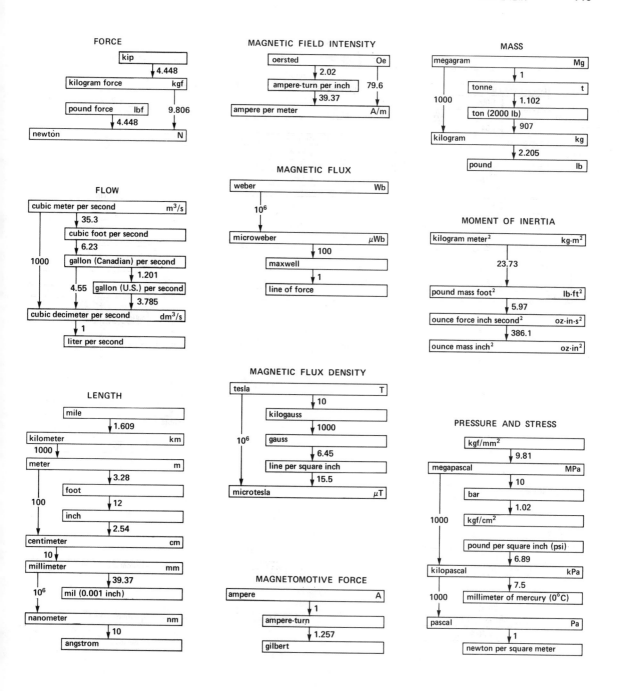

FORCE

kip
↓ 4.448
kilogram force   kgf
pound force   lbf   9.806
↓ 4.448
newton   N

MAGNETIC FIELD INTENSITY

oersted   Oe
↓ 2.02
ampere-turn per inch   79.6
↓ 39.37
ampere per meter   A/m

MASS

megagram   Mg
↓ 1
tonne   t
1000   ↓ 1.102
ton (2000 lb)
↓ 907
kilogram   kg
↓ 2.205
pound   lb

FLOW

cubic meter per second   m³/s
↓ 35.3
cubic foot per second
↓ 6.23
1000   gallon (Canadian) per second
↓ 1.201
4.55   gallon (U.S.) per second
↓ 3.785
cubic decimeter per second   dm³/s
↓ 1
liter per second

MAGNETIC FLUX

weber   Wb
↓ 10⁶
microweber   μWb
↓ 100
maxwell
↓ 1
line of force

MOMENT OF INERTIA

kilogram meter²   kg·m²
23.73
pound mass foot²   lb·ft²
↓ 5.97
ounce force inch second²   oz·in·s²
↓ 386.1
ounce mass inch²   oz·in²

LENGTH

mile
↓ 1.609
kilometer   km
1000 ↓
meter   m
↓ 3.28
foot
100   ↓ 12
inch
↓ 2.54
centimeter   cm
10 ↓
millimeter   mm
↓ 39.37
10⁶   mil (0.001 inch)
nanometer   nm
↓ 10
angstrom

MAGNETIC FLUX DENSITY

tesla   T
↓ 10
kilogauss
↓ 1000
10⁶   gauss
↓ 6.45
line per square inch
↓ 15.5
microtesla   μT

MAGNETOMOTIVE FORCE

ampere   A
↓ 1
ampere-turn
↓ 1.257
gilbert

PRESSURE AND STRESS

kgf/mm²
↓ 9.81
megapascal   MPa
↓ 10
bar
↓ 1.02
1000   kgf/cm²
pound per square inch (psi)
↓ 6.89
kilopascal   kPa
↓ 7.5
1000   millimeter of mercury (0°C)
pascal   Pa
↓ 1
newton per square meter

Example: Convert 580 psi to megapascals.
Solution: 580 psi = 580 (x 6.89) (÷ 1000) MPa = 4 MPa.

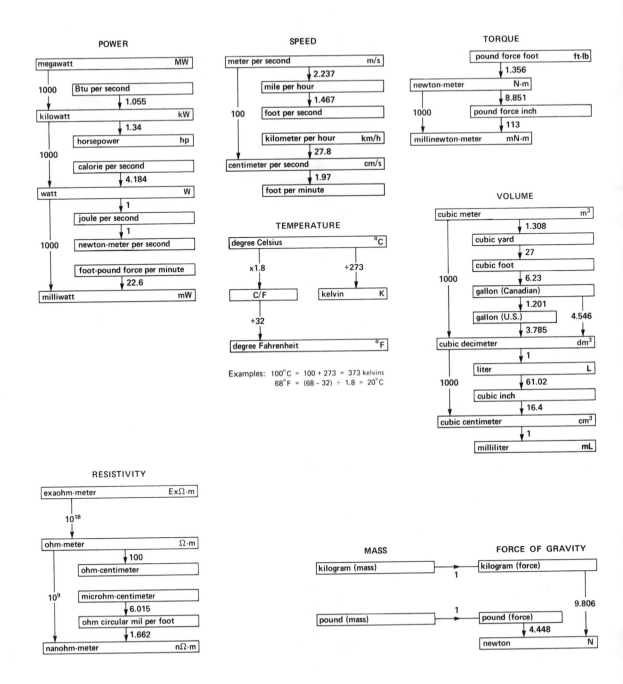

POWER

| megawatt | MW |

↓ 1000

| Btu per second |

↓ 1.055

| kilowatt | kW |

↓ 1.34

| horsepower | hp |

↓ 1000

| calorie per second |

↓ 4.184

| watt | W |

↓ 1

| joule per second |

↓ 1

| newton-meter per second |

↓ 1000

| foot-pound force per minute |

↓ 22.6

| milliwatt | mW |

SPEED

| meter per second | m/s |

↓ 2.237

| mile per hour |

↓ 1.467

| foot per second |

100

| kilometer per hour | km/h |

↓ 27.8

| centimeter per second | cm/s |

↓ 1.97

| foot per minute |

TEMPERATURE

| degree Celsius | °C |

×1.8   +273

| C/F | | kelvin | K |

+32

| degree Fahrenheit | °F |

Examples: $100°C = 100 + 273 = 373$ kelvins
$68°F = (68 - 32) \div 1.8 = 20°C$

TORQUE

| pound force foot | ft·lb |

↓ 1.356

| newton-meter | N·m |

↓ 8.851

| pound force inch |

↓ 113

1000

| millinewton-meter | mN·m |

VOLUME

| cubic meter | m³ |

↓ 1.308

| cubic yard |

↓ 27

| cubic foot |

↓ 6.23

1000

| gallon (Canadian) |

↓ 1.201

| gallon (U.S.) | 4.546

↓ 3.785

| cubic decimeter | dm³ |

↓ 1

| liter | L |

↓ 61.02

1000

| cubic inch |

↓ 16.4

| cubic centimeter | cm³ |

↓ 1

| milliliter | mL |

RESISTIVITY

| exaohm-meter | ExΩ·m |

↓ 10^{18}

| ohm-meter | Ω·m |

↓ 100

| ohm-centimeter |

10^9

| microhm-centimeter |

↓ 6.015

| ohm circular mil per foot |

↓ 1.662

| nanohm-meter | nΩ·m |

MASS   **FORCE OF GRAVITY**

| kilogram (mass) | → | kilogram (force) |
     1

| pound (mass) | → | pound (force) |
     1   ↓ 4.448   9.806

| newton | N |

Example: Calculate the force of gravity (in newtons) that acts on a mass of 9 lb.
Solution: 9 lbm → 9 (x 1) (x 4.448) N = 40 N.

TABLE AX1

PROPERTIES OF INSULATING MATERIALS					
Insulator	**Electrical Properties**		**Thermal Properties**		**Mechanical Properties**	**notes**
	dielectric strength	dielectric constant	max operating temperature	thermal conductivity	density	notes
	MV/m or kV/mm	ϵ_r	°C	W/(m·°C)	kg/m³	
dry air	3	1	2000	0.024	1.29	gas density is at 0°C 101 kPa
hydrogen	2.7	1	-	0.17	0.09	
nitrogen	3.5	1	-	0.024	1.25	
oxygen	3	1	-	0.025	1.43	
sulfur hexafluoride (SF₆)	30 MV/m at 400 kPa	1	-	0.014	6.6	
solid asbestos	1	-	1600	0.4	2000	synthetic liquid (restricted)
asbestos wool	1	-	1600	0.1	400	
askarel	12	4.5	120	-	1560	
epoxy	20	3.3	130	0.3	1600 to 2000	
glass	100	5 to 7	600	1.0	2500	
magnesium oxide	3	4	1400	2.4	-	(powder)
mica	40 to 240	7	500 to 1000	0.36	2800	
mineral oil	10	2.2	110	0.16	860	
mylar®	400	3	150	-	1380	a polyester
nylon	16	4.1	150	0.3	1140	a polyamide
paper (treated)	14	4 to 7	120	0.17	1100	
polyamide	40	3.7	100 to 180	0.3	1100	
polycarbonate	25	3.0	130	0.2	1200	
polyethylene	40	2.3	90	0.4	930	
polyimide	200	3.8	180 to 400	0.3	1100	
polyurethane	35	3.6	90	0.35	1210	
polyvinylchloride (PVC)	50	3.7	70	0.18	1390	
porcelain	4	6	1300	1.0	2400	
rubber	12 to 20	4	65	0.14	950	
silicon	10	-	250	0.3	1800 to 2800	
teflon	20	2	260	0.24	2200	

TABLE AX2

ELECTRICAL, MECHANICAL AND THERMAL PROPERTIES OF SOME COMMON CONDUCTORS (AND INSULATORS)

Material	Chemical symbol or composition	Electrical properties			Mechanical properties			Thermal properties		
		resistivity ρ		temp coeff	density	yield strength	ultimate strength	specific heat	thermal conductivity	melting point
		0°C $n\Omega \cdot m$	20°C $n\Omega \cdot m$	at 0°C (x 10⁻³)	kg/m³ or g/dm³	MPa	MPa	J/kg·°C	W/m·°C	°C
aluminum	Al	26.0	28.3	4.39	2703	21	62	960	218	660
brass	≈ 70% Cu, Zn	60.2	62.0	1.55	≈ 8300	124	370	370	143	960
carbon/ graphite	C	8000 to 30 000	—	≈ -0.3	≈ 2500	—	—	710	5.0	3600
constantan	54% Cu, 45% Ni, 1% Mn	500	500	-0.03	8900	—	—	410	22.6	1190
copper	Cu	15.88	17.24	4.27	8890	35	220	380	394	1083
gold	Au	22.7	24.4	3.65	19 300	—	69	130	296	1063
iron	Fe	88.1	101	7.34	7900	131	290	420	79.4	1535
lead	Pb	203	220	4.19	11 300	—	15	130	35	327
manganin	84% Cu, 4% Ni, 12% Mn	482	482	± 0.015	8410	—	—	—	20	1020
mercury	Hg	951	968	0.91	13 600	—	—	140	8.4	- 39
molybdenum	Mo	49.6	52.9	3.3	10 200	—	690	246	138	2620
monel	30% Cu, 69% Ni, 1% Fe	418	434	1.97	8800	530	690	530	25	1360
nichrome	80% Ni, 20% Cr	1080	1082	0.11	8400	—	690	430	11.2	1400
nickel	Ni	78.4	85.4	4.47	8900	200	500	460	90	1455
platinum	Pt	9.7	10.4	3.4	21 400	—	—	131	71	1773
silver	Ag	15.0	16.2	4.11	10 500	—	—	230	408	960
tungsten	W	49.6	55.1	5.5	19 300	—	3376	140	20	3410
zinc	Zn	55.3	59.7	4.0	7100	—	70	380	110	420
air	78% N₂, 21% O₂	—	—	—	1.29	—	—	994	0.024	—
hydrogen	H₂	—	—	—	0.09	—	—	14 200	0.17	—
pure water	H₂O	—	2.5 x 10¹⁴	—	1000	—	—	4180	0.58	0.0

TABLE AX3

PROPERTIES OF ROUND COPPER CONDUCTORS

Gauge number AWG/ B & S	Diameter of bare conductor		Cross section		Resistance $m\Omega/m$ or Ω/km		Weight g/m or kg/km	Typical diameter of insulated magnet wire used in relays, magnets, motors, transformers, etc.
	mm	mils	mm^2	cmils	$25°C$	$105°C$		
250MCM	12.7	500	126.6	250 000	0.138	0.181	1126	
4/0	11.7	460	107.4	212 000	0.164	0.214	953	
2/0	9.27	365	67.4	133 000	0.261	0.341	600	
1/0	8.26	325	53.5	105 600	0.328	0.429	475	
1	7.35	289	42.4	87 700	0.415	0.542	377	
2	6.54	258	33.6	66 400	0.522	0.683	300	
3	5.83	229	26.6	52 600	0.659	0.862	237	
4	5.18	204	21.1	41 600	0.833	1.09	187	
5	4.62	182	16.8	33 120	1.05	1.37	149	
6	4.11	162	13.30	26 240	1.32	1.73	118	
7	3.66	144	10.5	20 740	1.67	2.19	93.4	
8	3.25	128	8.30	16 380	2.12	2.90	73.8	mm
9	2.89	114	6.59	13 000	2.67	3.48	58.6	3.00
10	2.59	102	5.27	10 400	3.35	4.36	46.9	2.68
11	2.30	90.7	4.17	8 230	4.23	5.54	37.1	2.39
12	2.05	80.8	3.31	6 530	5.31	6.95	29.5	2.14
13	1.83	72.0	2.63	5 180	6.69	8.76	25.4	1.91
14	1.63	64.1	2.08	4 110	8.43	11.0	18.5	1.71
15	1.45	57.1	1.65	3 260	10.6	13.9	14.7	1.53
16	1.29	50.8	1.31	2 580	13.4	17.6	11.6	1.37
17	1.15	45.3	1.04	2 060	16.9	22.1	9.24	1.22
18	1.02	40.3	0.821	1 620	21.4	27.9	7.31	1.10
19	0.91	35.9	0.654	1 290	26.9	35.1	5.80	0.98
20	0.81	32.0	0.517	1 020	33.8	44.3	4.61	0.88
21	0.72	28.5	0.411	812	42.6	55.8	3.66	0.79
22	0.64	25.3	0.324	640	54.1	70.9	2.89	0.70
23	0.57	22.6	0.259	511	67.9	88.9	2.31	0.63
24	0.51	20.1	0.205	404	86.0	112	1.81	0.57
25	0.45	17.9	0.162	320	108	142	1.44	0.51
26	0.40	15.9	0.128	253	137	179	1.14	0.46
27	0.36	14.2	0.102	202	172	225	0.908	0.41
28	0.32	12.6	0.080	159	218	286	0.716	0.37
29	0.29	11.3	0.065	128	272	354	0.576	0.33
30	0.25	10.0	0.0507	100	348	456	0.451	0.29
31	0.23	8.9	0.0401	79.2	440	574	0.357	0.27
32	0.20	8.0	0.0324	64.0	541	709	0.289	0.24
33	0.18	7.1	0.0255	50.4	689	902	0.228	0.21
34	0.16	6.3	0.0201	39.7	873	1140	0.179	0.19
35	0.14	5.6	0.0159	31.4	1110	1450	0.141	0.17
36	0.13	5.0	0.0127	25.0	1390	1810	0.113	0.15
37	0.11	4.5	0.0103	20.3	1710	2230	0.091	0.14
38	0.10	4.0	0.0081	16.0	2170	2840	0.072	0.12
39	0.09	3.5	0.0062	12.3	2820	3690	0.055	0.11
40	0.08	3.1	0.0049	9.6	3610	4720	0.043	0.1

ANSWERS TO PROBLEMS

Note: The following numerical answers are usually rounded off to an accuracy of ±1%. Answers that are preceeded by the symbol ~ are accurate to ±5%.

CHAPTER 1. 7) MW 8) TJS 9) mPa 10) kHz 11) GJ
12) mA 13) μWb 14) cm 15) L 16) mg 17) μs 18) mK
19) mrad 21) mT 22) mm 23) r 24)S 33) mL 34) L/s or m^3/s
35) Hz 36) rad 37) Wb 38) kg/m^3 39) W 40) K or °C 41) kg
62) 11.95 yd^2 63) 126.9 mm^2 64) 2.549 in^2 65) 6.591 mm^2
66) 2.59 km^2 67) 76.77 Btu/s 68) 0.746 kW 69) 7.079 m^3
70) 13.56 × 10^6 μJ 71) 4.536 kgf 72) 0.93 T 73) 12 kilogauss
74) 1.417 kg 75) 6049.6 A/m 76) 3.107 miles 77) 288 000 C
78) 111.2 N 79) 11.34 kg 80) 6615 lb 81) 0.001 WbS 82) 8304 kg/
m^3 83) 67.7 mbar 84) 1.378 × 10^6 Pa 85) 482.3 × 10^3 N/m^2
86) 1.57 rad/s 87) 393 K 88) 366.3 K 89) 120 K 90) 1.67, 50,
0.33 91) 10.4 A, 231 Ω 92) 3.3 kΩ 93a) 402 93b) 0.076
93c) 16 m^2 93d) 64 m^3 93e) 93.75 93f) 3.24 × 10^5

CHAPTER 2. 2a) E$_{16}$ = −80 V 2b) E$_{25}$ = +80 V 2c) E$_{52}$ = 0
3) +20 V, 0 V, −30 V, +30 V 4) 20 V 5) 1.8 V 7) 320, 160, 112
8) 3840 A 9a) 960 N 9b) 960 N 9c) no 10b) 31.3 V, 141 V,
−200 V 11) 70.7 A 12a) 12 A 12b) 169.7 V 12c) 1440 W@
12d) 2880 W 13) 23 Hz 14) 1.67 ms 15a) I$_1$ lags I$_3$ by 60° 15b) I$_3$
lags I$_2$ by 90° 15c) E lags I$_1$ by 150° 16c) 2400 W, −800 W
17b) 120 V 18) A 19) no 20) −17.3 V, −17.3 V

CHAPTER 3. 1) 392 N 2) 2940 J 4) 60 N·m 5) 251 kW,
336 hp 6) 10.8 kW, 14.5 hp 7a) 100 kW, 134 hp 7b) 83.3%
7c) 68 256 Btu/h 8) 415.7 J 9a) 1096 J 9b) 3288 J 9c) 242 kJ
11b) 50 N·m 13) zero 14) 209 W 15) 209 W 16) 1.89 kW,
2.53 hp 17a) 13.1 N·m 17b) 88.8 kJ 17c) 2195 W 17d) 2400 W
18a) 1222 N·m 18b) 1332 N·m 19a) 475 N·m 19b) 4455 N
20) 3.04 MJ23) 47.8 r/min, 1156 ft·lbf

CHAPTER 4. 9a) 103 V 9b) 288 W 9c) 10.4 N·m
10a) E$_{NL}$ = 138 V 10b) polarity reverses 10c) voltage increases by-
less than 10% 11a) 5000 A 11b) 7800 A 12) 2 A
13) A = 18 V, B = 18 V, C = 18 V 14) at 90° E$_A$ = 20 V, at
120°E$_A$ = 18 V 16) 276.5 V 17a) 12 brush sets 17b) 150 A
19) E$_{XY}$ is (+) 20a) E$_{34}$ is (−) 21a) 292 V 21b) 0.436 T
21c) 530 μs 23a) 333 A 23b) 83.25 A

CHAPTER 5. 9a) 221 V 9b) 13 800 W 9c) 13.26 kW, 17.8 hp
10a) 1533 A 10b) 1.85 Ω 11a) 2975 A 11b) 2400 A 12) 144 V,
9.6 V 13a) 11.7 kW, 94% 13b) 10 A 13c) 8 mΩ, 223 V 13d) less
than 4 A 14a) 0.48 Ω 14b) 30 kW, 1.9 kW, 0 kW 15a) 0.96 Ω
15b) 45 kW,8.43 kW, 0 kW 16a) 17.8 kg/m^2 16b) 140 kJ 16c) 70 kJ

CHAPTER 6. 8) 180°C 9) 131.4 kW, 547 A 10) 22.8 hp
11) 882 A 12a) 28°C 12b) 88°C too hot 14) 82.9% 15) 10.4 kW,
14 hp 16) <155°C, <145°C, <129°C 17a) 139°C 17b) 108°C
18) 1 year 19a) 2.28 A/mm^2 19b) 13.4 W/kg 20a) 58.9 W/kg
20b) 985 cmil/A 21) 26.2 kW, 35hp 22) 64 years 23) ~768 h

CHAPTER 7. 2) 100 kVA 6) 64.3% 7) 667 kVA, 291 kvar
8) 4.34 kvar 9a) 1440 W 9b) 1440 VA 9c) 2880 W 9d) 1/120 s
10a) 1440 var 10b) 1440 VA 10c) 1440 W 10d) 1440 W 10e) 1/
240 s 11a) A is the active source 11b) C is the active source
11c) F is the reactive source 11d) J is the reactive source 11e) L
is the active source 11f) M is the reactive source 12) 7.2 A, 9.6 A
13) 0.72,2665 var 14a) 2765 W 14b) 745 var 14c) 2864 VA
14d) 11.9 A 14e) 96.5% 15a) 3.71 Ω 15b) 6.32 Ω 15c) 12.16 Ω
16a) 320 kW 16b) 240 kvar 17a) 1200 W 17b) 500 var
17c) 1300 VA 17d) 0.923 18a) 2 kW 18b) zero 19) 900 var
20a) 2765 W 20b) 3512 VA 20c) 0.787 21a) 3845 var
21b) 4000 var 21c) 769 W 21d) 784.5 VA 22a) 102 V
22b) 20.4 Ω 23a) 519.6 W and 300 var, both flowing from B to A
23b) 519.6 W from Dto C, 300 var from C to D

CHAPTER 8. 1) 4157 V 2a) −100, 0, 50, 50, −86.6 V 2b) +, 0,
−, −, +, 2c) 50, −86.6, −100,50, 86.6 V 3) yes 4a) 358 V
4b) 23.87 A 4c) 25.6 kW 5a) 694 A 5b) 13.2 kV 5c) 9.16 MW
5d) 27.5 MW 5e) 19 Ω 6a) a - b - c 6b) yes 7) 26 kVA
8) 120 V 9a) 13 kW 9b) 6.5 kW 10) single-phase 11a) 41.6 A
11b) 41.6 A 11c) 2.89 Ω 12a) 160 Ω 12b) 480 Ω 13) 0.9 Ω
14) 8 Ω 15a) 89.5% 15b) 62.3 kVA 15c) 37.2 kvar 15d) 80.2%

CHAPTER 9. 1a) 2 A 1b) 2.83 A 1c) 1415 A 1d) 0.9 mWb
2) 472 A, 0.3 mWb 4a) 360 V 4b) 30 A 4c) 18 A 4d) 10 800 W
4e) 10 800 W 5) 33.3 Ω 6) 400 V, 0.02 A

CHAPTER 10. 11) 120 V 12) 60 kV 13a) 110.4 V 13b) 0.353 A,
22.08 A 14) 2.42 kW, 11 A, 22 A 15) 50 A, 1250 A 16) 1.88 mWb
18a) 144 V 18b) 2.25 mWb 18c) 0.9 mWb 19a) zero 19b) 520 V
19c) additive 20a) short-circuit 23a) 72 Ω 23b) 0.115 Ω
24a) 126.4 V 24b) 5.22 A, 95 A 25a) 466.9 kW 25b) 466.9 kW,
99.1% 27) 9 mWb 29a) 84.1 Ω 29b) 3.04% 29c) 23.36 mΩ
29d) 3.04% 29e) 976 W 29f) 1.3%, 2.75% 30a) 1.9 Ω, 36.6 Ω
30b) 453.6 Ω 30c) 8.4% 31) 99.08% 32a) 28.9 W/kg 32b) 156 W/
kg

CHAPTER 11. 5) 300 turns 6) 0.2 V 7) 1300 kVA 8) H1-X2 or
H2-X1 together 9) 92 kVA 10) 13 mA 11a) 0.45 V
11b) 2.25 mV 11c) 250 A/5 A

CHAPTER 12. 2a) 20.8 A, 433 A 2b) 20.8 A, 250 A 3) 1506 A,
65 kA 4) 17.7 kA, 2175 A 5a) 1278 A 5b) yes 6a) delta-delta
6b) 577 A, 48.1 A 6c) 333 A, 27.8 A 7a) 236 kVA 7b) 1.89 kV
8a) no 8b) 433 kVA 9a) 347, 600, 347 V

CHAPTER 13. 7) no 10a) 360 r/min 10b) no 10c) 20
12) 936 A, 156 A, 15 A 15a) 600 r/min 15b) 564 r/min 16a) 3 V,
45 Hz 16b) 0.67 V, 10 Hz 16c) 1 V, 15 Hz 17a) 15 A, 90 A, 5.25 A
17b) 882 r/min, 812 N·m 18) 97.2 A 19a) 120 V, 30 Hz 19b) 60 V,
15 Hz 19c) 960 V, 240 Hz 20a) −8.66 A, 8.66 A, 0 20b) 86.6 A
21) 3, slot 1 to slot 8 22a) 23.56 m/s 22b) 3.3 V 22c) 196.35 mm
23a) 87.9% 23b) 2298 kW 23c) 34.5 kW 23d) 2251 kW, 30.3 kN·m,
96% 26) 1500 A, 75 N 27) 20 N 28) 38 kW 29a) 4.49 mΩ,
68.9 mΩ 29b) 1067 V, 40 Hz; 16 V, 0.6 Hz 29c) 1035 kvar
29d) 2.07 kW 29e) 50 kW 30a) 18.6 Mvar 30b) 670 kW
30c) 1498 kW 30d) zero 30e) 23.8 kN·m 31a) 400 V
31b) 508 mΩ, 314 kW 31c) 455 A 32) 264.5 mm 34) 100 mm,
120 kW

CHAPTER 14. 10) 78% 12a) 3 hp 12b) 54 or 56 13) 438, 280,
315 ft·lb; zero, 675, 765 r/min 14a) 3.8 kW 14b) 586 r/min
14c) 222.4 kW 14d) 5.3 kW 15a) 140 Hz 15b) 54 kW
15c) 23.8 kvar 15d) 65.6 A 15e) 100 hp 16a) 1109 A
16b) 434 kW 16c) 90.2 kW 16d) 5212 A, 83.5 kN·m
16e) 43.3 kN·m 17) 4.7°C 18a) 39.4 s 18b) 105 000 Btu
19a) 456 V, 1740 r/min 19b) ~16 hp 20a) design D 20b) same
21a) 54.5 mΩ 21b) 2.23 Ω 22a) ~1100 N·m 22b) 11.4 kg·m^2
23) 270 kJ 24a) 795 kg, 1509 lb·ft^2 24b) 810 r/min, 260 ft·lb
24c) 702 ft·lb 25a) 916 N·m 25b) 1.3 s 25c) 11.3 kJ 25d) 6.7 s
26a) 134 r/min 26b) 11:1 26c) 358 A 26d) 50 t 26e) 728 MJ
26f) 39 min 26g) 91 kW·h

CHAPTER 15. 2a) 5.05 Ω, 82° 2b) 811 r/min 2c) 45.4 A
2d) 332 N·m 3a) yes, 366 N·m 3b) no, as a brake, 35 N·m
4) 26.7 Ω, 202 Ω 5) 0.4 Ω, 0.5 Ω, 1.48 Ω, 268 N·m 6) 37.9 kN·m,
2.4 kN·m 7) 13.9 N·m, 35.5 N·m 8) 10.5 N·m, 42.6 N·m

CHAPTER 16. 3a) 20 poles 3b) 360 r/min 4) increased
6) 12 poles 7) 4 9a) 7500 V, 50 Hz 9b) 37.5 V, 0.25 Hz
12a) 150 A 12b) 50 A 13) 2400 V 14b) 0, 353, 600, 750, 600,
0 kW 15) 807 mm 16a) 145 Ω 16b) 192 Ω 16c) 144 Ω
16d) 10 A 16e) 1750 V 16f) 3.031 V 16g) 57.6 kW 16h) 36.9°
17) 16 kV 18a) 44.56° 18b) 2.475° 18c) 7.87 in 19a) 0.45 Ω
19b) 0.412 Ω 19c) 0.916 19d) 1.09 20a) 8130 kW 20b) 720 kW
20c) 24.23 MN·m 20d) 22.7°C 21) 0.409 T 22a) 126 MW
22c) 85.7% leading 23a) 228.5 MW 23b) 7230 A 24) 0 W, delivers
63.3 Mvar to the bus 25a) 236 r/min 25b) 108°, 1944° 26a) 40
26b) 180 26c) 1.5 26d) 1.73 in 26e) 3.7 Ω, 3565 W

718

CHAPTER 17.   7) 2300 hp   9a) 2217 kVA   9b) 90.2%
9c) 956 kvar   9d) 32 poles   11a) 333 A   11b) 3 Ω   12) 500 r/min
14a) 36.9°   14b) 2.16 MW   14c) 100%   14d) zero   15a) 300 A, 0°
15b) 720 kvar delivered   17a) 3457 kVA   17b) 289 A   17c) 5889 V
17d) 1.44°   17e) 1569 kvar   17f) 9150 hp   18a) 4741 V   18b) 32.8°
19a) 14.6 Ω, 1594 V   19b) 6807 ft·lb   19c) 142 A   20a) 4269 kW,
45.3 kN·m   20b) 1704 kW,36.2 kN·m   20c) 40.7 kN·m   20d) ~83 s

CHAPTER 18.   9a) 14 A, 14 A   9b) 33.8°   9c) 26.8 A   9d) 70.7%
10a) no   10b) probably not   12a) 50 Hz   12c) 4.26 J   13a) 8 mhp
13b) 55%   13c) 0.28   13d) 0.79 p.u., 1.06 p.u.   14a) 4.42 ft·lb
14b) 4.44 p.u.   14c) zero   14d) 2.5 p.u.   15a) 157.5 V   15b) 93.2°
16a) 0.577 A   16b) 86.6% lagging   16c) 30 W   16d) 37%
17a) 1.94 Ω   17b) 21.2 A, 84 V   17c) 3.2 N·m

CHAPTER 19.   4) 1152   6) 30°   8) true   9) 240 ms
10) 0.00175 in   11a) 144 W   11b) 78.5 W   12a) 30 mN·m
12b) 0.79 mhp   12c) 1.77 J   14) 437.5 r/min   15a) 1.5 ms
15b) 4.5 ms   15c) 0.23 A   16a) 7.2°   16b) 3.6°   16c) 1.8°
18a) 3000 r/min, 0.93 hp   18b) 13 ms   18c) 0.16 ms

CHAPTER 20.   10a) quadrant 1   10b) quadrant 4   11) 6.3 hp
12) 26 V, 7.5 Hz   13a) 40 A   13b) 120 A   13c) 40 A or 113 A
14a) 2 or 4   14b) 1 or 3 respectively   14c) 4 or 2 respectively
15) clockwise   16a) 2 min   16b) 5 s   18) 208 days   22) 28 s
23a) 18.85 kW   23b) 11.3 kW   23c) 30.15 kW   24) 1.57, 16,
12.57 kW   25a) 345 V, 45 Hz   25b) 633 V, 82.5 Hz   26) 38 V, 5 Hz
28a) 30 A   28b) 4.29 hp   28c) 63.7 N·m   28d) 56.4 ft·lb
29b) 10 Hz   30a) 834 ft·lb, 358 ft·lb   30b) 149 ft·lb   31a) 307 V,
40 Hz   31b) 67.3 kJ   31c) 29.9 kJ   32b) 29.5 ft·lb   33) 282 V

CHAPTER 21.   6a) 1620 V   6b) 200 A   6c) 600 A   7a) 3240 V
7b) 200 A   8a) 540 V   8b) 540 V   8c) 18 A   8d) 9 A   8e) 9720 W
9a) 150 V   9b) 3.75 A   10) 45 kW   11a) 0.72 kW   11b) 99.96%
12a) negative   12b) increasing   13a) 36 A   13b) 170 V
13c) >65 J   13d) 0.1 H   13e) 170 V   14a) 324 V   14b) 243 kW
14c) 750 A   14d) 5.55 ms   14e) 612 A   14f) zero   14g) 45.4 V
15a) 24 kW   15b) 24 kW   15c) 12 A   15d) 400 A   15e) 300 Hz
15f) 167 Ω   16a) zero   16b) 77.7°   16c) 102.3°   17a) 14 kV
17b) 6.3 MW   17c) 367 A   17d) 23.5 Mvar   18a) 40 kV
18b) −14.64 kV   18c) 54.64 kV   19a) 40 A   19b) 24 kW   19c) zero
19d) 100%   19e) zero   20a) 12.66 A   20b) 4.7 kW   20c) 70%
20d) 10.6 kVA   20e) 44%   20f) 63.5%   20g) 5.72 kvar
21) 6.25 mH   22a) 340 V   22b) 147°   22c) 30.4 A   23) 1/24 s
24a) 3.29 V·s   24b) 0.47 H   25a) 120 A   25b) 48 kJ   25c) 4 kW
26a) 120 A   26b) 80 A   26c) 200 A   26d) 160 A   27a) 155.2°
27b) 57.5 kW   27c) 57 kvar

CHAPTER 22.   3) 91 Hz   4a) 626 V   4b) 169 kW   4c) 90 A
4d) 220 hp   8a) 200 V   8b) 283 V   8e) 2 A   9a) 31.3°
9b) 5.1 kvar   9c) 28.6 A   9d) 113 V   10a) 85.1°   10b) 16.8 kvar
11a) 343 A   11b) 146°   12a) 300 V   12b)   315 V   13a) 150 A
13b) 75 V   13c) 60 V   14a) zero   14b) 150 A   14c) 15 V
15a) 100.7°   15b) 87.3°   16a) 232 A, 240 A   16b) 700 V
17a) 192 V   17b) 250 A   19a) 500 Hz   19b) −563 V   19c) 43.5°
19d) 73.7 A   19e) 360 Hz   19f) 3000 Hz   19g) 72.5%   19h) 60 A
20a) 44 V   20b) 44 V   20c) 14 Hz   20d) 14 Hz   20e) 220 W
20f) 2.5 N·m   21b) 2880 μF   22a) 24 V   22b) 97.8°   22c) 7340 var

CHAPTER 23.   3) 230 V, 30 Hz   4a) 1620 r/min   4b) 8.78 N·m
5) 600 r/min   10a) 8 Hz   10b) 42 V   11a) 50 ms   11b) 13 ms
12a) 8 N·m, 1080 r/min, 1.2 hp   12b) 3.5 N·m, 720 r/min, 0.35 hp
13a) 16.5 hp   13b) 39.5°   13c) 12.4 kvar   15a) 141 A   15b) 422 A
15c) 566 V   15d) 46.7 Hz   17a) 20.7 W   17b) 36.3 W   18) 184 V
19a) 3600 r/min   19b) 15 kW   19c) 337 V   19d) 7.6 Ω, 15 kW
20) 325 μs   21a) 115 V, 15 Hz   21b) 85 A   22) 31.8 A   23) 50 V,
15 Hz

CHAPTER 24.   7) 1176 MW   11a) 41 200 m³/s   11b) 40 400 MW
11c) 312 mi³   12) 15.6 h   13a) 60 MW, 50 MW   13b) 80 MW, 30 MW
14a) 5400 MW   14b) 2615 Mvar   14c) 9364 yd³/s   16a) 5715 tons
16b) ~57 000 tons   16c) 21.6 m³/s   17) 0.43 m³/s
18a) 1.86 × 10¹³ J, 1.76 × 10¹⁰ Btu   18b) 0.207 g   10a) 90.6 TW·h
19b) 10.34 GW   20) 1580 MW

CHAPTER 25.   7) ~2500   9) 750 A   10a) 5092 MW
10b) 151.8 MW   10c) ~3%   11a) 5700 V, 114 kW; 4500 V, 450 kW;
3000 V,600 kW; 1500 V, 450 kW   12) 0°   13a) 5992 V, 126 kW; 5692 V,
720 kW; 4243 V,1200 kW; 1897 V, 720 kW   14) E_R is 18.4° lagging α
15a) 6577 V, 9000 V   15b) 0°   16a) 5692 V   16b) no
17a) 240 kvar   17b) 139.2 A   17c) 290.6 kvar   17d) 50.6 kvar
17e) 802 kVA   17f) 5759 V   17g) 6251 V, 868 kW   18) 20.3°   19) 4 Ω,
2 Ω, 37.5 kΩ   21a) 421 MW   21b) 1074 A   21c) 148.8 Mvar
21d) 74.3 Mvar   22a) 22 Ω, 100 Ω, 1500 Ω   22b) 119 kV   22c) 45 A
22d) 9 Mvar   22e) 34.6 kW   23a) 648.5 A

CHAPTER 26.   3) 2 μΩ   6) 17.5 kV   8a) 48.8 kV   8b) zero
9a) 12 000 A   9b) 960 MW   9c) 4800 J   11) 5 and 7   12) 1.2 Ω,
3.18 mH   13a) 25 A, 35 A   13b) 10 A   13c) 0.5 A   14) 9.74 A
15a) not dangerous   15b) hazardous   17a) 115.6 A, 124.7 A, 31.6 A
17b) 2A   17c) 81.8% lagging   18a) 777 A   18b) 67.3 A, 86.8%
21b) 17.5 kA   21c) yes

CHAPTER 27.   2) $28.40   5) $1470   7) 4.33 MW
8a) 3975 GW·h   8b) 159 million dollars   9a) 72 kvar, 104 kVA
9b) 75 kW, 52 kvar   9c) 12.5%   10a) 243 kvar   10b) 32%
11a) 46.64 $/h   11b) $4205   12a) 30 cent/kW·h   12b) 2.4 cents
13) 6.45 cent/kW·h   15) 10 r/min   16) 1800 W   17a) meter reads
1 percent high   17b) no   18) ±5.6 kW·h

CHAPTER 28.   5a) ~45 kV   5b) ~492 A   6a) 150 MW
6b) 200 A   7a) 600 A, 57.6 MW   10a) 400 A   10b) 64 kV   11) 1800 A,
1000 Mvar   12) 45 kW   13) 4050 A   14a) 2.9 in²   14b) 13.38 Ω
14c) 43.4 MW   14d) 426 kV   14e) 94.7%   15a) 360 Hz, 720 Hz
15b) 9 Ω, 11 Ω   15c) 150 kV   16a) 17.7 A   16b) 20 kV   16c) 159 V

INDEX

– U – V –

– W – X – Y – Z –